Impact of Global Changes on Mountains

Responses and Adaptation

Impact of Global Changes on Mountains
Responses and Adaptation

Editors

Velma I. Grover
Faculty of Environmental Sciences
York University
Toronto, Ontario
Canada

Axel Borsdorf
Institute for Mountain Research: Man & Environment
Innsbruck, Tyrol
Austria

Jürgen Heinz Breuste
University Salzburg, Dept. Geography/Geology
Salzburg
Austria

Prakash Chandra Tiwari
Kumaun University
Nainital, Uttarakhand
India

Flavia Witkowski Frangetto
Brazilian Institute for Environmental Law
São Paulo/SP - CEP
Brazil

CRC Press
Taylor & Francis Group
Boca Raton London New York

CRC Press is an imprint of the
Taylor & Francis Group, an **informa** business
A SCIENCE PUBLISHERS BOOK

Cover Illustrations reproduced by kind courtesy of Dr. Velma I. Grover and Dr. Axel Borsdorf.

CRC Press
Taylor & Francis Group
6000 Broken Sound Parkway NW, Suite 300
Boca Raton, FL 33487-2742

First issued in paperback 2019

© 2015 by Taylor & Francis Group, LLC
CRC Press is an imprint of Taylor & Francis Group, an Informa business

No claim to original U.S. Government works

ISBN-13: 978-1-4822-0890-0 (hbk)
ISBN-13: 978-0-367-37790-8 (pbk)

Visit the Taylor & Francis Web site at
http://www.taylorandfrancis.com

and the CRC Press Web site at
http://www.crcpress.com

Preface

In every person's mind, the word 'mountains' evoke a different image. Mountains always bring forth images of height, waterfalls, tall green trees, sometimes glaciers and evoke the feelings and the sense of spiritual awakening and romanticism associated with nature. For most of us mountains have been sensationalized by the journalists who cover the climbers conquering the highest peaks in different continents or romanticized by movies! However, this book will go deep (without indulging in doing either of these two) into exploring the impact of global changes on high mountains. Avalanches in the Everest region have been in the news recently, with the new disaster depicted as the worst in history so far—prediction of such disasters is becoming more difficult with changing climate. The attempt of the authors in this book is to focus on what these global changes are and how these changes are impacting higher mountains in different regions: Africa, Americas (both North and South), Asia, and Europe.

Mountain regions encompass nearly 24% of the total land surface of the Earth. Mountains are highly critical from the view point of ecosystem services that sustain human society both in highlands and lowlands in large parts of the planet. Mountains also constitute the hot spots of biodiversity, and supports inhabitants of lowlands or high mountains—providing them with goods and services such as water, energy, lumber. Mountains are also known as the water towers of the world (since the headwaters for most of the major rivers are in one of the mountain ranges) that take care of freshwater needs of humanity. Mountains are an important part of human civilization in providing important resources needed for the survival of human kind. All the six continents have mountains in all the latitude zones and encompass some of the most spectacular landscapes. They also provide opportunities for recreation and spiritual renewal.

Currently, there are more than seven billion people on planet Earth. The planet has never been inhabited with so many people before and never with such extra-ordinary prosperity which is due to globalization. With an average per capita income of $12,000—meaning an economy of ninety trillion dollars—we live in a complex global society. Such a complex society has great impact on our lives whether in lowlands or up in the mountains and puts a pressure on natural resources. To fulfil the needs of the ever growing economy, various anthropogenic activities such as deforestation, expansion of urban areas and infrastructure, intensified agriculture, and climate change have changed the fragile and vulnerable mountain ecosystems. As a result, mountain regions of the world are passing through a process of rapid environmental changes, and exploitation and depletion of natural resources leading to ecological imbalances and economic unsustainability both in upland and lowland areas. Moreover, the changing climatic conditions have already stressed mountain ecosystems through higher mean annual temperatures and melting of glaciers and snow, altered precipitation patterns and hydrological disruptions, and more frequent and extreme weather events rendering the mountain communities and their economies more vulnerable to the long term impacts of climate change. It is not just the ecology of mountains which is changing, but the economic relations of mountain communities have also undergone transformations. Not too long ago, economies of most mountain communities depended on the use of resources at different altitudinal zones. Nowadays the effects of climate change and globalization are threatening the fragile ecosystems and livelihoods of people dependent on them. The aim of the book is to address some of these issues with case studies from different continents and to see how we can decelerate or even stop some of these changes by mitigation, new technologies or global, regional, and local policies.

Velma I. Grover
Axel Borsdorf
Jürgen Heinz Breuste
Prakash Chandra Tiwari
Flavia Witkowski Frangetto

Contents

SECTION 5: CASE STUDIES

SECTION 1

INTRODUCTION

1

Prelude: Mountains in an Uncertain World*

Jack D. Ives

Daunting rock pinnacles, knife-edged arêtes, cascading glaciers and explosive avalanches of blinding snow—and a team of grim-faced climbers risking everything in their arduous upward press to bag another summit—this is perhaps the conventional image conjured up by the word *mountains*. It is a vignette that might recall upper sections of the Alps, Andes or Himalaya. But these lofty regions represent only a small fraction of our total mountain terrain. Sadly, images of the heroic age are now too often supplanted by the reality of warfare, abused natural resources, impoverishment of mountain peoples and the environmental and economic disruption caused by climate change.

Mountains are found on every continent, from the equator to as close to the poles as land exists. As a single great ecosystem, they encompass the most extensive known array of landforms, climates, flora and fauna, as well as human cultures. From a geological point of view they comprise the most complex and dynamic of the Earth's underlying structures that are still in the process of formation as the numerous volcanoes and frequent earthquakes testify.

Adjunct Research Professor, Carleton University, Ottawa, Canada.
 Email: jack.ives@carleton.ca

* This chapter was prepared by merging and modifying two previously published items:
  1. A "commentary" co-authored by Bruno Messerli, *In*: H. Lewis-Jones ed. (2011).
  2. An Opinion column (Ives 2012: *A flood on The Mountain*), Ottawa Citizen, 10 June, 2012.

No surprise then, that it is hard to delimit mountains with a simple definition. Ask the people of northern England and Wales—perhaps shepherds at high pasture in the Lake District, or farmers at work in the shadow of Snowdonia's imposing crags—and they would all affirm that they live among mountains, yet altitudes there barely exceed a thousand metres. Peasants of southern Peru or Tibetan nomads likewise would be classed as mountain people; their local surroundings exceed 4,000 metres and yet may be as flat as the prairies of central Canada. The well known German mountain geographer of the 20th century, Carl Troll, described the highest parts of equatorial Indonesia as "high mountains without a high-mountain landscape"—unlike the remote, challenging, conventionally 'mountainous' *Hochgebirge* of popular imagination. It's clear that the diversity of the mountain world defies easy categorization.

Academics today, perforce, take a pragmatic approach to the problems of definition. It is broadly accepted that mountains occupy more than a fifth of the world's land surface, and that they provide the direct support-base for more than a 10th of humanity. When natural resources, including water, minerals, forests, grazing land and hydro-power are considered, more than half of humankind depends to some degree on the largesse of the mountains. To this must be added their pivotal role as repositories of cultural and biological diversity, their value as amenity and tourism assets, yet also their lamentable implication in a panoply of natural and human catastrophes.

We no longer have the luxury of viewing mountains simply as spectacular scenery or as goals to be attained by strenuous physical effort. They are certainly special regions, but they are places where communities live and have lived for centuries, and where environmental sustainability is a matter of survival not just to those who live there but to the many millions more for whom they are just a hazy blue line on the horizon.

In the far distant past when the earliest human societies were evolving, mountains were the object of veneration, inspiration and even fear; they had a profound influence on the emergence of many religions, including those that survive today. This spiritual approach to mountains fortunately persists and has been examined by Edwin Bernbaum in his outstanding work 'Sacred Mountains of the World,' where he argues for its essential relevance in any effort to achieve environmental stability (Bernbaum 1990).

With the dawn of the 18th-century European Enlightenment, fear began to ebb and the quest for inspiration led to the 'golden age' of alpinism, landscape art, poetry and the privileged tourism of the era. The identification of end points, or turning points, in the development of alpinism is highly subjective, but must include Alfred Wills's ascent of the Wetterhorn (1854) and Edward Whymper's ascent of the Matterhorn (1865), which encompass the golden age; the Aiguille du Grépon, first climbed in

1881 by Mummery, Bergener, and Venetz; the Austrian-German assault of Eiger's *Nordwand* in 1938; Maurice Herzog and Louis Lachenal's gruelling adventure on Annapurna in 1950, the first successful attempt on an 8,000 metre peak; or John Hunt's triumphant expedition in 1953 which put Edmund Hillary and Tenzing Norgay on the summit of Mount Everest. As in many other 'occupations', women entered the mountaineering scene relatively late. Vera Komarkova, on whose doctoral committee I served, was the first of her sex to climb an 8,000 metre peak (Annapurna 1978, and Cho Oyu six years later). Mountaineering had rapidly extended beyond the Alps to high mountains throughout the world, and remains a vital source of character formation, challenge and creative thought.

Today most people maintain that mountain peaks are never *conquered*— because (to paraphrase George Mallory's quip), they are still *there*, unfazed, unaccommodating and impersonal. We may climb, we may litter, but we always leave; that is, unless we succumb to altitude and leave our remains amongst the growing mounds of garbage. It is perhaps symbolic that Ed Hillary rated as his proudest accomplishment the building of schools and hospitals in the Khumbu. Nevertheless, electrifying feats of mountaineering still unfold and these may serve to provide inspiration if handled appropriately.

Tourism ushered in a new engagement between humans and mountains. It began in the Alps as an elitist diversion at the height of the British Empire and, with the rapid increase in accessibility during the 20th century, merged into mass tourism and trekking all over the world. In the early years following World War II this rapid growth of tourism was greeted with great anticipation as one likely solution to world poverty, particularly among mountain communities. This optimism has faded and the impacts have had mixed results at best. Nevertheless, the increased contact with hitherto isolated subsistent people has introduced a sharper awareness of alternatives to the frantic pace of life in the 'outer' world—an interest in regions characterized by the ingenuity, courage, persistence and dignity of mountain dwellers from whom we all have much to learn.

Exploration and scientific research unfolded as another dimension in our awareness of mountains, also linked chronologically to the development of alpinism and 19th-century elitist tourism in the Alps. The close association of mountaineering and research is exemplified by many of the early scholars: Horace-Bénédict de Saussure, Sveinn Pálsson, Alexander von Humboldt, Jean Louis Agassiz, John Tyndall, among others. Much of this early scientific development was motivated by intellectual curiosity, which certainly needs no justification, although imperial and national overtones were to emerge. Nevertheless, glaciology and tectonic geology unfolded as vital academic disciplines.

In recent years, while tourism and adventure among the high places have continued to flourish, a new approach defines the ambitions of those working hard to protect and preserve our mountain world. It might now be characterized, rather prosaically, as 'applied research'.

In the late 1960s concerns about environmental deterioration were beginning to surface from within various international agencies and non-governmental organizations. World-wide collaboration began to appear, especially following the United Nations 1972 Stockholm Conference on the Human Environment. A flashpoint was reached with the International Geographical Union's 1968 decision to recognize Carl Troll's contributions to mountain geography; he was granted his 'personal' Commission on High-Altitude Geoecology. The United Nations Educational, Scientific, and Cultural Organization (UNESCO) quickly perceived the utility of Troll's IGU commission for the development of its own mountain project within the new Man and the Biosphere (MAB) Programme; MAB Project 6 was launched in 1973 as a "study of the impact of human activities on mountain ecosystems". The IGU commission's name was subsequently changed to 'Mountain Geoecology and Sustainable Development'[1] and many of its objectives were incorporated into the newly created United Nations University project: 'Highland-Lowland Interactive Systems' in 1978.[2]

The quarterly journal *Mountain Research and Development* was founded in 1981 by the International Mountain Society. Its mission statement emphasizes the rapid progression from predominantly academic mountain research into the international political arena as 'applied mountain geoecology'. It called upon all of us, "to strive for a better balance between mountain environment, development of resources, and the well-being of mountain peoples". Today, this message is more important than ever.

The few of us involved at the onset of the current mountain political agenda felt we were voices crying in the wilderness. As the movement for appropriate development gained strength, however, it soon faced a quandary. At the beginning of the 1970s, the world's news media were suddenly inflamed with alarmist reports relating primarily to the Himalaya. The mountain paradigm of the day quickly became entrenched. There was an insistence that rapid population growth amongst 'ignorant' Himalayan farmers was leading to massive deforestation and construction of unstable agricultural terraces on steep slopes (see Photo 1). It was assumed that monsoon downpours were washing away these foolhardy constructions and were causing countless landslides and catastrophic flooding and siltation across Gangetic India and Bangladesh.

The beauty and scrupulously managed terraces of Nepal's Middle Mountains have been the object of criticism by many agency 'experts' as contributing to the *Himalayan devastation*. We were astounded by such prejudice and ignorance. Our UNU team quickly came to realize that, in

Photo 1 Nepal Middle Mountains. In places the Kakani ridge landscape has been so completely engineered by human hands through generations that no vestige of 'untouched nature' remains. This is a dry season view of *bari* (rainfed) terraces. We came to resent so-called Western experts characterizing such beautiful landscapes as wilful destruction of the environment by 'ignorant' peasants (1980).

many instances, they represented a marvel of landscape stability engineered under extremely difficult circumstances.

In the 1970s it was even predicted by the World Bank and other authoritative agencies, that all accessible forest cover in Nepal would be eliminated by 2000. We termed this alarmist process, for convenience, the 'Theory of Himalayan Environmental Degradation' and devoted all available research capacity at our disposal to determine the facts, eventually exposing its projections to exaggeration and misguided emotion.

The combined activities of the United Nations University, IGU, and International Mountain Society led to close contact with Maurice Strong who, in 1990, was appointed Secretary-General of the United Nations Conference on Environment and Development (UNCED), more popularly known as the Earth Summit, that convened in Rio de Janeiro in 1992. Together with the ardent support of Jean-François Giovannini and the Swiss Development Cooperation, the efforts of the now rapidly growing group of concerned scholars and institutions successfully stage-managed the insertion of the 'mountain chapter' into AGENDA 21 of the Rio Earth Summit.

Following Rio, mountain well-being moved from the back stage of world political and environmental attention to front and centre. This climaxed at the United Nations headquarters in New York on December 11, 2001 when the International Year of Mountains (2002) was launched. Under the adept leadership of the UN Food and Agricultural Organization (FAO), the designated 'lead agency' for the mountain chapter, mountain environmental and development problems began to receive widespread attention. Many NGOs were set up, with Bern University becoming a major focal point, and now December 11 is celebrated each year as the *Day of the Mountains*.

Although progress has slowed down in many regions, the 20th anniversary of the Rio Summit in 2012 offered an opportune juncture to regenerate momentum. Regrettably, the opportunity was squandered. The now greatly enlarged scope of mountain activism, including increasing numbers of indigenous mountain people and organizations, aided by the recent digital revolution, must renew their determination. Similarly, the world mountaineering community must come together as advocates and ambassadors for the mountains we all love—the mountains we all need (Ives 2013).

The reality of climate change is now undeniable—Earth *is* warming and the leading indicators have been decisively manifested in the mountains and polar regions. However, the urge to exaggerate and over-dramatize, to cry death and destruction, remains a serious problem that hampers effective progress and understanding. The Himalaya, once again, feature prominently in the news media, specifically regarding the contention that large numbers of melt-water lakes are forming on the surfaces and in front of retreating Himalayan glaciers and will inevitably burst, bringing death and destruction to hundreds of millions downstream. To this has been added the threat of the Ganges, and other major rivers of the region, withering to mere seasonal streams once the glaciers have all melted. Lack of good data is one of the primary constraints to proper evaluation, thus allowing these reckless forecasts to proliferate unchecked.

Mountains are among the least known and least understood areas of the world. Unreasonable and unsupportable statements propagated by the world-wide news media are dangerous, even immoral, yet complacency about the shared dangers and threats and the very real pressures that our mountain world faces must be avoided.

Our near-total lack of hydro-climatological information from vast glacierized regions such as the Himalaya and Andes needs to be corrected so that real evaluations can be made from a reliable database (Alford 2011). The countries concerned and the relevant international agencies are beginning to move ahead to fill this need.[3] The first tasks, however, must be to reduce the burden of conflict and to facilitate the involvement of mountain people in the management of their local resources and in development of relations with society at large. The ongoing problems of poverty and warfare are so formidable that they cannot be solved in the short term. But we must pursue them relentlessly; we must have *hope*.

It is perhaps appropriate at this point to revert to the latter-day evolution of mountaineering. This can be encompassed by reference to a single, regrettable tragedy.

On May 19, 2012, Shriya Shah-Klorfine died descending from the summit of Mount Everest where she had just planted the Maple Leaf of Canada, her adoptive homeland. Three other climbers died that day on the South Col route among the estimated 150 who mobbed a slender ropeway. The traffic jam at the Hillary Step is itself a huge risk factor, exhausting precious oxygen supplies, over-exposing climbers in the Death Zone above 8,000 metres, and delaying ascents until it is too late to descend safely. Shah-Klorfine's calamity was instantly spread by the news media across the length and breadth of Canada and is by now a familiar story comparable to many others; and yet, when it comes to life-threatening developments on Mount Everest, human bottlenecks at the Hillary Step, are just the tip of the iceberg.

People get sucked into the Everest vortex in diverse ways. I first experienced the tug of the Himalayan massif in the spring of 1953; I was an undergraduate, preparing to lead Nottingham University's first Arctic glacier research expedition while anxiously following the progress of the British Mount Everest expedition. The brilliant success of John Hunt, expedition leader, Edmund Hillary and Tenzing Norgay, perfectly timed to embellish the coronation of our new Queen Elizabeth heightened our resolve to persevere toward our own comparatively humble objectives. *Per aspera ad astra!*

The following autumn, I was invited to the triumphant Everest lecture at the Royal Albert Hall. There I was introduced to John Hunt (to become Lord Hunt of Llanfair Waterdine), Ed Hillary, and a cohort of Everest pioneers, including Noel Odell and Tom Longstaff. The dream of Everest and its heroes stayed with me until I eventually set eyes on The Mountain in 1979. By then, however, the Everest lustre was wearing thin.

After emigrating to Canada in 1954, I embarked on 12 years of research in the eastern Canadian Arctic and Subarctic, still an exotic destination for hard-core mountain and tundra aficionados. Later, as co-ordinator of the UNU mountain research undertaking, in close association with Bruno Messerli, I became engaged in mapping mountain hazards, both in the Middle Mountains of Nepal and in the Khumbu region below Everest. While environmental deterioration was not our direct concern, the negative impact of the burgeoning influx of mountain climbers and trekkers along the main trail to the Everest base camp was already apparent. Today, tens of thousands of trekkers throng the Khumbu each year and add their detritus to that of the commercialized attempts to push hundreds of would-be climbers to the summit, and this within a gazetted World Heritage Site. The northern approach to the summit from Chinese Tibet is witnessing comparable problems, a far cry from the situation prevailing during George Mallory's and Andrew Irving's legendary climb into oblivion in 1924.

As a member of the Sir Edmund Hillary Mountain Legacy Medal selection committee, which includes Peter Hillary, I can reveal that not one award has been made for mountaineering feats. While many of the medalists were originally climbers, their most remarkable achievements (like those of Sir Edmund) have been in the mitigation of threats to mountain environments, protection of traditional cultures, and alleviation of suffering in impoverished mountain communities. A few years before his death, Sir Edmund recommended to the Nepalese government that Everest be closed to climbing for several years. His advice was ignored.

Seven years later I can only second his recommendation and urge in addition that steps be taken to consider extensive reorganization of the Sagarmatha (Mt. Everest) National Park management plan. While trekking and mountaineering has become a vital part in the progressive emergence of the Khumbu Sherpa community toward a most appropriate higher standard of living, once international commercialism entered the scene a disturbing trend became established.

The title of a recent book by Michael Kodas: *High Crimes: The Fate of Everest in an Age of Greed,* alludes to a pervasive set of unethical and criminal factors: incompetent and vainglorious expedition leaders and climbers, staff and climbers who loot and vandalize critical equipment, and corrupt authorities who fail to maintain the standards on which the entire industry depends (Kodas 2008). It is time to stop glamorizing dubious *heroic* attempts to *conquer* mountains (Lewis-Jones 2011).

There is yet another issue that relates both to our original purpose in the Khumbu—the United Nations University project on mapping mountain hazards—and the way in which 'Everest-associated news' is purveyed by the media.

By 1985 our UNU research had brought us face to face with the phenomenon now widely known as Glacial Lake Outburst Floods (GLOFs). One occurred in 1985 during our survey of the Everest area, taking the lives of three Sherpas, and sweeping away houses, fields, a new hydroelectric

Photo 2. Nepal, Khumbu. Imja Lake from the air in 2007. The newly regarded major threat to outbreak of the lake is the course of its drainage through the end moraine via the series of small lakes (Photograph kindly supplied by Sharad P. Joshi, ICIMOD, Kathmandu).

Color image of this figure appears in the color plate section at the end of the book.

station, and every bridge for about 60 kilometres downstream. The first published assessment of the event and its tragic consequences was prepared for the International Centre for Integrated Mountain Development (ICIMOD) based in Kathmandu (Ives 1986). This led to identification of the Imja Glacier and its rapidly expanding supra-glacial lake a few kilometres south of Everest.

In 1956 there was no lake on the surface of the Imja Glacier; by 1985 there was a lake more than 1,000 metres long and 500 metres wide. Today that same lake exceeds 2.5 km in length. Between 2002 and 2012, its depth increased from 98 m to 116 m and its estimated volume from 35 million m^3 to 61.6 million m^3 (Watanabe et al. 2009, Byers et al. 2014).

Today Imja Lake, generally attributed to climate warming and glacier melt, has been characterized in the news media worldwide as the most dangerous glacial lake in the Himalaya. Should its apparently fragile end moraine dam be breached, a wall of water, mud and boulders is predicted to severely damage the small Sherpa village of Dingboche, breach the Everest base camp trail within three hours, and continue its devastation far downstream. Yet this scenario is disconcertingly under-reported by journalists concentrating on the dangers of climbing Everest as well as the environmental impacts of thousands of trekkers making their way to the base camp. Imja Lake has also become the object of projections of large scale catastrophe that are primarily based on emotion rather than substantiated scientific evaluation. Vital technical details are finally becoming available so that estimation of risk can be based upon scientifically generated facts rather than emotion and the journalistic desire to flash 'breaking news'. In addition, the local Sherpa communities are being incorporated into the research as key players (Byers et al. 2014).

Our UNU team became aware of the rapid growth and potential danger of Imja Lake by a curious coincidence. Professor Fritz Müller was a close friend from McGill University graduate student days in Montreal. In 1956 he was a member of the Swiss expedition to Everest and Lhotse, staying on to undertake the first glaciological and permafrost research in the region. His untimely death in 1980 resulted in part of his collection of Khumbu photographs being placed in my care. This included a 1956 panorama across the Imja Glacier looking on to the south face of Lhotse. In that year there was no lake on the surface of the Imja Glacier; by 1985 there was a lake more than 1,000 metres long and 500 metres wide. As stated above, the lake is many times larger today.[4]

The 1985 discovery led to further exploration by my own graduate students (and subsequently by their students) and others, research that has continued to the present (see end note 3). Even in 1985 we considered the risk of a glacial lake outburst although any realistic prediction of its timing and magnitude was not possible. Current, although still not sufficiently well-founded, predictions are that an outburst, should it occur, could be felt for up to 100 kilometres downstream. But this is only one of several hundred glacier lakes that are developing in the Himalaya and in many glacierized mountain regions world-wide.

How does this relate to mountaineering and trekking in the Everest region? Certainly any glacial lake outburst flood would likely cause extensive damage and loss of life; this must not be under-estimated. What is perhaps remarkable, however, is that such an event, depending on its timing, could overwhelm several hundred trekkers, mountaineers, porters and guides, certainly many times the intolerable total of lives already lost in the search for glory on The Mountain. Yet there is a curious disconnect in news media reportage. On one hand we read about the loss of life on Everest itself, overcrowding, heroism and serious environmental damage; on the other, there are grossly exaggerated reports that climate warming is causing formation of numerous glacier lakes throughout the Himalaya that are leading to the imminent threat of catastrophic outbursts, and consequent loss of millions of lives downstream—a particularly shameful case of uninformed exaggeration.

My personal response is that we do not have the necessary geotechnical data to predict if, or when Imja Lake, or any other glacial lake, will burst and so cause heavy loss of life and property. Nevertheless, news media reports, including spectacular and distorted videos (that win international prizes), continue to mislead us with false accounts.

There is surely a need for reassessment of the management of the Sagarmatha (Mt. Everest) National Park/World Heritage Site. There is also need for detailed geotechnical study of the stability of glacial lakes. This should include associated downstream land-use planning, restriction of new construction in potentially dangerous locations, and relocation of exposed sections of trekking routes in general and the route to the Everest base camp in particular. Climbing Everest will never be a walk in the park, but a lot can be done to make both climbing and trekking less dangerous to both participants and the host communities.

In conclusion, let us return to the more general overview of the first part of this account: we know, in the words of a former mayor of Narvik, that "the mountains wait". Theodor Broch was writing of the invasion of his town in 1940. The spiritual presence of the mountains of Arctic Norway inspired the local people to withstand five years of horror during the World War II (Broch 1942). Mountains, the world over, endure in all their stark beauty and mystery. Lifting their summits to the sky, they inspire and challenge us, prompting our determination and fortifying the courage of the generations who will follow us. Yet a concerted effort is also needed to help preserve in perpetuity this immense world heritage, vital for the well-being of us all.

References

Alford, D. 2011. Hydrology and Glaciers in the Upper Indus Basin. World Bank, technical report, forthcoming.

Bernbaum, E. 1990. Sacred Mountains of the World. Sierra Club Books, San Francisco.

Broch, T. 1942. The Mountains Wait. Webb Book Publ. Co., Saint Paul, USA.

Byers, A.C., D.C. McKinney, M.A. Somos-Valenzuela, T. Watanabe and D. Lamsal. 2013. Glacial Lakes of the Hongu Valley, Makalu-Barun National Park, Nepal. Natural Hazards 69: 115–139.

Byers, A.C., D. McKinney, S. Thakali and M. Somos-Valenzuela. 2014. Promoting Science-Based, Community-Driven Approaches to Climate Change Adaptation in Glaciated Mountain Ranges. Geography, IN Press, Sept. 2014.

Fujita, K. and T. Watanabe (eds.). 2012. Special issue of Global Environmental Research. Studies on the Recent Glacial Fluctuations, Glacial Lakes and Glacial Lake Outburst Floods in South Asian Mountains. Vol. 16, No.1, 122 pp.

ICIMOD. 2014. Glacier Status in Nepal and Decadal Change from 1980 to 2010 Based on Landsat Data. ICIMOD Research Report 2014/2, Kathmandu, Nepal.

Ives, J.D. Glacial Lake Outburst Floods and Risk Engineering in the Himalaya. Occl. Paper No. 5, International Centre for Integrated Mountain Development (ICIMOD), Kathmandu, Nepal.

Ives, J.D. 2013. Sustainable Mountain Development: Getting the facts right. Himalayan Assoc. Advancement Science, Kathmandu, Nepal (especially chapters 10 and 16).

Ives, J.D., R.B. Shrestha and P.K. Mool. 2010. Formation of glacial Lakes in the Hindu Kush-Himalaya and GLOF risk assessment. ICIMOD, Kathmandu, Nepal.

Kodas, M. 2008. High Crimes: The Fate of Everest in an Age of Greed. Hyperion, New York.

Lamsal, D., T. Sawagaki and T. Watanabe. 2011. Digital terrain modelling using Corona and Alos Prism data to investigate the distal part of Imja Glacier. Khumbu Himalaya, Nepal. Journ. Mountain Sci. 8: 390–402.
Lewis-Jones, H. (ed.). 2011. Mountain Heroes: Portraits of Adventure. Conway, in association with Polarworld, London.

Endnotes

1. The IGU commission was reconstituted in 2008 as 'Mountain Response to Global Change'.
2. To become 'Mountain Geoecology and Sustainable Development'.
3. In the last several years there has been a growing and highly effective response to filling this need. The International Centre for Integrated Mountain Development (ICIMOD), with greatly augmented funding, has been able to expand its research and coordination initiatives (ICIMOD 2014, Ives et al. 2010). Under the leadership of The Mountain Institute (West Virginia), with collaborating organizations, there has been extensive geophysical research on Imja Lake and neighbouring unstable lakes, with financial support from US AID and other organizations (Byers et al. 2013, 2014, Fujita and Watanabe 2012, Lamsal et al. 2011).
4. For much more detailed research on this topic see the following publications: Watanabe et al., Byers, etc.

2

Introduction and Road Map for Impact of Global Changes on High-Mountains

Velma I. Grover

INTRODUCTION

Mountain regions encompass nearly 24% of the total land surface of the Earth and is home to approximately 12% of the world's population, mountains are highly critical from the view point of ecosystem services that sustain human society both in highlands and lowlands in large parts of the planet. Mountains constitute the hot spots of biodiversity, and the mighty glaciers and forested mountain-ranges are the source of important rivers of the planet. Water discharge that builds up in these mountains is transported through the drainage system to the lowland where it sustains the economy and prosperity of people in numerous ways. Mountain regions with their super proportional discharge compared to the lowland, are therefore considered of significant hydrological importance. See box 1 for definitions of mountains. Mountain ecosystems provide various services and goods as well[1]:

1. Provisioning services—freshwater, fresh air, timber, food, renewable energy supply.
2. Regulating services—climate, water, air, erosion and natural hazard regulation, carbon sequestration.

Bryant Drake Guest Professor, Kobe College, Nishinomiya, Japan.
Email: Velmaigrover@yahoo.com

Box 1. Some definitions of Mountains and High Mountains

Axel Borsdorf

Mountains: We follow the UN Environmental Programme (UNEP, see Blyth et al. 2002: 74) definition. At least one of the following criteria must be relevant:

- Elevation of at least 2,500 m (8,200 ft);
- Elevation of at least 1,500 m (4,900 ft), with a slope greater than 2 degrees;
- Elevation of at least 1,000 m (3,300 ft), with a slope greater than 5 degrees;
- Elevation of at least 300 m (980 ft), with a 300 m (980 ft) elevation range within 7 km (4.3 mi).

According to this definition 24 percent of the global land area (35.8 million km²) can be classified as mountainous, that is 33% of Eurasia, 24% of North America, 19% of South America, and 14% of Africa. It can be estimated that in this area 12 percent of the global population are living, 25 percent live within or very close to mountain areas (see also Price et al. 2014: 5).

High mountains rise upon the Pleistocene snow line, formed by forces of glaciers, frost and solifluction. They extend above the natural timberline.

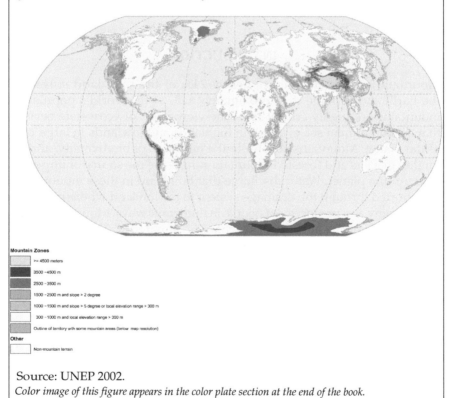

Mountain Zones

	>= 4500 meters
	3500 –4500 m
	2500 – 3500 m
	1500 –2500 m and slope > 2 degree
	1000 –1500 m and slope > 5 degree or local elevation range > 300 m
	300 –1000 m and local elevation range > 300 m
	Outline of territory with some mountain areas (below map resolution)

Other

	Non-mountain terrain

Source: UNEP 2002.

Color image of this figure appears in the color plate section at the end of the book.

Box 1. contd....

Box 1. contd.

Global Change can be seen as any change of the Earth system in the global scale on the continents, the oceans, the atmosphere, fauna and flora, the social and economic system, culture and civilization. It includes climate and the atmospheric circulation, ocean circulation, the carbon and nitrogen cycle, the water cycle (including glaciers and permafrost, sea ice and the sea level rise), resource use, energy, economy, transport, communication, technology, land use and—cover, urbanization, nutrition, demographic change and other cultural and environmental changes as biodiversity, pollution, health and more. In a more dense definition Global Change can be seen as a combination of Climate Change and Globalization.

Impact: We define impact as the collision of climate change and/or globalization to the environment, societies, cultures and/or economies of mountains.

Adaptation: In our book adaptation to Global Change is meant as any measure to minimize effects of Climate Change and Globalization.

References

Blyth, S., B. Groombridge, I. Lysenko, L. Miles and A. Newton. 2002. Mountain Watch. UNEP World Conservation Monitoring Centre, Cambridge, UK.
Price, M.F., AlC. Bryers, D.A. Friend, T. Kohler and L.W. Price (eds.). 2014. Mountain Geography. Physical and Human Dimensions. University of California Press. Berkeley, Los Angeles, London.

---

3. Cultural services—recreation, tourism, aesthetic value, cultural and spiritual heritage.
4. Supporting services—ecosystem functions, including energy and material flow, such as primary production, water and nutrient cycling, soil accumulation and provision of habitats.

Various anthropogenic activities such as deforestation, expansion of urban areas and infrastructure, intensified agriculture and climate change have changed the fragile and vulnerable mountain ecosystems. It is not just the ecology of mountains which is changing, but the economic relations of mountain communities have also undergone transformations. Not too long ago, economies of most mountain communities depended on use of resources at different altitudinal zones. However construction of new roads or in some cases (e.g., Nepal) availability of helicopters has connected remote isolated rural communities to the rest of the world. With easy accessibility tourism has increased quite a bit—to explore beautiful scenery, experience different cultures and for pilgrimage. Pilgrimage adds quite a bit of traffic in some areas, for example—9.3 million pilgrims arrive each year at Hardwar-Rishikesh, the entry point for pilgrimage in Garhwal (Martin Price, Mountains: globally important ecosystems). Influx of pilgrims leads to over-development in the regions and leads to major disasters like the one that happened in the summer of 2013 in India, washing out many religious places, hotels and of course people! Development has a negative impact on the fragile ecosystem due to deforestation which leads to floods, waste problems and others.

Because of all the above mentioned activities in the mountains and in response to globalizing economy with increased population dynamics, a variety of changes have emerged in the traditional resource use structure in high mountain areas in developed, developing as well as in underdeveloped regions of the world. As a result, mountain regions of the world are passing through a process of rapid environmental changes, and exploitation and depletion of natural resources leading to ecological imbalances and economic un-sustainability both in upland and lowland areas. Moreover, the changing climatic conditions have already stressed mountain ecosystems through higher mean annual temperatures and melting of glaciers and snow, altered precipitation patterns and hydrological disruptions, and more frequent and extreme weather events rendering the mountain communities and their economy more vulnerable to long term impacts of climate change. Besides, the recent food crisis followed by global economic recession have threatened the agricultural and food systems and livelihood security among mountain communities because of their subsistence economies, constraints of terrain, climate and resultant physical isolation and low productivity, vulnerability to natural risks, poor infrastructure, limited access to markets, higher cost of production, etc.

Our understanding about the problems of mountain regions and approach to their development has undergone drastic changes, during the recent years. Sustainable development of mountain regions assumed added significance in global context after Agenda 21 of Rio Summit 1992 and International Year of Mountain 2002. Nevertheless, mountains have long been marginalized from the view point of sustainable development of their resources and inhabitants and currently, mountain ecosystems as well as mountain communities are particularly threatened by global environment changes and emerging new international economic and political orders and the resultant problems of food and livelihood insecurity. In 2003, it was estimated that nearly 245 million rural-population living in world-mountains was vulnerable to food insecurity.

This book addresses these critical issues and looks at the ways to stop the downward spiral of resource degradation, rural poverty and food and livelihood insecurity in mountain regions, new and comprehensive approaches to mountain development are needed that identify sustainable resource development practices, how to strengthen local institutions and knowledge systems, and how to increase the resilience between mountain environment and their inhabitants. The book is divided into five sections:

beginning with impacts and risks of global changes on mountains, followed by drivers for change in mountains, adaptation frameworks, governance and legal issues in mountains followed by a few case studies to highlight the impacts of changes on mountains.

The prelude by **Jack Ives** gives an excellent introduction to the book and to the mountains, "They are certainly special regions, but they are places where communities live and have lived for centuries, and where environmental sustainability is a matter of survival not just to those who live there but for the many millions more for whom they are just a hazy blue line on the horizon."

DRIVERS FOR CHANGE AND IMPACT OF GLOBAL CHANGES ON HIGH MOUNTAINS

The only chapter in this section by **Borsdorf et al.** examines the impact of global changes on high mountains. As discussed by the authors, climate change and globalization meet complex man-environment systems in mountain regions. Whereas the chapter by Jack Ives serves a good introduction to majestic mountains, the chapter by Borsdorf et al gives good introduction to impacts and risks. The authors discuss the impact of climate change such as glacier—and permafrost retreat, water scarcity, soil erosion, as well as loss of biodiversity. Climate change thus impacts on ecosystem services for both the inhabitants of the mountain region as well as people living in the adjacent lowlands. Globalization processes also exert an ever faster impact on ecosystems, cultural landscape, agriculture, population structure, mobility, as well as the urbanization and marginalization of peripheral mountain regions. Societies are affected in socio-economic, political and institutional terms and require decision making at the regional and local level. Both subsystems interact in the sphere of land use and land management.

Global climate change and globalization have triggered dramatic changes in the Alps that are visible using the indicators discussed in the chapter. In the cultural landscape, persistent structures used to be at work for a long time, but the process that has its roots in the beginnings of industrialization is forcing accelerating dynamics onto them, speeded up further in recent decades as a result of globalization. Within the cultural space of the Alps, social and economic impact factors may dominate, but with each 'warm' year it becomes clearer that climate factors are gaining in significance. Climate change determines tourism just as much as the growth options for settlements and commercial areas, what is and is not a secure road and the routing of new roads and rail tracks. We may not yet perceive the full extent of this impact, yet the consequences are real and ever more significant.

Since Global Change affects man-environment systems with greatly varying intensity and speed in different mountain regions, comprehensive long-term observation and monitoring programs are necessary to capture it. As an example of a monitoring system, authors have discussed a global initiative: GLORIA. The challenges of climate change and globalization increasingly calls for target knowledge to be provided to decision makers in politics and the economy. It is vital to include stakeholders, not only in the creation of system knowledge but even more important in the assessment and transformation into target knowledge in truly trans-disciplinary style. For example, the authors have discussed projects such as TRIP and DIAMONT.

Borsdorf et al. have discussed the driver of change impacting high mountains and it is quite apparent that climate change has emerged as one of the major drivers transforming consistently the natural environment, society and economy of the mountains regions in all parts of the world, and mountains ecosystem being highly sensitive are extremely vulnerable to these changes. However, there is still a high degree of uncertainty about the trends and magnitude of climate change and its impacts on mountain systems. It is therefore highly imperative to improve our understanding of the trends of changes in temperature and variability in precipitation pattern at the local level through downscaling of regional climate models. This would require establishment of a comprehensive networks of hydro-meteorological monitoring stations in mountain areas across the world, particularly in the mountains of developing countries where currently, such monitoring is extremely lacking. Furthermore, a sharp focussed and comprehensive research on climate change impacts assessment, vulnerability and adaptation to climate change would be necessary at micro-regional scale.

ADAPTATION AND MITIGATION: STRATEGIES AND MODELS

As discussed by **Tiwari and Joshi** in the next chapter, an effective mechanism for the sharing of information, data, experience and knowledge generated from local, regional to international levels and international level transfer of knowledge would be crucial for better understanding of changing climatic conditions and evolving appropriate strategies for mitigation of climate change and responding to its impacts. Since mountains constitute headwaters of some of the largest trans-boundary basins on the Earth, it would also be indispensable to establish and strengthen international research collaboration, and develop international mechanisms on knowledge and data sharing. A regional geo-political cooperation framework among riparian countries is therefore highly crucial not only for evolving the framework of adaptation to climate change and improved governance of headwaters resources, but also for security and peace. In order to address the challenges posed by climate change, the mountain countries and regions should develop mountain specific adaptation and mitigation policies, programs, institutions and think tanks which would be necessary to enhance their resilience and ensure socio-economic and ecological sustainability in mountain areas. The authors have discussed some of these adaptation strategies in their chapter. For example, mountain communities through their traditional resource management practices contributed significantly towards preservation of forest and biodiversity, climate change mitigation through carbon sequestration, water conservation and preservation of cultural heritage and natural landscapes that provide a variety of ecosystem services and goods to considerably large populations in the downstream. In turn, the global community should contribute towards the conservation of natural ecosystem and improvement of the quality of life of mountain people by providing adequate incentives. Benefits of opportunities such as Reducing Emissions from Deforestation and Degradation (REDD) and Enhancement of Carbon Stocks (REDD+) under the United Nations Framework Convention on Climate Change (UNFCCC) which offer incentives for developing countries to reduce emissions from forested lands and invest in low carbon activities for their sustainable development, have also been discussed in this chapter. The forests conservation in mountains needs to be linked with climate change mitigation and adaptation, poverty alleviation and food and livelihood security of local people. Tiwari and Joshi have described some recent experiences of Forest *Panchayats* and lessons learned from Joint Forest Management (JFM) in the Indian Himalaya and Community Forestry (CF) in Nepal that can be replicated and used for institutionalizing forests and for their community oriented conservation and development. A considerably large proportion of population in mountain regions of developing countries depends for its livelihood on severely limited arable land symbolizing distress husbandry of land. Strategically, to preserve forests and biodiversity in the mountain regions, it is important to look beyond the traditional agricultural system and generation of rural employment opportunities in off-farm and non-traditional sectors in the mountains area of less developed countries. One of such examples can be promotion of local rural enterprise in different sectors of tourism.

The following chapter by **Monreal et al.** takes a discursive approach to explore the importance of adaptation to global climate change as a viable strategy in mountain regions. After describing the role of global change as a process that unfolds between interacting pairs of opposing forces/actors/trends and the reconciliation of which is the aim of recent thought in the study of social ecological systems. The authors have argued that mitigation and adaptation can be understood as part of this dialectic process. As other authors in the earlier chapters have discussed that changes in the mountainous regions not only impacts population/communities in the high mountains but also on lowlands. Monreal et al. also make a case for lowlands for a concerted effort on the part of lowland societies to support adaptation in mountain regions if for nothing else but their own self-interest. The authors have looked at the global significance of, and adaptation strategies for marginal places.

So far the chapters in the book have highlighted that the mountain ecosystems are most fragile and sensitive to climate change. Global climate estimates predict larger temperature rise in the mountainous regions even under conservative scenarios. Vulnerability of mountainous ecosystems is much higher because it is a repository of natural resources; water, forests and biodiversity, also it is home of the most poor people of world who are primarily dependent on the natural resources for their livelihood. Climate drivers like precipitation and temperature variability along with non-climatic drivers make the situation adverse and forces for reactive and planned adaption strategies. The operators and receptors of the adaptation are diverse and differ from place to place; hence the purpose of adaptation should be well defined for these regions. Adaptation models, tools and techniques need greater details of understanding to assess interaction, opportunities and challenges among available indicators. The chapter by **Joshi et al.** reviews available climate change adaptation frameworks and identifies their applicability for the Himalayan ecosystems. While Tiwari and Joshi discussed policies and strategies for adaptation, Joshi et al. have looked at different models and how information from these models can be used for adaptation policy making by decision makers. As discussed by Borsdorf et al., policy makers and decision makers look for information to make decisions around mitigation and adaptation to climate change, this chapter provides a comprehensive set of information for decision makers for developing region specific adaptive policy and researchers to analyze the importance of these, while recommending more models relevant to adaptive capacity to climate change.

LAWS AND GOVERNANCE

As we all know making laws and regulations is perhaps the easier part, but implementing them is the hard part (as rightly pointed out by **Frangetto**), to regulate mountains is probably more difficult than to climb a high mountain.

In this section (of the chapter) the focus is on governance and laws about mountains and how to appreciate the limits of nature rather than destroying the ecosystems. The section begins by an overview of international and national laws by **Leuzinger and da Silva** which looks at the development of 'soft law' at Rio followed by examples from Europe and Brazil. As discussed by Frangetto, Leuzinger and da Silva it will be interesting to have a global international accord designed specially to protect mountains.

The next part of this section by **Rosaio** looks at the Alpine Convention and if this can be a model for protecting trans-boundary areas of the mountains. The efforts of protection are also analyzed by checking what international organizations and research networks have been doing in regards to mountains.

After this analysis, **Vieria et al.** present a socio-environmental perspective of the ecological risks and disasters in hillsides and mountainous areas in Brazil because of the impact of global changes on mountains. As argued by the authors: The ecological disasters, either of natural or technological origin, can be understood as some of the greatest subjects of the current Environmental Law, for its aggravation before the climatic changes, intensification of the risk generation due to the fast technological development, but especially as a consequence of the environmental vulnerability generated by poverty, which contributes for a bigger exposition to the human rights violation. Probably the best at this point is to at least apply a 'precautionary principle' to prevent further damage to the mountainous ecosystem.

CASE STUDIES

The last section in this book looks at different case studies to see how specific mountains/mountainous regions are being impacted by the global changes discussed above. The first case study is from Africa. As discussed by **Evaristus,** some mountains such as those found in East Africa (e.g., Mt. Kenya, Mt. Kilimanjaro), appear like 'islands' rising above the surrounding plains. Mount Kenya like other mountains of the world is a zone where the signals of global change are quite apparent. Already there are indications that the glaciers on Mt. Kenya are receding and snow on its summits is disappearing. As the author has discussed in the chapter that although likely impacts of climate change on the mountain's tourism are unclear, yet the main issue is: Could climate change affect Mount Kenya's ecosystem so irreversibly that mountain tourism could become unsustainable?

Moving away from Africa to Asia, in the next chapter, **Chettri et al.** look at the impact of global changes on the Hindu Kush-Himalayas (HKH), one of the most dynamic ecosystems in the world with a rich and remarkable biodiversity. The region is endowed with a high level of endemism, diverse gene pools, species and ecosystems of global importance. As a result, the HKH have been highlighted in many global conservation prioritization strategies. However, this diverse ecosystem of the HKH is facing overarching threats from various drivers of changes including climate change. The ecosystems in the HKH are degrading mainly due to lack of incentive provisions for maintaining ecosystems and the goods and services provided by them. This is leading to development that is unsustainable including loss of biodiversity. Even the protected areas such as national parks, nature reserves and wildlife sanctuaries face tremendous pressures from external driving forces and communities living inside and outside of the HKH. It is a paradox that in spite of being rich in biodiversity the region is also home to poorest of the world and the most vulnerable in the face of climate change. So, there is a mounting challenge to balance conservation with development in the region. The authors of this chapter have documented the reconciling initiatives on maintaining ecosystem resilience through integrated conservation and development initiatives to address prevailing climate change challenges faced by the region with some evolving regional experiences.

Pathak in the next case study also looks at the HKH region and has observed the need for trans-boundary scientific cooperation to establish an effective communication between scientific community and policymakers to identify knowledge gaps for better understanding the complexities of the HKH region, and allow policy options based on appropriate scientific evidence. Since anthropogenic activities are the main engine of changing the HKH region, it is imperative for decision makers to revisit and redesign their growth and development strategy, along with new approaches apart from the ongoing adaptation and mitigation measures. All this however is not possible without bundling peace and security in the region. Since, HKH's environmental issues are trans-boundary, regional cooperation is imperative for peace and security and for sustainable development.

The next case study by Rawat describes the impact of climate change on community food and livelihood in the Lesser Himalaya in the Dabka watershed (a part of the Kosi Basin in the district of Nainital in India). The spatial distribution of climate throughout study area suggest three types of climatic zones, i.e., sub-tropical, temperate and moist temperate which are respectively favorable for mixed forest, pine forest and oak forest in the mountain ecosystem. The results of climate-informatics advocate that all these climatic zones are shifting towards higher altitudes due to global climate change and affecting the favorable conditions of the existing land-use pattern, e.g., decreasing the oak and pine forests. Consequently the high rates of forest degradation accelerated hydrological hazards during the monsoon and non-monsoon periods. The non-monsoon water-related hazards (i.e., decreasing underground water level, drying up of perennial springs and decreasing trends of stream water discharge) reducing irrigation facilities and decreasing the irrigated land by 2% each year from 1985 till 2012, whereas the monsoon hydrological hazards (i.e., flash flood, river-line flood, soil erosion and water induce landslide, etc.) degraded the 22% agricultural land each year through extreme hydro-meteorological events. Consequently deficit agricultural crop yields have led to food deficiency and the people are now adopting other occupations which ultimately is changing the existing rural livelihood pattern.

Sati explores the impact of global changes in Kewer Gahera sub-watershed in India. Just like in any other region, high population growth rate has also been observed in the Kewer Gadhera sub-watershed, increase in forestland and decrease in agricultural land was also observed during the last four decades. The author in this chapter has examined land-use/cover changes during the last four decades in the Kewer Gadhera. Sati has also described how large scale emigration and land abandonment have led to the land-use/cover changes in the region.

In the last case study about the HKH Tibetan plateau glaciers **Hasnain** takes a slightly different angle and has explored the geopolitics in the region and how international cooperation can be used to deal with impact of climate change on glaciers. From all the case studies discussing the HKH and Himalayan rivers, it is apparent that climate change is impacting the region and water source on which billions of people depend for water and their livelihoods. It has also become clear that it is a trans-boundary issue and keeping the sensitivity of relations between neighboring countries it is imperative that some kind of cooperation mechanism is developed to deal with the problems arising due to global changes in the region.

Globally, the glaciers are in a general state of retreat, most probably because of climatic warming. They often leave behind voids filled by melt water called glacial lakes which tend to burst because of internal instabilities in the natural moraine dams retaining the lakes (e.g., as a result of hydrostatic pressure, erosion from overtopping or internal structural failure) or as a result of an external trigger such as a rock or ice avalanche or even earthquake. In the last chapter with examples from the Himalayan region, **Manfred et al.** have taken the case study of GLOFs to describe the impact of climate change on high mountains (especially as they relate to melting glaciers). This is an important issue because of the damage it can cause to the settlements around the area.

Just like the HKH are important for water supply for millions of people, Sri Lanka's Central Highlands occupy a unique position among the main geographical zones of the country and has a diverse blend of most of the world's climatic features. The Central Highland area is also the watershed for 103 main rivers and more than 1,000 feeder streams joining the main rivers. This area is the heart of the entire country because of its important ecological conditions and as an economic driver for the whole country. However, the management of the sensitive ecosystems of the Sri Lankan Highlands is extremely fragile and destructive which started with the deforestation during the colonial period and is still going on. Climatic changes and the destruction of natural potential influence the whole country. **Breuste and Dissanayake** in this chapter have identified some of the strengths/weaknesses, opportunities and threats that have an impact on this area to identify the impacts of global social, political and environmental changes on Sri Lanka's Central Highlands. As suggested by the authors a strategy for sustainable utilization together with the preservation of natural potential must be developed and implemented to secure the values of the unique tropical highlands.

Moving from Asia to Europe the next chapter explores the post-disaster development in HighTatras National Park as an example of how natural and social systems imposed to global changes are becoming vulnerable to external as well as internal disturbances. Global climate changes may seriously affect the system's capacity to absorb external and internal disturbances and the adaptation capacity of natural and social systems without fundamental changes of their quality. In our study, the windstorm that devastated forest ecosystems of High Tatras Natural Park accelerated existing conflicts and institutional threats in resource management and created several social challenges for forest management and spatial development in the region. **Finka and Kluvankova** in this case study have explored 'Urbars' as a governance regime to cope with unpredictable disturbances and complexity of global changes. As discussed by the authors, the operation of urbars determine ecosystem dynamics and sustainable use of forest resources as an attribute of economic profit. In addition institutional structure of urbars increase internal system stability and reduces vulnerability against external shocks. Urbars are thus seen as more resilient than individual private or state property resource regimes.

The next few case studies focus on the Americas (both North and South). The chapter by **Barkin** examines the ingredients that went into developing constructive strategies to facilitate the survival of the hundreds of ethnic groups that continue to inhabit the highlands of Latin America. Without going into historical discussion of these developments, this presentation offers a suggestive examination of the concerted efforts by communities throughout the region to maintain their identities, to develop mechanisms to assure increasing degrees of autonomy in the face of intensifying efforts to integrate them into the ranks of the poorest people in national societies and global markets. Throughout Latin America mountain peoples are rediscovering and updating their traditional cosmologies along with their knowledge systems to develop unique proposals for harnessing their material, human and natural resources to improve their quality of life and ensure the protection of their ecosystems. The author has described selective local development initiatives but a lot of activities are being undertaken by millions of Mexicans and people elsewhere in the world implementing local development strategies on the margin of and in place of unsatisfactory market-based solutions. They are reclaiming cultural mechanisms for organizing productive structures responsive to local needs and strengthening traditional governance organs while creating a new generation of local cadre concerned with their societies' quality of life and the health of their ecosystems; in the process, they are transforming market relations with the outside world, replacing the commercial partners with fair trade institutions and other 'niche' marketers that protect them against unequal exchange.

Kaz et al. have discussed the impact of global changes on Canadian Rockies. In the Canadian Rockies, three major rivers (Fraser River, Columbia River and Saskatchewan River) provide freshwater to millions of Canadians for domestic, agricultural, industrial and power generation usage, as well as for tourism. The authors have looked at two case studies: the Cline River watershed in western Alberta and the Okanagan River watershed in the interior British Columbia. In both of these cases, impact of climate change on the seasonal variation of stream-flow is important in agriculture and power generation, as well as in tourism. The forest industry in British Columbia will be significantly impacted from the ecosystem redistribution of tree species, with many of the important conifer species quickly losing their habitats. A lot more research is required to understand how the interaction between the changes in meteorological variables under climate change and those factors which influence the biodiversity of the environment and the nature of human socio-economic activities.

Andean societies have always coped with environmental, economic, cultural and political changes. Often these exogenous forces have disturbed traditional livelihoods, required new adaptations, and also changed Andean landscapes and societies. At times, they also resulted in new economic outlooks, acculturations, migration patterns and different spatial and societal disparities. While the various agrarian reforms during the last half century attempted to achieve a more equitable land distribution and new opportunities for Sierra farmers in new colonization areas, the current impact of 'agro-capitalism' and market-orientation has created a situation, in which wealthy and powerful private and corporate stakeholders are the winners, with a majority of rural people remaining at the periphery of this form of development. Within the rural societies an intensified education and training (*capacitación*) and a mobilization and empowerment of people to remain 'in control' of their environment, and to find their own forms of resilience, adaptive strategies and development alternatives, will be imperative. However, these efforts have to be 'accompanied' and supported by an external financial, infrastructural, technical and political assistance. The ultimate objective of this partnership support has to be a sustainable and equitable improvement of the livelihoods for the majority of *campesinos,* rather than the often prevailing aim of maximizing outputs and profits for a small minority of stakeholders, without much concern for a responsible environmental stewardship. **Stadel** has suggested a conceptual model of 'Sustainable *Campesino* Communities' on the basis of favorable intrinsic and extrinsic influences.

Byrne et al. have discussed how climate change will affect the Rocky mountains of western North America. Climate change is real and ever present, and the role of each of us in changing the climate is also real and present. The complex terrestrial and aquatic ecosystems in the Rocky Mountains have evolved within the climate since the last great glaciation ended 10,000 years ago. Human induced climate change will bring weather extremes on both the short and long-term to the Rocky Mountains. The authors have discussed hydrological response to climate change, examined the impact of global changes on the ecosystem of the Rocky mountains: snow and the terrestrial ecosystem (there will be a greater variability in soil moisture condition in forests and valleys that can lead to longer and more intense droughts, more forest fires and some places will experience extreme flood events) and aquatic biota (aquatic species will need to adjust to mega floods, droughts, changing water temperatures and maybe challenged by invasive species invading their territory because of climate change). This is followed by some case studies of response to climate change in the Rocky Mountains (including glaciers and rare Alpine invertebrate, bull trout, freshwater algae) and the North Saskatchewan River. The authors have also looked at the adaptive management in the Rocky Mountains. The Rocky Mountains are a vast and complex region that is valuable both for resources and ecosystems. The Rockies cannot provide the valuable resources we need and treasure, particularly clean plentiful water, unless we protect and conserve mountain ecosystems. Hopefully the discussion in this chapter of the major changes ongoing in the Rocky Mountains due to climate change will help in understanding the impacts of climate change and add to the collective will in society to minimize this change in future.

In the next case study **Ginzo and Faggi** have assessed the information on regional vulnerabilities, mitigation and adaptation actions to plausible impacts of Climate Change on two mountain regions in Argentina: one in the Northwest (NOA region), and the other-in the West (Cuyo region) of the country. The NOA and Cuyo regions are closely related to the septentrional and the mid sections of the Argentinean Andean Ridge. As has been discussed by the authors the hydrologic regime in the areas of the Cuyo and NOA regions are most impacted by the changes in the Andean system of mountains and changes in precipitation and temperature due to global warming. Water stored in glaciers and in forested watersheds is expected to decrease in the future, thereby putting under stress the wellbeing of the regional population and ecosystems and, the population downstream which depends on these rivers. To cope with these threats the regions of NOA and Cuyo should take decisive and timely steps to increase their adaptability and resilience to a changing climate. As explained by the authors, in view of the foregoing each of NOA and Cuyo regions will need to implement context-specific policies of adaptation and mitigation to the extent of the expected climate change impacts. Most of those would be in line with recent conclusions from a relevant meeting of Latin American countries.[2]

Just like GLOFs are increasingly becoming a concern in the areas where glaciers are melting (discussed by Manfred et al. earlier in the book), mudslides are posing a huge challenge on hill/mountain sides as well. This becomes even more complicated when there are people illegally settled on these slopes. The authors in this chapter have looked at some technical issues such as stability of the slopes and social and environmental issues such as eroded hill sides, unplanned and illegal inhabitation of the fragile land in unstable housing, and socio-economic conditions of immigrants in urbanization process that lead to the disasters and its consequences. The authors have also looked at the way zones should be demarcated to mark the most risk-prone areas and how a monitoring system should be set up to save lives (at least till integrated plans can be put into place to prevent such disasters).

Conclusion

Mountain's ecosystem support about one-tenth of the human population. Mountains not only support the people living on it but also people on plains—since all the rivers originating from mountains carry water and sediments downstream supporting agriculture, fisheries and other livelihoods in addition to providing drinking water. Mountains are important part of global water cycle (since they store water in the form of snow or glaciers) and also centers of biological diversity.

As can be seen from the case studies, global changes are impacting mountainous ecosystems and livelihoods/lifestyles, economy and social structure of the communities. It is imperative that governments develop mitigation and adaptation strategies for these communities. As discussed in the various chapters in the book, mountain environments have been deemed essential to the survival of the global ecosystem in Chapter 13 of *Agenda 21*, the outcome document of the Rio 'Earth Summit' in 1992; by seven UNGA resolutions—including one in 2013; and by *The Future We Want*, the outcome document adopted at Rio+20 in 2012. However, the revised Zero Draft of the Sustainable Development Goals (new goals that will replace the

Millennium Development Goals) does not adequately reflect the vital role that mountains have in sustainable development. At this point, it is crucial that the targets be amended to ensure that mountains remain a priority through to 2030. There are some regional and global initiatives looking at furthering research and developing strategies for mountain communities. The Mountain Partnership sponsored by FAO play significant role in inter-connecting these various regional and global initiatives. Regional information networks need to be established which would act as effective learning and awareness generation forums between specialists, civil society organizations, and government agencies and to support capacity building activities through focussed education, training and research (as suggested by different authors such as Tiwari and Joshi, Pathak in the book). Further, in order to build resilient mountain social-ecological systems, the support and cooperation of civil society, including Non-Government Organisations (NGOs), Civil Society Organisations (CSOs), the private sector, educational and research institutions would be inevitable. These institutions play effective role in sensitizing policy planners, decision makers and society at large about the importance and significance of mountain ecosystems in sustaining global society as well as about their fragility, marginality, vulnerability and the emerging opportunities.

Endnotes

1. http://www.cbd.int/iyb/doc/prints/iyb-eu-biodiversitymessages-mountainecosystems-en.pdf.
2. Conclusions of a regional Technical Meeting on Climate Change impacts, adaptation and development in Mountain Regions. Organised by the Ministerio de Relaciones Exteriores de Chile, the Mountain Partnership, FAO, and the World Bank. 26–28 October 2011. Santiago de Chile.

SECTION 2

DRIVERS FOR CHANGE AND IMPACT OF GLOBAL CHANGES ON MOUNTAINS

3

Impacts and Risks of Global Change

Axel Borsdorf,[1,a,*] *Johann Stötter,*[2] *Georg Grabherr,*[3]
Oliver Bender,[1,b] *Carla Marchant*[4] and
Rafael Sánchez[5]

INTRODUCTION

This chapter analyzes the impacts and risks of climate variation and global change in mountain regions. Both involve highly sensitive ecosystems, providing services for the lowlands, and autochthonous ethnicities, traditional agricultural systems and regional traditions, still alive in mountain areas. While the impact of climate change is dramatic, its effects may be seen in a mid- to long-term dimension. In contrast, the effects of globalization have a much shorter time scale. The scientific analysis of both effects may help to develop adaptions strategies to secure sustainable regional developments in these important regions of the globe.

[1] Institute for Interdisciplinary Mountain Research, Austrian Academy of Sciences, Technikerstr. 21a, 6020 Innsbruck, Austria.
[a] Email: axel.borsdorf@oeaw.ac.at
[b] Email: oliver.bender@oeaw.ac.at
[2] Institute of Geography, University of Innsbruck, Austria, Innrain 52, 6020 Innsbruck. Email: hans.stoetter@uibk.ac.at
[3] GLORIA, Institute for Interdisciplinary Mountain Research, Austrian Academy of Sciences, Rennweg 14, 1030 Vienna, Austria.
[4] Institute of Geography, Universidad Austral de Chile, Valdivia. Email: carla.marchant@gmail.com
[5] Pontificia Universidad de Chile, Vicuna Mackenna 4850, Santiago Chile. Email: rafael.sancheza@gmail.com
* Corresponding author

Impacts of Climate Change

Climate warming and globalization have serious impacts on environment, space and society in mountain regions. Climate variation results in falling crop yields in many areas. However some mountain regions, like high latitude zones or in temperate climates may profit—to a certain degree and in some sectors—by climate warming, whereas harvests and yields will be reduced in many developing and even developed regions.

During the recent years, a variety of changes have emerged in the traditional resource use structure in high mountain areas in developed, developing as well as in underdeveloped regions of the world mainly in response to the globalizing economy and increased population dynamics. As a result, mountain regions of the world are passing through a process of rapid environmental changes, exploitation and depletion of natural resources leading to ecological imbalances and economic un-sustainability both in upland and lowland areas. Moreover, changing climatic conditions have already stressed mountain ecosystems through higher mean annual temperatures and melting of glaciers and snow, altered precipitation patterns and hydrological disruptions, and more frequent and extreme weather events rendering the mountain communities and their economy more vulnerable to long term impacts of climate change. Besides, the recent food crisis followed by global economic recession have threatened the agricultural and food systems and livelihood security on mountain communities because of their subsistence economies, constraints of terrain and climate and resultant physical isolation and low productivity, vulnerability to natural risks, poor infrastructure, limited access to markets, higher cost of production, etc. This has led to renewed concerns about the sustainability of mountain ecosystems and their inhabitants and calls for further investigation of the vulnerability of mountain communities with respect to ongoing process of global environmental change.

With the melting of inland icefields and glaciers and even the disappearance of small mountain glaciers both water and energy supply are endangered in several areas. To a certain degree, mountains may lose their function as water towers and energy providers. Specifically irrigation agriculture in the Mediterranean, semi-arid and arid climates is affected. As another consequence—at least for those mountain regions which are located near an ocean or, as islands, in an ocean—sea the level rise threatens coastal settlements and even some major cities on the shore. Alterations in marine ecosystems occur with evidence to the coastal environment and industry. With respect to mountains however, the terrestrial ecosystems are much more affected. Biodiversity is changing, many species will face extinction, invaders are immigrating and may enforce the process of suppression. Also, cover changes, and increases and decreases are to be observed.

Extreme weather events are increasing. This means a rising intensity of storms, extreme droughts, heat waves and forest fires in some regions and heavy precipitations and inudations in others. The precipitation variability is growing.

Environment

Glacier and permafrost smelting

The general climate warming since the end of the Little Ice Age ca. 150 years ago has left distinct traces in the cryosphere. In most high mountain areas across the world, glaciers have lost mass and surface area (Oerlemans 2005), permafrost thawing has increased thus causing mass movements (Haeberli et al. 2006), the zone of 'good winters', i.e., those with a continuous snow cover of 100 days, has withdrawn into higher altitudes (Breiling and Charamza 1999).

In the Alps, early studies of glaciers were triggered by catastrophic bursts of glacier lakes in the course of a cooling climate around the year 1600 (Nicolussi 1990). Since the Little Ice Age maximum around middle of the 19th century, Alpine glaciers have systematically been investigated (e.g., Schlagintweit and Schlagintweit 1850), since ca. 1890, changes in the length of the glaciers have been recorded annually (Patzelt 1970). In the second half of the 20th century, data on glacier mass balances have been collected by the World Glacier Monitoring Service. A global summary of changes in the cryosphere in general and the mountain glaciers in particular is included in the IPCC report (Lemke et al. 2007).

Since their Little Ice Age maximum extension around the middle of the 19th century, the glaciers of the Austrian Alps have lost at least 50% of their area (Gross 1987, Bender et al. 2011). The changes in glacier area are documented by country in glacier inventories (Abermann et al. 2009, Lambrecht and Kuhn 2007) largely based on remote sensing methods (Haeberli et al. 2007). Since the end of the last glacier-favouring period from the mid-1960s until the beginning of the 1980s, with some glacier advances, an intensive, more and more accelerating retreat of Alpine glaciers has been observed. Related phenomena, such as the formation of supra- and peri-glacial lakes require new strategies in glacier monitoring (Paul et al. 2007) and trigger research on basic mechanisms (Huss et al. 2007).

As consequence of the recent warming, the last two decades being the warmest since the beginning of instrumental measurements (Lemke et al. 2007), the glaciers of the Alps have undergone a tremendous development (UNEP and WGMS 2008). Due to negative mass balances, glaciers show losses both in volume and surface area from year to year. Hintereisferner, one of the best investigated glaciers in the Austrian Alps (e.g., Span et al. 1997), may be an example. Based on the worldwide unique airborne laserscan data set, it was shown that during the last 10 years, Hintereisferner with over 7 km² among the 10 largest glaciers in Austria, has lost more than 20% of its volume (Bollmann et al. 2011) (see Fig. 3.1).

Maximum annual losses were recorded in 2002/2003 and 2005/2006. The extreme summer temperatures of 2003 (see Schär et al. 2004) had an enormous impact on Hintereisferner (Fig. 3.2), resulting ca. 5% loss in the glaciological year 2002/2003 alone (Geist and Stötter 2007). As a further consequence of these extraordinary thermal conditions, a surplus of melt water at the basis of the ice-body caused a drastic leap of the glacier dynamics thus resulting in a doubling of the velocity of Hintereisferner (Bucher et al. 2006).

By studying long glaciological time series we obtain base lines for interpreting past climate in high mountain areas where only few direct recordings of weather stations are available (Vincent et al. 2004). Recent comparisons of glacier and climate data led to the development of models that allow an interpretation of glacier stands from the Holocene, documented by moraines, to extrapolate the climate of the time (e.g., Kerschner and Ivy-Ochs 2008).

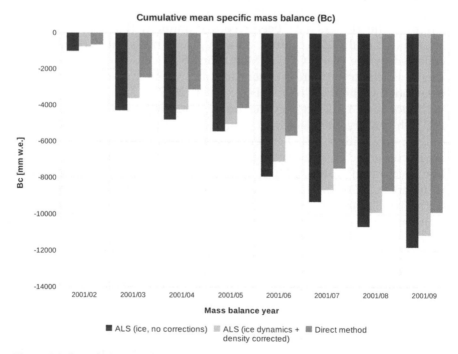

Figure 3.1. Cumulative mass balance of Hintereisferner for the glaciological years 2001/2002 to 2008/2009, elaboration by the authors.

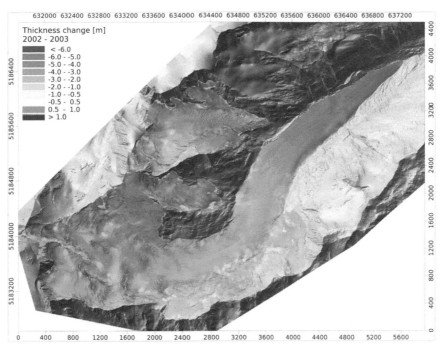

Figure 3.2. Surface elevation change of Hintereisferner for the glaciological year 2002/2003, derived from multi-temporal airborne laser scanner data. Blue indicates positive elevation changes (>0.3 m), white a 'stable' surface and yellow to red colours indicate negative elevation changes. For the respective period, the overall elevation change of Hintereisferner was −3.7 m, elaboration by the authors.

Color image of this figure appears in the color plate section at the end of the book.

While high latitude or polar permafrost is a significant driver of global warming due to increasing greenhouse gas emission, i.e., carbondioxide and methane that are released as consequence of its thawing, high altitude mountain permafrost is a sensitive indicator for climate change. Furthermore, its thaw bears considerable potential for natural hazards. After dramatic mass movement events in the Swiss Alps in summer 1987, Haeberli et al. (1990) were first in connecting climate change and thawing permafrost.

Since mountain permafrost controls both the stability and hydrologic behaviour of rock and debris slopes, its thaw and the related loss of stability bears a significant potential to cause natural hazards by facilitating mass movement and initiating slope instability processes (Stötter 1994, Stötter et al. 2012). Permafrost-related hazards include: (i) permafrost creep and the transport of material into debris flow zones, (ii) thaw settlement and frost heave, (iii) debris flow from permafrost due to increased depth of active layer, (iv) destabilization of frozen debris slopes as well as (v) rock-fall and rock avalanches from frozen rock faces (Kääb et al. 2006). Human life and settlements as well as infrastructure for high mountain tourism (huts, buildings, ski-lifts, hiking trails, climbing routes, etc.) and hydropower plants (buildings, dams, reservoirs) are potentially threatened by these events (Krainer et al. 2007).

The occurrence of permafrost as such does not present a danger, but permafrost is inherently a transitory phenomenon. Given this situation, detailed knowledge of the distribution of permafrost-related slope activity on a local scale is fundamental to any statement about the potential of permafrost hazards and for developing strategies by decision makers in natural hazard management (Monreal and Stötter 2010).

Recently the analysis of multi-temporal laserscan data sets, allowed a first quantification of area-wide permafrost melting (see Sailer et al. accepted). As a result of the general lowering of permafrost underlain high mountain areas, the losses of subsurface ice could thus be calculated. For the last decade melting rates average out at 5–10 cm per year can be observed (Fig. 3.3).

Impact on biodiversity

Initial indications of warming-induced migration of species, especially in the nival zone, date back to the beginning of the 20th century (Klebelsberg 1913). In the 1950s, Braun-Blanquet (1958) confirmed an increase in species richness on selected summits above 3000 m in the Rhaetian Alps. Grabherr et al. (1994) showed this to be a general trend in this region and beyond, even if some of the 25 summits studied, for which old and reliable records existed, presented no pronounced increase in species richness. Walther et al. (2005) reported that in the recent, very warm, decades the process of upward movement has accelerated. Meanwhile, effects of climate warming on alpine plants have been recorded for other mountain regions as well.

The current status of evidence-based climate impact research on mountain vegetation at cold-determined ecotones can be summarized as follows:

- the tree line ecotone has become denser (e.g., Urals; Moiseev and Shiyatov 2003; Fig. 3.4) and its upper limits are moving up; this trend had been reversed during cooler periods (1960s/1970s) as Kullmann (2007) showed for the tree line of birch forests in the Scandinavian Mountains;
- the total number of species on mountain summits, especially in the nival zone, has risen and implies an upwards move of species (e.g., Alps: Grabherr et al. 1994, Walther et al. 2005);

Figure 3.3. Surface elevation change of Reichenkar rock glacier (Austrian Alps), derived from multi-temporal airborne laser scanner data. Blue indicates positive elevation changes, white a 'stable' surface and yellow to red colours indicate negative elevation changes. Note: The area with blue colours are indicating horizontal motion, elaboration by the authors.

Color image of this figure appears in the color plate section at the end of the book.

- overall, reliable old species inventories are rare and missing for most of the high mountain areas; this makes it difficult to use high mountain areas as otherwise ideal sites for comparative studies of cold habitats across the world;
- life at the low-temperature limits, especially vascular plants, are excellent indicators for an ecological assessment of the impact of

Figure 3.4. Change in the treeline ecotone in a wilderness region of the Southern Urals in the previous century. The annual mean temperature increased by about 2°C with the period from 1919 to 1999. Photographs by Moiseev and Shiyatov, Yekaterinburg.

climate change, even if the vegetation reacts with some delay; this means that the responses of high mountain plants reflect trends of climate change rather than short-term vacillations;

- historical inventories are only usable if an exact localization is possible and a reliable identification of species combined with suitable quantitative measures;
- many high mountain areas are virtually free of human influence and even where there is some impact, there are always undisturbed and difficult to access places and peaks, i.e., high mountain areas are among the few regions on earth where the climate signal can be captured without interference.

These considerations are the basis for the monitoring network GLORIA, the Global Observation Research Initiative in Alpine Environments. It is established as a long-term programme and implemented across the world. It focuses on the well-documented indicatory value of alpine and nival plants and plant communities. The major results to date are:

Even the basic surveys in the GLORIA target regions provide new biogeographical and ecological findings. Some GLORIA teams have published their data and all data sets are compiled in the network's central database (see www.gloria.ac.at). These data represent a reliable record on alpine species for the selected summit areas from five continents. In remote or less studied mountain regions, e.g., in the Andes of southern Peru or in the high mountains of the Iran, hitherto unknown species may be discovered.

The data of the pilot project GLORIA Europe confirm the general hypothesis that phytodiversity (only vascular plants) at lower latitudes is higher than at high latitudes. The GLORIA summit floras at mid-latitudes (Alps, Pyrenees, Caucasus) deviate from this pattern by being the species-richest. The Mediterranean mountains (Sierra Nevada, Apennines, Lefka Ori [Crete]) are home to the most endemic species. For the target regions Sierra Nevada (Spain) and Hochschwab (Austria), Pauli et al. (2003) found that the proportion of endemic species increases with altitude. A generally warmer and drier climate (Sierra Nevada) should eventually lead to the loss of this unique mountain flora. Salick et al. (2009) and Grabherr (2009) estimate that this is also true for medicinal plants. In the Himalayas, Tibetan doctors use 76% of alpine species for medicinal purposes. However, Gottfried et al. (1999) have shown through spatially explicated modelling that microrefugia may support a longer survival. Some vegetation types, such as subalpine knee timber of *Pinus mugo* in the Limestone Alps, may be highly resistant against warming-driven impacts even in the longer term (100 years and more) (Dullinger et al. 2004).

The modern view of the altitudinal zonation of high mountain areas, which Humboldt already presented in a comparative approach, is a sequence of vegetation belts that are connected through narrower ecotones (Nagy and Grabherr 2009). By far the most conspicuous ecotone is the transition between closed montane forest and the treeless alpine zone. So far, attempts to describe this treeline ecotone in a general way have failed owing to the high level of individual forms dependent on the particular mountain area (Holtmeier 2009). Nor has the debate about the ecological causes of the tree line been concluded (Körner 2003, Butler et al. 2009).

Less distinct than the tree line ecotone is the transition from the alpine to the nival zone. Some authors understand this zone where the closed alpine grassland disintegrates into an open patchy vegetation as an altitudinal belt, the subnival zone; others see it as an alpine-nival ecotone (Nagy and Grabherr 2009). In the nival zone, the number of species is decreasing considerably (Grabherr et al. 1995, Körner 2003), but there are a number of species with their centre of distribution in this upper zone. These nival species are mainly characterized by a high tolerance of long snow cover, as Gottfried et al. (2002) were able to show through comparative samplings in the area of the GLORIA master station Schrankogel. Snow cover also prevents frost damage from cold spells in summer (Larcher et al. 2010).

On the whole, high mountain research has paid much less attention to the alpine-nival ecotone than to the tree line ecotone. A clear definition along abiotic and biotic criteria is still missing as is a detailed understanding of the processes involved in the dissolution of the vegetation and the exclusion of large numbers of alpine species in the nival zone.

A high frequency of monitoring cycles does not work in alpine vegetation. Alpine and nival species are long-lived perennials (Nagy and Grabherr 2009), some form clonal populations that can reach an age of several 1000 years (Grabherr 1997). It can be assumed that there is no massive inter-annual variability in the presence of the species as was confirmed by repeat photographs from permanent plots and by transplanting experiments in the Botanical Garden of Vienna (Holzinger and Grabherr 2009). A frequent reinvestigation of the plots would, moreover, cause damage by trampling. In 2008, a concerted repeat survey of the GLORIA Europe plots was carried out and the comparative data analysis is about to be completed. For the reasons quoted above and because of the generally moderate warming within the last seven years, dramatic impacts on a Europe-wide scale are not to be expected. On the GLORIA summits in the Dolomites, South Tyrol/Italy, however, species richness has increased by around 10% between 2001 and 2006 (Erschbamer et al. 2009).

Soil erosion

Loss of soil through either through denudation processes (soil erosion) on arable fields or shallow erosion in steep grasslands is a common problem in mountain areas. For instance, researchers point to an increase in soil erosion across the Alps within recent decades (Meusburger and Alewell 2008, Tasser et al. 2005, Geitner 2010). Soil erosion and nutrient losses are a severe problem in tropical agroecosystems (Maass et al. 1988), where—under demographic and economic pressure—shifting cultivation economies are substituted by permanent husbandry. Although deforestation, overgrazing and intensive agriculture are causing accelerated erosion, climate change phenomena inducing erosion (intensive rainfall, increasing variability of precipitations) are intensifying the process, as well as forest fires increase runoff and sediment yield rates, specifically in Mediterranean climates (Inbar et al. 1998).

The partial loss of soil by splash, rill, interrill and shallow erosion (Govers et al. 1999, Hessel et al. 2003) is accompanied by a degradation of its numerous ecosystem services. In addition to its performance in agriculture and forestry, the soil provides important functions as buffer and filter of pollutants from water and air, as carbon sink and as archive of natural and cultural history. Soils in the mountain regions provide such functions in varying degrees and take a long time to develop because of the limiting conditions in the sometimes extreme locations. Often the soils are shallow and more severely threatened by various morphodynamic processes (Fig. 3.5; Geitner 2007, Wiegand and Geitner 2010).

The Alpine Convention recognized the special situation of alpine soils and the threats to their performance. It included soils as a resource that deserves protection in the Soil Conservation Protocol and demanded concrete conservation measures (CIPRA 1998). Tasser et al. (2005) relate the apparent increase in soil erosion to the changes in land use as a result of changes in the socio-economic framework within the Alpine Space. Changes in agricultural use take the form of extensive management and

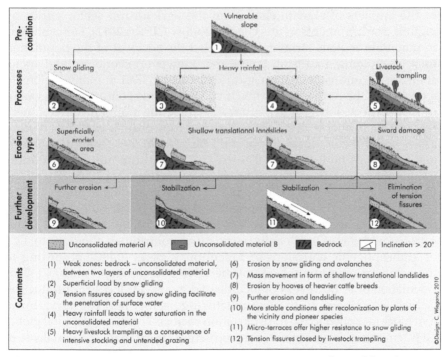

Figure 3.5. Simplified overview of process, erosion types and possible subsequent developments of shallow erosion in high mountains (Source: Wiegand and Geitner 2010: 81).

Color image of this figure appears in the color plate section at the end of the book.

abandonment of inaccessible and thus unprofitable mountain grasslands or of an intensification of easily accessible areas. Measures to manage areas extensively go along with a reduction of mountain pasture staff and a concomitant reduction of maintenance activities on the mountain grasslands. This leads to untended grazing if not the complete abandonment of mountain pasture activities, leaving these areas to natural succession processes. In contrast, more easily accessible areas are grazed more intensively or mown several times a year, increasing the yield through fertilization. In either case, the species composition of the vegetation cover changes, and with a certain time lag, affects the soil and its stability via changes in the litter composition (Tasser et al. 2005).

Land use

Agriculture

Agriculture is highly exposed to climate change, as farming activities directly depend on climatic conditions (Eitzinger et al. 2009). The severity of the impacts of climate change on the agricultural sector varies by regions (the arguments in this chapter follow CIPRA 2011). For centuries, in mountain regions farming has enabled the survival of the population thereby shaping a cultural landscape. However, economic and social changes, i.e., industrialization and urbanization, lead in many regions to a decline in the number of farms. If current trends continue to prevail, it must be expected that the agricultural sector in the mountain environment is shrinking further with the risk of depopulation of areas with poor natural assets and difficult access. Support for farming in these marginal areas is justified by its multifunctional roles, i.e., agriculture not only produces foodstuff but maintains the cultural landscape (Pruckner 2005). Therefore, financial support is granted in some countries at varying degrees.

It has to be said that global agricultural systems contribute substantially to climate change, primarily via emissions of methane (CH_4) and nitrous oxides (N_2O). Globally, agriculture contributes about 10 to 12% of anthropogenic emissions of greenhouse gases (GHG) according to the common reporting scheme of the UNFCCC (Smith et al. 2007). This calculation does however not include energy use for fertilizer production (which is assigned to the manufacturing sector) and for agricultural machines (ascribed to the transportation sector). Therefore, the total global GHG emissions from agricultural production are estimated to actually sum up to a much higher share (ITC and FiBL 2007). In addition, CO_2 emissions from agricultural soils are not included in the emission balance of the agricultural sector but in the land use, land-use change and forestry sector because they originate mainly from land-use changes such as deforestation.

Although agricultural lands generate very large fluxes of CO_2 to and from the atmosphere, the net flux of CO_2 is small (Smith et al. 2007).

Methane and nitrous oxides from agriculture contribute about 47 and 58% of total methane and nitrous oxide emissions, with a wide range of uncertainty, however. N_2O emissions from soils and CH_4 from enteric fermentation constitute the largest source, biomass burning, rice production and manure management account for the rest (Smith et al. 2007, CIPRA 2011: 8).

On the other hand climate change has a strong impact on agriculture. The effects shown in Fig. 3.6 differ between regions in occurrence, magnitude and impact.

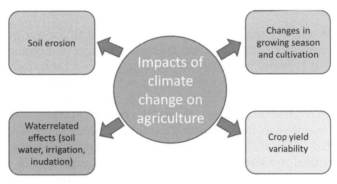

Figure 3.6. Climate change impacts on agriculture (modified from CIPRA 2011).

Soil erosion is one major impact of climate change. Specifically mountainous terrain is prone to soil erosion due to its steepness. Excess water due to intense or prolonged precipitation may cause tremendous damage to soil. Erosion is projected to aggravate with increases in precipitation amount and intensity (EEA 2008).

Ecosystems are strongly interlinked with the hydrological cycle that has already altered over the past several decades. The shrinking of glaciers, permafrost and snow cover, changes in precipitation patterns and increasing temperatures and evaporation will increase the competition for water by different sectors, in particular during the summer months when precipitation and run-off is reduced. The groundwater level is decreasing, and irrigation in semi-arid and arid zones is affected. F.i. in the Southern Alps the groundwater level dropped by 25% over the last 100 years (EEA 2009).

Increasing air temperatures are significantly affecting the duration of the growing season. In temperate- and cold humid zones the impact on plants is mainly reported as a trend towards an earlier start of growth of plants and grasslands in spring and its prolongation into autumn. This may enhance agricultural productivity as well as the change of cultivated crops.

As climatic conditions become more erratic (increase in frequency and scope of extreme events like floods, heat waves and severe droughts) new uncertainties in the future of the agricultural sector must be considered. More frequent drought could result in decreased productivity and declining quality. In permanent grassland, water scarcity might cause formation of gaps in the sward which can be colonized by weeds with negative implications for animal nutrition (Fuhrer et al. 2006). On the other hand heavy rainfall and floods have significant impacts on productivity and yield.

Tourism

Tourism is one of the largest and fastest growing economic sectors and of great importance for mountain regions. Tourism is obviously related to climate. It is therefore surprising that the tourism literature pays little attention to climate and climatic change (e.g., Witt and Witt 1995). It is equally surprising that the climate change impact literature pays little attention to tourism (e.g., Smith et al. 2001, Bourdeau 2008).

Since recent years the situation is changing. Three branches of literature have started to grow (Hamilton et al. 2003). Firstly, studies came up that build statistical models of the behaviour of certain groups of tourists as a function of weather and climate (e.g., Maddison 2001, Hamilton et al. 2003). Secondly, there are studies (e.g., Abegg 1996, UNWTO 2009) that relate the fates of particular tourist destinations to climate. Thirdly, there are studies that try to define indicators of the attractiveness to tourists of certain weather conditions (e.g., Matzarakis 2002). These three strands in the literature share a common deficit, namely the lack of a larger, global assessment of push and pull factors of international mountain tourism.

Climate defines not only the length and quality of tourism seasons, affects tourism operations, and influences environmental conditions that both attract and deter visitors but also determines the activities of tourists in their destinations. This is why the sector is considered to be highly climate sensitive. The effects of a changing climate have considerable impacts on tourism and travel businesses in mountain regions.

It has to be considered, that tourism and travel is also a vector of climate change, accounting for approximately five% of global carbon dioxide emissions. By 2035, under a 'business as usual' scenario, carbon dioxide emissions from global tourism are projected to increase by 130% (UNWTO 2009: 2).

Climate defines the length and quality of tourism seasons like hiking, summer and winter sports (e.g., Bender et al. 2007). Changes have considerable implications for the tourism flow. Mountain destinations are highly climate-dependent. It also directly affects various facets of tourism operations in mountains (drinking water, artificial snow production, heating and cooling, irrigation needs, pest management, evacuations and temporary closures due to floods, avalanches and other hazards). Furthermore tourism in mountain regions is based on a high quality natural environment which is to a high degree sensitive to climate variability (wildlife, biodiversity, surface water, glaciers). Last but not the least climate is a crucial determinant of tourist decision making and a key-driver of tourism demand at global and regional scales. Weather is an intrinsic component of the travel experience and influences tourist spending and holiday satisfaction (UNWTO 2009: 5).

Climate change increasingly threatens winter tourism, at first in lower-lying areas, later on possibly in all tourist destinations. If we assume a temperature rise of 4°C, only 30% of winter sports places will be able to guarantee snow (Abegg et al. 2007). This development will, however, be subject to great regional differences. It can be foreseen that the remaining winter sport destinations will be overcrowded in the future, whereas the decline of winter tourism will be highly problematically for others. In the European Alps, clear losers in this scenario are the lower-lying skiing areas with pistes not exceeding 2500 m a.s.l. (Alpine fringe, large parts of the Eastern Alps), while higher destinations with downhill runs above 3000 m (glaciated regions of the Western Alps and places in the Central Alps) are in a naturally more favourable position to cope.

In global competition, the destinations have to adapt to the changing behaviour and choices of tourists as well to customize their offer, specifically in winter, when climate change affect the snow availability. In the European Alps, about two decades ago, municipalities began to mitigate lack of snow by making artificial snow and have come to rely on snow canons more and more (Hahn 2004). Artificial snow makes the runs mechanically more robust and better adapted to developments in skiing equipment and to the rising standards of the visitors. Undoubtedly this must be seen as an example of successful development and implementation of adaptation strategies to the challenges of climate change.

During the Djerba International Conference on Climate Change and Tourism 2003 in the round table on tourism in mountainous regions, Richard Richardson described the effects of climate change on nature-based tourism in the Rocky Mountain National Park. As this park receives about 87% of its visitors (3 million a year) during the May to October period climate change already has positive effects on the demand, and it is estimated that there will be a visitation increase of 10–14% until 2020 (UNWTO 2003: 49). Bourdeau (2008) aspects that there might be a significant change in tourist's behaviour in the next decades: In summer former beach tourists will prefer the cooler climate in the mountains, whereas coastal destinations gain attractiveness in winter (Fig. 3.7). Abegg et al. (2008: 77) also predict

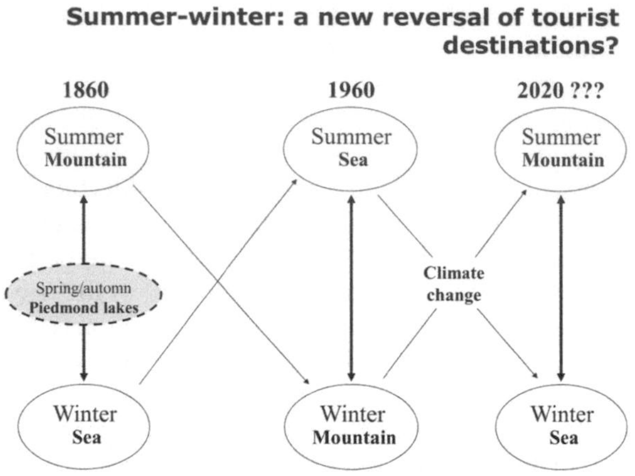

Summer-winter: a new reversal of tourist destinations?

Figure 3.7. Change in tourism by effects of climate change (Source: Bourdeau 2008).

that the negative impacts of climate change in Alpine tourism might be to a large part compensated by the positive impacts.

Therefore, highly developed winter destinations must diversify away from a narrow orientation on winter sports and towards sustainable tourism. For such a transition to be handled successfully, it is vital to create the right awareness (Kronberger et al. 2010). This will take new communication strategies as well as comprehensive participatory processes involving all stakeholders as well as tourist operators and the tourists themselves.

Water and energy provision

All the major rivers of the world have their headwaters in highlands and more than half of humanity relies on the freshwater that accumulates in mountain areas. Although they constitute a relatively small proportion of river basins, most of the river flow downstream originates in mountains, the proportion depending on the season (Liniger and Weingartner 1998). They provide a favourable temporal redistribution of precipitation and reduce the variability of flows in the adjacent lowlands. These 'water towers' are crucial to the welfare of humankind. As demand grows, the potential for conflict over the use of mountain water increases.

Mountain water resources are indispensable for fresh- and industrial water supply, irrigation, hydropower production and ecosystem services (Viviroli et al. 2003). The importance of mountains and their sensitivity to climate change was the subject of recent benchmark reports (Solomon et al. 2007, Bates et al. 2008, Stern 2007, WWAP 2009). Viviroli et al. (2011) analyzed the importance of the water supply of mountains based on detailed case studies.

Water, energy and climate change are inextricably linked, as energy production and use are sensitive to changes in the climate. Increasing temperatures will reduce consumption of energy for heating but increase energy used for cooling buildings. Global primary energy demand is projected to increase by over 50% by 2030. Freshwater withdrawals are predicted to increase by 50% in developing countries, and 18% in developed countries (UNEP 2007). The implications of climate change for energy supply are less clear than for energy demand. Mountains provide cheap and CO_2-neutral energy by their hydro-electrical potential, but also by other renewable energy production opportunities (solar, wind and geothermic energy). For example, in December 2008, the US Environmental Protection Agency announced an inter-agency agreement between the offices of Air and Water to collaborate on energy and climate efforts at water utilities.

Hydroenergy is by far the largest renewable source of electricity. Hydropower in mountainous regions is mostly dedicated to provide energy in times of high demand (pump storage plants). So there is a strong difference in hydroenergy production between different seasons (winter, summer, rainy and dry season). More and more hydropower serve as a storage facility for those energy sources having a weather dependent component (solar and wind energy).

Reservoirs store both water and energy and are becoming increasingly important for the management of climate change effects. Twenty nine percent of dams worldwide are used for hydropower and only 10% have hydropower as their main use (WBCSD 2006). Most of them are used for flood control, freshwater supply, irrigation, recreation or other purposes. Thus, water supply depends to a high degree on artificial water storage in mountains.

Impacts form climate change on regional and global hydrological systems is increasing, bringer higher levels of uncertainty and risk. Climate alteration has a significant impact on hydropower generation. However, the effects of solar radiation, evaporation, glacier retreat, run-off variations are not yet analyzed sufficiently (Borsdorf 2010).

Climate change effects on water and energy supply and demand will depend not only on climatic factors, but also on patterns of economic growth, land use, population growth and distribution, technological change and social and cultural trends that shape individual and institutional actions.

Urban habitat

A considerable proportion of the population in mountain regions is concentrated in large cities or even megacities. The causes of this dramatic growth can be traced back to overseas immigrations from the end of the 19th century, migration movements from nearby rural areas or even from other cities during the 20th century. Among the 40 urban agglomerations with more than five million inhabitants with more than 392 million inhabitants (Borsdorf and Coy 2009) 17 are located in mountainous environments, among those Mexico City, Guatemala City, Bogotá, Lima, Santiago and Rio de Janeiro are located in Latin America. Andean cities have received an enormous demographic burden which has provoked a macrocephalic effect in some countries, as in the case of Peru and Chile, where both Lima and Santiago de Chile, are 9 resp. 6 times bigger than the next biggest cities in terms of inhabitants ('primacy'). Apart from this huge mass of people we have to bear in mind that the largest and most important cities of the Andes represent more than two thirds of the productive capacity of their respective countries and are at the same time closely related to other national and international agglomerations.

It is evident, that the impact of climate change in such mountain megacities affect more people than in less densely populated areas. Heavy and long rainfalls cause catastrophic floods and in some cases mass movements, heat waves warm the urban agglomerations more than open land, droughts and wet years influence the water, energy and food supply. Increasing water scarcity is one of the major problems in near future to many megacities, even in mountainous areas, which are still function as water-towers.

Long high air pressure periods lead in agglomerations located in basins to the concentration of polluted air under a temperature inversion roof and may have in consequence health problems for the population. The warmer rivers are a source of pathogenic bacteria (Abraham 2011), the case of SARS in Asia a few years ago has highlighted the specific role that megacities as global hubs play in the spread of new diseases (Münchener Rück 2005).

Impacts of Globalization

Drivers of globalization

Whereas climate change effects have a long time scale, the impacts of globalization are much faster and in many cases even more effective than the natural processes. So the challenges for mountain regions cannot be analyzed without considering the socio-economic and cultural changes and the political framework. Mountains are regions where the manifold interdependencies between economic, political, social and cultural globalization processes and their effect are felt locally in many ways, as in most cases because of their peripherical location, inaccessibility of their inhabitants who were conserved in traditional culture, conserved traditional land- use methods and were—with exception to the global tourism and of mining—relatively separated from the global market. This is changing dramatically since the neoliberal paradigm got influence even on the most marginal regions of the globe and so also is influencing the mountain regions.

Neoliberalism as the dominant ideology of the age of globalization has been instrumental in the breakthrough of the principles of deregulation, privatization, flexibility, global competition, free transfer of goods and capitals. Against this background, economic policy everywhere, and also in mountainous regions, is moving towards a more market-oriented control of regional development. This means a shift in the constellations of policy actors. In many places, the private actors gain in influence at the expense of public authorities when it comes to the control and concrete design of regional development. Here, interdependencies with globalization also make themselves felt, as the real estate market and agriculture become an increasingly interesting business opportunity for globally acting real estate agents, tourism managers, planners and investors (Borsdorf and Coy 2009).

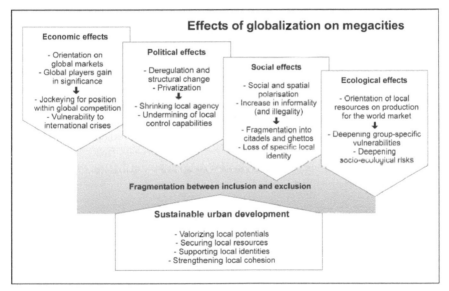

Figure 3.8. Effects of globalization on mountain agglomerations (Design: Martin Coy, source Borsdorf and Coy 2009).

The traditional population on one side suffers by these processes, on the other hand they adopt the new lifestyles and economic conditions. Globalized fast-food, music, vesture are mostly accepted, the fight for land rights, acceptance of regional cultures or indigenous nations, better prices for local products and acknowledgement to traditional land- use techniques cause many conflicts.

Amenity migration

Globalization—with free mobility of persons, goods and capital—brought new migrants to the mountain regions. They are looking for fresh air, healthy environment, tranquility and local culture—and in some cases spirituality— the amenities provided by mountain regions (Borsdorf 2009, Borsdorf and Hidalgo 2009, Borsdorf et al. 2012). Thus they may be called amenity migrants (Moss 2006) or lifestyle migrants (McIntyre 2009). Bourdeau (2008) interprets amenity migration even as part of post-touristic trends. Some authors estimate the migration reversal (from mountain exodus to mountain immigration) as a sustaining force for mountains and their cultures (Moss 2006), others look more critically to this new trend (Borsdorf et al. 2011). It started in the U.S., but is a wide spread phenomenon in many mountain regions of the world, and also includes international migration.

Amenity migration often is motivated by the search of a paradise on Earth. Among the senior migrants the desire of a retirement home is decisive, among the economical active people those professions are dominant, in which the place of work is arbitrary (medicals, lawyers, architects, etc.) or digital channels of communication can be used for work.

Although researchers only recently started to analyze the phenomenon, the literature is increasing rapidly. Amenity migration has been investigated in the Alps, the Scandes, the Carpathians, Rocky Mountains in the U.S. and Canada, the Andes, the Central American Cordillera, the Austrialian and New Zealandian Alps and the Philippines, among others (for a comprehensive overview see: Moss 2006 and Moss et al. 2008).

In a case study in the Chilean Andes based on their interests and behaviour different types of amenity migrants could be differentiated. Figure 3.9 gives an overview.

Impact on mountain tourism on local cultures

With summits, unique and rare flora and fauna, and a great variety of hill and mountain cultures, the tourism potential of mountains is very high. Tourism is one avenue where mountain specificities that are generally considered constraints to development—remoteness, difficult access, wilderness, insular cultures, and subsistence lifestyles—can be transformed into economic opportunities. The example of the Alps, where tourism already started in the beginning of the 19th century, demonstrated the strong impact on economic growth, overcoming the economic problems of the beginning of the industrial age. Being labour intensive, having relatively high multiplier effects, and requiring relatively low levels of capital and land investment, tourism can yield significant benefits in remote and rural areas where traditional livelihoods are under threat.

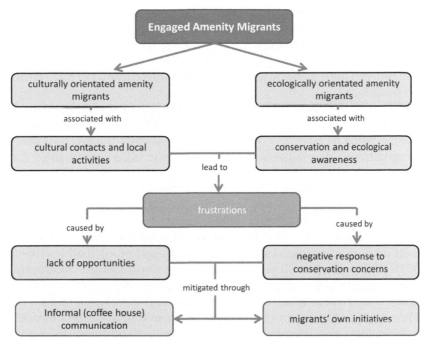

Figure 3.9. Classification, expectations and frustrations of amenity migrants in Pucón and Villarrica, Chile (Source: Borsdorf et al. 2012).

Tourism in manifold forms influences the local culture in the destinations. Culture may be defined as the complex composite of knowledge, beliefs, moral laws, conventions, customs, abilities and practices humans developed in their society (Tylor 1871). Today we may add the definition of White (1948): "Culture is a vast stream of tools, utensils, beliefs that are constantly interacting with each other, creating new combinations and synthesis. New elements are added constantly to the stream; obsolete traits drop out. The culture of today is but the cross section of this stream at the present moment, the resultant of the age-old process of interaction, selection, rejection, and accumulation that has preceded us."

In the 1970s and 1980s the socio-cultural effects of tourism in peripherical regions (like mountains) were discussed intensively (Krippendorf 1984) whereas today a more moderate perspective is dominant, regarding tourism as one factor among others which causes cultural change. There are four phases of cultural influence on the visited regions (Lüem 1985): In the beginning of tourism, tourists are perceived by the locals as idle rich and compared with the own situation, often characterized by relative poverty and hard work. The next step is an imitation effect which causes socio-cultural changes. The adoption of foreign cultural goods (clothing, music, drinks, food, etc.)—firstly by the young people—may lead to tensions and frustrations. This phase is followed by identification, which means the complete takeover of the foreign value systems and a weakening of the local culture. The last step is the acculturation to the dominant western civilization. This may be observed by a commercialization of local customs and hospitality.

Globalization has so far an enormous impact on mountain tourism all over the world, as travelling became easier (travel facilities, cheap air fares, increase of mobility). With globalization even the kinds of tourists changed from mountaineers and spa-tourists to hikers, sportsmen, skiers and young backpackers.

The Alps are among the earliest tourist destinations and from the beginning the focus has been on the landscape. With the advent of mass tourism in the 1960s, tourism and the leisure industry have become a major economic factor in the rural areas of the Alps (12% of jobs, 16% of GDP), albeit with great differences between regions and municipalities (Bätzing 2003). Within the last 30 years, a long-term trend towards skiing-based winter tourism has emerged for large parts of the Alps, with significantly higher added value than summer tourism. This development goes hand in hand with a knock-out competition between tourist destinations. Only municipalities that invest continuously in their tourism portfolio will be able to achieve growth in the future (Fig. 3.10).

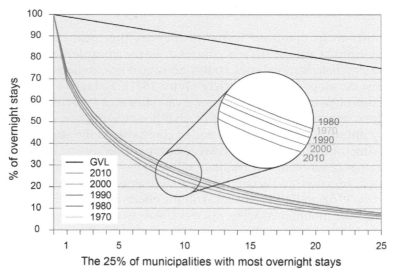

Figure 3.10. Increasing concentration of tourism in Austria (Lorenz Curve): GVL stands for the hypothesis that pernoctations are equally distributed to the municipalities. The distance to GVL demonstrates the concentration of tourism on some municipalities. © O: Bender 2011.

Color image of this figure appears in the color plate section at the end of the book.

Mining

Mountains offer manifold deposits of ore, minerals, and other raw materials. In the forelands often huge deposits of raw oil and gas are located. Whereas in some mountain regions (like the Alps) these deposits are mainly exhausted, others (like the Andes) are still exploited, and new deposits are detected. The mining sector causes severe damages to the environment and suffers by climate change, as in many cases water is needed, but the supply is more and more endangered by water scarcity and glacier retreat.

Even more the changes of the global economy affect the mining industry. Global players take over national mines and transfer the earnings to the centrals. They claim more and more land for future exploitation. This already caused social movements, struggling for the livelihood of farmers and other inhabitants of the affected areas (Bebbington et al. 2008, Bebbington 2012).

Agriculture

A variety of agricultural systems adapted to the specific environment exist in mountain regions. In temperate and arid climates forms of pastoral nomadism (alpine pasture farming, transhumance, nomadism) are frequently observed, in tropical mountains irrigation and rain-fed agriculture dominate the warm and temperate levels, and the higher levels are dominated by permanent stock farming. Frequently farmers use different levels. In many tropical mountains a community or an individual farmer possesses and utilizes agricultural areas from the foot of the mountains till the upper end of vegetation. In addition lower levels in some mountains show different forms of shifting cultivation. In temperate climates farmers use the upper stories for cattle grazing and dairy production in summer and have their homestead in the valleys. In arid or semiarid climates nomadism and transhumance also use different levels, according to temperature or rain seasons.

Economic globalization affects mountain areas in a similar manner to other peripheral regions. They now have to compete within the global market, as products cultivated under more favourable conditions enter even into remote markets. It can be said that under global competition mountain agriculture is becoming less and less competitive. This may be demonstrated by the example of the Alps. There an increasing abandonment of farms and marginal areas can be observed. In the Italian Alps in particular, we can observe a pronounced decline, while in the northern, German-speaking, Alpine countries and especially in Austria such processes are rare (cf. Borsdorf 2005, Borsdorf et al. 2010, Tappeiner et al. 2008). Here, the agrarian cultural landscape has largely been retained, even if it has been adapted in many places to allow modern utilization. This is linked to the mountain farming subsidies, which started several decades ago, but also to the fact that Austrian mountain farms are mostly run as a part-time concern, i.e., in combination with other economic activities, mainly in tourism. Moreover, agricultural production, especially in the Alps, has been oriented to high quality products, viz. the high proportion of organic farming and regional origin certified products (Bender 2010).

Despite the relatively stable situation of agriculture in Austria as compared to that of southern Alpine countries, recent decades did see considerable losses of cultural landscapes, leading to an expansion of wooded areas, particularly in steep mid-slope areas of the montane zone. Figures from the cadastre show that in the 1990s alone the wooded area increased by 5.5% (Borsdorf and Bender 2007). Mountain forests fulfil many functions for humans, particularly as protective forests and for timber production, but also as CO_2 sinks for climate protection. Forestry is increasingly making an effort to plant mixed forests suited to the individual location with varieties adapted to climate change. In the past, particularly near ore and salt processing sites in eastern Austria, fast growing fir trees were planted as monocultures. These are now being replaced by mixed forests. In many places, mountain forests are exposed not just to climate stress but also to additional factors such as game damage, immissions, pests, etc., which further increase the vulnerability of forest ecosystems to climate change. Overall, forestry in the Austrian Alps must be considered highly vulnerable to changes in climate (Kronberger et al. 2010).

At higher altitudes, more extensive mountain pasturing basically encourages a rise in the forest line, which has been held down artificially through grazing on land mostly cleared by burning. There are as yet no clear indications of tree stands rising solely as a result of changes in climate (Nicolussi et al. 2005, Wieser et al. 2009). The ecological conditions in the treeline ecotone are too complex and the response times of subalpine forest communities too long to draw direct conclusions (Borsdorf and Bender 2007).

Human habitat

The cultural landscape of mountain regions is unique and has a large variety of different forms and structures. Thus, many cultural heritage sites of the UNESCO are located in the mountains, in recent years even the immaterial heritage has been protected by this organization. However, large cultural regions are in danger by climate change, which affects water and electricity supply, air and water quality. Droughts and heavy precipitations put agriculture at risk. Mountain biosphere reserves, conceptualized as models of sustainable development face the challenges by implementing climate change adaptation technologies like organic farming, mixed cultures, and bioengineering. Good examples are the Río Piedras Basin in Southern Colombia (Borsdorf et al. 2011) or the livelihood studies in Northern Colombia (Marchant and Borsdorf 2013).

Under the pressure of globalization specifically the large agglomerations in mountain regions lose their individual shape and are transforming to globalized cities, not only in architecture, but also in the social space. Gated communities, increasing segregation, and fragmentation are indicators for this development (Borsdorf and Coy 2010).

Risks

Risks of climate change

Without doubt mountains are much more prone to risks than most other regions in the world (Stötter and Monreal 2010). At a first glance there is an increasing risk of dangerous feedbacks and abrupt, large-scale shifts in the climate system. Images of flooded valley bottoms, destroyed houses, eroded roads or cut-off villages may come into mind. These risks related to natural hazard processes, e.g., rock fall, debris flows, avalanches or floods, have been part of the specific human-environment system in mountains regions ever since settlement and intensive utilization of land began. In this sense they were perceived as part and parcel of the mountain environment (in the sense of a base disposition), infrequently topped by extreme events with the character of an existential threat (in the sense of a variable disposition). By learning to cope with these challenges throughout centuries, mountain societies have adapted to the specific natural hazard conditions of their specific environments.

In addition to these locally or regionally controlled interrelationships, nowadays global driving forces impact on human-environment systems in general, i.e., (i) global climate change, (ii) globalization, and (iii) resources scarcity. While the regional variations of climate change processes alter natural process dynamics, globalization processes cause new demographic, cultural, social and economic structures. As a consequence both base and variable disposition are superimposed by a trend of new forcings thus resulting in new dimensions of challenges to mountain societies never experienced before.

This story is more than a tale of minor relevance in an anyhow extreme environment. A fifth of the terrestrial surface is classified as mountains and roughly 12% of the world population lives in mountain areas. Moreover, these changing conditions affect nearly half of the humanity in the adjacent medium- and lower-watershed areas as they depend strongly on mountain-bound resources in one way or another (see Millennium Ecosystem Assessment 2005).

Ecosystem services, primarily water, hydropower, flood control and tourism, exceed the geographic limits of highlands through direct linkage with adjacent lowlands in catchments systems and for the extractive resources of mountains, such as timber and minerals exists global demand (Viviroli and Weingartner 2004).

Traditional risk concept

In order to address the linkage between natural hazard impact and exposed human systems and the intrinsic uncertainty of all future developments the idea of risk provides a conceptual framework with a high integration potential (e.g., Bohle and Glade 2008, Veulliet et al. 2009). In most risk concepts the sensitivity of the reacting system to the external impulse is determined by vulnerability and capacity, respective resilience, which as interacting and linking factors govern the dimension of risk and as a consequence the adaptability of the human-environment system.

Originally expressing the sensitivity of organisms to external impact in ecological systems, vulnerability can be seen much wider. In a (under) development context, vulnerability explains the degree to which an exposed social system is susceptible to harm due to perturbation or stress, and further the ability or lack thereof to cope, recover or fundamentally adapt (see Chambers 1989). Also in the interface study between natural hazards and human systems (Wisner et al. 2004), many approaches deal with dimensioning vulnerability to specific process magnitudes (Hollenstein et al. 2002, Thieken et al. 2005). In the context of global/regional warming, vulnerability may be understood as the degree to which a human-environment system is susceptible to, or unable to cope with, adverse effects of changing climatic conditions (see, e.g., Füssel and Klein 2006). Like vulnerability, the general idea of resilience, the ability of a system to withstand a shock-impact and to rebuild itself, has been adopted by different disciplines, thus evolving considerable understandings since it was first brought into discussion by Holling (1973).

While in natural systems, resilience stands for the capacity to tolerate disturbances without collapsing into a new state of the system controlled by a different set of conditions (Diamond 2005), resilience in the context of social systems is focussed on the added capacity of individuals to anticipate and to plan for the future (Watts and Bohle 1993). In modern interpretations, the concept of resilience is applied to social-ecological systems (Walker et al. 2004).

Both vulnerability and resilience have a dynamic character, as they are subject to temporal and spatial changes within the relationships between forcing and reacting systems (see Bohle and Glade 2008), thus being key pre-requisites for the understanding of human-environment systems in their attempt to adapt.

Risks related to natural hazards

In Central Europe in general and in the Alps in particular, a surprising clustering of extreme runoff events with a recurrence probability of one in 100 years and less is evidence that the frequency-magnitude relationship of natural hazard processes triggered by hydro-meteorological driving has undergone a marked change (first noted by Bader and Kunz 1999). The analysis of floods since the 1990s shows i) runoff maxima exceeding all measured records, and ii) the coupling of two or even three extreme flood events in independent river systems within a short period of time which is statistically highly unlikely (Stötter et al. 2009). Most prominent examples are the floods of both Bregenzer Ache and Lech in the years 1999, 2002 and 2005. The statistical probability of such an accumulation of extremes of approximately 1:30,000 highlights, how unlikely and thus how significant these events were.

These events may be seen as a clear indicator for a trend towards a more intensive precipitation—runoff relationship as a consequence of global/regional warming. Recent modelling of future scenarios supports the idea that due to seasonally differentiated warming with maxima from May to August and from November to February and marked increase of mean precipitation in the winter season will cause new patterns of the frequency-magnitude relationships. This new trend is additionally supported by the fact that higher winter temperatures will cause a tremendous rise of the snow line, which means that a much higher portion of winter precipitation will fall as rain (see Beniston 2003).

As major consequences within in mountain human-environment systems, this development of precipitation-runoff relationships means a major threat to the natural hazard management system. As this is based on the acceptance of defined protection limits (goals) and remaining risks, it does only provide measures against, e.g., floods, which have so far been understood as 100 years events but may now and in future be expected to occur more frequently. The extreme socio-economic changes since the mid 20th century, resulting in population growth and rapidly increasing numbers of houses and other values further contribute to this new challenge.

Risks related to global climate change—global change

In the Fourth Assessment Report the Intergovernmental Panel Climate Change (IPCC) stated that most of the observed global warming over the last 50 years has very likely (>90 to 99% probability, see Manning et al. 2004, IPCC 2005) been caused by greenhouse gas forcing, while it is cited to be very unlikely (1 to 10% probability) that it is due to known natural external causes alone (Hegerl et al. 2007). These findings altered the perception of climate change fundamentally and consequently led to a worldwide acceptance that global warming and dependent changes of further climate elements as well as multiple effects on nature and society are no longer disputable—they have become a fact (see, e.g., Oreskes 2004). Prior to the World Climate Conference held in Copenhagen in December 2009 a group of IPCC authors highlighted that in recent years multiple evidence has been produced for even more drastic warming in the 21st century.

Due to the complex topography as well as specific and spatially intensive variability of human-environmental sub-systems, mountains tend to become regions affected by Global Climate Change far beyond average. In some mountain areas, it can be shown that warming trends and anomalies are elevation dependent, where increasing temperature has a steeper positive gradient in higher altitudes (e.g., in the Alps, see Böhm 2009).

The impact of intensified climate change on the natural mountain environment has become especially apparent in the shrinking water storages of the cryosphere, i.e., snow and ice cover (see, e.g., Lemke et al. 2007, UNEP and WGMS 2008). Being key components of the hydrological cycle this causes further radical changes of the seasonal character (regimes) and amount of runoff of mountains and adjacent lowlands (see, e.g., Viviroli et al. 2007, Bates et al. 2008). In fact mountains are the source for 50% of the world's rivers (Beniston 2003). The Himalaya and Hindu Kush alone feed the Indus, Ganges, Brahmaputra, Irrawaddy, Salween, Mekong, Tarim, Yangtse and Yellow Rivers, the Alps supply the Rhine, Po, Rhône and Danube tributaries. Mountains are i) water pumps, which extract moisture from the atmosphere through the orographic uplift of air masses and they are ii) water towers due to their water storage capacities in glaciers, permafrost, snow, soil and groundwater. Much of the inter- and intra-annual variation of discharge is compensated by discharge from mountains. In semi-arid areas mountain discharge accounts for 50–90%, in extreme cases (e.g., Nile, Colorado, Rio Negro) for more than 95% of the total river discharge (Viviroli et al. 2007). Roughly 23% of China's 1.3 billion people depend on glacier discharge from the Himalayas. The Alps supply a significant proportion of freshwater for the population of Europe (Braun et al. 2000). This water means freshwater, irrigation and hydropower.

Due to much of the cryosphere being at a temperature close to 0°C, mountain regions are highly sensitive indicators of climate change. This is manifested by 7000 km² of mountain glaciers having disappeared in the last four decades of the 20th century. The European glacier extent decreased 30–40% during the 20th century (Haeberli and Beniston 1998, Lamprecht and Kuhn 2007) and further 30–50% of glacier mass may be lost by 2100 (Maisch et al. 1999). These changes in the cryosphere will have significant repercussions in the hydrological cycle and alter availability of water and seasonality of run-off regimes (Ellenrieder et al. 2007). After a period of increased discharge due to melting, the compensatory discharge of melt water will wane as glaciers disappear. Coupled with changing seasonality of precipitation, with less rainfall in summer and more liquid precipitation in winter this may lead to severe water shortage due to exhausted water stores. The cryosphere and linked hydrological cycle in mountain regions is most severely affected by the impacts of a warmer climate and feedbacks are transferred to other resource areas, i.e., hydro power.

The extreme summer of 2003 in Europe (see Schär et al. 2004) may give a glimpse of potential future consequences of regional warming to mountain water cycles. Due to the extreme dry conditions melt water runoff from Alpine glaciers could hardly compensate the water deficit in the foreland river systems, e.g., the upper Danube, where it caused lowest recorded water levels since more than a century with multiple economic losses due to very limited river trade or reduced production of electricity at water power plants along the Danube and its tributaries (EEA 2009).

As at the end of this century (2071–2100), about every second summer could be as warm (or even warmer) and as dry (or even drier) as the summer of 2003 (Schär et al. 2004), periods of minimum discharge like in 2003 are expected to become more frequent (see, e.g., Mauser et al. 2008). Due to the fact that glacial and snow melt waters will no longer compensate the missing precipitation, it is very likely that the consequences will be more serious than in 2003.

Besides this specific reaction to global climate change impacts, it is the unique spatial situation of mountain areas' natural, i.e., meteorological, hydrological, vegetation, geomorphological conditions that change dramatically over relatively short distances. Consequently boundaries between these systems experience drastic shifts due to climate warming or changing precipitation. These extraordinary spatial variations of environmental resources mean a tremendous challenge to societies in mountain areas. Due to the limited utilizable space there are rather restricted alternatives to the specialized economic situation.

Although there is a common pattern of Global Climate Change challenges to mountain areas worldwide (e.g., melting of the cryosphere, increase of natural hazards), it has to be stated that due to the position in the global circulation system and due to the specific state of development vulnerability and resilience/coping capacities vary strongly from one to another mountain region.

Since the 19th century mountain regions have become attractive destinations for tourism and recreation. Today both play a key role in mountain economies. International tourism has increased 25-fold in the second half of the 20th century and mountain regions take an increasing share of possible destinations (Beniston 2003). In fact with 336–370 million overnight stays, i.e., 11% worldwide, the Alps are the number one tourist destination in Europe (Bätzing 2003).

Urban metropolises—a risky habitat

The demographic concentration in huge urban agglomerations is a matter of risk per se (Kraas 2003, Münchener Rück 2005). Political, social or economic conflicts often arise from those human accumulations. Natural hazards, scarcity of water, energy, pests, and other threats have catastrophic dimensions. Emissions and immissions produced by industry, car traffic and household pollution have more victims than elsewhere. Megacities in mountainous areas deserve the label of future 'risk hotspots' (Borsdorf and Coy 2009).

This raises questions about the opportunities for and limits of control and development of these megacities in terms of sustainability. In their specific local differentiation, they will be strongly influenced by the tension between global impacts on local development and the opportunities for implementing location-specific responses. These questions will be dealt in Chapter 4 of this book.

The potential impact of environmental change can affect the development of megacities directly and indirectly. Droughts, floods, desertification, soil degradation and the environmental and resource conflicts that these disasters trigger are push factors which will continue to cause people to leave their rural home regions in search of a better future in the megacities. In scenarios which take into account future climate wars megacities even serve as havens for environmental and conflict refugees (Borsdorf and Coy 2009).

To demonstrate the impact of climate change and natural processes to large urban agglomerations we will take Andean cities as an example. They are located in areas that are extremely sensitive to the emergence of natural hazards (Sánchez 2010). Between 1950 and 2009, more than 650 disasters hit the countries of the Andean region. These catastrophes had both natural and socio-natural antecedents, with floods (38%) and earthquakes (20%) the most frequent phenomena that directly or indirectly affected the main metropolitan areas (Fig. 3.5).

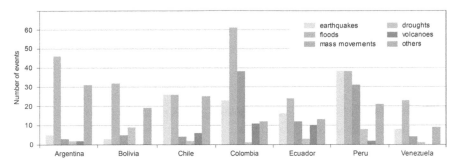

Figure 3.11. Natural and socio-natural disasters in the countries of the Andean region, 1950–2009 (Source: The International Disaster Database 2010).

Color image of this figure appears in the color plate section at the end of the book.

In this sub-region, a number of natural factors converge that allow the development of specific phenomena of a certain magnitude and intensity. In the East Pacific region, the continental crust of South America heaves and rises above the oceanic crust, triggering earthquakes and volcanic eruptions. The American west coast is part of the so-called Ring of Fire, which is one of the most seismic areas of the world and releases more than 80% of the global energy produced by seismic waves. It is worth noting that a number of very serious earthquakes have taken place in this region, such as the ones in Chile on May 22, 1960 and February 27, 2010 with a magnitude of 9.5 resp. 8.8 on the Richter scale, and the earthquake in Ecuador in 1906, which also reached 8.8 on the Richter scale. Even though most volcanoes are located at a reasonable distance from large cities and mostly affect small villages and rural and aboriginal settlements, some eruptions have had catastrophic consequences, as for example in 1859, when the volcano Pichincha (4794 m) destroyed the city of Quito.

Another important assortment of extreme phenomena is that of hydro-meteorological hazards. The variability of rains and droughts is closely linked with the phenomenon known as El Niño-Southern Oscillation (ENSO). During the El Niño phase, the trade winds weaken, permitting the warmer water of the Equatorial Pacific Ocean to spread towards Ecuador, raising the temperature of the Central and East Pacific Oceans by one or two degrees. This encourages the formation of clouds and more intense precipitation, often leading to catastrophic floods as in 1940, 1972, 1982, 1987, 1992, 1997, 2000 and 2005. This meteorological phenomenon finds its counterpart in La Niña, characterized by cooler than usual surface water temperatures of the Equatorial Pacific Ocean and producing contrary effects to those of El Niño, i.e., droughts (1946, 1968, 1980, 1995 and 1998).

Modernization has led to the development of new technologies and goods which, in the course of their production, utilization and eventual disposal, generate huge amounts of toxic residuals that are difficult to assimilate by nature and alter or reduce the well-being of the population. This situation is constantly evolving, not only as a consequence of population growth and of the accelerated urban expansion, but also as a direct result of the economic and technological developments, which raise consumption levels and the demand for satisfaction of the ever-increasing and complex requirements of the people.

The principal aspects of urban contamination in the Andean sub-region are manifold: urban over-expansion characterized by considerable difficulties in regulating land use and the ever-increasing distances travelled by car, a notable population and economic growth and the accompanying increase in energy use, together with erosion and biogenic factors.

Latin America is not a major producer of greenhouse gases, in particular of carbon dioxide (CO_2). In spite of this, there is evidence of a continuous growth, around 2.4% p.a., in the emission of this pollutant (CEPAL 2007). In urban areas in particular, this situation has led to an increasing atmospheric contamination with worrying effects on the health, productivity and quality of life in the cities.

People are also exposed to contamination in the comfort of their homes when they burn biomass. Given its affordable price, wood is the most accessible and the cheapest source of energy for impoverished communities. This accounts not only for a greater contamination but also signifies a deterioration of the ecotones adjacent to cities. Even though wood is a renewable source of energy, limited knowledge of silvicultural techniques (degradation) leads to overexploitation and loss of this natural resource (deforestation).

Another element that contributes to environmental degradation is the production, handling and final disposal of both household and industrial waste, which in urban areas is often carried out without control by the authorities.

The institutional set-up of the countries in the Andean sub-region is characterized by fissures and gaps, which means that not all public institutions are politically stable or economically and socially valued. Not all of them are known, accepted or practiced by the society. Also, without exception, these public institutions present a partial territoriality as the state is unable to cover the entire nation, exhibiting an inability to ensure the validity of the territory as a joint field of articulation, jurisdiction, and regulations. Thus the stability of the state is in question. It is unable to maintain and reproduce itself under a set order of principles (Campero and León 1999).

It is therefore common for never-ending crises in these countries (usually called 'crises of legitimacy') to be generated by the collision of opposing projects. The low institutionalization of administrative processes and the political and institutional instability of the higher levels of the public organizations fail to ensure a strengthening process of the democracies as a political regime (mistrust of the institutions linked to political power, such as the judiciary, parties and the congress). This state of affairs facilitates and fosters the development of organized crime, corruption, influence peddling (also known as 'influence trafficking') and a lack of transparency (Transparency International 2009).

The elements discussed above reveal themselves most strikingly in urban centres as these are a reflection of social processes engendered by the various strategic actors and their exercise of power. There seems to be evidence that uncontrolled urbanization combined with weak institutions greatly erodes social trust and promotes the increase of violence, creating a lack of confidence in cities.

Open risk concept

Risk research provides the conceptual framework for investigating uncertain impacts of climate change on the environment and society on regional and local scales (Schneider et al. 2007, UNEP 2007), whatever consequences, negative or positive, may occur. As a consequence, risk has to be understood as an indicator for an open and uncertain future bearing options for both positive and negative outcomes. Thus the often negatively connoted concept of risk is superseded by a neutral understanding allowing for both good risk, i.e., an opportunity to be grasped, and bad risk in the classical sense of a negative option to be avoided (Campbell and Vuolteenaho 2004, Stötter and Coy 2008). Since risk analysis helps to understand the likelihood of occurrence as well as the magnitude of an anticipated impact it constitutes the primary decision-making tool for development and deployment of adequate adaptation measures as highlighted in the research concept of the alpS-Centre for Climate Change Adaptation Technologies.

There may be many different attitudes towards risk, but without doubt common to all definitions are the core aspects of: i) future-orientation, and ii) uncertainty. In Global Change research risk has to be understood as an open and uncertain future bearing both options of positive and negative outcome, thus interpreted as good risk and bad risk (Stötter and Coy 2008).

All future climate change driven developments of human-environment systems in mountain regions may generally be understood as pointing in one of two directions, which means either an improvement or a deterioration of the accustomed situation. Consequently all adaptation activities have to aim in both directions, either to moderate specific harm or to exploit beneficial opportunities, thus corresponding to the principle idea of minimizing bad risks and optimizing good risks.

Due to the spatially varying character of effects of climate change, all adaptation activities, no matter if they have anticipatory, autonomous or planned character, have to be especially designed to respond to climate change impacts by meeting the demands of sustainability goals on the local or regional level. In this sense, it must be a primary goal to develop solutions, which enable to retain the present quality of life nearly unchanged, respectively to further develop living conditions in areas with current deficits in development. Again, good and bad risks have to be distinguished in order to be able to define respective needs and development goals, which meet the normative standards of sustainability.

Need for Action

The costs of heavy and urgent action to avoid or reduce serious impacts of climate change are considerably less than the damages by the consequences of temperature and precipitation alteration. Although still in discussion, there is a human influence on climate change, however even with strong actions to mitigate greenhouse gas emission adaptation must be a crucial part of development strategy. Therefore policy requires an urgent and international action, pricing for damages from greenhouse gases, supporting innovative technologies, combating soil erosion and deforestation, minder energy consumption and improve insulation of buildings, and to secure water supply, agriculture, energy production and tourism.

Conclusion

Climate change and globalization meet complex man-environment systems in mountain regions. Glacier and permafrost retreat, water scarcity, soil erosion and land-use change, as well as loss of biodiversity are consequences of climate change on the regional scale of the mountain area. Climate change thus impacts on ecosystem services, not just for the societies in the mountains but also for the inhabitants of the adjacent lowlands. Globalization processes also exert an ever faster impact on ecosystems. Even more dramatic is their effect on the cultural landscape, on agriculture, population structure, mobility, as well as the urbanization and marginalization of peripheral mountain regions. Societies are affected in socio-economic, political and institutional terms and require decision making at the regional and local level. Both subsystems interact in the sphere of land use and land management (Fig. 3.6).

Since Global Change affects man-environment systems with greatly varying intensity and speed in different mountain regions, comprehensive long-term observation and monitoring programmes are necessary to

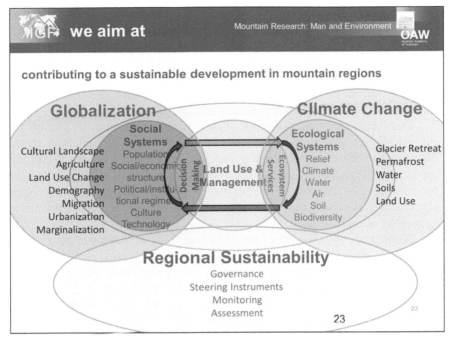

Figure 3.12. Man-environment system in mountain regions (modified by the authors from: GLP, www.globallandproject.org).

Color image of this figure appears in the color plate section at the end of the book.

capture it. One such programme is the global GLORIA initiative with its high-mountain ecology monitoring instruments (Grabherr et al. 2010, Pauli et al. 2007). It is not enough to study individual system elements of the natural- and the anthroposphere (single discipline approach) or the interaction between subsystems or even the entire man-environment system (interdisciplinary approach). When all is said and done, such analyses are always restricted to the production of system knowledge.

The challenges of climate change and globalization increasingly call for target knowledge to be provided to decision makers in politics and the economy. Everybody uses key terms such as sustainability, preserving biodiversity, safeguarding ecosystem services, disaster protection, establishing protected areas, ensuring economic capability, competitiveness and social coherence as things to strive for. These are the expressions of a heightened awareness of social and political responsibility, but as a rule they are not very specific and their implementation is not without conflict. Therefore it is vital to include stakeholders, not only in the creation of system knowledge but even more importantly in the assessment and transformation into target knowledge in truly a transdisciplinary style. The mountain.TRIP project is making a valuable contribution to the dissemination of system and target knowledge to a variety of recipients (stakeholders and general populace) by facilitating the sharing of insights (Braun 2010). Another important aspect is creating and disseminating management knowledge at different levels and across different regions. The DIAMONT project produced exemplary databases on steering instruments and best practices (Borsdorf et al. 2010).

Tourist advertising may talk about 'time standing still' but global climate change and globalization have triggered dramatic changes in the Alps that can be made visible using the indicators mentioned. In the cultural landscape, persistent structures used to be at work for a long time, but the process that has its roots in the beginnings of industrialization is forcing accelerating dynamics onto them, speeded up further in recent decades as a result of globalization.

Within the cultural space of the Alps, social and economic impact factors may dominate (cf. Slaymaker 2001), but with each 'warm' year it becomes clearer that climate factors are gaining in significance. Climate change determines tourism just as much as the growth options for settlements and commercial areas, what is and is not a secure road and the routing of new roads and rail tracks. We may not yet perceive the full extent of this impact, yet the consequences are real and ever more significant.

However: "When wind increase, some build up walls, others set sails" is a Chinese saying. Change is a common experience in human life and in many cases changes content risks and cause adaptations. The next chapter therefore will deal with adaptation strategies and technologies.

References Cited

Abegg, B. 1996. Klimaänderung und Tourismus. Klimafolgenforschung am Beispiel des Wintertourismus in den Schweizer Alpen. Vdf Hochschulverlag, ETH, Zurich.

Abegg, B., S. Agrawala, F. Crick and A. de Montfalcon. 2007. Climate change impacts and adaptation in winter tourism. pp. 25–60. *In*: S. Agrawala (ed.). Climate Change in the European Alps: Adapting Winter Tourism and Natural Hazards Management. OECD, Paris.

Abegg, B., R. Bürki and H. Elsasser. 2008. Climate change and tourism in the Alps. pp. 73–80. *In*: A. Borsdorf, J. Stötter and E. Veulliet (eds.). Managing Alpine Future. IGF-Forschungsberichte 2. Austrian Academy of Sciences Press, Vienna.

Abermann, J., A. Lambrecht, A. Fischer and M. Kuhn. 2009. Quantifying changes and trends in glacier area and volume in the Austrian Ötztal Alps (1969–1997–2006). The Cryosphere 3(2): 205–215.

Abraham, W.-R. 2011. Megacities as sources for pathogenic bacteria in rivers and their fate downstream. International Journal of Microbiology, Article ID 798292, 13 pages. doi:10.1155/2011/798292.

Bader, S. and P. Kunz. 1999. Klimarisiken: Herausforderungen für die Schweiz. Schlussbericht NFP 31. Zürich.

Bates, B.C., Z.W. Kundzewicz, S. Wu and J.P. Palutikof (eds.). 2008. Climate change and water. IPCC Technical Paper 6. Geneva.

Bätzing, W. 2003. Die Alpen. Geschichte und Zukunft einer europäischen Kulturlandschaft (2nd edition), Beck, München.

Bebbington, A.J. (ed.). 2012. Social Conflict, Economic Development and Extractive Industry. Evidence from South America. Routledge, London.

Bebbington, A.J., J. Bury, D. Humphreys Bebbington, J. Kingan, J.P. Muñoz and M. Scurrah. 2008. Mining and social movements: struggles over livelihood and rural territorial development in the Andes. MWPI Working Paper 33. Manchester.

Bender, O. 2010. Entstehung, Entwicklung und Ende der alpinen Bergbauernkultur. pp. 113–137. *In*: H. Heller (ed.). Über das Entstehen und die Endlichkeit physischer Prozesse, biologischer Arten und menschlicher Kulturen, LIT, Wien, Berlin.

Bender, O., A. Borsdorf, A. Fischer and J. Stötter. 2011. Mountains under climate and global change conditions—research results in the Alps. pp. 403–422. *In*: J. Blanco and H. Kheradmand (eds.). Climate Change—Geophysical Foundations and Ecological Effects. InTech, Rijeka.

Bender, O., K.P. Schumacher and D. Stein. 2007. Tourism and seasonality in Central Europe. pp. 181–214. *In*: H. Palang, H. Sooväli and A. Printsmann (eds.). Seasonal Landscapes. Springer Landscape Series, 7. Heidelberg, New York.

Beniston, M. 2003. Climatic change in mountain regions: a review of possible impacts. Climatic Change 59(1): 5–31.

Bohle, H.-G. and T. Glade. 2008. Vulnerabilitätskonzepte in Sozial- und Naturwissenschaften. pp. 99–119. *In*: C. Felgentreff and T. Glade (eds.). Naturrisiken und Sozialkatastrophen. Spektrum Akademischer Verlag, Heidelberg.

Böhm, R. 2009. Klimarekonstruktion in der instrumentellen Periode—Probleme und Lösungen für den Großraum Alpen. pp. 145–164. *In*: R. Psenner, R. Lackner and A. Borsdorf (eds.). Alpine Space—Man and Environment Vol. 6, Innsbruck University Press, Innsbruck.

Bollmann, E., R. Sailer, C. Briese, J. Stötter and P. Fritzmann. 2011. Potential of airborne laser scanning for geomorphologic feature and process detection and quntification in high alpine mountains. Zeitschrift für Geomorphologie, Supplementary Issues 55(2): 83–104.

Borsdorf, A. (ed.). 2005. Das neue Bild Österreichs. Strukturen und Entwicklungen im Alpenraum und den Vorländern, ÖAW, Wien.

Borsdorf, A. 2009. Amenity migration in rural mountain areas. Die ERDE 140(3). 225–228.

Borsdorf, A. 2010. The hydroelectrical potential of North-Western Patagonia—balancing economic development and ecological protection. pp. 154–161. *In*: A. Borsdorf, G. Grabherr, K. Heinrich, B. Scott and J. Stötter (eds.). Challenges for Mountain Regions—Tackling Complexity. Böhlau, Vienna.

Borsdorf, A. and O. Bender. 2007. Kulturlandschaftsverlust durch Verbuschung und Verwaldung im subalpinen und hochmontanen Höhenstockwerk: Die Folgen des klimatischen und sozioökonomischen Wandels. pp. 29–50. *In*: Alpine Kulturlandschaft im Wandel. Hugo Penz zum 65. Geburtstag, Innsbrucker Geographische Gesellschaft, Innsbruck.

Borsdorf, A. and M. Coy. 2009. Megacities and global change. Case studies form Latin America. Die ERDE 140(4): 341–353.

Borsdorf, A. and R. Hidalgo. 2009. Searching for fresh air, tranquility and rural culture in the mountains: a new lifestyle for Chileans? Die ERDE 140(3): 275–292.

Borsdorf, A., F. Borsdorf and L.A. Ortega. 2011. Towards climate change adaptation, sustainable development and conflict resolution—the Cinturón Andino Biosphere Reserve in Southern Colombia. Eco.mont—Journal on Mountain Protected Areas Research and Management 3(2): 43–48.

Borsdorf, A., R. Hidalgo and H. Zunino. 2011. Amenity Migranten im Seengebiet Chiles. Auf der Suche nach Natur und Spiritualität. Grazer Schriften der Geographie und Raumforschung 46: 9–48.

Borsdorf, A., R. Hidalgo and H. Zunino. 2012. Amenity migration: a comparative study of the Italian Alps and the Chilean Andes. Journal of Sustainability Education. Available at http://www.jsedimensions.org/wordpress/content/amenity-migration-a-comparative-study-of-the-italian-alps-and-the-chil-ean-andes_2012_03/.

Borsdorf, A., U. Tappeiner and E. Tasser. 2010. Mapping the Alps. pp. 186–191. *In*: A. Borsdorf, G. Grabherr, K. Heinrich, B. Scott and J. Stötter (eds.). Challenges for mountain regions. Tackling Complexity. Vienna.

Bourdeau, P. 2008. The Alps in the age of new style tourism: between diversification and post-tourism? pp. 81–86. *In*: A. Borsdorf, J. Stötter and E. Veulliet (eds.). Managing Alpine Future. IGF-Forschungsberichte 2. Austrian Academy of Sciences Press, Vienna.

Braun, F. 2010. Closing the gap between science and practice. mountain.TRIP—an EU project coordinated by IGF. pp. 204–207. *In*: A. Borsdorf, G. Grabherr, K. Heinrich, B. Scott and J. Stötter (eds.). Challenges for mountain regions. Tackling Complexity. Böhlau, Wien.

Braun, L., M. Weber and M. Schulz. 2000. Consequences of climate change for runoff from Alpine regions. Annals of Glaciology 31(1): 19–25.

Braun-Blanquet, J. 1958. Über die obersten Grenzen pflanzlichen Lebens im Gipfelbereich des Schweizerischen Nationalparks. Kommision der Schweizerischen Naturforschenden Gesellschaft zur wissenschaftlichen Erforschung der Nationalparks 6: 119–142.

Breiling, M. and P. Charamza. 1999. The impact of global warming on winter tourism and skiing: a regionalised model for Austrian snow conditions. Regional Environmental Change 1(1): 4–14.

Bucher, K., T. Geist and J. Stötter. 2006. Ableitung der horizontalen Gletscherbewegung aus multitemporalen Laserscanning-Daten. Fallbeispiel: Hintereisferner/Ötztaler Alpen. pp. 277–286. *In*: J. Strobl, T. Blaschke and G. Griesebner (eds.). Angewandte Geoinformatik. Salzburg.

Butler, C.R., G.P. Malanson, S.J. Walsh and D.B. Fagre (eds.). 2009. The Changing Alpine Treeline. Developments in Earth Surface Processes 12. Elsevier, Amsterdam, London.

Campbell, J.Y. and T. Vuolteenaho. 2004. Bad Beta, Good Beta. American Economic Review 94: 1249–1275.

Campero, G. and F. León. 1999. Crisis y legitimidad de las reformas institucionales. pp. 19–10. *In*: Naciones Unidas, CEPAL, División de Desarrollo Social (ed.). América Latina y las crisis. Serie Políticas Sociales 33. Santiago de Chile.

CEPAL (Comisión Económica para América Latina). 2007. Anuario Estadístico. Santiago de Chile.

Chambers, R. 1989. Editorial Introduction: Vulnerability, Coping and Policy. IDS Bulletin 20(2): 1–7.

CIPRA. 1998. Protocol on the implementation of the Alpine Convention of 1991 in the field of soil conservation. Bled, Schaan.

CIPRA. 2011. Agriculture in climate change. A background report of CIPRA. Compact 2. CIPRA. Schaan, Liechtenstein.

Diamond, J.M. 2005. Collapse: How Societies Choose to Fail or Succeed. New York.

Dullinger, S., T. Dirnböck and G. Grabherr. 2004. Modelling climate change-driven treeline shifts: relative effects of temperature increase, dispersal and invisibility. Journal of Ecology 92: 241–252.

EEA (European Energy Agency). 2008. Impacts of Europe's changing climate—2008 indicator-based assessment. EEA, Copenhagen.

EEA (European Environment Agency). 2009. Regional climate change and adaptation. The Alps facing the challenge of changing water resources. EEA, Copenhagen.

Eitzinger, J., K.C. Kersebaum and H. Formayer. 2009. Landwirtschaft im Klimawandel. Auswirkungen und Anpassungsstrategien für die Land- und Forstwirtschaft in Mitteleuropa, Agri Media.

Ellenrieder, T., L.N. Braun and M. Weber. 2004. Reconstruction of mass balance and runoff of Vernagtferner from 1895 to 1915. Zeitschrift für Gletscherkunde und Glazialgeologie 38(2): 165–178.

Erschbamer, B., T. Kiebacher, M. Mallaun and P. Unterluggauer. 2009. Short-term signals of climate change along an altitudinal gradient in the South Alps. Plant Ecology 202(1): 79–89.

Fuhrer, J., M. Beniston, A. Fischlin, Ch. Frei, S. Goyette, C. Jasper and C. Pfister. 2006. Climate risks and their impact on agriculture and forests in Switzerland. Climatic Change 79(1-2): 79–102.

Füssel, H.-M. and R.J.T. Klein. 2006. Climate change vulnerability assessments: an evolution of conceptual thinking. Climatic Change 75: 301–329.

Geist, T. and J. Stötter. 2007. Documentation of glacier surface elevation change with multitemporal airborne laser scanner data-case study: Hintereisferner and Kesselwandferner, Tyrol, Austria. Zeitschrift für Gletscherkunde und Glazialgeologie 4: 77–106.

Geitner, C. 2007. Böden in den Alpen—Ausgewählte Aspekte zur Vielfalt und Bedeutung einer wenig beachteten Ressource. pp. 56–62. In: A. Borsdorf and G. Grabherr (eds.). Internationale Gebirgsforschung. IGF-Forschungsberichte 1. Austrian Academy of Sciences Press, Vienna.

Geitner, C. 2010. Soils as archives of natural and cultural history; examples from the Eastern Alps. pp. 68–75. In: A. Borsdorf, G. Grabherr, K. Heinrich, B. Scott and J. Stötter (eds.). Challenges for Mountain Regions. Tackling Complexity. Böhlau, Vienna.

Gottfried, M., H. Pauli, K. Reiter and G. Grabherr. 1999. A fine-scaled predictive model for changes in species distribution patterns of high mountain plants induced by climate warming. Diversity and Distributions 5: 241–251.

Gottfried, M., H. Pauli, K. Reiter and G. Grabherr. 2002. Potential effects of climate change on alpine and nival plants in the Alps. pp. 213–223. In: C. Körner and E. Spehn (eds.). Mountain Biodiversity: A Global Assessment. Parthenon Publishing, New York.

Govers, G., D.A. Lobb and T.A. Quine. 1999. Preface—Tillage erosion and translocaton: emergence of a new paradigm in soil erosion research. Soil Tillage Research 51: 167–174.

Grabherr, G. 1997. The high-mountain ecosystems of the Alps. pp. 97–121. In: F.E. Wielgolaski (ed.). Polar and Alpine Tundra. Ecosystems of the World 3. Amsterdam.

Grabherr, G. 2009. Biodiversity in the high ranges of the Alps: Ethnobotanical and climate change perspectives. Global Environmental Change 19(2): 167–172.

Grabherr, G., M. Gottfried, A. Gruber and H. Pauli. 1995. Patterns and current changes in alpine plant diversity. pp. 167–182. In: F.S. Chapin and C. Körner (eds.). Arctic and Alpine Biodiversity. Ecological Studies 113. Springer, Berlin, Heidelberg.

Grabherr, G., M. Gottfried and H. Pauli. 1994. Climate effects on mountain plants. Nature (369): 448.

Grabherr, G., H. Pauli and M. Gottfried. 2010. A worldwide observation of effects on climate change on mountain ecosystems. pp. 48–57. In: A. Borsdorf, G. Grabherr, K. Heinrich, B. Scott and J. Stötter (eds.). Challenges for Mountain Regions. Tackling Complexity. Böhlau, Vienna.

Gross, G. 1987. Der Flächenverlust der Gletscher in Österreich 1850–1920–1969. Zeitschrift für Gletscherkunde und Glazialgeologie 23(2): 131–141.

Haeberli, W., D. Rickenmann, M. Zimmermann and U. Roesli. 1990. Investigation of 1987 debris flows in the Swiss Alps: general concept and geophysical soundings. Hydrology in Mountainous Regions 194: 303–310.

Haeberli, W. and M. Beniston. 1998. Climate change and its impacts on glaciers and permafrost in the Alps. Ambio 27: 258–265.

Haeberli, W., C. Goudong, A.P. Gorbunov and S.A. Harris. 2006. Mountain permafrost and climatic change. Permafrost and Periglacial Processes 4(2): 165–174.

Haeberli, W., M. Hölzle, F. Paul and M. Zemp. 2007. Integrated monitoring of mountain glaciers as key indicators of global climate change: the European Alps. Annals of Glaciology 46(1): 150–160.

Hahn, F. 2004. Künstliche Beschneiung im Alpenraum. CIPRA, Available from: http://www.cipra.org/de/alpmedia/publikationen/2709/.

Hamilton, J.M., D.J. Maddison and R.S.J. Tol. 2003. Climate change and international tourism: a simulation study. Centre for Marine and Climate Research Working Paper FNU-31, Hamburg.

Hegerl, G.C., F.W. Zwiers, P. Braconnot, N.P. Gillett, Y. Luo, J.A. Marengo Orsini, N. Nicholls, J.E. Prenner and P.A. Stott. 2007. Understanding and attributing climate change. *In:* Solomon, S. D. Qin, M. Manning, Z. Chen, M. Marquis, K.B. Averyt, M. Tignor and H.L. Miller (Hrsg.). Climate Change 2007: The Physical Science Basis. Contribution of Working Group I to the Fourth Assessment Report of the Intergovernmental Panel Climate Change, Cambridge.

Hessel, R., V. Jettn and G.H. Zhang. 2003. Estimating manning's for steep slopes. Catena 54: 77–91.

Hollenstein, K., O. Bieri and J. Stückelberger. 2002. Modellierung der Vulnerability von Schadenobjekten gegenüber Naturgefahrenprozessen. Zürich.

Holling, C.S. 1973. Resilience and stability of ecological systems. Annual Review of Ecology and Systematics 4: 1–23.

Holtmeier, F.-K. 2009. Mountain Timberlines, 2nd ed. Springer, Berlin.

Holzinger, B. and G. Grabherr. 2010. What happens when alpine plants are exposed to lowland climate? Verhandlungen der Zoologisch-Botanischen Gesellschaft Österreich 146: 139–150.

Huss, M., A. Bauder, M. Werder, M. Funk and R. Hock. 2007. Glacier-dammed lake outburst events of Gornersee, Switzerland. Journal of Glaciology 53(181): 189–200.

Inbar, M., M. Tamir and L. Wittenberg. 1998. Runoff and erosion processes after a forest fire in Mount Carmel, a Mediterranean area. Geomorphology 24: 17–33.

IPCC—Intergovernmental Panel on Climate Change. 2005. Guidance Notes for Lead Authors of the IPCC Fourth Assessment Report on Addressing Uncertainties. Cambridge.

ITC and FiBL. 2007. Organic Farming and Climate Change. International Trade Centre UNCTAD/WTO and Research Institute of Organic Agriculture (FiBL). Doc. No. MDS-08-152.E, ITC, Geneva.

Kääb, A., C. Huggel and L. Fischer. 2006. Remote Sensing Technologies for Monitoring Climate Change Impacts on Glacier- and Permafrost-Related Hazards. 2006 ECI Conference on Geohazards. http://services.bepress.com/eci/geohazards/2.

Kerschner, H. and S. Ivy-Ochs. 2008. Palaeoclimate from glaciers: examples from the Eastern Alps during the Alpine Lateglacial and early Holocene. Global and Planetary Change 60(1-2): 58–71.

Klebelsberg, R. 1913. Das Vordringen der Hochgebirgsvegetation in den Tiroler Alpen. Österreichische Botanische Zeitschrift 177-187: 241–254.

Konz, N., D. Baenninger, M. Konz, M. Nearing and C. Alewell. 2010: Process identification of soil erosion in steep mountain regions. Hydrology and Earth System Sciences 14: 675–686.

Körner, C. 2003. Alpine Plant Life. Functional Plant Ecology of High Mountain Ecosystems, 2nd ed. Springer, Berlin.

Kraas, F. 2003. Megacities as Global Risk Areas. -Petermanns Geographische Mitteilungen 147(4): 6–15.

Krainer, K., W. Mostler and C. Spötl. 2007. Discharge from active rock glaciers, Austrian Alps: a stable isotope approach. Austrian Journal of Earth Sciences 100: 102–112.

Krippendorf, J. 1984. Die Ferienmenschen: Für ein neues Verständnis von Freizeit und Reisen. dtv. München.Kronberger, B.M. Balas and A. Prutsch (Red.). 2010. Auf dem Weg zu einer nationalen Anpassungsstrategie, Policy Paper, Bundesministerium für Land- und Forstwirtschaft, Umwelt und Wasserwirtschaft (BMLFUW), 13.04.2011, Available from: http://umwelt.lebensministerium.at/filemanager/download/68173/.

Kullmann, L. 2007. Modern climate change and shifting ecological states of the subalpine/alpine landscape in the Swedish Scandes. Geoöko 28: 187–221.

Lambrecht, A. and M. Kuhn. 2007. Glacier changes in the Austrian Alps during the last three decades, derived from the new Austrian glacier inventory. Annals of Glaciology 46(1): 177–184.

Larcher, W., C. Kainmüller and J. Wagner. 2010. Survival types of high mountain plants under extreme temperatures. Flora 205(1): 3–18.

Lemke, P., J. Ren, R.B. Alley, I. Allison, J. Carrasco, G. Flato Y. Fujii, G. Kaser, P. Mote, R.H. Thomas and T. Zhang. 2007. Observations: changes in snow, ice and frozen ground. pp. 338–383. *In*: S. Solomon, D. Qin, M. Manning, Z. Chen, M. Marquis, K.B. Averyt, M. Tignor and H.L. Miller (eds.). Climate Change 2007. The Physical Science Basis. Contribution of Working Group I to the Fourth Assessment Report of the Intergovernmental Panel on Climate Change, Cambridge University Press, Cambridge, UK.

Liniger, H. and R. Weingartner. 1998. Mountains and freshwater supply. FAO Corporate Document Repository. Unasylva 195. http://www.fao.org/docrep/w9300e/w9300e08. htm, accessed 31 October 2011.

Lüem, T. 1985. Sozio-kulturelle Auswirkungen des Tourismus in Entwicklungsländern. Ein Beitrag zur Problematik des Vergleiches von touristischen Implikationen auf verschiedenartige Kulturräume der Dritten Welt. Zentralstelle der Studentenschaft. Zurich.

Maass, J.M., C.F. Jordan and J. Sarukhan. 1988. Soil erosion and nutrient losses in seasonal tropical agroecosystems under various management techniques. Journal of Applied Ecology 25: 595–607.

Maddison, D. 2001. In search of warmer climates? The impact of climate change on flows of British tourists. pp. 53–76. *In*: D. Maddison (ed.). The Amenity Value of the Global Climate: Earthscan, London.

Maisch, M., D. Vonder Mühll and M. Hoezle. 1999. Die Gletscher der Schweizer Alpen. Gletscherhochstand 1850, aktuelle Vergletscherung, Gletscherschwund-Szenarien. Schlussbericht NFP 31. Zürich.

Manning, M.R., M. Petit, D. Easterling, J. Murphy, A. Patwardhan, H.-H. Rogner, R. Swart and G. Yohe (eds.). 2004. IPCC Workshop on Describing Scientific Uncertainties in Climate Change to Support Analysis of Risk and of Options: Workshop report. Intergovernmental Panel on Climate Change (IPCC). Geneva.

Marchant, C. and A. Borsdorf. 2013. Protected areas in Northern Colombia—on track to sustainable development) Eco.mont—Journal on Protected Mountain Areas Research and Management 5(2): 5–14.

Matzarakis, A. 2002. Examples of climate and tourism research for tourism demands, 15th Conference on Biometeorology and Aerobiology joint with the International Congress on Biometeorology. 27, October to 1. November 2002, Kansas City, Missouri 391–392.

Mauser, W., M. Muerth and S. Stöber. 2008. Climate change scenarios of low-flow conditions and hydro power production in the upper Danube River basin. IOP Conferences Series: Earth and Environmental Science 6. Available at: http://iopscience.iop.org/1755-1315/6/29/292018 (accessed: 11/08/2010).

McIntyre, N. 2009. Rethinking amenity migration: Integrating mobility, lifestyle and social-ecological systems. Die ERDE 140, 3: 229–250.

Meusburger, K. and C. Alewell. 2008. Impacts of anthropogenic and environmental factors on the occurrence of shallow landslides in an alpine catchment (Urseren Valley, Switzerland). Natural Hazards and Earth System Sciences 8: 509–520.

Millennium Ecosystem Assessment. 2005. Available at http://www.maweb.org/en/index. aspx.

Moiseev, P.A. and S.G. Shiyatov. 2003. Vegetation dynamics at the treeline ecotone in the Ural highlands, Russia. pp. 423–435. *In*: L. Nagy, G. Grabherr, C. Körner and D.B.A. Thompson (eds.). Alpine Biodiversity in Europe. Ecological Studies 167, Berlin.

Monreal, M. and J. Stötter. 2010. Alpine permafrost: a rock glacier inventory of South Tyrol based on laser scanning data. pp. 40–47. *In*: A. Borsdorf, G. Grabherr, K. Heinrich, B. Scott and J. Stötter (eds.). Challenges for Mountain Regions. Tackling Complexity. Böhlau, Vienna.

Moss, L.A.G. (ed.). 2006. The Amenity Migrants. Seeking and Sustaining Mountains and their Cultures. Cobi, Wallingford.

Moss, L.A.G., R.S. Glorioso and A. Krause (eds.). 2008. Understanding and Managing Amenity-Led Migration in Mountain Regions. The Banff Center. Banff, Alberta, Canada.

Münchener Rück (ed.) 2005. Megastädte- Megarisiken.—Trends und Herausforderungen für Versicherung und Risikomanagement.—München.

Nagy, L. and G. Grabherr. 2009. The Biology of Alpine Habitats. University of Oxford Press. Oxford.

Nicolussi, K. 1990. Bilddokumente zur Geschichte des Vernagtferners im 17. Jahrhundert. Zeitschrift für Gletscherkunde und Glazialgeologie 26, 2: 97–119.

Nicolussi, K., M. Kaufmann, G. Patzelt, J. van der Plicht and A. Thurner. 2005. Holocene tree-line variability in the Kauner Valley, Central Eastern Alps, indicated by dendrochronological analysis of living trees and subfossil logs. Vegetation History and Archaeobotany 14(3): 221–234.

Oerlemans, J. 2005. Extracting a climate signal from 169 glacier records. Science 308(5722): 675–677.

Oreskes, N. 2004. Beyond the ivory tower—The scientific consensus on climate change. Science 306: 1686.

Patzelt, G. 1970. Die Längenmessungen an den Gletschern der österreichischen Ostalpen 1890–1969. Zeitschrift für Gletscherkunde und Glazialgeologie 6(1-2): 151–159.

Paul, F., A. Kääb and W. Haeberli. 2007. Recent glacier changes in the Alps observed by satellite: Consequences for future monitoring strategies. Global and Planetary Change 56(1-2): 111–122.

Pauli, H., M. Gottfried, T. Dirnböck, S. Dullinger and G. Grabherr. 2003. Assessing the long-term dynamics of endemic plants at summit habitats. pp. 195–207. In: L. Nagy et al. (eds.). Alpine Biodiversity in Europe. Ecological Studies 167. Springer, Berlin, Heidelberg.

Pauli, H., M. Gottfried, D. Hohenwallner, K. Reiter and G. Grabherr (eds.). 2004. The GLORIA Field Manual—Multi Summit Approach. European Commission DG Research. Luxembourg.

Pauli, H., M. Gottfried, K. Reiter, C. Klettner and G. Grabherr. 2007. Signals of range expansions and contractions of vascular plants in the high Alps: observations (1994–2004) at the GLORIA master site Schrankogel, Tyrol, Austria. Global Change Biology 13(1): 147–156.

Pruckner, G. 2005. Non-governmental approaches for the provision of non-commodity outputs and the reduction of negative effects of agriculture. Agritourism and landscape conservation program in Austria. pp. 57–62. In: OECD (ed.). Multifunctionality of Agriculture. OECD, Paris.

Salick, J., Z. Fang and A. Byg. 2009. Eastern Himalayan alpine plant ecology, Tibetan ethnobotany, and climate change. Global Environmental Change 19(2): 147–155.

Sánchez, R. 2010. Risks in the Andean metropolises. pp. 102–109. In: A. Borsdorf, G. Grabherr, K. Heinrich, B. Scott and J. Stötter (eds.). Challenges for Mountain Regions. Tackling Complexity. Vienna Wien. Böhlau.

Schär, C., P.L. Vidale, D. Lüthi, C. Frei, C. Häberli, M.A. Liniger and C. Appenzeller. 2004. The role of increasing temperature variability in European summer heatwaves. Nature 427: 332–336.

Schlagintweit, H. and A. Schlagintweit. 1850. Untersuchungen über die physicalische Geographie der Alpen in ihren Beziehungen zu den Phaenomenen der Gletscher, zur Geologie, Meteorologie und Pflanzengeographie. Barth, Leipzig.

Schneider, S.H., S. Semenov, A. Patwardhan, I. Burton, C.H.D. Magadza, M. Oppenheimer, A.B. Pittock, A. Rahman, J.B. Smith, A. Suarez and F. Yamin. 2007. Assessing key vulnerabilities and the risk from climate change. In: M.L. Parry, O.F. Canziani, J.F. Palutikof, J. van der Linden and C.E. Hanson (eds.). Climate Change 2007: Impacts, Adaptation and Vulnerability, Contribution of Working Group II to the Fourth Assessment Report of the Intergovernmental Panel on Climate Change, Cambridge 779–810.

Slaymaker, O. 2001. Why so much concern about climate change and so little attention to land use change. The Canadian Geographer 45(1): 71–78.

Smith, J.B., H.-J. Schellnhuber, M.M.Q. Mirza, S. Fankhauser, R. Leemans, E. Lin, L. Ogallo, B. Pittock, R.G. Richels, C. Rosenzweig, R.S.J. Tol, J.P. Weyant and G.W. Yohe. 2001. 21'Vulnerability to climate change and reasons for concern: a synthesis. pp. 913–967. *In*: J.J. McCarthy, O.F. Canziani, N.A. Leary, D.J. Dokken and K.S. White (eds.). Climate Change. Impacts, Adaptation, and Vulnerability. Cambridge University Press, Cambridge.

Smith, P., D. Martino, Z. Cai, D. Gwary, H. Janzen, P. Kumar, B. McCarl, S. Orgle, F. O'Mara, C. Rice, B. Scholes and O. Sirotenko. 2007. Agriculture. pp. 498–540. *In*: B. Metz, O.R. Davidson, P.R. Bosch, R. Dave and L.A. Meyer (eds.). Climate Change: Mitigation. Contribution of Working Group III to the Fourth Assessment Report of the Intergovernmental Panel on Climate Change. Cambridge University Press, Cambridge, United Kingdom and New York.

Solomon, S., D. Qin, M. Manning, Z. Chen, M. Marquis, K.B. Averyt, M. Tignor and H.L. Miller (eds.). Climate Change. 2007. The Physical Science Basis. Contribution of Working Group I to the Fourth Assessment Report of the Intergovernmental Panel on Climate Change, Cambridge University Press, Cambridge, UK and New York.

Span, N., M. Kuhn and H. Schneider. 1997. 100 years of ice dynamics of Hintereisferner, Central Alps, Austria, 1894–1994. Annals of Glaciology 24: 297–302.

Stern, N. 2007. The Economics of Climate Change. The Stern Review. Cambridge University Press, Cambridge, UK.

Stötter, J. 1994. Veränderungen der Kryosphäre in Vergangenheit und Zukunft sowie Folgeerscheinungen—Untersuchungen in ausgewählten Hochgebirgsräumen im Vinschgau (Südtirol). Habilitation Thesis, Ludwig-Maximilian-Universität München.

Stötter J. and M. Coy. 2008. Forschungsschwerpunkt Globaler Wandel—regionale Nachhaltigkeit. *In*: Innsbrucker Geographische Gesellschaft (eds.). Jahresbericht 2007. Innsbruck.

Stötter, J. and M. Monreal. 2010. Mountains at risk. pp. 86–93. *In*: A. Borsdorf, G. Grabherr, K. Heinrich, B. Scott and J. Stötter (eds.). Challenges for Mountain Regions—Tackling Complexity. Vienna.

Stötter, J., H. Weck-Hannemann and E. Veulliet. 2009. Global change and natural hazards, new strategies. pp. 1–34. *In*: E. Veulliet, J. Stötter and H. Weck-Hannemann (eds.). Sustainable Hazard Management in Alpine Environments. Springer, Dordrecht, Heidelberg, London, New York, Berlin.

Tappeiner, U., A. Borsdorf and E. Tasser (eds.). 2008. Alpenatlas Atlas des Alpes Atlante delle Alpi—Atlas Alp—Mapping the Alps. Spektrum Akademischer Verlag, Heidelberg.

Tasser, E., U. Tappeiner and A. Cernusca. 2005. Ecological effects of land-use changes in the European Alps. pp. 409–420. *In*: U.M. Huber, H. Bugmann and M. Reasoner (eds.). Global Change and Mountain Regions. Springer, Berlin.

Thieken, A.H., M. Müller, H. Kreibich and B. Merz. 2005. Flood damage and influencing factors: New insights from the August 2002 flood in Germany. Water Resources Research, 41, 12, W12430.

Transparency International. 2009. Global Corruption Report 2009. Berlin. Available at: http://www.transparency.org/publications/gcr (accessed: 11/08/2010).

Tylor, E.b. 1924[1871]. Primate Culture. 2 volumes. 7th edition. Brentano's, New York.

UNEP—United Nations Environment Programme. 2007. Global Environment Outlook—environment for development (GEO-4). Valletta.

UNEP and WGMS. 2008. Global Glacier Changes: facts and figures.

UNEP and WGMS. 2008. Global Glacier Changes: facts and figures. Zürich. UNWTO (United Nations World Tourism Organization) 2003. Climate change and tourism. Proceedings of the 1st International Conference on Climate Change and Tourism, Djerba 9–11 April 2003. Djerba.

UNWTO (United Nations World Tourism Organization). 2009. From Davos to Copenhagen and beyond: advancing tourism's response to climate change. UNWTO Background Paper. Madrid.

Veulliet, E., J. Stötter and H. Weck-Hannemann (eds.). 2009. Sustainable Natural Hazard Management in Alpine Environments. Springer, Heidelberg.

Vincent, C., G. Kappenberger, F. Valla, A. Bauder, M. Funk and E. Le Meur. 2004. Ice ablation as evidence of climate change in the Alps over the 20th century. Journal of Geophysical Research - Atmospheres 109. D10104, doi:10.1029/2003JD003857.

Viviroli, D., R. Weingartner and B. Messerli. 2003. Assessing the hydrological significance of the world's mountains, Mountain Research and Development 23: 32–40.

Viviroli, D. and R. Weingartner. 2004. The hydrological significance of mountains: from regional to global scale. Hydrology and Earth System Sciences 8(6): 1017–1030.

Viviroli, D., H.H. Dürr, B. Messerli, M. Meybeck and R. Weingartner. 2007. Mountains of the world, water towers for humanity: Typology, mapping, and global significance. Water Resources Research 43(7): 1–13.

Viviroli, D., D.R. Archer, W. Buytaert, H.J. Fowler, G.B. Greenwood, A.F. Hamlet, Y. Huang, G. Koblotschnig, M.I. Litaor, J.I. López Moreno, S. Lorentz, B. Schäderl, H. Schreier, K. Schwaiger, M. Vuille and R. Woods. 2011. Climate change and mountain water resources: overview and recommendations for research, management and policy. Hydrology and Earth Systems Sciences 15: 471–504.

Walker, B., C.S. Holling, S.R. Carpenter and A. Kinzig. 2004. Resilience, adaptability and transformability in social–ecological systems. Ecology and Society 9(2): 5.

Walther, G.-R., S. Beißner and C.A. Burga. 2005. Trends in the upward shift of alpine plants. Journal of Vegetation Science 16(5): 541–548.

Watts, M. and G. Bohle. 1993. The space of vulnerability: The causal structure of hunger and famine. Progress in Human Geography 17(1): 43–67.

WBCD (World Business Council for Sustainable Development). 2009. Water, energy and climate change. A contribution from the business community. Geneva.

WBCSD (World Business Council for Sustainable Development). 2006. Powering a sustainable future: An agenda concerted action. Facts and Trends. Geneva.

White, L.A. 1948. Man's control over civilization. An anthropocentric illusion. Scientific Monthly 66(3): 235–247.

Wiegand, C. and C. Geitner. 2010. Shallow erosion in grassland areas in the Alps. What we know and what we need to investigate further. pp. 76–83. In: A. Borsdorf, K. Heinrich, B. Scott and J. Stötter (eds.). Challenges for Mountain Regions. Tackling Complexity. Böhlau, Vienna.

Wieser, G., R. Matyssek, R. Luzian, P. Zwerger, P. Pindur, W. Oberhuber and A. Gruber. 2009. Effects of atmospheric and climate change at the timberline of the Central European Alps. Annals of Forest Science, Vol. 66, No. 4, (June 2009), pp. 402–412, ISSN 1286-4560.

Wisner, B., P.M. Blaikie and T. Cannon. 2004. At Risk. Natural Hazards, People's Vulnerability and Disasters. 2nd ed. Routledge, London.

White, L.A. 1948. Man's control over civilization. An anthropocentric illusion. Scientific Monthly 66(3): 235–247.

Witt, S.F. and C.A. Witt. 1995. Forecasting tourism demand: A review of empirical research. International Journal of Forecasting 11: 447–475.

WWAP (World Water Assessment Programme). 2009. The United Nations World Water Development Report 3: Water in a Changing World, UNESCO, Paris, France and Earthscan, London.

SECTION 3

ADAPTATION AND MITIGATION: STRATEGIES AND MODELS

4

Global Change and Mountains: Consequences, Responses and Opportunities

Prakash C. Tiwari[1,]* and *Bhagwati Joshi*[2]

INTRODUCTION

Mountain regions which encompass nearly 24% of the total land surface of the Earth (UNEP-WCMC 2002) and constitute homes for approximately 12% of the world's population (Huddleston et al. 2003) in addition to nearly 14% global population living in their foothills and adjoining lowlands are highly critical from the view point of marginality, environmental sensitivity, climate change, constraints of terrain, geographical inaccessibility and less infrastructural development (Meybeck et al. 2001). Mountains constitute the sources of a variety of ecosystems services, including water, biodiversity, soils, natural beauty, recreational opportunities, wilderness and cultural diversity which sustains the livelihood and economy of large population both in mountains and their vast lowlands. Mountain headwaters provide freshwater to approximately half of the world population inhabiting the large river basins located far away from mountains (Viviroli et al. 2007). The largest trans-boundary river systems of the planet have their origin in high mountains, and mountains have still the largest proportion of the

[1] Professor of Geography, Kumaon University, Nainital – 263002, Uttarakhand, India.
  Email: pctiwari@yahoo.com
[2] Assistant Professor of Geography, Government Post Graduate College, Rudrapur (U.S. Nagar), Uttarakhand, India.
  Email: bhawanatiwari@yahoo.com
* Corresponding author

world's forests which not only constitute global biodiversity hot spots and the pool of genetic resources, but they also regulate and modify climatic conditions and contribute towards mitigating global warming through serving as carbon sinks (ICIMOD 2010). Mountain agriculture and farming systems constitute the principal source of food and livelihood for about half a billion population. The indigenous communities inhabiting mountain regions since time immemorial have evolved diversity of cultures that comprise traditional knowledge, resource development and environmental conservation practices, agricultural and food systems, adaptation and coping mechanism, languages, customs, traditions, costumes, conventions and rituals which have immense relevance and practical significance in environmental restoration, climate change adaptation and ensuring sustained resource productivity in mountain ecosystems (ICIMOD 2010).

But, mountains have long been marginalized from the view point of sustainable development of their resources and inhabitants. However, our understanding about the problems of mountain regions and approach to their development has undergone drastic changes, during the recent years. Currently, mountain ecosystems as well as mountain communities are particularly threatened by the ongoing processes of environmental global change, population dynamics and globalizing economy and resultant exploitation of mountain resources (Borsdorf et al. 2010). During the recent years, a variety of changes have emerged in the traditional resource use structure in high mountain areas, particularly in developing and underdeveloped regions of the world mainly in response to changing global economic order, transforming political systems, rapid urban growth, increased demographic pressure and resultant increased demand and exploitation of natural resources. As a result, mountain regions of the world are passing through a process of rapid environmental, socio-economic and cultural transformation and exploitation and depletion of their natural resources leading to ecological imbalances and economic un-sustainability both in upland and lowland areas (Haigh et al. 2002, Tiwari 2000).

Moreover, the changing climatic conditions have already stressed mountain ecosystems through higher mean annual temperatures and melting of glaciers and snow, altered precipitation patterns and hydrological disruptions, and more frequent and extreme weather events. In this context, climate change acts as an additional stress which can multiply existing development deficits and may also reverse the process of socio-economic development in mountain regions particularly in underdeveloped and developing countries (UNDP 2010). Mountain people, who have contributed the least to global greenhouse gas (GHG) emissions, and helped significantly in mitigating climate change through promoting carbon sequestration by preserving the largest proportion of forests on the planet are likely to be the worst affected by long-term impacts of climate change (ICIMOD

2010). Further, these changes are likely to undermine the inherent capacity of indigenous mountain communities to respond and adapt to changing environmental conditions including climate change. Besides, the recent food crisis followed by global economic recession has adversely affected the food and livelihood security of mountain communities because of their subsistence economies, constraints of terrain and climate and resultant physical isolation and low productivity, vulnerability to natural risks, poor infrastructure, limited access to markets, higher cost of production and poor employment and livelihood opportunities. It has been estimated that a large proportion of food insecure population now live in mountain regions of the world (Huddleston et al. 2003, FAO 2008a, 2008b).

It is therefore high time that national as well as international policy and decision-making agencies and organizations must realize the significance of mountain ecosystems is sustaining world population, and evolve a framework for the sustainable development of mountain regions and well being of their inhabitants. The world community should come forward to compensate mountain people for the value of the ecosystem services provided by the mountain regions, and facilitate them in the protection of their fragile environment and conservation of natural resources and take benefit from emerging opportunities of globalization. This is particularly imperative in the context of the United Nations Framework Convention on Climate Change (UNFCCC) for providing monetary benefits to mountain inhabitants for their contribution towards mitigating climate change through conservation of forests. The developed countries should line up to act as important resource centres for capacity building of developing countries by providing knowledge, technology and expertise to mountain regions.

The Mountains

Although mountains constitute very significant phenomena on the Earth's land surface with varying altitudinal ranges and heterogeneity of landscape and terrain characteristics and diversity ecosystems, but so far, the mountains of the planet have not been defined very vividly and in a well acceptable manner (Mahat 2006). However, mountains can be characterized by their altitudinal progression across different climatic and vegetation zones, physiographic characteristics and landforms types. "Mountains include all areas of a marked relief with significant ecological differences and slopes which are susceptible to natural hazards and human activities" (Mahat 2006). The mountains and highlands constitute a large part of the planet as nearly 46.7% of the total area of all the continents is above 500 m; 26.9% above 1000 m; and 11.1% is over 2000 m (Mountain Agenda 1997).

United Nations Environment Programme—World Conservation Monitoring Centre (UNEP-WCMC) characterized mountain area as a land mass with slope and elevation which includes both mountains and hills (UNEP-WCMC 2000). However, in several cases mountains are separated from hills on the basis of their higher elevations, steeper slopes and lower temperatures. Food and Agricultural Organization (FAO) of the United Nations has defined mountains as areas with high altitudes and characteristic topography, and all areas above 2500 m elevation have been included under mountains. Further, the areas having an altitude between 300 m and 2500 m can also be categorized as mountain regions if they are marked with steep slopes or wide range of elevation in a small spatial unit (FAO 2002). However, highland plateaus having an elevation below 2500 m that lack surface slope and/or local elevation rate are therefore not included under mountains. But hills marked with low altitudes, exhibiting difficult terrain conditions have been classified as mountains (FAO 2002).

The UNEP-WCMC classified world mountains into six categories ranging from Class-1 to Class-6 mountains taking into account elevation from the mean sea level and surface slope. The elevations higher than 2500 m from the mean sea level have been categorized as mountain areas irrespective of the magnitude of slope and topographic characteristics (Table 4.1 and Fig. 4.1). The Class-1 mountains are those areas which are situated in an altitudinal range of 300 m and 1000 m with a Local Elevation Range (LER) of 300 m of 5 km (i.e., characterized with 5° slope); Class-2 Mountains have an elevation from 1000 m to 1500 m with slope equal to or more than 5° or a Local Elevation Range 300 m of 5 km; Class-3 Mountains should be located between the altitude of 1500 m and 2500 with a surface slope equal to and above 2°; Class-4 mountains area with elevation between 2500 and 3500; Class-5 with altitudinal range between 3500 m and 4500 m; and

Table 4.1. Mountains of the World (Source: Huddleston et al. 2003).

Sub Region	Mountain Area Category-wise (in thousand km²)						
	Class-1	Class-2	Class-3	Class-4	Class-5	Class-6	Total
Asia & Pacific	2731	1151	1219	759	853	1581	8294
Latin America & Caribbean	1412	730	812	443	585	155	4136
Near East & North America	857	752	798	223	37	13	2681
Sub Saharan Africa	921	668	438	94	5	–	2125
Total Developing Countries	5921	3301	3268	1518	1479	1748	17237
Total Transition Countries	3353	1129	546	156	93	27	5305
Total Developed Countries	3263	1573	1296	698	12	–	6842
Total World	12538	6003	5110	2372	1585	1776	29384

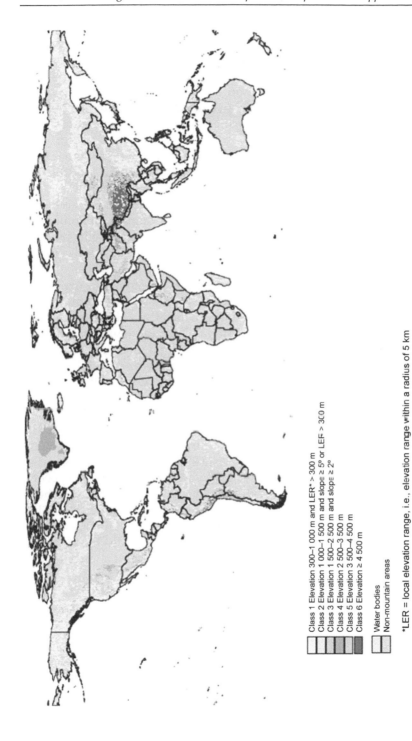

Class 1 Elevation 300–1 000 m and LER* > 300 m
Class 2 Elevation 1 000–1 500 m and slope ≥ 5° or LER > 300 m
Class 3 Elevation 1 500–2 500 m and slope ≥ 2°
Class 4 Elevation 2 500–3 500 m
Class 5 Elevation 3 500–4 500 m
Class 6 Elevation ≥ 4 500 m

Water bodies
Non-mountain areas

*LER = local elevation range, i.e., elevation range within a radius of 5 km

Figure 4.1. World Mountains (Source: Huddleston et al. 2003).

Class-6 mountains should have altitude above 4500 m. It is interesting to observe that this classification of mountain does not consider regions below the altitude of 2500 m, high plateaus and large intermountain valleys as mountain areas, despite their showing several features of mountain ecology and being interconnected with mountain ecosystems. However, areas located just at an altitude of 300 m above mean sea level but marked with more than 2° of slope have been classified as mountain areas from Class-1 to Class-3 Mountains (Table 4.1).

As per the UNEP-WCMC classification of the world's mountains, nearly 29 million km² area which accounts for about 22% of the total geographical land surface of the world is under different types of mountains. The maximum proportion of World Mountains (22 million km²) is located in developing countries and only a small proportion (5 million km²) is situated in transition countries (Table 4.1). Asia and the Pacific region of the developing world shares 28%, developed countries 23% and countries in transition have 18% of the world's mountains. The sub-regions of developing and transition countries in which the mountain regions are extend over more than 1 million km² include East Asia, the Commonwealth of Independent States (CIS), South America, the Near East, Southeast Asia and Oceania, South Asia and East Africa, and mountains in these sub-regions account for approximately 90% of the total mountain areas of the developing and transition countries (Table 4.1). However, Huddleston et al. (2003) have classified those countries as mountainous where one-third of their population was living in mountain areas and/or in which mountainous terrain encompass more than one-third of the total geographical area.

However, taking into the proportion of total geographical area the sub-regions in developing and transition countries, East Asia, North America (Mexico), Central America, Southeast Asia and Oceania and the Near East respectively account for 50%, 45%, 41%, 35% and 34% areas of the respective sub-region. Interestingly, 80% of the world's mountains have an altitude 2500 m and almost 50% mountains regions have an elevation below 1000 m. The high mountains in the planet extend across the Himalaya, Andes mountain ranges and in the Tibetan plateau. Geologically, the Himalaya, Andes, Alps and Rockies are considered as the youngest and tectonically active mountain ranges of the world, and owing to their high elevation, steep slope and geo-tectonic instability these mountains are characterized with high environmental fragility as well as socio-economic vulnerability. Nearly 90% of the mountain population live in developing or transition countries (Huddleston et al. 2003); 50% in the Asia-Pacific region and one-third live in China. About 30% of total world's mountain population lives in urban centres even though during the recent years the mountain regions of heavily populated developing countries have been experiencing rapid urban growth (Hassan et al. 2005). Out of the total population living in the mountain regions more than 50% live below 1000 m, and over 70% live below 1500 m, whereas less than 10% global mountain population live in the mountain areas located above 2500 m (Huddleston et al. 2003). However, distribution of the population living in mountain areas varies remarkably from one region to another (Table 4.2). The Hindu Kush Himalayan (HKH) mountains ranges extending across the countries of Bhutan, Nepal, Afghanistan, India, Pakistan, the Tibetan Plateau and the Andes mountains in Bolivia, northern Chile, Ecuador, Peru are virtually the only mountain regions which are inhibited by humans above an elevation of 4500 m even though the population living in these high mountains account for merely less than 1% of total global mountain population (Huddleston et al. 2003).

Table 4.2. Mountain Population and its Vulnerability (Source: Huddleston and Ataman 2003).

Sub Region	Mountain Area, Mountain Population & its Vulnerability				
	Mountain Area (thousand km²)	Mountain Area as Share of Total Land Area (%)	Mountain Population (Thousand Persons)	Mountain Population as Share of Total Population (%)	Vulnerable Rural Population (Thousand Persons)
Asia & Pacific	8294	41	333070	11	139999
Latin America & Caribbean	4136	20	112421	22	32906
Near East & North America	2681	22	97239	25	29738
Sub Saharan Africa	2125	09	87980	15	30874
Total Developing Countries	17237	23	630710	14	233517
Total Transition Countries	5305	23	32082	08	11200

Mountain Ecosystem Services

The Millennium Ecosystem Assessment (MA) was carried out between 2001 and 2005 to assess the consequences of ecosystem change for human well-being and to establish the scientific basis for actions needed to improve the conservation and sustainable utilization of ecosystems and their contributions to human well-being (Millennium Ecosystem Assessment 2005a). MA defines ecosystem as a dynamic complex of biotic components of nature that include plant, animal, and microorganism communities and the nonliving environment interacting as a functional unit, and also as "the benefits people obtain from ecosystems" (MA) (Millennium Ecosystem Assessment 2005a). The MA deals with the full range of relatively undisturbed ecosystems, such as natural forests and landscapes with mixed patterns of human use, and also ecosystems intensively managed and modified by humans, such as agricultural land and urban areas. Ecosystems are natural assets that provide a wide range of services and products that sustain humanity across the planet (Millennium Ecosystem Assessment 2005a). Mountains support many different ecosystems and provide key resources and services for human sustenance far away from mountains as most of the goods and services provided by mountains have their origin in the headwaters, the beneficiaries of these services are mostly in the lowlands (Beniston 2005). Highland and lowland ecosystems are thus highly interactive and inter-dependent in terms of ecology and economy as well as in social and political perspectives. The mountain population have contributed significantly to the conservation and protection of these ecosystem goods and services with their indigenous knowledge and traditional resource management practices. In view of this, mountain people need to given some tangible incentives and adequate compensation for their most sincere efforts towards supporting the sustainability of large lowland population. The goods and services provided by mountain ecosystems can be divided into the following three primary categories (Hassan et al. 2005, UNEP-WCMC 2002)):

(i) Supporting Services which maintain the essential natural conditions for all forms of life on the Earth and are intangible and do not have an explicit market value. These services mainly include soil formation, photosynthesis and nutrient cycling. (ii) Provisioning Services that provide means of livelihoods and the economy by supplying various natural resources and products, such as, food, water, timber and fibre. Undoubtedly, the most important goods provided by the mountain ecosystem is water, and therefore mountains are often known as 'water towers' for the world (Viviroli 2007, UNEP-WCMC 2002) (Table 4.3 and 4.4). Mountains slopes and valleys have the capacity to store a large amount of freshwater in the form of glaciers, snow and ice; as well as in the form of groundwater and lakes. Almost all the principal rivers of the world, besides a large number of others have their sources in mountains (Messerli and Ives 1997). It has been observed that more than half of the total geographical area of world's mountains is highly crucial for supplying freshwater to a large population inhabiting the lowlands. The freshwater of Hindu Kush Himalayan (HKH) mountain ranges is used by more than 200 million people living in the region and by 1.3 billion people living in the 10 downstream river basins of South and East Asia (Viviroli et al. 2007). Besides water, mountains are endowed with high biological and agricultural diversity that includes variety of food and fibre crops and medicinal plants (UNEP-WCMC 2002). (iii) Regulating Services regulate climate, flood and disease, maintain water

Table 4.3. Principal River Basins of Hindu Kush Himalayan Mountains (Source: ICIMOD 2009).

River	Annual Mean Discharge (m³/ Second)	% of Glacial Melt in River Flow	Basin Area (km²)	Population Density (Person/ km²)	Population (× 0000)	Water Availability (m³/person/ year)
Amu Darya	1376	Not Available	534739	39	20855	2081
Brahmaputra	21261	12	651335	182	118543	5656
Ganges	12037	09	1016124	401	407466	932
Indus	5533	50	1081718	165	178483	978
Irrawaddy	8024	Not Available	413710	79	32683	7742
Mekong	9001	07	805604	71	57198	4963
Salween	1494	09	271914	22	5982	7876
Tarim	1262	50	1152448	07	8067	4933
Yangtze	18811	18	1722193	214	368549	2465
Yellow	1438	02	944970	156	147415	308

Table 4.4. Glaciers and Glaciated Area in Major Basins of Hindu Kush Himalaya (Source: ICIMOD 2009).

Basin	Basin area within HKH (km²)	Number of Glaciers	Glaciated Area (km²)	Estimated Ice Reserves (km²)	Average Glacier Size (km²)
Amu Darya	166686	3277	2566	162.61	0.78
Indus	555450	18495	21193	2696.05	1.15
Ganges	244806	7963	9012	790.53	1.13
Brahmaputra	432480	11497	14020	1302.63	1.22
Irrawaddy	202745	133	35	1.29	0.27
Salween	211122	2113	1352	87.69	0.64
Mekong	138876	482	235	10.68	0.49
Yangtze	565102	1661	1660	121.40	1.00
Yellow	250540	189	137	9.24	0.73
Tarim (Interior)	26729	1091	2310	378.64	2.12
Qinghai-Tibetan Interior	909824	7351	7535	563.10	1.02
Total	3705721	54252	60054	2126.85	1.11

quality and recycle wastes. Mountains play a vital role in maintaining and supporting a healthy and safe environment and climatic conditions for the survival of human beings as well as for other living organism. Mountains contribute significantly in maintaining the hydrological cycle through purification and retention of rainwater in the form of groundwater, ice and snow, as well as in lakes and streams. Mountain ecosystems play an important role in atmospheric circulation and climate regulation and act as a pool for the storage carbon and soil nutrients. The mountain slopes consisting of nearly 28% of the world's forests, provide mechanical support to landscape and hence protect fragile slopes against land degradation and slope instability (IPCC 2007a,b). (iv) Cultural Services that provide opportunities for recreation and education and spiritual and aesthetic inspiration (World Resource Institute 2005). The mountain regions exhibit cultural and ethnological diversity including spiritual traits and rich traditional ecological knowledge, which besides strengthening the adaptive capacity of mountain communities to global change also provide fascinating attractions to a variety of visitors (Bernbaum 1997).

During the last 50 years human interventions have transformed natural ecosystems more rapidly and extensively than in any comparable period of time in human history on the Earth. This was primarily to fulfil rapidly growing demands of natural resources, such as, arable land, food, freshwater, timber, grazing areas and fuel which resulted into a considerable and largely irreversible loss of biodiversity and ecosystem services on Earth. Although these changes in the natural system have contributed significantly towards betterment of the quality of human life, but they also caused rapid exploitation of natural resources, substantial loss of ecosystem services and imbalanced economic growth in many parts of the planet. The depletion of natural resources and continued disruption and loss of ecosystem services may adversely affect the process of attaining the Millennium Development Goals (MDGs) (Millennium Ecosystem Assessment 2005a). The challenge of restoration of ecosystem services and conservation of environment, while meeting increasing demands of natural resources would require significant changes in policy planning and decision-making processes from local to international levels. There are a number of possibilities and opportunities that may be used to restore certain types of ecosystem goods and services in such a way that they provide positive synergies with other ecosystem services (Millennium Ecosystem Assessment 2005a).

Marginality and Vulnerability of Mountains and their Inhabitants

Mountains include some of the most fragile ecosystems on the planet (ICIMOD 2010) as they are highly sensitive to a variety of changes caused by natural as well as by anthropogenic factors primarily owing to their high altitude, steepness of terrain, high rainfall, thin and shallow soils, sensitive flora and fauna, climatic complexities and geodynamic instability (Sonesson and Messerli 2002) (Fig. 4.2). The drivers of changes range from a variety on natural hazards and disasters including, degradation of land, slope instability and landslides, volcanic eruptions and seismic activities, floods and flesh floods, cloud bursts, avalanches, droughts to a series

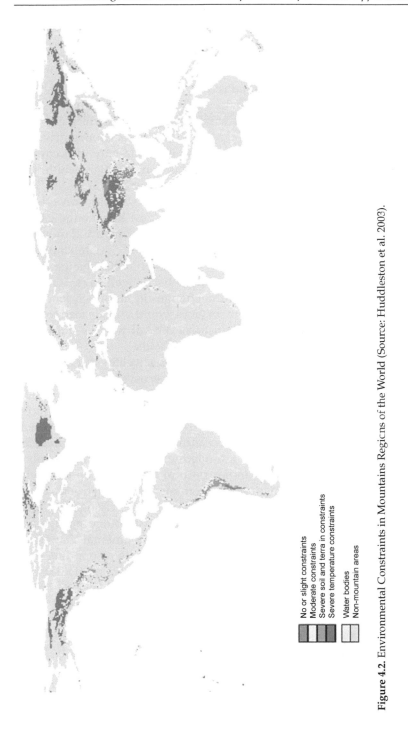

Figure 4.2. Environmental Constraints in Mountains Regions of the World (Source: Huddleston et al. 2003).

of human induced processes such as, population growth and resultant intensification of land use and depletion of natural resources, forest fires, globalization, urbanization, changing economic policies and orders; political marginalization and cultural dilution. Moreover, changing climatic conditions have stressed the mountain environments through, higher mean annual temperatures and melting of glaciers and snow, altered precipitation patterns, and more frequent and extreme weather events which are likely intensify the impacts of other natural as well as socio-economic drivers of change and may cause substantial decrease in availability of ecosystem services, and consequently increase the proportion of water, health, food and livelihood of insecure people in large part of their lowlands.

However, despite exhaustive debate and discussion on climate change and its possible drivers, still there exist vast knowledge gaps in understanding the crucial linkages between climate change and mountain ecosystems. This indicates that the exact impacts of climate change on mountains and its connection with other drivers of global change are yet to be fully understood and interpreted at global as well as regional and local levels. However, it is clear that the changes in mountain landscape and resultant disruption of ecosystem services would directly lead to social unsustainability and economic insecurity of 30% of world's population living in mountains as well as in their vast lowlands.

This will have particularly severe impacts on the livelihood and economy of mountains and their densely populated plains ecosystems in developing countries dependent primarily on subsistence crop farming (Tiwari 2000). Mountains exhibit a very high level of ethno-culturally complexity which is reflected in diversity of languages, dialects, food habits, costumes, traditions and economic systems. The mountain communities and their culture is a vast depository of rich traditional knowledge which they have evolved through experimentation with complex natural conditions, and applying in the conservation, management and governance of natural resources and adapting to impacts of global change including rapidly changing climatic conditions in mountains. The traditional community resource management practices and traditional coping mechanism to environmental changes are now being considered highly relevant and significant in responding to the impacts of global change mountain environments, and particularly for low-intensity production systems at high altitude (UNEP-WCMC 2002).

This clearly indicates that mountain communities, particularly in developing and underdeveloped countries are highly vulnerable to the impacts of global change (UNEP-WCMC 2002) (Table 4.2). The driving forces of mountain people's vulnerability include unbalanced poverty levels, declining agricultural productivity and increasing food insecurity, poor health conditions, high dependency on natural resources, socio-economic and political marginalization and limited livelihood options. The potency of these factors is likely to be further exacerbated by the process of global change especially in least developed countries. It has been indicated that the impacts of climate change on economic development will be unevenly distributed around the globe (Sanderson and Islam 2009). In order to improve rural livelihood in mountains and reduce poverty, it would be highly imperative to address the issue of climate change and its relationship with economic development through adaptation measures.

Approximately, 2.1 billion people amounting to 40% of the rural population of developing countries live in Less Favoured Areas (LFA) which are primarily defined as low productivity regions mainly due geo-environmental constraints (Fig. 4.2), lack of access to infrastructure and market and socio-economic and political marginalization (Ruben et al. 2004). Nearly, 90% of the mountain-population lives in developing or transition countries; 50% in Asia Pacific region and one third of population of mountains resides in China. Although mountains particularly those situated in developing countries, have been passing through a process of rapid urbanization, yet majority of mountain population (70%) across the planet live in rural areas. A considerably large proportion of the population living in mountain regions of developing countries primarily depends on subsistence agriculture for its food and livelihood even-though the availability of arable land is severely limited and crop productivity is poor. The Food and Agricultural Organization (FAO) of United Nations has identified more than 75% of the land surface of the world's mountains as unsuitable or marginally suitable for practicing agriculture. The mountains due to constraints of terrain and climate and lack of access infrastructure and essential services and resultant low level of socio-economic development largely overlap with LFA (Pender and Hazell 2000, Messerli and Ives 1997). As a result, the mountain ecosystems and their inhabitants are highly vulnerable to the long-term impacts of global change particularly in the developing world (Table 4.2). Although mountain areas are rich in natural resources, such as biodiversity, water and scenic beauty, but they often present several challenges because of their subsistence economy, fragile environment, physical isolation, inadequate access to markets and inputs, low resource productivity and resultant vulnerability to risks of a variety of natural hazards and disasters.

Mountain regions, particularly in developing countries are largely inhibited by poor people (IFAD 2001). The mountain communities are subject to prevailing poverty and lower levels of socio-economic development than those in their adjoining lowland areas in developing countries. The main drivers of mountain poverty include fragility of ecosystems; remoteness; poor accessibility and marginalization of mountain communities from the mainstream; lack of equity in terms of access to basic facilities such as health care, education and physical infrastructure, as well as to markets, political marginalization; lack of employment opportunities; and proneness to natural disasters (Tiwari and Joshi 2012b, Jodha 1992, ICIMOD 2010). It has been observed that the proportion of poor and vulnerable people increases with elevation, and the ongoing process of globalization seems to have further strengthened poverty imbalances between highlands and lowlands (Hassan et al. 2005, Huddleston et al. 2003). There are indications that poverty inequality between mountain people and those living in other areas is currently increasing in Nepal (ICIMOD 2010).

Approximately, 271 million people which accounts for nearly 40% of the mountain population in developing and transition countries have been estimated to be highly vulnerable to food insecurity; of which about 50% are surviving under chronic hunger. Out of the total of 245 million mountain rural population identified as vulnerable food insecurity as much as 87% live below 2500 m above mean sea level which constitute the most densely populated transact of mountains specifically in Asia and Latin America (Tiwari and Joshi 2012b). Although in higher mountains regions the number of vulnerable rural population is comparatively small, but that accounts for nearly 70% of the population living in high mountains above 2500 m. The lack of balanced diets reduces the supply of micronutrients in mountain people which results in a very high level of malnutrition among the local population (Huddleston et al. 2003). Furthermore, due to very limited and no access to primary health-care and sanitation facilities and increasing food deficiencies the maternal and infant mortality rates have been observed very high in mountain regions, particularly in developing and underdeveloped countries (FAO 2008).

High dependency on natural resources and increasing marginalization are some of the important factors for the prevailing poverty, food and livelihood insecurity, poor community health, in mountains of developing countries which are further increasing their vulnerability to long-term impacts of global change (Tiwari and Joshi 2012b). The mountain population is traditionally dependent on natural resources and agriculture constitutes the prime source of rural livelihood in mountain environment (Huddleston et al. 2003). At lower mountains, intensive mixed farming systems are the most common form of agriculture, however, due to environmental constraints such as unfavourable climatic conditions, poor-quality or shallow soils, and sloping terrain, productivity is generally low and the produces are not competitive in the global market (Huddleston et al. 2003). Whereas, at higher elevations pastoralism is the most common livelihood option as due to rising altitude, declining temperature and steeper slopes restrict agricultural practices.

Global Change and Mountains

Globalization and climate change are interrelated and interdependent drivers of global change. However, global change studies in mountains so far dominated by climate change, and the impact of globalization have not received the desired focus (Kohler and Maselli 2009, Borsdorf et al. 2008, Jandl et al. 2009). Climate change has long-term impacts on ecosystem services both in mountains and their downstream, and the process of globalization also exerts a sharp impact on ecosystems and an even more severe effect on the socio-economic landscape including agriculture, population structure and its dynamics, as well as the urbanization and marginalization of peripheral mountain communities (Borsdorf et al. 2010). The ongoing global discussion on responses at different spatial scale that might take place under the processes of future global change is generally focussed on fragile environments, such as the Arctic, coastal regions and mountains (ACIA 2004, IPCC 2007a,b). Mountains consisting of a vast repository of a variety of ecosystem services and goods constitute one of the most fragile environments on the planet and are considered as being highly sensitive to global change (Diaz et al. 2003).

The drivers of global change and the process of transformation in mountains are highly complex and multifaceted; hence the understanding of current responses and projecting future changes in these ecosystems is exceedingly difficult (Loeffler et al. 2011). People living in mountain areas are exposed to a series of environmental and non-environmental stressors which are interconnected and have serious implications on the livelihood and quality of life of mountain communities. These stressors include population growth and processes of socio-economic development which are linked to increasing demand of goods and services and globalization that leads to depletion of natural resources, such as, land, water, forests, minerals and biodiversity (ICIMOD 2010). The global change is referred to the changes caused by both natural and anthropogenic processes and encompass, among other factors, climate change, land-use cover change, industrialization, urbanization, and changes in atmospheric chemistry (Goudie and Cuff 2002). Becker and Bugmann (2001) have categorized global environmental changes affecting mountain ecosystems into two groups: (i) systemic changes that operate at a global scale (such as climate change) and (ii) cumulative changes caused by processes at a local scale but that are globally pervasive (such as land-use cover change).

The environmental and economic changes and their impacts are parts of invariable processes operating in the mountain environment. However, during the recent years, the magnitude and rate of these changes and their impacts on natural and social systems have become severe with serious consequences for the sustainability of mountain communities as well as large populations living in their catchments in the downstream. Currently, mountain ecosystems are changing more rapidly than at any time in human history on the Earth. The long-term impacts of changes caused by deleterious anthropogenic interventions, such as intensification of land use and overexploitation of natural resources are much more deleterious for mountain communities than the affects of natural processes, such as volcanic and seismic events, landslides and flooding that devastate large parts of mountain ecosystems every year (Messerli and Hurni 2000, FAO 2002b, Pratt and Shilling 2002). Currently, mountain ecosystems and their inhabitants are exposed to a variety of drivers of change including globalization; economic policies; and increasing pressure on land and mountain resources due to economic growth and changes in population and lifestyle (ICIMOD 2010). The major drivers of global change on the world's mountains mainly include climate change, population dynamics, economic globalization, land-use change, urbanization (Loeffler et al. 2011, Millennium Ecosystem Assessment 2005b). The Andes, covering 33% of the area of the Andean countries, are vital for the livelihoods of the majority of the region's population and the countries' economies. However, increasing pressure, fuelled by growing population numbers, changes in land use, unsustainable exploitation of resources, and climate change, could have far-reaching negative impacts on ecosystem goods and services.

Climate Change

Global Climate Models (GCMs) have improved our understanding about climate change and its possible impacts in mountains. The mountains across the world, particularly the Himalaya and Tibetan Plateau, have shown consistent trends in overall warming during the past 100 years and the temperature rise in mountains is faster than in plains (Yao et al. 2006, Du et al. 2004). A number of studies indicated that the increase in temperature in the Himalaya has been much higher than the global average of 0.74°C over the last 100 years (IPCC 2007a, Du et al. 2004). The Himalaya has experienced warming between 0.6°–0.15°C per decade during the last three decades (Ouyang 2012). The Himalayan ranges in Nepal recorded a temperature increase of 0.6°C per decade between 1977 and 2000. It was observed that the temperature rise in Nepal and Tibet has been progressively higher with elevation (ICIMOD 2007). Global climate change is expected to exacerbate the impacts of other drivers of change on the mountains. However, the exact impacts of climate change on mountain ecosystems, and the interlink-ages with other drivers of global change are yet to be investigated and properly understood (ICIMOD 2010). The mountain regions across the world have experienced the effects of global warming which are resulting in the retreat of glaciers, decrease in permafrost, changes in the seasonality of runoff, upward shifts of the vegetation line, and changes in the biodiversity of alpine lakes and streams. The long-term impacts of these changes are not only of direct relevance to the high mountains, but they will have serious implications for downstream regions. Mountain regions have experienced warming of above-average level in the 20th century (IPCC 2007a,b). The rise in temperature in mountain regions has affected mountain environments and ecological processes across the globe. In case of the Himalaya, the progressive warming at higher altitudes has been three times higher in comparison to global average (Eriksson et al. 2008, Xu 2008).

Warming projections for the current century indicate that the mountain systems of the world are expected to be warmer in the 21st century although the rise in temperatures will be disproportionate and in varying magnitude depending both on the altitude and latitude of mountains all over the planet (IPCC 2007a,b). IPCC Special Reports on Emissions Scenarios (SRES) indicates that the highest temperature rise is expected in high and medium-latitude mountains, including the high-latitude mountains of Asia, North America and Europe and mid-latitude mountains of Asia compared to the mountain ranges of Australia and New Zealand (Rial et al. 2004, Burkett et al. 2005, Nogués-Bravo et al. 2007, Wookey et al. 2009). The high-latitude mountains of Asia are expected to experience the greatest increase in temperature, whereas tropical and mid-latitude mountains in Africa and South America are expected to be less warm with latitudinal gradient of temperature change (Nogues-Bravo et al. 2007). Depending on the different emissions scenario the average temperature in the mountain regions across the globe has been estimated to rise between 2.1°C and 3.2°C by 2055 (Nogues-Bravo et al. 2007), which is 2–3 times higher than the temperature recorded during the 20th century in different mountain regions of the Earth (Pepin and Seidel 2005).

Mountain ecosystems are expected to react very sharply to climate change (IPCC 2007a), with both natural and social systems being influenced all over the world (Beniston 2000, 2003, 2005, 2006). These changes can be irreversible if they are allowed to go beyond the critical threshold limits (Beniston 2003). Despite long-term predictions about changes in climatic conditions are not practically possible, but scenarios can be used to indicate potential changes (Nogués-Bravo et al. 2007). Climate change affects all spheres of the planet and as well as almost all components of the anthroposphere. However, the changes in the cryosphere are most clearly visible (Borsdorf et al. 2010). The climate change has shown the most remarkable and significant impact on the glaciers which are retreating fast since the last century. The trend of glaciers melting is global and rapid which may lead to the de-glaciation of large parts of many mountain ranges in the world in the next decades (UNEP WGMS 2002). The Alps are among the most intensely studied mountains on the planet where an intensified retreat of Alpine glaciers has been observed during the previous century (Borsdorf et al. 2010). The Alpine glaciers have reduced by nearly 35% between 1850 and the 1970s, and as much as by 22% by 2000 (UNEP-WGMS 2008). The volume of glaciers in Canadian Rockies has decreased a minimum of 25% during the last century (Luckman and Kavanagh 2000). In the world's highest and largest mountain system—Himalayas, the rate of retreat of glaciers has been relatively faster than the world average and they are thinning at the magnitude of 0.3–1 m/year (Dyurgerov and Meier 2005). Further, it was investigated that the Himalayan glaciers are losing mass faster than European glaciers but slower than those in Alaska (Inman 2010). The rate of retreat of the Gangotri—the largest glacier of Himalaya —has been three times higher than the rate at which it melted during the preceding 200 years (Bhandari and Nijampurkar 1988, Tiwari 1972, Tiwari and Jangpangi 1962, Mukhophadayay 2006, ICIMOD 2007, IPCC 2007a,b, Rawat 2009, Srivastava 2003). In the Nepal Himalaya, a large number of glaciers have been observed to be receding rapidly (Seko et al. 1998, Kadota et al. 2000, Fujita et al. 2001). The study indicated that 6% glaciers in the headwaters of Tamor and Dudh Koshi basins in the Nepal Himalaya have decreased during 1970–2000 (Ouyang 2012). As much as 82% glaciers in western China have receded during the last half century (Liu et al. 2006, Kang et al. 2010) and as much as 5.5% glaciers receded during 1980–2005 (Ouyang 2012). Further, the area under glaciers in the Tibetan Plateau has decreased by about 11.5% over the last 40 years (CNCCC 2007). Similar trends in the responses of glaciers and snow-packed areas have been observed in South African mountains where about 85% of the total ice volume of the plateau glaciers of Mount Kilimanjaro in Africa vanished between 1912 and 2000 (Thompson et al. 2009). In general, de-glaciation is accelerating in mountains, and consequently, most snow and ice caps across the world are shrinking at alarming rates. However, exceptionally and interestingly, some of the high altitude glaciers in the Karakorom ranges in Asia have been found either advancing or stable as in the 1970s many of these glaciers showed evidence of short-term thickening and expansion (Hewitt et al. 1989). Out of 42 studied glaciers in Karakoram region, 58% are either advancing or stable and 42% retreating (Scherler et al. 2011, Hewitt 2005). Inman (2010) observed that the remote glaciers of the Himalayan mountains have been a subject of much controversy, but without adequate research. Out of about 12000 to 15000 glaciers in the Himalaya and nearly 5000 in Karakorum, very few have so far been monitored on the ground. However, he maintains that Himalayan glaciers have been losing mass, with noticeably greater loss in the past decade (Inman 2010). In all other mountain regions including the Himalaya, Hindukush, West Kunlun Shan, the glaciers are mostly retreating, and the study has observed that the debris cover has a significant influence on glacier terminus dynamics (Scherler et al.

2011). In North American mountains, it was observed that a large number of glaciers were continuously retreating and thinning, and the unconsolidated and unstable sediment is exposed and mobilized into rivers, which causes aggradation downstream. As a result, the rate of sedimentation in most of the river-beds has increased from 7–13 cm up to 1.8 m per decade leading to more frequent catastrophic shifts in the courses of river channels (ASPEN International Mountain Foundation 2012).

The glaciers have been found retreating in all Andean countries over the last three decades with the complete disappearance of a large number of glaciers. The average rate of retreat of glaciers in Cordillera Blanca of Peru has been observed to be 26% between 1970 and 2003, in Ecuador 27% during 1997–2006, 87% reduction in the Mérida Cordillera in Venezuela in the last 50 years, and 2 to 5% annual decrease in glaciers in Colombia over the last 10 years. The annual rate of contribution to sea-level rise from the Patagonian ice-fields has doubled during 2000–2005 in comparison to 1975–2000. The area of small glaciers has reduced by 50% in the Argentinean Tierra del Fuego. The Northern Patagonian Ice Field in South America lost nearly 140 km^2 of its snow covered area during 1942 and 2001.

These observed changes in the cryosphere are transforming the hydrological regimes of mountain headwaters and disrupting drainage systems all across the basins. Consequently, the regime of water resources in mountain headwaters is likely to change rapidly, with respect to discharge, volumes and availability (Tiwari 2000, 2008, Tiwari and Joshi 2012a, 2005, Bandyopadhyay et al. 2002, Viriroli et al. 2003). In the Himalaya, as many as 35% of the present glaciers will disappear and runoff will increase with peak flow attaining between 2030 and 2050 with a projected temperature increase of 2°C by 2050 (Qin 2002). The impacts of retreating glaciers on river-flow is expected to be greater, in small, more highly glaciated headwaters in both the western and eastern Himalayan mountains under the uniform warming scenario of +0.06°C per year. Water discharge in most of the glacier-fed sub-catchments (glaciation ≥50%) is likely to attain peak flow of 150 and 170% of initial flow around 2050 and 2070 in the western and eastern Himalayan ranges respectively before decreasing until the respective glaciers vanish in 2086 and 2109 (Rees and Collins 2006). The western Himalayan glaciers are expected to retreat in the next 50 years causing increase of Indus River flows, and then the glacial water-reservoirs will be empty, resulting in decrease of flows by up to 30 to 40% over the subsequent 50 years (World Bank 2005). At the same time, a large number of mountain communities have been severely threatened by the diminishing or disappearance of the small ice-masses and snow fields on which they depend for their water supplies (ICIMOD 2007). It is most likely that most of the mountain regions and their lowlands face a catastrophic water scarcity by the 2050s resulting from population growth, climatic change and the increase of water demand (Oki 2003).

The Himalaya and Tibetan Plateau—the 'Water Tower of South Asia' sustains livelihood and economy of approximately 200 million people. The observed changes in headwater ecosystems are likely to have significant implications on water availability for the mountain as well as downstream population (Stern et al. 2006). The continued melting of glaciers, snow and ice cover will adversely affect the supply of water leading to severe water crisis and potential conflicts in large part of the world. The Himalaya, for example, constitutes headwaters of some of the largest trans-boundary basins of planet (e.g., Amu Darya, Brahmaputra, Ganges, Indus, Irrawaddy, Mekong, Salween, Tarim, Yangtze and Yellow River basins) that sustain one-fourth global population dependent primarily on subsistence agriculture in Pakistan, India, Nepal, Bhutan, China and Bangladesh (Table 4.3, 4.4 and Fig. 4.3). Climate change has stressed hydrological regimes of Himalayan headwaters through higher mean annual temperatures, melting of glaciers and altered precipitation patterns causing substantial decrease in water availability. This may increase the proportion of health, food and livelihood of the insecure population in South Asia which includes some of the poorest people of the world with access to less than 5% of the planet's freshwater resources. This will have enormous regional implications for fundamental human endeavours ranging from poverty alleviation to environmental sustainability and climate change adaptation, and even to human security

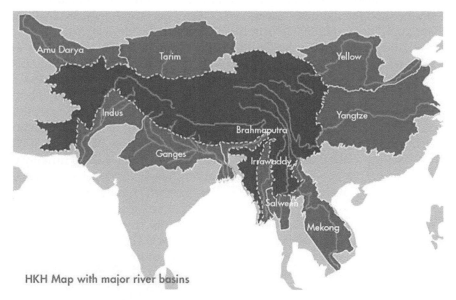

HKH Map with major river basins

Figure 4.3. Hindu Kush Himalaya and Tibetan Plateau with Major River Basins (Source: ICIMOD).

Color image of this figure appears in the color plate section at the end of the book.

(Stern et al. 2006). However, in view of the complexity of inter-linkages between the climatic and non-climatic drivers influencing the mountain freshwater resources and very high improbability regarding changes in precipitation patterns it is extremely difficult to project future changes in mountain freshwater ecosystems.

In the Andes, clean water becomes a critical resource for future sustainable development, taking into consideration the generalized scarcity of resources in the country. Climate change has already decreased the annual average rainfall by 50% in the southernmost section of Chilean Patagonia, mean and minimum temperatures are increasing, snowlines are rising and glaciers are retreating (CONDESAN 2011). There are already serious concerns about decreasing availability of freshwater and reduced supply of hydropower in cities, such as La Paz, Lima and Quito (CONDESAN 2011). The glacier runoff is highly significant for the supply of freshwater in Peru and Bolivia (Buytaert et al. 2011) where the decreasing glacier runoff is affecting the prime agricultural areas in many of the Andean valleys, particularly in Cordillera Norte and in Central Argentina and Chile (CONDESAN 2011). Decreases in water levels in rivers originating in mountains in the provinces of Río Negro and Neuquén, probably due to reductions in snowfall in the Andes, have already led to 40% reductions in hydroelectricity generation (República de Argentina 2007) (Table 4.5).

Mountain ecosystems sustain about 50% of the global biodiversity, and support approximately half of the biodiversity hotspots of the planet (Hassan et al. 2005). Furthermore, nearly 28% of the geographical area of the world's terrestrial protected areas has their home in the mountains (Kollmair et al. 2005). Nonetheless, mountain biological species are adapted to specific altitudinal zones and microclimatic conditions; hence they are highly sensitive to climate variability (Hassan et al. 2005, IPCC 2007a,b). The high altitude, steep terrain, the compression of climatic zones, and landscape fragmentation, increases the biological richness in terms of both species diversity and endemism in mountain ecosystems (IPCC 2007a,b, Körner 2009). The observed trends of changes in temperature are expected to push the vegetation belts upward in higher elevations and the geographical ranges of species may advance northward in the northern hemisphere changing the species composition of communities (Nogues-Bravo et al. 2007, Singh et al. 2010). The advancing unfolding, blossoming, and ripening in the leaves and fruit of wild plants; and of hibernation, migration, and breeding of wildlife have been observed in mountain regions (ICIMOD 2007). In China, the phenology of events has become two to four days earlier than observed in 1980s (Zheng et al. 2002). The climate change also increases the risk of extinction for species with narrow geographic or climatic distributions and also disruption of existing communities. A large number of endemic plant species are unable to respond appropriately to

Table 4.5. Expected Climate Change Impacts in the Andean Region of Countries (Source: National UNFCCC Communications).

Country	Expected Climate Change Impacts
Argentina	▪ Less snowfall in mountains (affecting hydroelectric production, and water availability for irrigation). ▪ Reduction in rainfall in mountains (trends recorded since last century) ▪ Warming of 1°C (greater demand for water in agriculture due to greater evapo-transpiration
Bolivia	▪ Greater concentration of rainfall with less days of rain and more intense flooding ▪ Greater frequency of frost ▪ Greater frequency of hail (destruction of crops) ▪ Longer periods without rain (greater need for irrigation, reduction in hydroelectric energy) ▪ Retreat of glaciers
Chile	▪ Decrease in rainfall from north to centre of country (reduced agricultural yields), and increases on altiplano and further south (increase in suitable climate for grasslands and yields) ▪ Decrease in frosts, milder spring temperatures (improve conditions for temperate fruit growing), but colder winter temperatures
Colombia	▪ Transition from semi-humid to semi-arid climate in mountain regions
Ecuador	▪ Reduction in rainfall (affecting hydroelectric production), conversion to grasslands in some agriculture regions ▪ Reduction in glacier areas
Peru	▪ Increase in rainfall north, central mountains, decrease in rainfall further south ▪ Increase in temperature in all mountain regions ▪ Drastic reduction in areas of glaciers, or disappearance (affecting tourism)
Venezuela	▪ Increase in areas with less than four months of rain per year ▪ Relocation of tourism to higher elevations

rapidly changing climatic conditions (McCarty 2001). In the Alpine zone the upward movement of treeline and encroachment of woody vegetation has been observed. In the eastern Himalaya, the treeline is moving upward at a rate of 5–10 m per decade (Baker and Moseley 2007). However, some mountain species, such as, territorial animals, late successional plant species, species with small, restricted populations, and species confined to summits are likely to face extinction (Körner 2009).

There is widespread agreement that global warming is associated with the most severe fluctuations, particularly in combination with intensified monsoon circulations that changes the frequency and magnitude of extreme weather events and causes severe natural hazards and disasters. Mountain ecosystems are highly vulnerable to a variety of natural hazards and disasters which now being triggered by rapidly changing climatic conditions. The changes in temperature and precipitation pattern and melting of glaciers and snow alter the frequencies, distribution and magnitudes of natural hazards and disasters in mountains. Mountains are typically exposed to a variety of natural hazards and their multiple impacts, and the incidences and magnitude of hazards and extreme events, such as, floods, windstorms, droughts, cloud bursts and flash floods will increase with increasing temperature and changes in precipitation pattern (Kohler and Maselli 2009, IPCC 2007a,b). The global warming is to intensify the hydrological cycle in mountain watersheds changing the frequency, intensity and severity of floods and droughts in mountains as well as in their lowlands (Beniston 2005). The high intensity precipitation could trigger flash floods, slope failure and landslides in mountainous terrains having severe impacts on the natural and socio-economic sustainability of fragile mountain ecosystems especially in the tropics and at higher latitudes where an increase in overall precipitation is expected. The more intense precipitation has increased the incidences as well as the spatial scale of floods in the Central Europe during the past few decades. Since the 1990s, an increase in extreme events has been recorded in several river systems with a probability of recurrence in such short succession in large parts of Europe (Borsdorf et al. 2008). On the contrary, the risks of droughts are likely to increase in most subtropical and mid-latitude regions. The most dangerous natural disasters in high mountains are associated with the direct consequences of changes in the cryosphere. The changes in cryosphere have led to the formation of a large number of glacial lakes in the Himalaya, and many of these high altitude lakes are potentially dangerous. The resulting Glacial Lake Out Flows (GLOFs) can cause catastrophic flooding in lowlands damaging human lives, property, forests, agricultural land, crops and infrastructure. The Himalaya in Nepal has recorded 25 GLOFs over the last 70 years (Mool 2001, Yamada 1998). In the Hindu Kush-Himalayan region, some 204 critical glacial lakes bearing the high risk of breaching have been identified (Ives et al. 2010). It has been observed that liquid precipitation in combination with excessive melt-water often triggers flash floods and debris flows. In Karakoram ranges, the catastrophic rockslides have a substantial influence on glaciers and have triggered glacial surges which are specific hazards in the Karakoram and Pamir mountains (Hewitt 2005). In the Andes, the frequency of extreme climate or weather events in Andean countries, such as flooding, extreme temperatures, landslides, droughts and wildfires have increased by almost 40% in the period 2001–2010 (CEPAL 2011) (Table 4.5).

As discussed earlier, mountains consisting of fragile ecosystems are highly vulnerable to the impacts of these changes that may cause substantial loss of forests, biodiversity and natural habitats, decrease in availability of water for drinking, sanitation and food production, and consequently an increased proportion of water, health, food and livelihood of the insecure population in the entire region. Moreover, the changing climatic conditions are also expected to increase both the severity and frequency of natural disasters in fragile mountains (IPCC 2007a,b, ICIMOD 2007a). Climate change is particularly threatening sustainable development, especially poverty alleviation and livelihood improvement programmes, in the mountains as critical livelihood resources such as agricultural crops, storage of food and seeds, and agricultural land have come under increasing threats of climate change triggered natural risks particularly in developing countries. The mountain communities are expected to face more stern impacts in future due to the likelihood of more and more frequent occurrences of extreme events, and their economic development prospects are coming under increased risks of natural calamities. The long-term impacts of climate change may thus further widen the existing socio-economic inequalities between highland and lowland communities. Tourism is emerging as one the most important opportunities in mountains all over the world. However, like agriculture, the tourism industry is extremely susceptible to changes in climatic and other environmental conditions. Picturesque and pristine natural landscapes, twisting rivers, glaciers, diversity of flora and fauna and the range of protected areas are some of the major tourist attractions in mountains. Climate change and the resultant increase in extreme events, fast melting of glaciers, diminishing water flow in streams and rivers have already threatened the tourism industry both in developing and developed countries. Winter skiing resorts in the Alps have made heavy investments into adapting to higher temperatures over the past several years, and in view of this the resort located at lower elevations would no longer be competitive (IPCC 2007a,b).

It is likely that food insecurity could increase more rapidly with climate change, and community health conditions, particularly the sensitive segment of society, such as the old people, infants and children, pregnant women, and the chronically sick may further deteriorate. The declining rainfall and decreasing number of rainy days and the reduced availability of water and increasing incidences and severity of natural disasters may result in crop failure spread of crop and livestock pests and vector-borne diseases to higher elevations with rising temperatures. This may further undermine the sustainability and wellbeing of mountain communities in poor regions. Moreover, the traditional livestock rearing and pastoral practices are becoming increasingly vulnerable due to population growth and the resultant land-use intensifications at higher elevations, and also owing to the impacts of more frequent and severe droughts, and the breakdown of traditional trade routes and patterns of exchange (Huddleston et al. 2003).

Land Use and Land Cover Change

Besides climate change, changes in land use and land cover are now strongly considered as one of the powerful drivers of global change affecting mountain environments (Bugmann et al. 2007, Zierl and Bugmann

2007, Batllori and Gutierrez 2008). During the recent years, the process of land use and land cover change has become the most important driver of landscape change across the globe (Goudie et al. 2002). Further, it is anticipated that anthropogenic interventions and resultant land-use changes will become increasingly dominant in the 21st century. The integrity of the argument depends on defining and comparing effects at specific temporal (century) and spatial (landscape) scales (Löffler et al. 2011). The economic globalization and population growth are likely to have far reaching impacts on the natural ecosystems as well as on human sustainability in mountains. In many mountain regions of the world there have been drastic changes in land-use patterns during the last few decades. The mountain areas in least developed countries—from Jamaica to Nepal—mountain forests are depleted as population increases in the lowlands, forcing poorer people into the mountains, where they are left with no option but to cultivate marginal land for their livelihood (Löffler et al. 2011).

Economic development often results in land-use changes with consequent degradation of ecosystem services. These changes may lead to environmental degradation through over-exploitation of natural resources and resultant intensifications of land use on fragile slopes in densely populated mountains of developing countries (Beniston 2000, 2003), while the mountain regions of developed countries are expected to experience extensification and reforestation (OcCC 2007). The changes in land-use pattern can also disrupt the circulation of sensible and latent heat, carbon-dioxide, nutrients and pollutants in the ecosystem by altering their exchange processes (Tasser et al. 2005). The European Alps have experienced dramatic transformation of changes in land use and land cover over the last decades mainly due to the abandonment of low productive and less accessible agricultural land and the agricultural intensification of high productive and easily accessible areas (Tasser et al. 2005, Börst 2006, Giupponi et al. 2006, Tasser et al. 2005, Lambin et al. 1999). In many cases, these land-use changes have been derived by advancement in farming technology, such as mechanized harvesting of hay or introduction of improved breeds of grazing cattle. In the European Alps 16% of all farmland has been abandoned during 1980 and 1990, and agricultural is now being practiced as a secondary source of income in another 70% farms. It was observed that a minimum 20% to a maximum of 70% farmland has been abandoned in the Alps during the recent past (Lambin et al. 1999).

Land-use changes are now being considered as one of the major driving forces transforming the natural landscape and affecting the ecosystem function and dynamics. These changes in ecosystem structure and function are causing great loss of biodiversity and disrupting biogeochemical cycles and hydrological processes in the Alps (Borsdorf et al. 2010). In the Himalaya, owing to constraints of terrain and climate biomass based subsistence agriculture constitutes the main source of rural livelihood. More than 75% population of the region depends on this traditional subsistence agriculture even though the availability of arable land is severely limited and agricultural productivity is considerably poor (Tiwari and Joshi 2007). This traditional agriculture is interlinked with forests and pastures and flow of biomass energy from forests to agro-ecosystem is mediated through livestock (Singh et al. 1984). During the recent past, a variety of changes have emerged in the traditional resource use structure mainly in response to growing population pressure and resultant increased demand of natural resources, such as, arable land, grazing areas, fodder, fuel wood, etc. and increasing socio-economic and political marginalization in the region (Tiwari 2010, Palni et al. 1998). This facilitated and also compelled people to utilize the critical natural resources, such as, land, water and forests beyond their ecological carrying capacity. Large-scale deforestation, mining and quarrying, extension of cultivation, excessive grazing, rapid urban growth and development of tourism contributed significantly to the depletion of natural resources (Ives 1989, Tiwari 1995, 2000, 2008, 2010, Tiwari and Joshi 1997, 2005, 2007, 2009). Besides, the fast expansion of road linkages has facilitated the rapid urbanization, emergence and growth of rural service centres and increased access to markets. A large proportion of arable land is being encroached upon by growing urbanization and expansion of infrastructure, services and economic activities in the region (Tiwari and Joshi 2011, Tiwari and Joshi 2005, Tiwari 2007). The process of economic development in mountains usually leads to a weakening of traditional cultures that have provided the foundation for local sustainability.

Moreover, there is a regional shift from traditional crop farming and animal husbandry system to village-based production of fruits, vegetables, flowers and milk for sale both in the nearby and far-off markets, in the villages situated in the influence zone of urban centres and market places, and along and near roads (Singh et al. 1984). This has a large impact on the traditional resource development process and land-use pattern. As a result, critical natural resources, such as, land, water, forests, biodiversity, pastures, etc., have depleted steadily and significantly leading to their conversion into degraded and wastelands in the region during the last 30 years. These land-use changes have an unprecedented adverse impact on basic ecosystem services, particularly, water, biomass, soil-nutrients leading to decline in productivity of rural ecosystem and undermining food and livelihood insecurities in the Himalaya (Palni et al. 1998, Ives 1989, Tiwari 2000, 2002, Bisht and Tiwari 1996, Haigh 2002). Besides, these rapid land-use changes, particularly conversion of forests into degraded and wasteland are also contributing to ongoing climate changes (Tiwari 2010). The land-use intensifications have disrupted the hydrological regimes of the Himalayan headwaters (Tiwari and Joshi 2012a, Tiwari 2008, Ives 1989). The studies carried out in the region revealed that the amount of surface runoff from cultivated and barren lands is much higher compared to the amount of runoff from other categories of land, particularly, forests and horticulture (Tiwari 1995, 2000, 2008, Rawat 2009).

The large-scale depletion of forest resources is causing great damage to the underground water resources by reducing the water-generating capacity of land to springs and streams in the region. Since, a large proportion of the rainfall is lost through surface run-off without replenishing the groundwater reserves, a large number of springs that support a variety of life-sustaining activities are drying up fast in the region. The water resources of the region are diminishing and depleting fast owing to the rapid land-use changes and resultant reductions in groundwater recharge (Tiwari and Joshi 2012a, Valdiya and Bartarua 1991, Tiwari 1995, 2000, Bisht and Tiwari 1996, Wasson et al. 2008, Jianchu et al. 2008). In the Himalaya, these hydrological imbalances are discernible in terms of:

1. The long-term decreasing trend of stream discharge (Tiwari and Joshi 2012a, Rawat 2009, Tiwari 2008);
2. Diminishing discharge and drying up of springs (Valdiya and Bartarya 1991, Rawat 2009, Tiwari 1995, 2000, 2002, 2008, 2010, Tiwari and Joshi 2012a, Tiwari and Joshi 2005, 2007, 2009);
3. Human impacts on surface run-off flow systems and channel network capacity (Tiwari and Jangpangi 1962, Rawat 2009).

In addition to above-mentioned hydrological implications, the natural risks of rapid land-use changes in the Himalaya are also clearly discernible in terms of the dwindling capacity of lakes situated in and around the urban areas through their accelerated silting and pollution in the region (Khanka and Jalal 1984, Rawat 2009). Bathymetric investigations of Bhimtal and Nainital Lakes situated in the Kumaon Himalaya revealed that the capacity of these important lakes has respectively decreased by 5494 m³ and 14150 m³ during the last 100–110 years due to rapid siltation of the lakebeds. The annual average rate of siltation in the Nainital Lake was 65.32 m³ (Khanka and Jalal 1984, Hashimi et al. 1994). Most of the lakes of the region are heavily infested by weeds and invaded marshy conditions (Valdiya and Bartarya 1991).

Rapid commercial and residential land development throughout the North American Rocky Mountains has caused significant, but poorly assessed, ecological and social effects (Theobald et al. 1996). British Columbia is a mountainous province, and most of its geographical area lies in high Rocky and coastal mountain ranges which are among the youngest and recent mountains systems of the Earth. Consequently, these vast mountains tracts are geologically instable and ecologically vulnerable (Robinson 1991). In the Canadian Rockies and coastal mountains in British Columbia the indiscriminate deforestation (200000 ha/year) has been bringing about major land-use changes for the last two centuries. This has adversely affected the hydrological cycle of watersheds resulting in the destruction of fish and wildlife habitats, accelerated soil-erosion and landslides, pollution of water sources and loss of biodiversity and wilderness, in the entire province. As a result, the forest resources have depleted and eroded steadily and significantly bringing about rapid changes in the natural environment and biodiversity of the region.

In the Andes Mountains large scale conversion of forests into pastures and cultivated land has been taking place for the last century. The pressures on natural resources in the Andes result from expanding populations, expanding agricultural areas and intensity, and increasing mineral extraction (Romero et al. 2009). Studies indicated that intensive grazing is disrupting the Andean hydrological cycle by reducing interception and transpiration and increasing runoff (CONDESAN 2011). In Chilean Andes, hundreds of thousands of hectares of rainforest and shrubs were burned at the beginning of the 20th century in an attempt to introduce livestock and agriculture in unsuitable lands, resulting in degradation of natural ecosystem through accelerated soil erosion, rapid loss of biodiversity and depletion of water resources (Romero et al. 2009). These changes also resulted in conflicts between public and private use of resources all over Chile after 30 years of extreme application of liberal economic and political ideas (Romero et al. 2009). Current changes in land use are responsible for growing pressures on natural ecosystems in the Andes. This is substantiated by the fact that at the regional level South America suffered the largest net loss of forest between 1990 and 2010, at about 0.6 hectares per year above Africa for the period 2000–2010 (FAO 2010). Between 1990 and 2010, forest extent for the whole of the seven Andean countries decreased by 239,110 km^2 registering a decrease of 35 to 38% of country area covered in forest (FAO 2010). In the northern Andes excluding Chile and Argentina, transformed ecosystems were found to the extent of 22% of the area for the period 2000–2003 which varies from 3% of transformed areas in Bolivia to as much as 58% in Colombia. The transformed natural landscape is actually much larger in the northern part of the continent in comparison to the Central Andes (Romero et al. 2009). In the European Alps, mountain agriculture is becoming less and less competitive given the difficult conditions of climate and terrain (Borsdorf et al. 2008).

Urban Growth in High Mountains

Urban development has emerged as one of the most important global change drivers influencing the mountain regions not only in North America and Europe but also in developing countries in Asia, Africa and Latin

America where mountain regions offer high value human habitat and where urban growth has been rapid and mostly unplanned. Urbanization as a general term covers a variety of dynamics. Large cities proximate to the mountains expand into the mountains. Smaller towns grow through amenity migration by distant immigrants while others grow by offering new economic or human development options to proximate rural migrants. Some grow as gateways to mass tourism. Yet others grow with ongoing economic development and penetration of global economies. While in all cases, population densities increase and urban footprints expand, the nature of the population and the footprint vary as do the longer-term development trajectory. Urbanization has impacts, both ecological and social, within the urban footprint and beyond. Expanding urban areas consume and pollute resources, and frequently put inhabitants at risk. But urbanization also contributes significantly to not only economic growth through the improvement of infrastructure, the development of tourism and the generation of employment opportunities, but also increased community sustainability by strengthening social services, particularly, education, health, communication, etc. At the same time, rapid urbanization alters land use far beyond the urban periphery and unlocks the world's mountains, particularly their remote and comparatively inaccessible areas for exploitation of their natural resources by growing global markets. As urbanization is to a large extent at least theoretically under policy control, it is conceivable that the mix of costs and benefits attributable to urbanization, and the distribution of those costs and benefits across the many social groups inhabiting these areas can be influenced by policy.

Our mountains across the planet are rapidly urbanizing. In the Andes the proportion of urban population has increased tremendously during the last few decades. Out of the total population of Andean countries 69 and 91% now living in urban areas compared to 55 to 87% in 1990 (Rumero et al. 2009, CEPAL 2004). Bolivia and Ecuador have emerged as the most urbanized countries in the continents, with difference of 13 and 15 percentage points, respectively (CONDESAN 2011). The largest urban centres in the Andean countries are evenly distributed inside and outside the Andean region. However, many situated outside, such as Lima—the most populated city in the Andean countries, depend heavily on the Andes for natural resources, particularly water. The environmental and socio-economic impacts of rapid urban growth in the Andes have reached an alarming level (CONDESAN 2011). Chile has experienced a more rapid, persistent and comprehensive process of economic development in the last 20 years that resulted in fast urban growth in high mountains as well as in their pediment zones. As a result, 87% of Chile population lived in urban areas in 2002 which is having severe impacts on mountainous ecosystems (CONDESAN 2011).

In Santiago and its surrounding areas 19500 ha of productive agricultural land was lost due to urbanization between 1989 and 2003. Furthermore, the transformation of 6789 ha of land with high vegetation productivity, 4224 ha of land with high soil moisture content, and 6654 ha with high concentrations of biomass caused enormous loss to natural ecosystem of the fragile mountains. The growth of imperviousness areas must be considered one of the most devastating environmental impacts of urbanization in semi-arid Mediterranean Andean landscapes. Urbanization is responsible for increasing surface runoff, decreasing groundwater recharge, and polluted water sources. It was also observed that the process of urbanization is adversely affecting ecosystem services, such as groundwater recharge. This situation is severely affecting the ecosystem in Mediterranean area where average annual rainfall is less than 300 mm which falls in less than 20 days annually, and the region faces severe drought that occurs at least once every 10 years (CONDESN 2011). The increasing urbanization is also contributing towards warming through creating large urban heat islands. These heat islands are affecting the cooler piedmont zones that have traditionally acted as a source of airflow which cleans the heavily polluted city atmosphere during the night (Romero 2004).

The expansion of urbanized centres is threatening the very last natural relics in the European Alps. Many Alpine valleys have already lost most of their biodiversity value due to urban sprawl. In the Alps, the initial process of the evolution of settlement started in elevated areas of the most accessible alpine valleys which offered the best conditions for housing and agriculture. Later, with the growth and gradual spread of urban settlements in the side valleys, the catchments of the easily reachable Rhone, Rhine, Inn, and Adige Rivers have already lost most of their biodiversity prosperity. Natural habitats and ecosystems in the valley bottoms—riverbeds, floodplain forests, wetlands, and alpine steppes are now disappearing with expanding urbanization. The development of transport infrastructure associated with this urban growth is a major barrier for many species in the Alps, preventing the establishment of ecological networks. Currently, the Alps are experiencing a new level of urbanization with the average living space occupied by a person has doubled since 1950. Although the population is growing at a very slow rate, but there has been a sharp rise in the number of residential properties. The continual expansion of cities, towns, tourist and skiing resorts are now threatening even the more remote and inaccessible areas in the Alps. Tourist complexes are significantly contributing to the rising level of urbanization in the sensitive mountains. The urban expansion coupled with the growth of population has reduced the permeability of land surfaces and, hence, the time of concentration of floods, in the absence of mitigating solutions in the design of urban drainage systems (urbanization and floods in Italian Alps). Similarly, the Rockies which are intensified by the increasing urbanization, growing demands for water, and altered precipitation and stream-flow patterns driven by climate change.

In the Himalaya, with the growth of population and infrastructure, particularly increasing road connectivity, the region has experienced rapid urban growth during the recent past. The fast expansion of road linkages has facilitated the rapid urbanization, emergence and growth of rural service centres and increased access to markets. This is clearly indicated by fast growing urban population in the newly created Himalayan state of Uttarakhand in India where the urban population increased from 16.36% of the total in 1971 to 20.7% in 1981, 22.97% in 1991, and 25.59 in 2001. As per Census of India (2001), the state registered 56.38% growth of urban population during 1971–1981, however, decadal urban population growth declined slightly during 1981–1991 (42.20%) and 1991–2001 (32.81%). Nevertheless, growth of the urban population in Uttarakhand during 1971–1981 and 1981–1991 was much higher than the national decadal growth of urban population in India (46.39 and 36.24% respectively). Moreover, there is a regional shift from traditional crop farming and animal husbandry system to village-based production of fruits, vegetables, flowers and milk for sale both in the nearby and far-off urban markets, in the villages situated in the influence zone of urban centres and market places, and along and near roads. This has a large impact on the traditional resource development process and land-use pattern.

The rapid growth of urban settlements in the high Himalayan Mountains, particularly in tectonically alive and ecologically fragile Lesser Himalayan ranges has been resulting in the depletion and destruction of nature as well as increased incidence and severity of natural risks, such as, slope failures, disruption of natural drainage and water pollution, degradation of forests, etc. within the urban ecosystems as well as in their surrounding areas. More recently, comparatively less accessible areas of the region are also being affected by process of fast urbanization mainly owing to the extension of road network, development of horticulture, gradual shift from primary resource development practices to secondary and tertiary sectors, and the growth of domestic tourism through the publicity and marketing of new tourist sites. Consequently, there has been tremendous increase in size, area, number and complexity of urban settlements in the region resulting in the expansion of urban land use (i.e., expansion of urban land use in surrounding agricultural zone, forests and rural environments) in urban fringe areas as well as intensity of land use (i.e., increase in the density of covered area, density of building, density of population, increase in the height of buildings, increase in the volume of traffic on roads and increase in the consumption of energy and water, etc.) within the towns (Tiwari and Joshi 2011, Tiwari 2008, 2007).

The natural risks of this unplanned urban growth are now clearly discernible in most of the urban centres and their surroundings, in the densely populated Lesser Himalayan ranges of the newly carved mountain state of Uttarakhand, such as, Nainital, Mussoorie, Pauri, etc. It is expected that the urban growth cannot be stopped or reduced but can be steered in a more sustainable way by a proper integrated land-use management. The urban development in the region is also having long-term impacts on the fragile ecosystem and environment of the urban fringe areas consisting of natural forests, wildlife habitats and water sources including, lakes, streams and natural springs and agricultural land. The natural components of the urban fringe zone are being degraded and depleted steadily and significantly through the expansion of urban land use, deforestation, habitat destruction, mining of aggregate material for construction, waste and sewage disposal, and facilitating changes in the traditional land use and resource management practices by the multiplier effect of urban growth (Tiwari 2008, 2007, Tiwari and Joshi 2011, 2005).

A large proportion of cultivated land and other areas are being encroached upon by the process of rapid urbanization and expansion of infrastructure, services and economic activities in the Himalaya, every year (Tiwari and Joshi 2011, 2005). A study indicated that the most densely settled and rapidly growing urban centres of Uttarakhand Himalaya have been fast intruding upon productive agricultural land in their surrounding rural regions. This has caused huge transformation of cultivated land within urban centres as well as in their peri-urban zones leading to land-use intensifications. Rural areas surrounding these urban centres have lost their prime agricultural land ranging between 4.71% to as much as 12.97% due expansion of urban land use in urban fringe during the 30 years. The loss of fertile agricultural land, decrease in supply of biomass manure and reduction of irrigated area caused by the ruin of forests and development of urban structure contributed 19 to 55% decline in agricultural productivity in 10 urban zones. Consequently, rural settlements situated on the fringe of these urban complexes are currently facing food deficit between 65 and 95%. Undoubtedly, urbanization has contributed significantly to socio-economic betterment of the region through development of infrastructure, generation of employment opportunities in various emerging sectors, such as, tourism, trade, services, etc. However, depletion of forest resources and decline in agricultural economy decreased off-farm employment opportunities in traditional forestry and agricultural sectors. As a result, rural households in peri-urban zones have lost 7 to 12% of their food purchasing power. This will have long-term impacts on local food security affecting particularly the poor and socially marginalized communities constituting nearly 75% of total population (Tiwari and Joshi 2011, Tiwari 2008).

Resource Extraction

Mountains are considered as the largest reservoir of natural resources, such as, water, forests, biodiversity, minerals and hydropower. Hydropower generation is one of the major resource extraction activities in all mountain regions of the world (Millennium Ecosystem Assessment 2005b). A large number of hydropower generation projects have inundated large areas under reservoirs depleting land, water, forests and biodiversity in mountains including in the higher Himalaya. This brings the severe risks of seismic events and dam failure besides the threat catastrophic flooding in the mountains as well as in their vast lowlands. In addition, the large hydropower projects have displaced millions of people and ruined their culture, traditions, customs, indigenous knowledge and traditional resource utilization system worldwide (Millennium Ecosystem Assessment 2005b). Nevertheless, the lion's share of the benefit generated by such projects is enjoyed by the people far away from the mountains with no sentiments for mountain environments and their poor people. In the Himalaya, a large number of hydropower generation projects have now become highly controversial on environmental, socio-economic as well as on cultural grounds.

The mountains regions share the largest proportion of forests in many countries, and therefore constitute one of the most valuable natural resources in mountains. Besides providing fodder, fuel and timber, and raw material for different forest-based industries; forests constitute one of the prime sources of livelihood and employment to a considerably large proportion of population both in the developed and developing regions across the planet (Singh 2007, Tiwari and Joshi 2001). The forest resources of the mountain regions are now depleting rapidly mainly due to their over-exploitation in many countries. The unabated loss of forest cover not only adversely affecting the mountain ecosystems services, water, biodiversity, soils, etc. but also depriving a large number of forest-dependent communities of their traditional livelihood in both developed and developing countries. It is very difficult to estimate the losses of ecosystem services as they are not priced or traded in markets. However, it has been observed that such losses often overshadow the economic return coming out of logging and timber. The studies indicated that the value of alternative uses exceeds the value of logging by more than 50% (Hamilton 1996). In the Indian Himalaya, the commercial exploitation of forests caused massive loss natural ecosystem and rural livelihood during the last century which resulted in community movement—'*Chipko Movement*'—for protection of forests against economic utilization, and followed by prohibition on green felling at and above 1000 m elevation across the country. However, depletion of forest resources is still continuing for construction of roads, dam, public building and also for fuel-wood and fodder extraction from forests by local people. During the recent years, the forest ecosystem in the Himalaya have come under increased biotic stress mainly due to high population growth, rapid urbanization, industrial growth and tourism development. As a result, the forest resources have depleted and degraded steadily and significantly leading to their conversion into degraded and waste lands in many parts of the region. Besides, construction of road, dams, mining and quarrying, and extension of cultivation and grazing areas have contributed significantly towards depletion of forests resources in the Himalaya. Moreover, climate change have stressed the Himalayan forest ecosystem through higher temperatures, altered precipitation patterns, more frequent and extreme weather events, and forest fires disrupting ecosystem services, depleting biodiversity and decreasing forest productivity (Singh 2007). These changes in forest ecosystem are undermining agricultural and food systems, and increasing the vulnerability of millions of poor people to water, health, food and livelihood insecurity in large part of South Asia mainly dependent of subsistence agriculture (Tiwari and Joshi 2012b).

The mountainous province of British Columbia with a forest area of 60.30 million ha, contains as much as 49.20% of natural forests of Canada. The forests of British Columbia have sustained the economy and culture of the hundreds of traditional forest-dependent communities across the province, and the timber they provide has been a driving force of the provincial economy for more than a century. A broad range of values are associated with British Columbia's forests—biodiversity, community stability, gathering, geo-climatic, grazing, heritage, hunting, recreation, science and education, socio-political, spiritual and aesthetic, timber, tourism, water, etc. (British Columbia, Ministry of Forests 1995). But, traditionally, timber has been considered as the only important value of forests in British Columbia. It has been widely proclaimed that 50 cents out of every dollar spent in British Columbia is generated by the forest industry. Even today, with increased emphasis on multiple use and protection of other values associated with the forests, the timber remains the most important forest resource, and the commercial forests sector is the dominant user of this natural resource (Commission on Resources and Environment 1995).

The commercial forestry in British Columbia has disrupted the fragile ecosystem of the mountain watersheds through reduced water recharge, increased surface run-off, accelerated erosion, and destruction of wilderness and wildlife, in the province. The indiscriminate clear cutting of natural forests over large areas and construction of logging roads for the transportation of timber from remote forests to main road and rail heads have promoted the processes of slope failure and mass movement resulting in massive landslides and mudslides along the fragile mountain slopes. The effects of poor logging and road building practices are clearly visible along the steep wet slopes of fragile mountains in the region. It was observed that it takes about seven to 20 years for the big roots of tree stumps to rot out and massive landslides to begin to occur on steep slopes. But, steep slope logging, still legal under British Columbia's Forest Practices Code, is happening all over the province, particularly along ecologically vulnerable Pacific coast. Hundreds of miles of shoreline have been degraded by the clear-cut, which have turned some of the most beautiful cruising coastlines in the world into some of the despoiled ones (Power 1995). However, it was investigated that the overexploitation of forest resources has declined employment in the forestry sector despite continued increase timber harvest (Schwindt and Heaps 1997).

In addition to large number of rural communities that depend on commercial forestry sector, a significant number of people, particularly, the native population still depends of the extraction of primary forest resources in British Columbia. Due to intensive logging and massive loss of biodiversity the sustainability of rural communities is collapsing through the depletion of primary forest resources that support traditional economic activities, such as gathering, grazing, fishing, hunting, ranching, etc. Consequently, a large number of native families inhabiting the forests of British Columbia are now being forced to leave their traditional forest-based means of livelihood, and migrate out of their homes in search of alternative means of livelihood. With the deprivation of traditional economic activities, the native communities are also losing their culture and indigenous knowledge developed through several thousand years, which has great implications to community sustainability and conservation and management of natural resources.

Mining activities have often devastating impacts on the local environment and causes displacement of indigenous people living in the immediate area (Pratt 2001). The pollution and toxic wastes produced by mines have still more severe environmental problems. Mines are now therefore emerging as the major source of toxic pollution which is contaminating large areas both in upstream and downstream, particularly in developing countries. Although mineral extraction contributes a relatively small part of global GDP, but mineral revenues are often significant in mountainous countries. For example, the mineral resources of Andes are incredibly important for the economies of the all seven mountainous countries of South America with the presence of globally significant reserves metals and minerals in these mountains (CEPAL 2004). Minerals account for 45 and 49% national export of Chile and Peru. The ecologically sensitive mountainous regions like Paramos in the north and Puno in the south are now increasingly affected by mining activities and emerging as a major threat to fragile mountain systems (Ministerio De Medio Ambiente 2010). Large-scale mining has overstrained steep and fragile terrain, polluted water sources, damaged aquatic ecosystem and other wildlife habitats and caused slope instability, landslides, and degradation of land around the mining areas. The impacts of mining, particularly pollution, opening access to undisturbed areas and changes in local social dynamics and economies, are reported often widespread and reaching all across the Andean mountains. Furthermore, resource-use conflicts, especially water, are becoming increasingly common, and are expected to become more severe as climate change affects water availability. In Andean mountains of Argentina and Chile large mining projects have caused depletion of glaciers.

Owing to inaccessibility of terrain, remoteness, lack of infrastructure and strong community opposition, the mining activities are not very common in the Himalayan mountains. The ruling of the Apex Courts of India and Nepal against mining in the fragile Himalayan mountains contributed significantly towards bringing awareness to the risks of mining in these fragile mountains (Bandyopadhyay and Shiva 1985). However, extraction of lime-stone, building-stone, magnesite has already caused devastating effect in many parts of the Indian Himalaya during the last century. Nevertheless China has expanded its mineral exploration and extraction activities in the western and north western parts of country.

Tourism in High Mountains

Tourism is now one of the fastest growing industries in the world with direct as well as indirect impacts on, natural ecosystem, economy, society, culture, perception and attitude of people. Tourism and recreation are major industries in many mountain regions. Tourism and recreation are both affected by global change, and at the same time they also constitute a form of global change themselves. The improvement in transportation system has now opened access to increasing number of tourists to almost all mountain regions of the world. Tyrol in Austrian Alps, for example, has experienced a remarkable spatial expansion of settlements, commercial and transport infrastructure, primarily caused by the shift from a largely agrarian society to a service society shaped by tourism (Borsdorf et al. 2010). The Alps are among the earliest nature based tourist destinations. Tourism and the leisure industry have become a major economic factor in the rural areas of the Alps with the advent of mass tourism in the 1960s as tourism was contributing 12% employment and as much as 16% of GDP (Bätzing 2003). During the last three decades, a long-term trend towards skiing-based winter tourism has emerged with significantly higher added value than summer tourism in the Alps (Borsdorf et al. 2010). However, tourism activities are very unevenly distributed over the mountain regions.

As mentioned above, tourism is one of the fastest growing industries in the mountains of the world with direct as well as indirect impacts on, natural ecosystem, economy, society, culture, perception and attitude of people. During the recent years, mountains have emerged as the second most popular tourist destinations after coastal regions in the world, and they share as much as 15–20% of the global tourism market (FAO 2005). One the one hand, the rapidly growing tourism industry has contributed significantly to world economy and played an important role in alleviating poverty through providing economic opportunities for local people in both developed and developing regions of the world, while on the other hand increasing tourism has created a series of environmental problems ranging from land-use changes, deforestation, loss of biodiversity, destruction of natural habitats, pollution, etc., and resource-use conflicts among different stakeholders. These changes have threatened the very basis of tourism in many mountain areas, particularly in developing countries where tourism is still mostly unplanned and unregulated. Further, it is expected that the growth of tourism would increase the demand of water and energy besides creating resource-use conflicts with other sectors of economy, and it may also create conflicts among various competitive resource users and with conservation practices in all mountain regions of the world.

The mountain tourism is highly diverse, involving a variety of seasonal activities, and at the same time it has become highly competitive and sensitive to political and economic changes. Hence, the benefits of increasing tourism are very unevenly distributed and unpredictable over the mountains regions across the world, for instance, in the Alps 40% of communities have no tourism and only 10% have major tourist infrastructure. The changing climatic conditions have severely threatened winter tourism in all tourist destinations in the Alps, a rise of 4°C temperature is expected to reduce the availability winter sport sites to 30% (Abegg et al. 2007).

One the one hand, the rapidly growing tourism industry has contributed significantly to world economy and played important role in alleviating poverty through providing economic opportunities for local people in both developed and developing regions of the world (Kruk et al. 2007), while on the other hand increasing tourism has created a series of environmental problems ranging from land-use changes, deforestation, loss of biodiversity, destruction of natural habitats, pollution, etc., and resource-use conflicts among different stakeholders. These changes have threatened the very basis of tourism in many parts of the planet, particularly in developing countries where tourism is still mostly unplanned and unregulated. Moreover, it is expected that the growth of tourism would increase the demand of water and energy besides creating resource-use conflicts with other sectors of economy, and it may also create conflicts among various competitive resource users and with conservation practices in sensitive mountain ecosystems which are already under stress of various other drivers of change.

Despite encouraging projections about the growth of global tourism industry by the World Travel Organization (WTO), and a wider offer than ever, such as, sports, heritage and wellness, mountain tourism is experiencing uncertainty, stress and crisis across the world. Since, climate is one of the crucial factors affecting the tourist, locale, connectivity, requirement of water and energy and various tourism stakeholders directly as well as indirectly; the nature-based tourism in the mountain regions is highly vulnerable to change in climatic conditions. Although the direct as well indirect impacts of climate change on tourism in mountains would vary widely according to geographical location, tourist settings, infrastructure and tourist seasons, yet keeping in view the inaccessibility, fragility, marginality, diversity and niche of the region the climate change may bring considerable uncertainties in tourism economy, particularly the mountains regions of developing countries, such as the Himalaya and Andes (Jodha et al. 2002, Dyurgerov and Meier 2005). In the European Alps, climate and geo-cultural changes are causing great uncertainty and crisis in tourism industry. In this context, climate change is likely to act as an indicator for structural contradictions and weaknesses of alpine tourism as well as booster for cultural, geographic and economic revolution in the tourism industry in the Alps.

The changing global economic order is having wide ranging and long-term implications to tourism growth and development across the world. Keeping in view the foreseen uncertainties in world economy these impacts are likely to intensify in the near future and affect tourism industry in mountains. The economic recession that crumpled the global economy during 2008–2012 has unprecedented and depressing impact on all sectors of economy including tourism across the world. The tourism industry in the European Alps is currently facing challenges by major global changes such as, economic instability and market uncertainty and consequent loss of shares in the tourist market in the Alpine countries throughout Europe; competition from other tourist destinations, the growing economic and territorial divide between large and small resorts, new recreational practices, the ageing of the tourist population, demands for environmental quality, the changed notion of resort, the social issue of seasonal work, the need for huge new investments against the background of a reduction of public funding, and risk management (UNWTO 2010). However, the growth in international tourism recorded an increasing trend at the end of 2009 (UNWTO 2010). For example, in part of the Indian Himalaya, the global financial crisis did not have much adverse impact on tourism. This is clearly brought out by the trends of tourist growth in the region during 2008–2009 when the region recorded a tourist growth of more than 186% during 2000–2009, even though the region has not been able to attract foreign tourists in significant number during the period. But, it is particularly interesting to note that during the period of global financial crisis (2008–2009) the region registered respectively 11.73 and 41.81 growth in domestic and foreign tourist arrivals. However, as in some other Himalayan states of India the fast globalizing economy and the ongoing process of economic liberalization have now unlocked the valuable mountain resources to fast extending international markets. As a result, tourism resources, assets and products have become highly vulnerable to exploitation by large national as well as multinational tourism enterprises depriving local people, particularly poor and marginalized sections of society of the benefits of tourism development in the region (Tiwari 2012).

Further, the lack of structural and institutional mechanism, inappropriate and fault process of development planning, absence of land-use policies and regulation and dearth of political will particularly in developing countries are sharpening the adverse impacts of global change in mountains as important indirect drivers (Rodgers 2002). For instance, the sectoral fragmentation of institutional responsibilities, increasing political interference, over reliance on technocrats and bureaucrats at the planning and implementation stages, non-involvement of local communities, in the process of decision making and over exploitation of natural resources resulted in the failure of decades of mountain water conservation initiatives (Pratt and Shilling 2002). Furthermore, a series of other indirect drivers, such as lack of public awareness and lack of valuation of natural resources and ecosystem services are increasing the vulnerability of mountain communities to the long-term impacts of global change, particularly in less developed countries. The well informed and socio-politically empowered local communities have proved to be a key to managing the impacts of global changes imposed on them from outside (Dhar 1997).

Downstream Impacts of Changes in Mountains

The consequences of global change in mountains have always serious implications for downstream ecosystems and human sustainability as and the impacts of climate and other changes on mountain ecosystem reach far beyond the mountain areas (Hassan et al. 2005). The disruptions in mountain ecosystems and the resultant loss of mountains services and goods due to climate change and other drivers of change will affect the food and livelihood security of hundreds of millions of people in mountains and a much higher number in their lowlands (Viviroli et al. 2003). These changes would particularly affect the supply and availability of water both in highlands and lowlands. There are clear indications that the reduction of water supplies during the dry season associated with melting of glacier and snow and reduced groundwater recharge will affect up to one-sixth of the world's population in South Asia, over a quarter of a billion people in China, and up to 50 million people in the Andes. The Andes, covering 33%

geographical of the area of the Andean countries, are highly crucial for the livelihoods and economies on the Andean countries. However, increasing population pressure, changes in land use, unsustainable exploitation of resources, and climate change are now having far-reaching adverse impacts on ecosystem goods and services of the Andean countries (Stern et al. 2006). Declining flows of water from mountains will decrease the availability of freshwater for drinking, sanitation and food production, and consequently will have serious implications for water, health, food and livelihood security of downstream communities dependent on subsistence agriculture. Increasing water stress is expected to lead to political as well as social conflicts, especially in arid and semi-arid areas which are highly dependent on mountain water and particularly in trans-boundary river basins, such as South Asia where as many as seven largest rivers of the consentient have their sources in the mighty glaciers of the Hindu-Kush Himalayan mountains. The recent investigations indicated that Himalayan river basins in China, Bangladesh, India and Nepal will face massive water depletion losing almost 275 billion cubic metres of annual renewable water in the next two decades. Depletion of water resources and resultant water scarcity and effects of desertification and soil erosion would decline food production in China and India by as much as 50% by 2050, and consequently both the countries will need to import more than 200 to 300 million tonnes of wheat and rice within the next 20 years. These changes in the regime of water resources are likely to displace millions of people leading to mass migration and increased conflicts within and between countries (Himalaya Post 2010). This will have enormous regional implications for fundamental human endeavours ranging from poverty alleviation to environmental sustainability and climate change adaptation, and even to human security in a large part of the world. The increase in air temperature is likely to expand malaria and other diseases to higher altitudes in both Africa and Latin America.

Global Change and Emerging Opportunities in Mountains

Changing climatic conditions, particularly changes in temperature and precipitation patterns and other processes of global changes have increased the vulnerability of mountain communities to various natural and socio-economic risks and posed serious environmental as well as developmental threats on the one hand, and created several opportunities and possibilities for the sustainable development of mountain regions and well being of their people on the other. The emerging opportunities include conservation of biodiversity and genetic resources, growing demand for high-value mountain niche products, such as, eco-tourism, hydro-energy generation, carbon trading, compensation for ecosystem services and production of high value fruits, flowers, vegetables and medicinal plants. The mountains provide very little scope for the development of multiple livelihood options to mountain communities other than subsistence agriculture, primarily due to constraints of terrain and climate and the resultant inaccessibility. However, tourism is now emerging as major livelihood option in mountain areas in both the developed and the developing world. Tourism is now a major source of employment and foreign exchange in the developing countries, and it may also check the increasing rate of rural outmigration of the educated and entreprenuering population. At the same time, with rising temperatures mountain destinations in the subtropics and tropics attracting increasing numbers of visitors, particularly during summer months (Hoermann and Kollmair 2009).

Mountain systems consisting nearly 50% of global biodiversity hotspots sustain and support half of the world's biological diversity and genetic resources of which some are very rare. With rising temperatures upward shifts of vegetation belts to higher elevations and northward advances in it is expected that the geographical ranges of species may shift upward with increase in global temperature in the northern hemisphere. These changes are likely to bring a variety of changes in the composition, structure and spatial distribution of biological resources creating a set of opportunities for their conservation and sustainable development in high mountain ecosystems across the planet. Thus mountain ecosystems would be able to provide necessary protection to a variety of species which may face extinction in lowlands (Körner 2009).

The drivers of global change have contributed significantly towards increasing awareness and improving the understanding of ecosystem goods and services flowing down from mountains and called for sustainable development of mountain ecosystems and creating new livelihood opportunities for mountain communities through restoration of ecosystem services. The emerging opportunities for sustainable development in mountains include the growing demands for mountains as popular destinations for recreation. Furthermore, globalization has provided mountain communities with a set of new economic opportunities in the production of high value mountain products, such as, freshwater , fruits, nuts, off-season vegetables, flowers, honey, dairy products, and cosmetic, aromatic and medicinal plants. However, the value added by mountain dwellers will likely remain proportionally small unless local processing replaces the export of raw produce (Jodha 2002). These trends are likely to accelerate as market forces gain primacy. The rise in global demand for mountain herbs and other organic and non-timber forest products is leading to over-extraction. Climate change may also provide new opportunities in the agricultural sector by increasing the length of growing seasons for certain crops or the possibility of growing crops at higher altitudes, providing shelter for rare and endangered species of plants and animals, generation of hydro-power and potential carbon sink. The value addition to raw produce through local processing would be necessary for ensuring the maximum benefit of these emerging opportunities should go to the mountain communities. However, these changes may lead to the over-exploitation of mountain natural resources (Jodha 2002).

It is now well understood that the carbon sequestration and conservation potential of the forests is much higher than total carbon sequestration capacity of all other terrestrial ecosystems of the planet. The forests of the world account for 90% of the annual carbon flux between the atmosphere and the land surface of the Earth (Gupta 2007). This has further improved our understanding of the importance of forests to the global environment, and has influenced forest management policy decisions around the world during the recent years. Over the past decades, forests have emerged as a major consideration in global discussions on sustainable development. Since, the United Nation's Earth Submit held in Rio de Janeiro, Brazil in 1992, remarkable progress has been made in advancing the worldwide consensus on the protection, conservation and sustainable development of forest resources (Tiwari and Joshi 2001). At the same time, the globalizing economy has increased the demand of forests in various economic sectors, and thus unlocked the valuable and rich forest resources of remote and inaccessible regions, such as high mountains, for their exploitation, degradation and depletion. However, the Clean Development Mechanism (CDM) of The Kyoto Protocol gives more emphasis on afforestation and reforestation programmes, and conservation and management of existing forests are not included under the Clean Development Mechanism (CDM) (Geoffrey 2005). Since, a significant proportion of the population is currently involved in the conservation and management of their forests, particularly in the developing countries, the incorporation of forest management in the Kyoto Protocol or alternatively voluntary carbon market would benefit a large number of communities in developing countries (Climate Community and Biodiversity Alliance 2005).

Economic globalization, participatory resource management systems, decentralized governance mechanism, public-private partnership based infrastructure development, growing urbanization, development of information technology and communication system have sensitized mountain communities for capturing the potential of these drivers of transformation. Fast emerging economies, particularly in South and East Asia and Latin America have also increased flow of resources and thereby accelerated the pace of economic growth in mountains. As result, mountain communities which suffered from marginalization and underdevelopment have been able to attract the attention of national as well as international agencies for their sustainable development. Despite prevailing uncertainty on comprehensive climate change impact scenario it is clear that mountains ecosystems will be essential building blocks for long term sustainable global development. Now it is the responsibility of mountain-countries and those countries with sizeable areas under mountain eco-system to build and improve their respective individual capabilities to take up the challenge to collaborate in order to benefit from these opportunities of global change.

Mountains in International Sustainable Development Agenda

Mountains have long been marginalized from the view point of sustainable development of their resources and inhabitants. However, our understanding about the problems of mountain regions and approach to their development has undergone drastic changes, during the recent years (ICIMOD 2010). The first UN Conference on the 'Human Environment' held in Stockholm in June 1972 figured the subsequent international mountain conferences. This was followed by the pioneering publication of famous report 'The Limits to Growth' by Club of Rome's (Meadows et al. 1972) which initiated research in globalization through bringing into focus the projections of high populations increase and growth of economy and technology in the world. Over the next years a large number of international conferences took place in different mountain regions of the world which was followed by the establishment of the International Centre for Integrated Mountain Development (ICIMOD) involving all eight countries of the Hindu Kush-Himalaya (HKH) in Kathmandu, Nepal in 1983, the African Mountain Association in Ethiopia in 1986, and Andean Mountain Association in 1991 in Chile. This facilitated developing countries to support the proposal of a mountain chapter in the Agenda 21 at the third preparatory conference for Rio in 1991, and consequently, 'increased awareness and improved understanding of the effects of climate change and globalization on world' mountains (Borsdorf et al. 2010).

The significance of mountain social-ecological systems was acknowledged for the first time on a global scale in Agenda 21 of the United Nations Conference on Environment and Development (UNCED), held in Rio de Janeiro, Brazil, in 1992. Chapter 13 titled 'Managing Fragile Ecosystems: Sustainable Mountain Development' of Agenda 21 recommends two priority programmes for the sustainable development of the world mountains. These include: (i) generating and strengthening knowledge about the ecology and sustainable development of mountain ecosystems; and (ii) promoting integrated watershed development and livelihood opportunities in mountains. The Food and Agriculture Organization (FAO) of the United Nations was assigned with the responsibility of facilitating and reporting of the implementation of these two programmes.

This resulted in several specific initiatives by different governments of mountainous countries, international organizations, NGOs and scientific agencies across the world during the decade following the UN Conference on Environment and Development. Establishment of Mountain Forum in 1995 was one the important initiatives taken for a global network for information exchange, mutual support, and advocacy towards equitable and ecologically sustainable mountain development and conservation of mountains ecosystem across the world. This was followed by designating 2002 as the International Year of Mountains (IYM) by United Nations General Assembly which was followed by Global Mountain Summit in Bishkek, Kyrgyz Republic in the same year. The International Year of Mountains provided a great opportunity for raising awareness about the importance of mountains for sustainability of the global environment. This resulted in the manifestation of several new initiatives, including the Adelboden Group which lead to the creation of Sustainable Agriculture and Rural Development in Mountains (SARD-M), Global Change in Mountain Regions (GLOCHAMORE) and the Mountain Research Initiative (MRI). Further, Mountain Partnership was launched at the World Summit on Sustainable Development held in Johannesburg in 2002 to promote, strengthen and facilitate closer collaboration between governments, civil society organizations, inter-government agencies, and the private sector toward achieving sustainable mountain development. All these initiatives and actions have been quite successful in raising awareness of the importance of mountains, and some of them also initiated successful interventions for sustainable mountain development.

On the other hand, in spite of all these successful initiatives and actions and the UN General Assembly continued emphasis on the sustainable mountain development, the mountain regions of the world have never received the desired consideration in the global development agenda. One of the important reasons for mountains not receiving adequate attention was the international developmental agenda is always dominated by the sustainable development agenda, such as the Millennium Development Goals (MDGs) and the Poverty Reduction Strategy Papers (PRSP), which were largely implemented as national programmes ignoring the ecology and specific developmental requirements of mountain areas. Furthermore, the outline of United Nations Framework Convention on Climate Change (UNFCCC) still lacks a mountain perspective, largely because of substantial knowledge gaps and an uncoordinated approach by the countries that are most affected by climate change in their mountains. In view of this, several critical issues related with sustainable mountain development, particularly, management of water resources; conservation of biological and cultural diversity; infrastructure development, access to health services and markets; proper recognition and valuation of mountain ecosystem services and the aesthetic, recreational and spiritual significance of mountains need to be fully addressed (Sonesson and Messerli 2002).

The Second UN Conference on Environment and Development (Rio+20) did not adequately and satisfactorily address the critical issues related to sustainable mountain development. However, mountains have been as one of the important thematic areas under the framework for action and follow-up under Para 210, 211 and 212 which are appended as follows (United Nations 2012):

210. We recognize that the benefits derived from mountain regions are essential for sustainable development. Mountain ecosystems play a crucial role in providing water resources to a large portion of the world's population; fragile mountain ecosystems are particularly vulnerable to the adverse impacts of climate change, deforestation and forest degradation, land use change, land degradation and natural disasters; and mountain glaciers around the world are retreating and getting thinner, with increasing impacts on the environment and human well-being.

211. We further recognize that mountains are often home to communities, including indigenous peoples and local communities, who have developed sustainable uses of mountain resources. These communities are, however, often marginalized, and we therefore stress that continued effort will be required to address poverty, food security and nutrition, social exclusion and environmental degradation in these areas. We invite States to strengthen cooperative action with effective involvement and sharing of experience of all relevant stakeholders, by strengthening existing arrangements, agreements and centres of excellence for sustainable mountain development, as well as exploring new arrangements and agreements, as appropriate.

212. We call for greater efforts towards the conservation of mountain ecosystems, including their biodiversity. We encourage States to adopt a long-term vision and holistic approaches, including through incorporating mountain-specific policies into national sustainable development strategies, which could include, inter alia, poverty reduction plans and programmes for mountain areas, particularly in developing countries. In this regard, we call for international support for sustainable mountain development in developing countries.

It is therefore highly imperative to place sustainable mountain development in developing countries and the increasing socio-economic vulnerability of their population at the centre of climate change mitigation and adaptation strategy. Further, in view of increasing demands for freshwater and hydro-power and other mountains ecosystem services, it would also necessary to realize the need of integrated framework for addressing upstream-downstream interlink-ages, as well as comprehensive trans-boundary river-basin management approaches. The emerging need of adaptation to climate change, particularly in developing countries, calls for further redefining the global sustainable development agenda focusing on eco-region based integrated development. With the rapid retreat of mountains glaciers across the planet and its potential serious long-term impacts on the freshwater ecosystem and availability of water resources in large parts of the world, the mountains regions have suddenly gained global attention, which need to be seized into concrete actions and effective programmes for sustainable development of mountain regions and improvement of the livelihood of their inhabitants, particularly in developing countries.

Conclusion

Mountains of the planet despite their marginality and remoteness are changing rapidly under increasing pressure from globalization and climate change. These changes have unlocked the natural resources of mountains for exploitation without serious concern for sustainable development of mountain and wellbeing of their inhabitants. Consequently, mountain ecosystems as well as mountain communities, specifically in developing and underdeveloped countries are particularly threatened by a series of drivers and processes of global change and emerging new international economic and political orders. However, we are experiencing an emergence of responsiveness of the ecological significance of mountain systems and their importance for the sustainability of global community, particularly after the United Nations Conference on Sustainable Development—the Rios Earth Summit in 1992. As a result, our understanding about the dilemmas of mountain regions and approach to their development has undergone drastic changes, during the last two decades. Climate change has emerged as one of the major drivers transforming consistently the natural environment, society and economy of the mountains regions in all parts of the world, and mountains ecosystem being highly sensitive are extremely vulnerable to these changes. However, there is still a high degree of uncertainty about the trends and magnitude of climate change and its impacts on mountain systems. It is therefore highly imperative to improve our understanding of the trends of changes in temperature and variability in precipitation pattern at the local level through downscaling of regional climate models. This would require establishment of a comprehensive networks of hydro-meteorological monitoring stations in mountain areas across the world, particularly in the mountains of developing countries where currently, such monitoring is extremely lacking. Furthermore, a sharp focussed and comprehensive research on climate change impacts assessment, vulnerability and adaptation to climate change would be necessary at micro-regional scale.

An effective mechanism for the sharing of information, data, experience and knowledge generated from local, regional to international levels and international level transfer of knowledge would be crucial for better understanding of changing climatic conditions and evolving appropriate strategies for mitigation of climate change and responding to its impacts in an amicable manner. Since mountains constitute headwaters of some of the largest trans-boundary basins on the earth, it would also be indispensable to establish and strengthen international research collaboration, and develop international mechanisms on knowledge and data sharing. A regional geo-political cooperation framework among riparian countries is therefore highly crucial not only for evolving a framework of adaptation to climate change and improved governance of headwaters resources, but also for security and peace in the entire world. The international conventions, initiatives and organization can play an effective role in initiating trans-boundary climate adaptation diplomacy in different mountain regions. All mountain regions across the world are currently facing common threats from climate change and economic globalization in the backdrop of the similar constraints of terrain, fragility; geographical isolation and socio-political marginalization. In many developing countries, policy decisions and planning processes have plains perspective largely ignoring the disproportionate vulnerability of social-ecological systems of mountains. Hence, effective mountain-specific policies need to be designed not only at the national level but also at global scale, as these challenges are independent of national territories. In order to address the challenges posed by climate change, the mountain countries and regions should develop mountain specific adaptation and mitigation policies, programmes, institutions and think tank which would be necessary to enhance their resilience and ensure socio-economic and ecological sustainability in mountain areas.

Mountain communities through their traditional resource management practices contributed significantly towards preservation of forest and biodiversity, climate change mitigation through carbon sequestration, water conservation and preservation of cultural heritage and natural landscapes that provide a variety of ecosystem services and goods to a considerably large population in the downstream. In turn, the global community should contribute towards the conservation of natural ecosystem and improvement of the quality of life of mountain people by providing adequate incentives for these high value services flowing down from mountains. Moreover, mountain areas, especially those situated in subtropical and tropical zones have contributed the least to global greenhouse gas emissions. The mountain inhabitants therefore need to be supported in their sincere efforts to adapt to the challenges and facilitated to be benefited from emerging opportunities of global change. Payment for environmental services (PES) can pave the way for rewarding mountain communities for the critical services they provide. Payment for environmental services would help in preventing further depletion of mountain natural resources and restoration of ecosystem services which in turn will make them more resilient to long-term impacts of climate change.

Reducing Emissions from Deforestation and Degradation (REDD) and Enhancement of Carbon Stocks (REDD+) are other important opportunities under the United Nations Framework Convention on Climate Change (UNFCCC) which offer incentives for developing countries to reduce emissions from forested lands and invest in low carbon activities for their sustainable development. Mountains accounting for nearly 28% of the world's forests bear a huge potential for carbon storage and sequestration and are therefore in a privileged position to attract such funds for climate change mitigation bear a huge potential for carbon storage and sequestration and are therefore in a privileged position to attract such funds for climate change mitigation. Hence, the mountain forest-ecosystems are one of the most vital mainstays of Green Economy and in attaining the goals of both REDD and REDD+. The forests conservation in mountains therefore needs to be linked with climate change mitigation and adaptation, poverty alleviation and food and livelihood security of local people. In view of this, the recent experiences of Forest *Panchayats* and lessons learned from Joint Forest Management (JFM) in the Indian Himalaya and Community Forestry (CF) in Nepal can best be replicated and used for institutionalizing forests and for their community-oriented conservation and development. This would provide mountain people with the opportunity of getting involved in global carbon credit process and enhancing their quality of lives through reduction of poverty, improvement of livelihood and restoration of ecosystem services. In addition, several ecological benefits, such as conservation of biodiversity, increased forest productivity, erosion control, slope stabilization and hydrological restoration, could be obtained by integrating participatory forest management with existing carbon markets.

A considerably large proportion of population in mountain regions of developing countries depends for its livelihood on severely limited arable land symbolizing distress husbandry of land. Paradoxically, on the other hand, nearly large forest area put together with water-bodies, high altitude pastures, etc. characterized with charismatic landscapes, natural splendor, variety of flora and fauna, enthralling wilderness and rich biodiversity have so far been utilized to provide livelihood to a small percentage of the rural population. The situation therefore calls for looking beyond the traditional agricultural system and generation of rural employment opportunities in off-farm and non-traditional sectors in the mountains area of less developed countries. This brings out the fact very clearly that restoration of ecosystem services through sustainable utilization and conservation of critical natural resources, such as, land, water, forests, and biodiversity, and generating economically viable options of rural livelihood other than agriculture, and ensuring food security would constitute the critical components of the process of climate change adaptation in mountain areas of less developed countries. The strategy should have a wider scope for the generation of off-farm livelihood employment opportunities particularly through the promotion of local rural enterprise in different sectors of tourism. Hence, tourism presents a great potential for the inclusive development of the region.

However, agriculture will remain as one of the important economic activities in the mountains of developing countries, and thus will constitute one of the core components of overall climate change adaptation strategy in the mountains in times to come. This is primarily because, agriculture is not merely an important economic activity and fundamental source of livelihood of local rural communities, but also constitutes an integral part of their culture, history and traditions, and an invaluable treasure of traditional ecological knowledge required for adapting to climate change. Secondly, mountains have some of the highly productive and agriculturally prosperous valleys and mid-slopes which still have the potential of contributing towards food as well as livelihood security. Thirdly, the potential of varying agro-climatic zones from valleys to higher elevation can be utilized for growing a variety of crops and producing seasonal as well as off-season agricultural products. Fourthly, for making tourism ecologically conducive, economically viable and pro-poor livelihood and adaptation strategy needs to be linked with local agricultural and food systems. Lastly, integration of tourism with local production system will create local viable market for the agricultural products and thus make local agriculture economically viable. In view of this, 'Ecological Tourism' which is now popularly known as 'Ecotourism' has immense potential to be developed as potential adaptation strategy to climate change in mountain areas of less developed countries. Ecotourism capitalizing upon both the socio-cultural and biophysical strength of the mountain landscape would contribute significantly towards securing viable alternative livelihood opportunities, particularly for the poor and marginalized mountain communities in the mountains.

Regional initiatives, including the Mountain Research Imitative (MRI), Mountain Forum (MF), Alpine Convention (AC) and the recent global Mountain Initiative (MI) floated by the Government Nepal need to be strengthened. The Mountain Partnership sponsored by FAO plays significant role in inter-connecting these various regional and global initiatives. Regional information networks need to be established which would act as effective learning and awareness generation forums between specialists, civil society organizations, and government agencies and to support capacity building activities through focussed education, training and research. Further, in order to build resilient mountain social-ecological systems, the support and cooperation of civil society, including Non-Government Organizations (NGOs), Civil Society Organizations (CSOs), the private sector, educational and research institutions would be inevitable. These institutions play effective role in sensitizing policy planners, decision makers and society at large about the importance and significance of mountain eco-systems in sustaining global society as well as about their fragility, marginality, vulnerability and the emerging opportunities.

Acknowledgement

The authors are grateful to the International Centre for Integrated Mountain Development (ICIMOD), Kathmandu, Nepal and Food and Agricultural Organization (FAO), Rome for having kindly granted permission to reproduce a number of maps, diagrams, tables and other information from their different publications in this paper.

References

Abegg, B., S. Agrawala, F. Crick and A. de Montfalcon. 2007. Climate change impacts and adaptation in winter tourism. pp. 25–60. *In*: Climate Change in the European Alps: Adapting Winter Tourism and Natural Hazards Management. OECD, Paris.

ACIA. 2004. Impacts of a Warming Arctic: Arctic Climate Impact Assessment. Cambridge.

ASPEN International Mountain Foundation. 2012. Sustainable Mountain Development: North American Report—Report Submitted for Rio+20, ASPEN International Mountain Foundation.

Baker, B.B. and R.K. Moseley. 2007. Advancing tree-line and retreating glaciers: implications for conservation in Yunnan, P.R. China. Arctic, Antarctic and Alpine Research 39(2): 200–209.

Bandyopadhyay, J. and V. Shiva. 1985. Conflict over limestone quarrying in Doon Valley, Himalaya. Environmental Conservation 12(2): 131–139.

Bätzing, W. 2003. Die Alpen. Geschichte und Zukunft einer europäischen Kulturlandschaft (2nd edition), Beck, ISBN 978-3-406-50185-0, München.

Batllori, E. and E. Gutierrez. 2008. Regional tree line dynamics in response to global change in the Pyrenees. Journal of Ecology 96: 1275–1288. DOI: 10.1111/j.1365-2745.2008.01429.x.

Becker, A. and H. Bugmann. 2001. Global change and mountain regions. IGBP Report 49, Stockholm, pp. 88.

Beniston, M. 2000. Environmental Change in Mountains and Uplands. Arnold/Hodder and Stoughton/Chapman and Hall, London.

Beniston, M. 2003. Climatic change in mountain regions: a review of possible impacts. Climatic Change 59: 5–31. DOI: 10.1023/A:1024458411589.

Beniston, M. 2006. Mountain weather and climate: a general overview and a focus on climatic change in the Alps. Hydrobiologia 562: 3–16. DOI: 10.1007/s10750-005-1802-0.

Beniston, M. 2005. Mountain climates and climatic change: an overview of processes focusing on the European Alps. Pure and Applied Geophysics 162: 1587–1606.

Bernbaum, E. 1998. Sacred Mountains of the World. University of California, Press, Berkeley, Los Angeles, London, pp. 291.

Bhandari, N. and S. Nijampurkar. 1988. Himalayan Glaciers: our neglected Water Banks. pp. 76–92. *In*: S.K. Chadha (ed.). Himalayas: Ecology and Environment. Mittal Publications, New Delhi.

Bisht, B.S. and P.C. Tiwari. 1996. Land use planning for sustainable resource development in Kumaon Lesser Himalaya: a study of Gomti watershed. Inl. J. Sust. Dev. Ecol. 3: 23–34.

Borsdorf, A., U. Tappeiner and E. Tasser. 2010. Mapping the Alps. pp. 186–191. *In*: A. Borsdorf, G. Grabherr, K. Heinrich, B. Scott and J. Stötter (eds.). Challenges for Mountain Regions. Tackling Complexity. Böhlau, Vienna.

Borsdorf, A. and V. Braun. 2008. The European and Global Dimension of Mountain Research —An Overview. Revue de Géographie Alpine 96: 4, pp. 117–129, ISSN 1760-7426.

Börst, U. 2006. Nachhaltige Entwicklung im Hochgebirge. Eine Systemanalyse von Mensch-Umwelt-Szenarien im Lötschental (Zentral-Alpen). Diss. Bonn. URN: urn:nbn:de:hbz:5N-07103.

British Columbia, Ministry of Forests. 1995. Forest, Range and Recreation Analysis. Crown Publications, Victoria.

Bugmann, H. and C. Pfister. 2000. Impacts of inter-annual climate variability on past and future forest composition. Regional Environmental Change 1: 112–125. DOI: 10.1007/s101130000015.

Burkett, V.R., D.A. Wilcox, R. Stottlemyer, W. Barrow, D. Fagre, J. Baron, J. Price, J.L. Nielsen, C.D. Allen, D.L. Peterson, G. Ruggerone and T. Doyle. 2005. Nonlinear dynamics in ecosystem response to climatic change: case studies and policy implications. *In*: Ecological Complexity. 2: 357–394. DOI: 10.1016/j. ecocom.2005.04.01.

Buytaert, W., F. Cuesta-Camacho and C. Tobón. 2011. Potential impacts of climate change on the environmental services of humid tropical alpine regions. Global Ecology and Biogeography 20: 19–33.

Census of India. 2001. Census of India 2001, the Registrar General Census, Government of India, New Delhi.

CEPAL. 2011. Climate Change: A Regional Perspective. Report. UN Economic Commission for Latin America and the Caribbean (ECLAC), IDB.

CEPAL. 2004. Renewable energy sources in Latin America and the Caribbean: Situation and Policy Proposals. UN Economic Commission for Latin America and the Caribbean (ECLAC).

Climate Community and Biodiversity Alliance. 2005. Climate, Community and Biodiversity Project Design Standards (First Edition). Climate Community & Biodiversity Alliance, Washington DC. May 2005.

CNCCC. 2007. China national report on climate change 2007 (in Chinese). China National Committee on Climate Change, Beijing.

Commission on Resources and Environment. 1995. British Columbia's Sustainability Strategy. Report to the Legislative Assembly, 1994–95.

CONDESAN. 2011. 20 years of Sustainable Mountain Development in the Andes—from Rio 1992 to 2012 and beyond—Draft Regional Report, prepared for Lucerne World Mountain Conference.

Dhar, U. (ed.). 1997. Himalayan Biodiversity—An Action Plan. The G.B. Pant Institute of Himalayan Environment and Development, Almora, India.

Diaz, H., M. Grosjean and L. Graumlich. 2003. Climate variability and change in high elevation regions: past, present and future. *In*: Climatic Change. 59: 1–4. DOI: 10.1023/A:1024416227887.

Du, M.Y., S. Kawashima, S. Yonemura, X.Z. Zhang and S.B. Chen. 2004. Mutual influence between human activities and climate change in the Tibetan plateau during recent years. Global and Planetary Change 41: 241–249.

Dyurgerov, M.D. and M.F. Meier. 2005. Glaciers and Changing Earth System: A 2004 Snapshot. Boulder (Colorado): Institute of Arctic and Alpine Research, University of Colorado, p. 117.

Eriksson, M., J. Fang and J. Dekens. 2008. How does climate affect human health in the Hindu Kush-Himalaya region? Regional Health Forum 12(1): 11–15.

Food and Agricultural Organization. 2010. Global Forest Resources Assessment 2010. Food and Agriculture Organization of the United Nations, Rome.

Food and Agricultural Organization. 2008a. Soaring Food Prices: Facts, Perspectives, Impacts and Actions Required, High-level Conference on World Food Security: The Challenges of Climate Change and Bio Energy. Hlc/08/Inf/1, Rome, 3–5 June 2008.

Food and Agricultural Organization. 2008b. Food Security in Mountains—High time for action. Brochure of the International Mountain Day 2008. http://www.mountaineering.ie/documentbank/uploads/IMD08%20brochure.pdf.

Food and Agricultural Organization. 2005. Mountain Tourism: making it work for the poor. Rome.

Food and Agricultural Organization. 2002a. International Year of the Mountains. Food and Agriculture Organisation of the United Nations, Rome.

Food and Agricultural Organization. 2002b. Land-water linkages in rural watersheds. Land and Water Bulletin 9. Food and Agriculture Organisation of the United Nations, Rome.

Fujita, K., T. Kadota, B. Rana, R.B. Shrestha and Y. Ageta. 2001. Shrinkage of Glacier AX010 in Shorong Region, Nepal Himalayas, in the 1990s. In Bulletin of Glaciological Research 18: 51–54.

Geoffrey, H. and K. Conrad. 2005. A Solution to Climate Change in the World's Rainforests. Financial Times, November 29, 2005, 20: 40.

Giupponi, C., M. Ramazin, E. Sturaro and S. Fuser. 2006. Climate and land use changes, biodiversity and agri-environmental measures in the Belluno province, Italy. *In:* Environmental Science & Policy 9: 163–173. DOI: 10.1016/j.envsci.2005.11.007.

Goudie, A.S. and D.J. Cuff (eds.). 2002. Encyclopedia of Global Change, Environmental Change and Human Society. Oxford University Press, Oxford.

Gupta, M.K. 2007. Promoting Self Sufficiency Through Carbon Credits from Conservation and Management of Forests. Unpublished Masters Research paper. Submitted to the faculty of Clark University, Worcester, Massachusetts, Department of International Development, Community and Environment (IDCE).

Haigh, M. 2002. Headwater control: integrating land and livelihoods, paper presented at the International conference on Sustainable Development of Headwater Resources. United Nation's International University, Nairobi, Kenya, September, 2002.

Hamilton, G. 1996. Victoria Close to Forest Industry Conflict. Suzuki Foundation Claims, Sun Forestry Reporter, June 21 1997, Vancouver.

Hassan, R., R. Scholes and N. Ash (eds.). 2005. Ecosystems and human well-being: Current state and trends. Volume 1: Findings of the condition and trends, Working Group of the Millennium Ecosystem Assessment. Island Press, Washington.

Hashmi, N.H., M.C. Pathak, P. Jauhari, R.R. Nair, A.K. Sharma, S.S. Bhakuni, M.K.S. Bisht and K.S. Valdiya. 1994. Bathymetric Study of the Neo-tectonic Naini Lake in Outer Kumaon Himalaya. Journal of Geological Society of India 41: 91–104.

Hewitt, K. 2005. The Karakoram Anomaly? Glacier Expansion and the 'Elevation Effects,' Karakoram Himalaya. Mountain Research and Development 25(4): 332–340.

Hewitt, K., C.P. Wake, G.J. Young and C. David. 1989. 'Hydrological investigation at Biafo Glacier, Karakoram Himalaya: an important source of water for the Indus River'. Ann. Glaciol. 13: 103–108.

Himalaya Post. 2010. http://archive.wn.com/2010/06/29/1400/himalayapost/.

Hoermann, B. and M. Kollmai. 2009. Labour migration and remittances in the Hindu Kush Himalayan region. ICIMOD, Kathmandu, Nepal.

Huddleston, B., E. Ataman and L. d'Ostiani. 2003. Towards a GIS-based analysis of mountain environments and populations. Environment and Natural Resources Working Paper No. 10. Food and Agriculture Organization of the United Nations, Rome.

ICIMOD. 2010. Mountains of the World–Ecosystem Services in a Time of Global and Climate Change: Seizing Opportunities—Meeting Challenges. Framework paper prepared for the Mountain Initiative of the Government of Nepal by ICIMOD and the Government of Nepal, Ministry of Environment.

ICIMOD. 2009. The Changing Himalayas: Impact of Climate Change on Water Resources and Livelihoods in the Greater Himalayas. ICIMOD, Kathmandu, Nepal.

ICIMOD. 2007. Melting Himalayas: regional challenges and local impacts of climate change on mountain ecosystems and livelihoods. ICIMOD, Kathmandu, Nepal.

IFAD. 2001. Rural Poverty: The Challenge of Ending Rural Poverty Report. Oxford University Press, Oxford.

Inman, M. 2010. Settling the Science on Himalayan Glaciers. Nature. http://www.nature.com/climate/2010/1003/full/climate.2010.19.html March 2, 2010.

IPCC. 2007a. Climate change 2007: The scientific basis. Working Group I contribution to the Intergovernmental Panel on Climate Change Fourth Assessment Report. Cambridge University Press, Cambridge.

IPCC. 2007b. Climate change 2007: Impacts, adaptation and vulnerability. Working Group II contribution to the Intergovernmental Panel on Climate Change Fourth Assessment Report. Cambridge University Press, Cambridge.

Ives, J.D., R.B. Shrestha and P.K. Mool. 2010. Formation of glacial lakes in the Hindu Kush-Himalayas and GLOF risk assessment. ICIMOD, Kathmandu, Nepal.

Ives, J.D. 1989. Deforestation in the Himalaya: the cause of increased flooding in Bangladesh and Northern India. Land Use Policy 6: 187–193.

Jandl, R., A. Borsdorf, H. Van Miegroet, R. Lackner and R. Psenner (eds.). 2009. Global Change and Sustainable Development in Mountain Regions. COST Strategic Workshop, Innsbruck University Press, Innsbruck.

Jianchu, X., R. Sharma, J. Fang and Y. Xu. 2008. Critical linkages between land-use transition and human health in the Himalayan region. Environment International 34(2): 239–247.

Jodha, N.S. 2002. Rapid Globalisation and Fragile Mountains: Sustainability and Livelihood Security Implications in Himalayas. Final narrative report of the research planning project submitted to the MacArthur Foundation, ICIMOD, Kathmandu, Nepal.

Jodha, N.S. 1992. Mountain perspective and sustainability, a framework for development. pp. 41–82. In: N.S. Jodha, M. Banskora and T. Partap (eds.). Sustainable Mountain Agriculture. Oxford & IBM Publishing, New Delhi

Jodha, N.S., B. Bhadra, N.R. Khanal and J. Richter. 2002. Poverty: issues and options in mountain areas. pp. 1–31. In: N.S. Jodha, B. Bhadra, N.R. Khanal and J. Richter (eds.). Proceedings of the International Conference on Poverty Alleviation in Mountain Areas of China 11–15 November 2002, ICIMOD, Kathmandu, Nepal.

Kadola, T., K. Seko, T. Aoki, S. Iwata and S. Yamaguchi. 2000. Shrinkage of the Khumbu glacier, east Nepal from 1978 to 1995. In Debris Covered Glaciers, IAHS Publication No. 264: 235–243. IAHS, Wallingford, UK.

Kang, S., Y. Xu, Q. You, W. Fl'ugel and T. Yao. 2010. Review of climate and cryospheric change in the Tibetan Plateau. Environmental Research Letters 5: 1–8.

Khanka, L.S. and D.S. Jalal. 1984. Morphological Analysis of Lake Naini Tal in Kumaon Himalaya. Geogrl. Rev. India 46: 64–69.

Kohler, T. and D. Maselli (eds.). 2009. Mountains and climate change—From understanding to action. Geographica Bernensia and Swiss Agency for Development and Cooperation, Berne.

Kollmair, M., G.S. Gurung, K. Hurni and D. Maselli. 2005. Mountains: Special places to be protected? An analysis of worldwide nature conservation efforts in mountains. International Journal of Biodiversity Science and Management 1: 181–189.

Körner, C. 2009. Conservation of mountain biodiversity in the context of climate change. In: Proceedings of the International Mountain Biodiversity Conference, Kathmandu, 16–18 November 2008, ICIMOD, Kathmandu, Nepal.

Kruk, E. 2009. Tourism in the Himalaya—Mountains of Opportunities in a Changing Climate. Position Paper, ICIMOD, Kathmandu, Nepal.

Lambin, E.F., X. Baulies, N. Bockstael, G. Fischer, T. Krug, R. Leemans, E.F. Moran, R.R. Rindfuss, Y. Sato, D. Skole, B.L. Turner and C. Vogel. 1999. Land-use and land-cover change (LUCC): Implementation strategy. IGBP Report No. 48, IHDP Report No. 10, Stockholm, Bonn.

Liu, S.Y., Y.J. Ding, J. Li, D.H. Shangguan and Y. Zhang. 2006. Glaciers in response to recent climate warming in Western China. Quaternary Sciences 26(5): 762–771.

Löffler, J., K. Anschlag, B. Baker, Oliver-D Finch, B. Diekkrüger, D. Wundram, B. Schröder, P. Pape and A. Lundberg. 2011. Mountain Ecosystem Response to Global Change. Erdkunde 65(2): 189–213.

Luckman, B. and T. Kavanagh. 2000. Impact of climate fluctuations on mountain environments in the Canadian Rockies. Ambio 29: 371–380.

Mahat, T. J. 2006. Issues in Sustainable Mountain Development: The Himalayan Experience. "Hamro Kalpabrikshya", 17 (189), a monthly magazine of the Department of Forest (Ministry of Forest and Soil Conservation), Government of Nepal, Kathmandu, NEPAL. Unedited version is available from http://www.mtnforum.org/oldocs/486.doc.

McCarty, J.P. 2001. Ecological consequences of recent climate change. Conservation Biology 15(2): 320–331.

Messerli, B. and H. Hurni. 2000. African Mountains and Highlands: Problems and Perspectives. African Mountains Association, Walsworth Press, Missouri.

Messerli, B. and J.D. Ives (eds.). 1997. Mountains of the World—A Global Priority. A contribution to Chapter 13 of Agenda 21. Parthenon, New York.

Meybeck, M., P. Green and C. Vörösmarty. 2001. A new typology for mountains and other relief classes: an application to global continental water resources and population distribution. Mountain Research and Development 21: 34–45.

Millennium Ecosystem Assessment. 2005a. Ecosystems and Human Well-being: Synthesis. Island Press, Washington, DC.

Millennium Ecosystem Assessment. 2005b. Ecosystems and Human Well-being: Current State and Trends: Findings of the Condition and Trends Working Group, Mountain Systems. Island Press, Washington, DC.

Ministerio De Medio Ambiente. 2010. Perú y el Cambio Climático. Segunda Comunicación Nacional ante la Convención Marco de las Naciones Unidas sobre Cambio Climático.

Mool, P. 2001. Glacial Lakes and Glacial Lake Outburst Floods. ICIMOD Mountain Development Profile, MDP 2.Kathmandu: ICIMOD.

Mountain Agenda. 1997. Mountains of the World: Challenges for the 21st Century. United Nations. Switzerland.

Mukhophadayay, S.C. 2006. Glaciers and water source appraisal of Himalaya with special reference to Central Himalaya, Uttaranchal. pp. 106–119. In: M.S.S. Rawat (ed.). Resource Appraisal, Technology Application, and Environmental Challenges in Central Himalaya, Vol. 1. Transmedia, Srinagar, Uttarakhand.

Nogués-Bravo, D., M.B. Araújo, M.P. Errea and J.P. Martinéz-Rica. 2007. Exposure of global mountain systems to climate warming during the 21st century. Global Environmental Change 17: 420–428. DOI: 10.1016/j.gloenvcha.2006.11.007.

OcCC and ProClim. 2007. Climate change and Switzerland 2050—Expected impacts on environment, society and economy, OcCC Secretariat, Bern, Switzerland.

Oki, T. 2003. Global Water Resources Assessment under Climatic Change in 2050 using TRIP'. Water Resources Systems—Water Availability and Global Change (Proceedings of symposium HS02a held during IUGG 2003 in Sapporo, July 2003). IAHS Publ. No. 280. Fontainebleau (France): IAHS.

Palni, L.M.S., R.K. Maikhuri and K.S. Rao. 1998. Conservation of the Himalayan agro-ecosystem: issues and priorities. pp. 253–290. In: Ecological Co-operation for Biodiversity Conservation in the Himalaya. United Nations Development Programme, New York.

Pender, J. and P. Hazell (eds.). 2000. Promoting Sustainable Development in Less-Favored Areas. 2020 Vision, Focus 4. Washington, D.C.: International Food Policy Research Institute.

Pepin, N.C. and D.J. Seidel. 2005. A global comparison of surface and free-air temperatures at high elevations. Journal of Geophysical Research 110: D03104.

Power, T.M. 1995. Economic Wellbeing and Environmental Protection in the Pacific Northwest: A Consensus Report. Pacific Northwest Economists, December 1995, mimeo.

Pratt, D.J. 2001. Corporations, Communities, and Conservation. California Management Review, Vol. 43, No. 3, Haas School of Business, University of California Press, Berkeley.

Pratt, D.J. and J. Shilling. 2002. High Time for Mountains. Background paper prepared for the World Development Report 2002/2003. World Bank, Washington DC.

Qin, D.H. 2002. Glacier Inventory of China (Maps). Xi'an: Xi'an Cartographic Publishing House, Science Press, Beijing, China.

Ouyang, H. 2012. Climate Changes and Water Resources Management in the HKH region: Strategy and Implementation. Paper Presented at ICIMOD-MAIRS joint International Workshop on Climate Change Impacts on Water/Land and Adaptation Strategies in the Tibet-Himalayan Region, Pokhara, Nepal, 27–29 June, 2012.

Rawat, J.S. 2009. Saving Himalayan Rivers: developing spring sanctuaries in headwater regions. pp. 41–69. In: B.L. Shah (ed.). Natural Resource Conservation in Uttarakhand. Ankit Prakshan, Haldwani.

Rees, G.H. and D.N. Collins. 2006. Regional differences in response of flow in glacier-fed Himalayan rivers to climate warming. Hydrological Processes 20: 2157–2167.

República de Argentina. 2007. 2da Comunicación Nacional de la República Argentina a la Convención Marco de las Naciones Unidas sobre Cambio Climático.

Rial, J.A., R.A. Pielke, M. Beniston, M. Claussen, J. Canadell, P. Cox, H. Held, N. De Noblet-Ducourde, R. Prinn, J.F. Reynolds and J.D. Salas. 2004. Nonlinearities, feedbacks and critical thresholds within the earth's climate system. In: Climatic Change. 65: 11–38. DOI: 10.1023/B:CLIM.0000037493.89489.3f.

Rodgers, W.A. 2002. Development contradictions. pp. 230–231. In: Kilimanjaro: The Story of a Mountain. National Geographic Press, Washington, DC.

Robinson, J.L. 1991. Geographical regions of British Columbia. pp. 19–40. In: Paul M. Koroscil (ed.). British Columbia: Geographical Essays. Department of Geography, Simon Fraser University, Burnaby, British Columbia.

Romero, H., P. Smith and A. Vasquez. 2009. Global Changes and Economic Globalization in the Andes - Challenges for Developing Nations. Alpine Space—Man & Environment, Vol. 7: 'Global Change and Sustainable Development in Mountain Regions'. Innsbruck University Press, Innsbruck, Austria.

Romero, H. and F. Ordenes. 2004. Emerging urbanization in Southern Andes: environmental impact of urban sprawl in Santiago De Chile on the Andean Piedmont. Mountain Research and Development 24(3): 195–199.

Ruben, R., J. Pender and A. Kuyvenhoven (eds.). 2004. Sustainable Poverty Reduction in Less Favoured Areas: Problems, Options, Strategies. CAB International, Oxfordshire, UK.

Sanderson, J. and S.M.N. Islam. 2009. Climate Change and Economic Development—SEA Regional Modelling and Analysis. Palgrave Macmillan, Basingstoke, UK.

Scherler, D., B. Bookhagen and M. Strecker. 2011. Spatially variable response of Himalayan glaciers to climate change affected by debris cover. Nature Geoscience 4: 156–159.

Schwindt, R. and T. Heaps. 1996. Chopping up the Money Tree: Distributing the Wealth from British Columbia's Forests. Report to the David Suzuki Foundation, Vancouver, Canada.

Seko, K., H. Yabuki, M. Nakawo, A. Sakai, T. Kadota and Y. Yamada. 1998. Changing surface features of Khumbu glacier, Nepal Himalayas revealed by SPOT images. Bulletin of Glaciological Research 16: 33–41.

Singh, S.P. 2007. Himalayan Forest Ecosystem Services: Incorporating in National Accounting. Central Himalayan Environment Association, Nainital, Uttarakhand, India.

Singh, S.P., V. Singh and M. Skutsch. 2010. Rapid warming in the Himalayas: ecosystem responses and development options. Climate Change and Development 2: 1–13.

Singh, J.S., U. Pandey and A.K. Tiwari. 1984. Man and forests: a central Himalayan case study. Ambio 13: 80–87.

Sonesson, M. and B. Messerli (eds.). 2002. The Abisko Agenda: Research for Mountain Area Development. Ambio Special Report 11. Royal Swedish Academy of Sciences, Stockholm.

Srivastava, D. 2003. Recession of Gangotri glacier. *In*: D. Srivastava, K.R. Gupta and S. Mukerji (eds.). Proceedings of a Workshop on Gangotri Glacier, Lucknow, 26–28 March 2003, special Publication No. 80, Geological Survey of India, New Delhi.

Stern, N., S. Peters, V. Bakhshi, A. Bowen, C. Cameron, S. Catovsky, D. Crane, S. Cruickshank, S. Dietz, N. Edmonson, S.L. Garbett, L. Hamid, G. Hoffman, D. Ingram, B. Jones, N. Patmore, H. Radcliffe, R. Sathiyarajah, M. Stock, C. Taylor, T. Vernon, H. Wanjie and D. Zenghelis. 2006. Stern Review: The Economics of Climate Change. HM Treasury, London.

Tasser, E., U. Tappeiner and A. Cernusca. 2005. Ecological effects of land-use changes in the European Alps. *In*: U.M. Huber, H.K.M. Bugmann and M.A. Reasoner (eds.). Global Change and Mountain Regions: An Overview of Current Knowledge, Advances in Global Change Research, Name of Publisher, Dordrecht 23: 409–420.

Theobald, D.M., H. Gosnell and W.E. Riebsame. 1996. Land use and landscape change in the Colorado mountains II: a case study of the East River Valley. Mountain Research and Development 16(4): 407–418.

Thompson, L.G., H.H. Brecher, E. Mosley-Thompson, D.R. Hardy and B.G. Mark. 2009. Glacier loss on Kilimanjaro continues unabated. Proceedings of the National Academy of Science of the United States of America 107(35): 1–6.

Tiwari, A.P. 1972. Recent changes in the position of the snout of the Pindari Glacier, Kumaon Himalaya, Almora District, UP India. pp. 1144–1149. *In*: Proceedings of the symposium on role of snow and ice in hydrology. UNESCO-WHO-IAHS, Canada.

Tiwari, A.P. and B.S. Jangpangi. 1962. The retreat of the snout of the Pindari Glaciers. pp. 245–248. *In*: Proceedings of the symposium of Obergurage. International Association of Hydrological Sciences.

Tiwari, P.C. 2011. Tourism Trends and Potential: in Indian Part of Kailash Sacred Landscape. Draft Report, ICIMOD International (China, India & Nepal) Collaborative Project on 'Kailash Sacred Landscape Conservation Initiative Developing a Tran-boundary Framework for Conservation and Sustainable Development', ICIMOD, Kathmandu, Nepal.

Tiwari, P.C. 2010. Land use changes and conservation of water resources in Himalayan headwaters. pp. 170–174. *In*: Proceedings of the 2nd Indo-German Conference on Research for Sustainability: Energy & Land Use, Bonn, Germany.

Tiwari, P.C. 2008. Land use changes in Himalaya and their impacts on environment, society and economy: a study of the Lake Region in Kumaon Himalaya, India. Advances in Atmospheric Sciences (an International Journal of Chinese Academy of Sciences, Beijing) 25(6): 1029–1042.

Tiwari, P.C. 2007. Urbanization and environmental changes in Himalaya: a study of the Lake region of District Nainital in Kumaon Himalaya, India, International Working Paper Series ISSN 1935–9160, Urbanization & Global Environmental Change (UGEC), International Human Dimension Programme (IHDP), Working Paper 07–05, pp. 1–19.

Tiwari, P.C. 2000. Land use changes in Himalaya and their impact on the plains ecosystem: need for sustainable land use. Land Use Policy 17: 101–111.

Tiwari, P.C. 1995. Natural Resources and Sustainable Development in Himalaya. Shree Almora Book Depot, Almora, India.

Tiwari, P.C. and B. Joshi. 2012a. Environmental changes and sustainable development of water resources in the Himalayan headwaters of India. International Journal of Water Resource Management 26(4): 26: 883–907. DOI 10.1007/s11269-011-9825-y.

Tiwari, P.C. and B. Joshi. 2012b. Natural & socio-economic drivers of food security in Himalaya. International Journal of Food Security 4(2): 195–207. DOI 10.1007/s12571-012-0178-z.

Tiwari, P.C. and B. Joshi. 2011. Urban Growth and Food Security in Himalaya, International Working Paper Series, Urbanization & Global Environmental Change (UGEC), View Point, International Human Dimension Programme (IHDP) 1(5): 20–23.

Tiwari, P.C. and B. Joshi. 2009. Resource utilization pattern and rural livelihood in Nanda Devi biosphere reserve buffer zone villages, Uttarakhand Himalaya, India. Ecomont: International Journal of Protected Areas 1(1): 25–32.

Tiwari, P.C. and B. Joshi. 2007. Rehabilitation and management of wasteland in the Himalayan headwaters: an experimental study of Kosi headwater in Kumaon lesser Himalayas in India. International Journal of World Association for Soil and Water Conservation J2, pp. 39–62.

Tiwari, P.C. and B. Joshi. 2005. Environmental changes and status of water resources in Kumaon Himalaya. pp. 109–123. *In*: J. Libor, M. Haigh and H. Prasad (eds.). Sustainable Management of Headwater Resources: Research from Africa and Asia. United Nations University, Tokyo.

Tiwari, P.C. and B. Joshi. 2001. Integrated resource management in forestry using remote sensing and GIS. pp. 242–258. *In*: A.S. Rawat (ed.). Forest History of the Mountain Regions of the World, Proceedings of the International Forest History Conference.

Tiwari, P.C. and B. Joshi. 1997. Wildlife in the Himalayan Foothills: Conservation and Management. Indus Publishing Company, New Delhi.

UNDP. 2010. Summary of implications from the East Asia and South Asia consultations: Asia Pacific human development report on climate change. Colombo: UNDP Asia Pacific Regional Centre, Human Development Report Unit.

UNEP-WCMC. 2002. Mountain Watch: Environmental change and sustainable development in mountains. UNEP, Nairobi. www.unep-wcmc.org/mountains/mountainwatchreport/ (accessed 11 March 2012).

UNEP-WGMS. 2008. Global glacier changes: Facts and figures. UNEP, Nairobi http://www.grid.unep.ch/glaciers/ (accessed 22 June 2010).

UN. 2012. Outcome of the Conference, Agenda Item—10. Rio+20: United Nations Conference on Sustainable Development, Rio de Janeiro, Brazil, June 2012.

UNWTO. 2009. International Tourism on Track for a Rebound after an Exceptionally Challenging 2009. Report, Madrid, Spain.

Valdiya, K.S. and S.K. Bartarya. 1991. Hydrological studies of springs in the catchment of Gaula River, Kumaon Lesser Himalaya, India. Mountain Research and Development 11. 17–25.

Viviroli, D., H.H. Dürr, B. Messerli, M. Meybeck and R. Weingartner. 2007. Mountains of the world, water towers for humanity: typology, mapping, and global significance. Water Resource Research 43: W07447.

Viviroli, D., R. Weingartner and B. Messerli. 2003. Assessing the hydrological significance of the world's mountains. Mountain Research and Development 23: 32–40.

Wasson, R.J., N. Juyal, M. Jaiswal, M. McCulloch, M.M. Sarin, V. Jain, P. Srivastava and A.K. Singhvi. 2008. The mountain-lowland debate: deforestation and sediment transport in the upper Ganga catchment. Journal of Environmental Management 88(1): 53–61.

Wookey, P.A., R. Aerts, R.D. Bardgett, F. Baptist, K.A. Brathen, J.H.C. Cornelissen, J.L. Gough, I.P. Hartley, D.W. Hopkins, S. Lavorel and G.R. Shaver. 2009. Ecosystem feedbacks and cascade processes: understanding their role in the responses of arctic and alpine ecosystems to environmental change. *In*: Global Change Biology. 15: 1153–1172. DOI: 10.1111/j.1365-2486.2008.01801.x.

World Resources Institute. 2005. Millennium Ecosystem Assessment—Ecosystems and Human Well-being: Synthesis. Island Press, Washington, DC.

World Bank. 2005. Pakistan's Water Economy Running Dry: Pakistan Water Strategy, World Bank Report 2006.

Xu, J.C. 2008. The highlands: a shared water tower in a changing climate and changing Asia. Working Paper No. 67. Beijing: World Agro-forestry Centre, ICRAF-China.

Yamada, T. 1998. Monitoring of Glacier Lake and its outburst floods in Nepal Himalaya. *In*: Japanese Society of Snow and Ice, Monograph No. 1. Japanese Society of Snow and Ice, Tokyo.

Yao, T., K. Duan, B. Xu, N., Wang, X. Guo and X. Yang. 2008. Ice core precipitation record in Central Tibetan plateau since AD 1600. Climate of the Past Discuss 4: 233–248.

Zheng, J.Y., Q.S. Ge and Z.X. Hao. 2002. Impact of Climate Warming on Plant Phenology during Recent 40 years in China. In Scientific Bulletin 47(20): 1582–1587 (in Chinese).

Zierl, B. and H. Bugmann. 2005. Global change impacts on hydrological processes in alpine catchments. *In*: Water Resources Research 41, W02028. DOI: 10.1029/2004WR003447.

5

Adaptation to Climate Change in Mountain Regions: Global Significance of Marginal Places

Matthias Monreal[1,*] and *Johann Stötter*[2]

INTRODUCTION

This chapter takes a discursive approach to explore the importance of adaptation to global climate change as a viable strategy in mountain regions. It sets out by trying to understand global change as a process that unfolds between interacting pairs of opposing forces/actors/trends, the reconciliation of which is the aim of recent thought in the study of social ecological systems. It is argued that mitigation and adaptation can be understood as part of this dialectic process. After presenting the particular environmental and societal circumstances in mountain regions, their vulnerability and their importance for lowlands the case is made for a concerted effort on the part of lowland societies to support adaptation in mountain regions if for nothing else but their own self-interest.

---

[1] Institute of Geography University of Innsbruck Innrain 52A-6020 Innsbruck Austria.
  Email: matthias.monreal@uibk.ac.at
[2] Institute for Geography, University of Innsbruck, Innrain 52, A-6020 Innsbruck, Austria.
  Email: hans.stoetter@uibk.ac.at
* Corresponding author

Global Change—A Dialectic Process

The striking appeal 'think globally, act locally' has long been the maxim of grassroots movements and countless mainly environmentally concerned initiatives. Its message is twofold, reminding us that local action must be embedded within a global awareness, but also that in order to achieve global results, local engagement is necessary. In its simplicity this appeal perfectly captures the tensions inherent when thinking about global change. It seems, in fact, that global change can very well be understood in terms of such sets of polar opposites that go beyond a mere differentiation in terms of scale, i.e., local vs. global.

First, there exists a principle dichotomy of attitude towards change. Global change is on the one hand, driven by forces that promote change by concentrating on the integration of the globally diverse (e.g., online social networks). On the other hand, global change is affected by forces that oppose change by concentrating on the preservation of the distinct (e.g., religious fundamentalism). Second, there is also a dichotomy of the actors involved in global change. They are at one end, individuals who drive and experience global change through their work and travel, their preferences and tastes, in other words through the personal choices they make. On the other end, there are institutions that result from and aim to handle global change. These are intergovernmental organizations concerned with politics, trade, finances, health, etc., but also multi-national companies, religious groups, research groups, etc. A further dichotomy lies in the type of system that is affected. There is the cultural system, which thus far has attracted much of the focus of discussions on global change. Change within these cultural system(s) is what is commonly understood by globalization. There is also the natural system that undergoes change on a global scale, through resource scarcity and pollution, biodiversity loss, desertification and land-use change, but most comprehensively through global climate change. The various motivations for actions in the context of global change do also line up between two polar extremes. There is the motivation of self-interest leading to a behaviour that seeks to produce maximum personal gain in the sense of the rational choice theory. Opposing that, there is the socially-minded, globally aware, altruistic mind-set that informs and motivates a multitude of social activities.

It seems feasible to understand the dynamics of global change in the form of a dialectic process between such opposing forces. A deeper understanding and active engagement within the processes of global change means attempting reconciliation and the discovery of complementarity within these tensions. The emergence of new scientific 'disciplines', such as human-environment systems, social and political ecology, cybernetics, etc. that focus on the connectedness of subsystems is a consequence of such an understanding and provides the theoretical background for action. The aim is to transcend what superficially seems like divergent trends. The study of global climate change is no exception and features its own dichotomy, in the two principal, distinct but complementary, strategies which apply in order to answer its challenges: mitigation and adaptation.

Mitigation and Adaptation

Because it is not always easy to allocate a specific activity to either one of the two strategies, it is particularly important to develop a clear and unequivocal understanding of the terms as they are used within the context of the climate change discourse. In fact the often contrary use of the terms in everyday language as well as in other scientific disciplines complicates the matter. Particularly climate change adaptation runs danger to become a tautological truth since almost everything can be argued to be somehow an adaptation to climate change. As we will see by the end of the chapter, climate change adaptation in mountain regions (but also elsewhere) addresses a specific situation that calls for specific types of actions that must be integrated but not confused with climate change mitigation efforts. With no doubt, both the artificial production of snow and the introduction of Douglas fir are successful measures to adapt to changing climate in the Alps, with the former having negative mitigation outcome (additional energy demand) (Schmidt et al. 2012), and the latter acting as enhanced carbon sink and thus contributing to mitigation efforts (van Loo et al. 2012).

Various definitions for climate change mitigation and adaptation that differ in nuances rather than conflicting in their core content have been put forward by researchers and practitioners (OECD 2006). In an attempt to present a distillation of these, mitigation can be defined as measures that decrease radiative forcing that leads to global warming by reducing sources and enhancing sinks of greenhouse gases. Adaptation are measures that decrease the vulnerability or enhance the capacity to deal with direct effects of climate variability and climate change in order to avoid harm or reap its benefits.

Mitigation rarely occurs by coincidence. When switching to renewable energies, developing new mobility concepts, promoting afforestation and energy efficient building in most cases mitigation is the stated aim of the activity rather than a side effect. These activities assume reflection and willingness to further the goals of mitigation, to prevent our climate from further change. Adaptation on the other hand can happen unconsciously, both by nature and by action. It can range from spontaneous, chaotic and haphazard to deliberately planned and implemented (for detailed treatment see (Smit and Wandel 2006, Smit et al. 2000)). Before all adaptation is a 'natural' process inherent in the constant change and co-evolution of nature and culture that plays out with or without our deliberation (Guillet et al. 1983). Because adaptation is what happens anyway, it not always leads to desirable outcomes. By contrast to evolutionary adaptation, in planned climate change adaptation such maladaptations (see Carey et al. (2012) for an interesting case study on unintended effects of adaptation) can be avoided prior to their emergence through foresight and planning in a holistic approach. Above all, adaptations that bear negative mitigation consequences need to be avoided. Energy intensive air-conditioning as a reaction to warmer temperatures serves as the stereotypical negative example of a maladaptation that solves one problem (overheated buildings) and exacerbates a range of others (resource scarcity, urban heat islands, operational overheads and climate change).

Over the last decade or so it has become increasingly clear that all efforts to prevent dangerous levels of global warming are too late. This finding should by no means deflate continued efforts of emissions reduction, but it means that global climate change is inevitable and that adaptation to its impacts will and must occur. Since the 4th IPCC report it is clear that there is now a common understanding that climate change mitigation must be complemented by strategies to adapt to its impacts and that in many cases the latter is the more urgent call (Parry et al. 2007). This paradigm shift from a focus on mitigation to a concentration of efforts in climate change adaptation brought with it a growing role of the social sciences to the field of climate change that was hitherto a realm of natural sciences alone.

Adaptation to climate change needs a strong involvement of political sciences, sociology, anthropology, development studies and the range of emerging 'hybrid disciplines' mentioned above. The importance of contextualizing climate change within societal processes is pivotal and has been understood as such particularly within an integrative geography that spans physical and human dimensions. Carey et al. (2012) observe that,

> "geographers and many other scientists and social scientists studying climate change readily recognize the benefits from the intertwined analysis of coupled natural and human systems that go beyond token consideration of the 'social' dimensions",

and further that

> "this intertwined, dialectic relationship that plays out among multiple human and non-human actors is precisely why scholars have recently devoted so much attention to technonatures, social-ecological systems, coupled natural and human systems, hybrid landscapes, Actor Network Theory, and the hydro-social cycle."

Climate change adaptation is foremost about human decision making and social processes. Notions of behavioural psychology, communication, culture and power relations need to be considered for all adaptation measures. This is not to say that adaptation measures do not often have a technological/infrastructural core. On the contrary, a coastal dyke to prevent intrusion of flood waters or a drainage system that prevents glacier lake outburst floods clearly are very tangible adaptation measures. What it means is that the successful implementation of such measures depends on local actors embedded in local cultures, social networks and power relations, but also on decisions driven by political hierarchies that maybe are removed from the reality of local (mountain) societies. Without their consideration the danger of unintended effects is considerable (Carey et al. 2012).

To some extent the distinct but complementary strategies of climate change mitigation and climate change adaptation map to the dichotomies laid out in the preceding section. The central actors in climate change mitigation are governance and policy relevant institutions that take on global responsibility. It is their objective to prevent the global (natural) climate system from further change. The central actors in climate change adaptation are individuals and communities that act in their self-interest and promote and engage in change of their socio-economic conduct (their culture) to accommodate locally precipitating effects of climate change.

After looking at the specific characteristics of mountain regions and the ways in which they are affected by climate change it will be argued later that the complex of attributes associated with climate change adaptation bears particular relevance for mountain regions.

Climate Change and Mountain Regions

To continue the juxtaposing theme of this chapter it may be tempting to compare mountain societies with those of small island states. Both seem to occupy opposite ends of the terrestrial surface of the globe: islands are to the coastal context what mountains are to the continental context. In many ways they share defining characteristics such as an endemic ecology (difficult access, cultural distinctness and economic hardship (see, e.g., Messerli and Ives 1997, Baldacchino 2007). Like mountain regions, small island states are strongly affected by global climate change. In the long term sea level rise will lead to their disappearance altogether. The drama and utter devastation that this inevitability means for their inhabitants is hard to comprehend. More tragic still that this drama largely remains solely with the people whose life it destroys: small island states have little relevance for the rest of the world therefore there is little reason to get involved beyond the role of a spectator. Their populations are small, they are by and large economically and often politically marginalized and the existence of their above sea level natural habitat bears little impact on main land ecosystems. Small island states are as disjunct from the rest of the world as it gets.

This is not true in the case of mountain societies. Though mountain populations and societies are often remote and in many ways isolated, they are by no means disconnected from the low lands. It is through their connection with the lowlands that the fate of mountain regions and the way mountain regions adapt to climate change will continuously be interwoven with the fate of the adjacent lowlands. Therefore climate change in mountain regions is a common concern for a large part of humanity.

Mountain region characteristics

Before discussing the implications of climate change in mountain regions for lowland societies it is important to look at the defining characteristics. What is it that makes mountain regions and the human-environment systems they host particularly vulnerable to global climate change?

Mountains and mountain ranges are distinct features of the Earth's surface that share common characteristics. Defined as a "substantial elevation of the Earth's crust above sea level" they result in "localized disruptions to climate, drainage, soils, plants and animals" (Andrew Goudie, The Encyclopedic Dictionary of Physical Geography). Depending on the sources quoted (Xu et al. 2009, Beniston 2003, Millennium Ecosystem Assessment 2005) 20–25% of the terrestrial surface can be classified as mountains, being home to between 12 and 26% of the world population and providing resources to 40–50% of the world population. In mountain regions land resources are sparse due to steep slopes, low temperatures, poor soils, general inaccessibility and susceptibility to natural hazards (Becker and Bugmann 2001).

Maybe most defining characteristic of mountain areas is that natural (i.e., meteorological, hydrological, biological/ecological, geomorphological) conditions change strongly over short distances. The decisive factor in this change is not horizontal distance to the poles, i.e., latitude, but vertical distance, i.e., elevation. Bioclimatic changes mimic those that are usually associated with moving toward higher latitudes which results in a distinct vertical zonation of plant and animal life (Goudie 2000). Some understand culturally specific societies as a response to these bioclimatic spatial patterns (Guillet et al. 1983). Accommodating so much diversity within confined horizontal spaces generates a volatile situation with enormous potential for disruptions should general framework conditions alter (Xu et al. 2009). And indeed global climate change does alter these framework conditions. Due to general warming and/or changing precipitation, the boundaries between vertically organized systems experience drastic shifts. The entailing extraordinary changes in the distribution of environmental resources imply tremendous challenges to the ecology but also to highly specialized societies in mountain areas.

Exacerbating the effect of the vertical change gradient is the circumstance that total available space becomes less along this gradient. If one moves upwards in mountains, less and less area is available for ecosystems and societies that depend on them. In all mountains the drive upwards on the slopes ultimately ends with a summit. This means that not only do temperature-dependent systems need to be able to rapidly move upwards they are also running out of space. For example the snow leopard habitat in the Himalaya has found to be affected by climate change induced shifts of the tree-line. As a result of the upwards shift a recent study (Forrest et al. 2012) has shown that the snow leopard habitat might shrink by 30%.

The principal repercussions of the vertical distribution of ecosystems in mountain regions, i.e., volatility to change and diminishing available space as one moves upwards, provide the physical basis for an understanding of the effect of climate change and the way its impacts precipitate in mountain regions and the societies they host.

Impact of climate change on the natural environment

It is expected that with continued climate change increases in mean annual air temperature are considerably higher in mountain regions than elsewhere. Given that the benchmark for dangerous global warming of 2°C increase of global average air temperature seems increasingly likely to be exceeded (e.g., Allison et al. 2009) this is alarming for all mountain dwellers. For example temperatures on the Tibetan Plateau are expected to rise substantially more (Rupa et al. 2006), which would have catastrophic effects for the Greater Himalayan people and ecosystems (Xu et al. 2009, Anderson and Bowe 2008, Hansen et al. 2008, Solomon et al. 2009). Indeed it can already empirically be shown that in some mountain areas warming trends and anomalies are elevation dependent (Giorgi et al. 1997, Matulla et al. 2004), and temperatures increase more rapidly in higher altitudes (e.g., in the Alps). That means that on top of the more fragile situation of vertically organized ecosystems as described above, the changes in one of the principal impacts of climate change, i.e., temperature, are more pronounced in mountain regions. Extremes in impact and vulnerability, in driver and in capacity to respond, combine.

Mountain regions are highly sensitive indicators of climate change. This is manifested by 7,000 km² of mountain glaciers that have disappeared in the last four decades of the 20th century (Nogués-Bravo et al. 2007). The European glacier extent decreased 30–40% during the 20th century (Haeberli and Beniston 1998, Lamprecht and Kuhn 2007) and further 30–50% of glacier mass may be lost by 2100 (Haeberli 1999, Maisch et al. 1999, Zemp et al. 2006). These changes in the cryosphere will have significant repercussions in the hydrological cycle and alter availability of water and seasonality of runoff regimes (Ellenrieder et al. 2007, Braun and Weber 2007, Hagg et al. 2007). In dry seasons, this may lead to severe water shortage due to exhausted water stores with feedbacks to other resources, e.g., hydro power (IPCC 2008). Xu et al. (2009) discuss how the cascading effects of rising temperatures and impacts on the cryosphere affect "water availability (amounts, seasonality), biodiversity (endemic species, predator–prey relations), ecosystem boundary shifts (tree-line movements, high-elevation ecosystem changes), and global feedbacks (monsoonal shifts, loss of soil carbon)."

Fragile and sensitive and often endemic mountain ecosystems with their exceptionally high biodiversity (Grabherr et al. 2005) are particularly affected by direct and cascading effects of rising temperatures. The expected upward migration with the entailing shrinkage of available area per ecotone will lead to an intensified encroachment and competition between species. Together with the high degree of adeptness of mountain ecosystems and their dependence on temperatures and melt water this will lead to species loss (Grabherr et al. 2004). The impacts on mountain ecosystems will have a profound knock-on effects on mountain communities.

Impacts of climate change on mountain societies

In the same way as a high degree of adaptation is required from mountain flora and fauna in order to occupy their respective niches in a marginal ecosystem, land-use in marginal areas for human habitation is highly adapted (Stadel 1991). This particular vulnerability of mountain regions to impacts of Global Climate Change represents a bottleneck that affects human-environment systems almost everywhere. Due to limited utilizable land there are rather restricted alternatives to the specialized economic situation.

A common trait among the enormous cultural diversity of mountain societies around the world is that they are highly specialized and adapted to the harsh reality of their natural environment. Some proponents of a vertical control on the formation of mountain cultures (Brush 1977, Webster 1971, Forman 1978) argue that ecological zonation leads to a specific resource exploitation strategy that in turn determines cultural expressions of a society. Others accept environmental constrains (low-yield, fragility) and vertical zonation as playing a major role but not as ultimately determining factor in mountain cultures (Guillet et al. 1983). Clearly though verticality and ecological zonation as well as low productivity, considerable hazard potential and mobility constraints inform strategies of environmental adaptation and resource exploitation. The latter can range from exploitation of one vertical zone to increasing exchange and supplementation from other zones (Brush 1977, Werge 1979). Alpwirtschaft, common throughout the

European Alps but also in the Andes and the Himalaya, is an example of an adaptive strategy that is based on agropastoral transhumance. The seasonal movement of people and animals in vertical space shapes much of the regional cultural identity and has brought about its own social institutions (Rhoades and Thompson 1975, Netting 1982).

Drastic changes of the distribution, available space and interconnection of ecological zones that do in a variety of ways underlie various subsistence strategies has direct implications on mountain societies meeting their economic and cultural needs. It is important to note that once the intricate link to either a directly inhabited or otherwise utilized vertical zone is broken, not only the economic resource base but the entire fabric of a culture may be threatened. In this sense the demise of mountain farming (Bergbauerntum) serves as an example for an impact of economic globalization from the highly developed economies of the European Alps (e.g., Lahn-Gärtner 2007). With mountain farming being no longer economically viable the perceived degradation from self-sustained farmers to subsidy dependent de facto government employed landscapers leads to a breakdown of cultural identity.

In many cases the existential threat results directly from global climate change induced changes of the natural environment and resource base. For example Carey et al. (2012) point out that

> "societies that depend on snowmelt and glacier run off from the Andes and Himalayas to the Rockies and Alps are some of the most vulnerable to these fluctuations or reductions of freshwater supplies. In the tropical Andes in particular, future glacier shrinkage will diminish water sources, especially during the May September dry season, which will affect the export agriculture economy, indigenous people's subsistence food production, urban drinking water, industries, and hydroelectricity generation that accounts for up to 80% of Andean energy supplies. Retreating glaciers have also caused disastrous glacial lake outburst floods (GLOFs) in mountain ranges worldwide but particularly in Peru; as climate continues to change, the threat of these floods will persist."

Land resources in mountains are increasingly exposed to natural hazards, i.e., mass movements due to thawing of permafrost (Stötter et al. 1996, Kääb et al. 2007), or the heat wave in 2003 (Stott et al. 2003, Schär et al. 2003) and the floods in 2002 and 2005 (Habersack and Moser 2003, Godina et al. 2006). Recently Jäger et al. (2008) developed a method to compute the fraction of attributable risk related to climate damages by a Bayesian filtering approach. In case of the alpine summer heat wave this fraction amounted to 90%. Trends in floods and droughts are difficult to assess due to the coarseness of GCMs.

Although there is a common pattern of Global Climate Change challenges to mountain areas worldwide (e.g., melting of the cryosphere, increase of natural hazards), vulnerability and resilience/coping capacities vary strongly from one to another mountain region. This is on the one hand, due to their respective position within the global circulation system and on the other hand, due to the specific state of development, political and cultural context of their populations. While in the Alps adaptation may aim at continuation of a tourism-based economy and the resulting quality of life, in mountain regions like the Andes or the Himalaya the growing water scarcity presents an existential threat of pure survival.

Since the 19th century the Alps have become attractive destinations for tourism and recreation which today play a key role in mountain economies (Smeral 2000). Mountain regions take an increasing share of possible destinations (Beniston 2003). In fact, with 11% of the world overnight stays the Alps are the number one tourist destination in Europe (Siegrist 1998, Bätzing 2003). Mountain tourism will be particularly affected in winter season. Ski tourism requires 100 days of continuous snow cover over 30 cm, which may put low altitude resorts out of business (Breiling 1997, Abegg et al. 2007, Steiger 2008).

As global climate change proceeds and landscape composition and vegetation become affected the Alpine image of a Heidi-land of lush meadows, alpine creeks and dark forest may disappear. This not only threatens a cultural identity, but without being able to cater to this deeply ingrained image of what the European Alps are about (imagine a Lederhosen-clad farmer bringing his cattle down from the summer pastures through olive groves, or a Bierfest in midst vineyards), they may lose much of their attraction as summer destinations as well.

The reason for the particular vulnerability of mountain regions ultimately lies in the particularly intertwined human-environment systems that prevail. It is a further systemic connection to the adjacent lowlands that ought to give mountain regions special attention.

The lowland connection

The marginality of mountain human-environment systems makes them particularly vulnerable to global climate change. But as mentioned above, the most important distinction between mountains and islands in the global climate change context is that mountain vulnerability does not remain within the confines of their geographic limits. Mountain regions' vulnerability is exported to the lowlands. And the main reason for that is the fact that mountains are 'water towers'.

Mountains are water pumps which extract moisture from the atmosphere through the orographic uplift of air masses and they are water towers due to their water storage capacities in glaciers, permafrost, snow, soil and groundwater (Liniger et al. 1998). The 10 largest Asian rivers, namely the Amu Darya, Indus, Ganges, Brahmaputra, Irrawaddy, Salween, Mekong, Yangtze, Yellow and Tarim all originate in the Himalaya. They provide water for approximately 1.3 billion people (Xu et al. 2009). Melting snow and ice contribute about 70% of summer flow in the main Ganges, Indus, Tarim, and Kabul Rivers during the shoulder seasons, that is before and after precipitation from the summer monsoon (Singh et al. 2006, Barnett et al. 2005). The contribution of glacial melt to flow in the Inner Asian rivers is even greater (Yao et al. 2004, Xu et al. 2004, Chen et al. 2006). Indus River irrigation systems in Pakistan depend on snowmelt and glacial melt from the eastern Hindu Kush, Karakoram, and western Himalayas for about 50% of total runoff (Winiger et al. 2005). In western China, about 12% of total discharge is glacial melt runoff, providing water for 25% of the total Chinese population in the dry season (Li et al. 2008, Xu 2008). The Alps supply a significant proportion of water for the population of Europe (Baumgartner et al. 1983, Braun et al. 2000). This water means freshwater, irrigation and hydropower. The energy aspect of mountain waters based on water abundance and topography must not be underestimated in comparison to freshwater and irrigation uses. Hydropower constitutes a significant part of the energy supply in many mountain regions, i.e., hydropower covers about 60% of the Austrian and nearly 100% of the energy demand of Norway. Worldwide mountains are the source for 50% of all rivers (Beniston 2003). Much of the inter- and intra-annual variation of discharge is compensated by discharge from mountains. In semi-arid areas mountain discharge accounts for 50–90%, in extreme cases (e.g., Nile, Colorado) for more than 95% of the total river discharge (Viviroli et al. 2007).

A direct knock-on effect of glacier retreat and thawing of permafrost is the resultant destabilization of mountain slopes and thus endangerment of life and infrastructure in the valleys (Stoffel et al. 2005) while its cascading effects precipitate via river runoff, water availability and hazards, pollination and phenology, prey-predator relations, endemism and extinction, shifting tree-line, ecosystem composition to society, economy and culture. Carey et al. (2012) point out that "social conflicts in response to real and perceived water shortages have already emerged in Peru's most glaciated mountain range, the Cordillera Blanca".

The ecosystem services that mountains provide go beyond water and include hydropower, flood control, mining and tourism. For many exists a direct linkage with adjacent lowlands in catchments systems and for their extractive resources exists global demand (Löffler 2004, Viviroli and Weingartner 2004). It has been estimated that all together mountains provide resources (Viviroli and Weingartner 2004) and ecosystem services to about half of humanity (Becker and Bugmann 2001). It is this central role in the global human-environment system that gives mountain areas their exceptional relevance.

The particular case for adaptation

The urgent case for adaptation in mountain regions comprises three arguments. First, the rate of change in which mountain human-environment systems are affected outpaces any process of natural adaptation. This is due to the particular circumstances laid out above. Second, mountain societies need to adapt to the drastic changes that their natural resource base will undergo if they are to survive. This is due to their high specialization and dependence on their natural environment. Third, adaptation will have to occur for the sake of the low-lands. This is due to the inseparable fate of mountain regions and their adjacent lowlands that results from a multitude of links most importantly that of water. Here it is particularly an integrated watershed management and natural hazard prevention must that take centre stage. However, this cannot be achieved in isolation, but only if mountain human-environment systems are understood in their complex interactions and fragility and are targeted as a whole in attempts to strengthen adaptive capacities and reduce their vulnerability.

As prerequisite for guided, deliberate adaptations (either as a change in livelihood and resource use, or in form of technological/infrastructural measures), intricate knowledge of mountain human-environment systems needs to be established in order to produce adaptive capacity and governance structures which in turn need to be in place for adequate adaptation decision making (Smit and Wandel 2006). Given that mountain regions are already marginal in economic and political terms it is unlikely that sufficient efforts to this end can be made by them alone. An active role in governance, financial support and provision of resources must be taken on by the economic and political centres of the lowlands in order to help mountain communities adapt; for their sake but also in order to manage the knock-on effects that climate change in mountain regions has on the lowlands themselves. Further impetus therefore is given in the sense of a moral obligation. Since the marginal economies of mountain regions have contributed little to anthropogenic climate change, and since they are among the most vulnerable to its impacts there is duty for the lowlands to engage that goes beyond self-interest and is a matter of climate justice.

Conclusions

At the beginning of this chapter the polar extremes of various aspects of global change were roughly aligned and grouped accordingly. It is clear that these are not alternative pathways in the sense of mutually exclusivity.

On the contrary, these various aspects of global change, the driving of it and its prevention, the local and the global processes, the individual and institutional actions all occur simultaneously. They may be mutually reinforcing sometimes and neutralizing their effect at other times. Only the establishment of sound knowledge of the various processes and forces at play embedded within a holistic approach can help steering these processes to a maximum of common benefit. 'Think globally, act locally' transcends the inherent tensions of global change. In the context of climate change it is an appeal for both, mitigation and adaptation. Only with the global good in mind will mitigation work and only when acting in accordance with local needs will marginal human-environment systems ensure their survival.

References Cited

Abegg, B., S. Agrawala, F. Crick and A. de Montfalcon. 2007. Climate change impacts and adaptation in winter tourism. pp. 25–60. *In*: S. Agrawala (ed.). Climate Change in the European Alps: Adapting Winter Tourism and Natural Hazards Management. OECD, Paris.

Allison, N.L., R.A. Bindoff, P.M. Bindschadler, P.M. Cox, N. de Noblet, M.H. England, J.E. Francis, N. Gruber, A.M. Haywood, D.J. Karoly, G. Kaser, C. Le Quéré, T.M. Lenton, M.E. Mann, B.I. McNeil, A.J. Pitman, S. Rahmstorf, E. Rignot, H.J. Schellnhuber, S.H. Schneider, S.C. Sherwood, R.C.J. Somerville, K. Steffen, E.J. Steig, M. Visbeck and A.J. Weaver. 2009. The Copenhagen Diagnosis: Updating the World on the Latest Climate Science. Sydney.

Anderson, K. and A. Bows. 2008. Reframing the climate change challenge in light of post-2000 emissions trends. Transactions of the Royal Academy-A: Mathematical, Physical and Engineering Sciences 366: 3863–3882.

Baldacchino, G. 2007. Introducing a World of Islands. pp. 1–29. *In*: G. Baldacchino (ed.). A World of Islands: An Island Studies Reader. Malta and Canada, Agenda and Institute of Island Studies.

Barnett, T.P., J.C. Adam and D.P. Lettenmaier. 2005. Potential impacts of a warming climate on water availability in a snow-dominated region. Nature 438: 303–309.

Bätzing, W. 2003. Die Alpen: Geschichte und Zukunft einer europäischen Kulturlandschaft, 2. aktual. u. neu konzip. Aufl., München.

Baumgartner, A., E. Reichel and G. Weber. 1983. Der Wasserhaushalt der Alpen, München.

Becker, A. and H. Bugmann (eds.). 2001. Global Change in Mountain Regions, International Geosphere-Biosphere Programme (IGBP) Report 49, Stockholm.

Beniston, M. 2003. Climatic change in mountain regions: A review of possible impacts. Climatic Change 59: 5–31.

Braun, L., M. Weber and M. Schulz. 2000. Consequences of climate change for runoff from Alpine regions. Annals of Glaciology 31: 19–25.

Braun, L.N. and M. Weber. 2007. Gletscher: Wasserkreislauf und Wasserspende. pp. 48–55. *In*: Bundesministerium für Umwelt, Naturschutz und Reaktorsicherheit (Hg.): Klimawandel in den Alpen: Fakten—Folgen—Anpassung, Berlin.

Breiling, M., P. Charamza and O. Skage. 1997. Klimasensibilität österreichischer Bezirke mit besonderer Berücksichtigung des Wintertourismus, Rapport 1, Alnarp.

Brush, S. 1976. Introduction to the proceedings of the symposium on cultural adaptation of mountain ecosystems. Human Ecology 4: 125–134.

Carey, M., A. French and E. O'Brien. 2012. Unintended effects of technology on climate change adaptation: an historical analysis of water conflicts below Andean Glaciers. Journal of Historical Geography 38: 181–191.

Chen, Y.N., K. Takeuchi, C.C. Xu, Y.P. Chen and Z.X. Xu. 2006. Regional climate change and its effects on river runoff in the Tarim basin, China. Hydrological Process 20: 2207–2216.

Ellenrieder, T., L.N. Braun and M. Weber. 2004. Reconstruction of mass balance and run-off of Vernagtferner from 1895 to 1915. Zeitschrift für Gletscherkunde und Glazialgeologie 38: 165–178.

Forman, S.H. 1978. The future value of the 'verticality' concept: implications and possible applications in the Andes. Actes du XLII Congres International des Americanistes, Paris 4: 234–256.

Forrest, J.L., E. Wikramanayake, R. Shrestha, G. Areendran, K. Gyeltshen, A. Maheshwari, S. Mazumdar, R. Naidoo, G. Jung Thapa and K. Thapa. 2012. Conservation and climate change: assessing the vulnerability of snow leopard habitat to treeline shift in the Himalaya. Biological Conservation 150: 129–135.

Giorgi, F. and R. Francisco. 2000. Evaluating uncertainties in the prediction of regional climate change. Geophysical Research Letters 27: 1295–1298.

Godina, R., P. Lalk, P. Lorenz, G. Müller and V. Weilguni. 2006. Das Hochwasser in Österreich vom 21. bis 25. August 2005 – Beschreibung der hydrologischen Situation. Wien.

Goudie, A.S. and D.S.G. Thomas. 2000. The Encyclopedic Dictionary of Physical Geography. Blackwell, Oxford.

Grabherr, G., A. Björnsen, J.P. Gurung, W. Dedieu, W. Haeberli, D. Hohenwallner, A.F. Lotter, L. Nagy, H. Pauli and R. Psenner. 2005. Long-term environmental observations in mountain biosphere reserves: Recommendations from the EU GLOCHAMORE project. Mountain Research and Development 25: 376–382.

Grabherr, G., H. Pauli, D. Hohenwallner, M. Gottfried, C. Klettner and K. Reiter. 2004. GLORIA (the Global Observation Research Initiative in Alpine Environments): alpine vegetation and climate change. pp. 109–114. In: C. Lee and T. Schaaf (eds.). Global Change Research in Mountain Biosphere Reserves—Proceedings of the International Launching. Geneva.

Guillet, D., R.A. Godoy, C.E. Guksch, J. Kawakita, T.F. Love, M. Matter and B.S. Orlove. 1983. Toward a cultural ecology of mountains: the Central Andes and the Himalayas compared. Current Anthropology 24: 561–574.

Habersack, H. and A. Moser (eds.). 2003. Ereignisdokumentation Hochwasser August 2002, Plattform Hochwasser Hochwasser 02/2003, ZENAR—Zentrum für Naturgefahren und Risikomanagement, Universität für Bodenkultur (BOKU), Wien.

Haeberli, W. and M. Beniston. 1998. Climate change and its impacts on glaciers and permafrost in the Alps. Ambio 27: 258–265.

Haeberli, W. 1999. Eisschwund und Naturgefahren im Hochgebirge. pp. 8–11. In: Fachtagung Naturgefahren: Gletscherund Permafrost. CENAT/VAW-ETH Zürich.

Hagg, W., L.N. Braun, M. Kuhn and T.I. Nesgaard. 2007. Modelling of hydrological response to climate change in glacierized Central Asian catchments. Journal of Hydrology 33: 40–53.

Hansen, J., M. Sato, P. Kharechal, D. Beerling, R. Berner, V. Masson-Delmotte, M. Pagani, M. Raymo, D.L. Royer and J.C. Zachos. 2008. Target atmospheric CO2: where should humans aim? The Open Atmospheric Science Journal 2: 217–231.

Intergovernmental Panel on Climate Change (IPCC). 2008. Technical Paper on Climate Change and Water, on: <http://www.ipcc.ch/pdf/technicalpapers/climate-change-water-en.pdf/>, (30.09.2008).

IPCC AR4. 2007. Contribution of Working Groups I, II and III to the Fourth Assessment Report of the Intergovernmental Panel on Climate Change. pp. 104. Core Writing Team, R.K. Pachauri and A. Reisinger (eds.). IPCC, Geneva, Switzerland.

Jäger, C.C., J. Krause, A. Haas, R. Klein and K. Hasselmann. 2008. A method for computing the fraction of attributable risk related to climate damages. Risk Analysis 28: 815–823.

Kääb, A., M. Charle, B. Raup and C. Schneider. 2007. Climate change impacts on mountain glaciers and permafrost. Global and Planetary Change 56: vi–ix.

Lahn Gärtner, E. 2007. Untersuchungen zur längerfristigen Entwicklung und räumlichen Differenzierung der Landwirtschaft in Österreich mit besonderer Berücksichtigung des

Alpenraumes : Grundzüge regionaler Veränderungen der land- und forstwirtschaftlichen Bevölkerung und Betriebe sowie der Rinderhaltung, Innsbruck.

Lambrecht, A. and M. Kuhn. 2007. Glacier changes in the Austrian Alps during the last three decades, derived from the new Austrian glacier inventory. Annals of Glaciology 46: 177–184.

Li, X., G. Cheng, H. Jin, S. Kang, T. Che, R. Jin, L. Wu and Z. Nan. 2008. Cryospheric change in China. Global and Planetary Change 62: 210–218.

Liniger, H., R. Weingartner and M. Grosjean. 1998. Mountains of the World: Water Towers for the 21st Century, Centre for Development and Environment, Bern.

Löffler, J. 2004. Degradation of high mountain ecosystems in northern Europe. Journal of Mountain Science 2: 97–115.

Maisch, M., A. Wipf, B. Denneler, J. Battaglia and C. Benz. 1999. Die Gletscher der Schweizer Alpen. Gletscherhochstand 1850—Aktuelle Vergletscherung—Gletscherschwund-Szenarien. Schlussbericht NFP31, Zürich.

Matulla, C., H. Formayer, P. Haas and H. Kromp-Kolb. 2004. Mögliche Klimatrends in Österreich in der ersten Hälfte des 21. Jahrhunderts. Österreichische Wasser- und Abfallwirtschaft 6: 1–9.

Messerli, B. and J.D. Ives (eds.). Mountains of the World: A Global Priority. Parthenon, Carnforth.

Messerli, B., D. Viviroli and R. Weingartner. 2004. Mountains of the world: vulnerable water towers for the 21st century. Ambio 13: 29–34.

Mutke, J. and W. Barthlott. 2005. Patterns of vascular plant diversity at continental to global scales. *In*: I. Friis and H. Balslev (eds.). Plant Diversity and Complexity patterns—Local, Regional and Global Dimensions. The Royal Danish Academy of Sciences and Letters, Copenhagen, Biologiske Skrifter 55: 521–537.

Netting, R.McC. 1972. Of men and meadows: strategies of alpine land use. Anthropological Quarterly 45: 132–144.

Nogués-Bravo, D., M.B. Araújo, M.P. Errea and J.P. Martínez-Rica. 2006. Exposure of global mountain systems to climate warming during the 21st century. Global Environmental Change 17: 420–428.

OECD. 2006. Adaptation to Climate Change: Key Terms, Paris.

Parry, M.L., O.F. Canziani, J.P. Palutikof, P.J. van der Linden and C.E. Hanson (eds.). 2007. Contribution of Working Group II to the Fourth Assessment Report of the Intergovernmental Panel on Climate Change, Cambridge and New York.

Rhoades, R.E. and S.I. Thompson. 1975. Adaptive strategies in alpine environments: beyond ecological particularism. American Ethnologist 2: 535–51.

Rupa, K.K., A.K. Sahai, K.K. Krishna, S.K. Patwardhan, P.K. Mishra, J.V. Revadkar, K. Kamala and G.B. Pant. 2006. High resolution climate change scenario for India for the 21st century. Current Science 90: 334–345.

Salick, J., A. Byg, A. Amend, B. Gunn, W. Law and H. Schmidt. 2006. Tibetan Medicine Plurality. Economic Botany 60: 227–253.

Schär, C., P.L. Vidale, D. Lüthi, C. Frei, C. Häberli, M.A. Liniger and C. Appenzeller. 2004. The role of increasing temperature variability in European summer heatwaves. Nature 427: 332–336.

Schmidt, P., R. Steiger and A. Matzarakis. 2012. Artificial snowmaking possibilities and climate change based on regional climate modeling in the Southern Black Forest. Meteorologische Zeitschrift 21: 167–172.

Siegrist, D. 1998. Europa im Wandel—die Alpen im Wandel. pp. 11–22. *In*: Internationale Alpenschutzkommission (CIPRA) (eds.). Wer Geld hat und ein Auto…. Nahversorgung—Öffentlicher Verkehr—Öffentliche Dienste für die Bevölkerung im Alpenraum—wie lange noch? Wien.

Singh, P., U.K. Haritashiya, N. Kumar and Y. Singh. 2006. Hydrological characteristics of the Gangotri Glacier, Central Himalayas, India. Journal of hydrology 327: 55–67.

Smeral, E. 2000. Wirtschaftliche Rolle des Tourismus in den Alpen. Maßnahmen zur Verbesserung der Wettbewerbsposition. pp. 49–60. *In*: A. Sartoris (ed.). Alpentourismus, CIPRA, Schaan.

Smit, B., I. Burton, R. Klein and J. Wandel. 2000. An anatomy of adaptation to climate change and variability. Climatic Change 45: 223–251.

Smit, B. and J. Wandel. 2006. Adaptation, adaptive capacity and vulnerability. Global Environmental Change 16: 282–292.

Solomon, S., G.K. Plattner, R. Knutti and P. Friedlingsstein. 2009. Irreversible climate change due to carbon dioxide emissions. Proceedings of the National Academy of Sciences of the United States of America 106: 1704–1709.

Stadel, C. 1991. Altitudinal belts in the Tropical Andes: their ecology and human utilization. Yearbook. Conference of Latin Americanist Geographers 17/18: 45–60.

Steiger, R. 2008. The impact of poor winter seasons on ski tourism and the role of snowmaking as an adaptation strategy. pp. 95–102. *In*: A. Borsdorf, J. Stötter and E. Veulliet (eds.). Managing Alpine Future. Proceedings of the Innsbruck Conference 15–17 October 2007 (IGF-Forschungsberichte 3), Wien.

Stoffel, M., I. Lièvre, D. Conus, M.A. Grichting, H. Raetzo, H.W. Gärtner and M. Monbaron. 2005. 400 years of debris flow activity and triggering weather conditions: Ritigraben, Valais, Switzerland. Arctic, Antarctic and Alpine Research 37: 387–395.

Stott, P.A., D.A. Stone and M.R. Allen. 2004. Human contribution to the European heatwave of 2003. Nature 432: 610–614.

Stötter, J., M. Maukisch, J. Simstich and K. Belitz. 1996. Auswirkungen des zeitlich/räumlichen Wandels derPermafrostverteilung im Suldental (Ortlergebiet) auf das Gefährdungspotential durch Erosionsprozesse imLockermaterial pp. 447–457. Interpraevent, 1, Klagenfurt.

Van Loo, M., W. Hintsteiner and H. Hasenauer. 2012. Douglasie: Zukunftpassiertjetzt, begannaberschongestern. Forstzeitung 10: 20–21.

Viviroli, D., H.H. Dürr, B. Messerli, M. Meybeck and R. Weingartner. 2007. Mountains of the world, water towers for humanity: Typology, mapping, and global significance. Water Resources Research 43: 1–13.

Viviroli, D. and R. Weingartner. 2004. The hydrological significance of mountains: from regional to global scale. Hydrology and Earth System Sciences 8: 1017–1030.

Webster, S. 1971. An indigenous Quechua community in exploitation of multiple ecological zones. Actas y memorias del XXXIX Congreso Internacional de Americanistas 3: 174–83. Lima.

Werge, R.W. 1979. The agricultural strategy of rural households in three ecological zones of the Central Andes. International Potato Center, Social Science Unit, Working Paper 1979-4.

Winiger, M., M. Gumpert and H. Yamout. 2005. Karakoram-Hindu Kush-Western Himalaya: assessing high-altitude water resources. Hydrological Processes 19: 2329–2338.

Xu, J.C., R.E. Grumbine, A. Shrestha, M. Eriksson, X. Yang, Y. Wang and A. Wilkes. 2009. Las Himalaya Se Derriten: Efectos En Cascada Del Cambio Climático Sobre El Agua, La Biodiversidad y Los Medios De Vida.Conservation Biology 23: 520–530.

Xu, J.C., Y. Yang, Z.Q. Li, N. Tashi, R. Sharma and J. Fang. 2008. Understanding land use, livelihoods, and health transitions among Tibetan Nomads: a case from Gangga Township, Dingri County, Tibetan Autonomous Region of China. EcoHealth 5: 104–114.

Xu, Z.X., Y.N. Chen and J.Y. Li. 2004. Impacts of climate change on water resources in the Tarim River basin. Water Resources Management 18: 439–458.

Yao, T., Y. Wang, S. Liu, J. Pu and Y. Shen. 2004. Recent glacial retreat in high Asia in China and its impact on water resource in northwest China. Science in China Series D Earth Sciences 47: 1065–1075.

Zemp, M., W. Haeberli, M. Hoelzle and F. Paul. 2006. Alpine glaciers to disappear within decades? Geophysical Research Letters 33.

Adaptation Frameworks for Climate Change—Eloquent to Himalayan Ecosystems

P.K. Joshi,* Kamna Sachdeva and A.K. Joshi

INTRODUCTION

Global climate change has substantial impacts on the Himalayan ecosystems and its resources (Barnett et al. 2005). For centuries, the region is also known to face formidable environmental and livelihood challenges in its effort to safeguard its valuable natural resources including land, water and forests (Grabherr et al. 1994, 1995, Zobel and Singh 1997, Pauli et al. 2003, Kazakis et al. 2006). Furthermore, the region is highly fragile and has been subject to variety of natural hazards like earthquakes, landslides, could bursts, floods, and flash floods, forest fires, etc. (Ives 1987, Dortch et al. 2009). These all formidable challenges are enhanced in light of climate change as per the projections done so far. There is evidence of prominent increases in the intensity and/or frequency of many extreme weather events such as intense rainfall, Glacier Lake Outburst Floods (GLOFs), snow avalanches, tropical cyclones, prolonged dry spells, thunderstorms, and severe dust storms in the region and adjoining landscapes (Cruz et al. 2007, Nandargi and Dhar 2012). Many of times, impacts of such natural disasters results multiple stresses ranging from ambient environment to socio-economic configuration. Such impacts are hunger and susceptibility health disorders,

Department of Natural Resources, TERI University, New Delhi 110 070, India.
* Corresponding author: pkjoshi27@hotmail.com

loss of resources, assets, property and livelihoods, affecting human survival and well-being and resulting imbalance in the daily life of local community.

The impacts of climate change are already being experienced world-wide. As per the Intergovernmental Panel on Climate Change (IPCC), climate change is unequivocal and will affect everyone. It is expected to have a disproportionate effect on communities living in poverty in developing countries and ecosystems or regions more sensitive to precipitation and temperature changes (MEA 2005). Nature of the Himalayan region, sensitive and difficult landscapes with sparsely poor infrastructure makes is more vulnerable. It is a matter of great concern as the region has more snow, ice and glaciers than any other region in the world outside the poles (Barnett et al. 2005). The Himalayan glaciers are source of nine major river (Yangtze, Ganges, Yamuna, Brahmaputra, Irrawady, Mekong, Indus, Tarim) basins in South Asia, of which only the Gangetic basin supports more that 500 million people. Over 200 million people live in the mountains, valleys, and hills of the Hind Kush-Himalayan (HKH) region and around 1.3 billion people, a fifth of the world's population live in the basins downstream. All told, an estimated 3 billion people benefit from the water and other services (Immerzeel et al. 2010). Recognizing the importance of mountains as ecosystems of decisive significance and essential for sustainable development, many national and international efforts have been started. Despite all these efforts, the Himalayan ecosystems are facing enormous pressures from various climate change drivers and natural disasters.

As we see climate has already been changing and its adverse impacts are evident, we cannot stop this change in the short order. Hence adaption is the only way to deal the momentum of physical and social processes of climate change. As we know the pace of climate change determines the severity of its impacts hence responses like mitigation and adaptation are necessary for slowing the human-induced climate change and stresses. The developing world has more problems and challenges hence are more vulnerable. To know who and what is vulnerable and to understand locally available adaptation strategies, it is important to understand the dynamics (socio-economic) of any region for that matter specific framework for adaption can be developed by modifying a few general well accepted lessons like: (i) recognize the problem and respond early, (ii) mainstreaming the adaption with development goals, (iii) knowledge and awareness, (iv) use of locally oriented strategies, (v) identification of co-benefits like protection of natural resources, health benefits and others, and (vi) involvement of people who are at risk. All the adaptation frameworks involve these lessons but modifications as per the region can be done. Definitely for the Himalayan ecosystem, more careful strategies are needed because combined pressures of climate and non-climatic drivers makes it more vulnerable.

This chapter reviews available climate change adaptation frameworks and identifies their applicability for the Himalayan ecosystems. The authors expect this will provide a comprehensive set of information to the decision makers for developing region specific adaptive policy and researchers to analyze the importance of these while recommending more models relevant to adaptive capacity to climate change.

Vulnerable Himalayan Ecosystem

Mountains occupy a quarter of the global land surface area (about 25 million km^2) and homes 12% of the world's population (GTOS 2008). In these regions, livelihoods are highly dependent on climate-sensitive resources. For example a vast area of the HKH region consists of rangelands, where 30 million people depend on livestock for their livelihoods. Similarly, in different zones of Indian Himalayas, communities are dependent on forests, alpine grasslands and also snow covered landscapes. A growing number of studies from these regions suggest that in many areas more than one-third of total cultivated land has been abandoned (Khanal and Watanabe 2006) due to such factors. Principally the agriculture is in these regions is categorized as subsidence, of which up to 90% is rain-fed and accounts for 70% of regional employment and livelihoods. These communities already struggle to cope with the current state of water availability, soil fertility and production rates. The projected extreme weather events and climate variability would further enhance the vulnerability in these regions. Thus there is a direct linkage between climate variables ecosystem services and human beings (Fig. 6.1). Himalayan regions, including the majority of developing countries are in tropical and sub-tropical regions, areas predicted to be seriously affected by the impacts of climate change. The Himalayas are expected to warm by 2.7°C/ 3.8°C according to B1/A1F1 scenarios (Nogues-Bravo et al. 2006). This is compounded by the fact that developing countries are often less able to cope with adverse climate impacts because of poor economic conditions and is exacerbated further by the impacts of environmental change (Grabherr et al. 1994, Körner 1999).

In the recent past decades, many natural disaster-related deaths occurred in the developing countries and 90% of all natural disasters were climate, weather, geology and water related (Reddy and Assenza 2009). Indeed, continuing land pressures, rapid urban development and settlement of the impoverished in exposed low-lying areas greatly influence flood risks, health and livelihoods. More than 40% of the world's floods takes place in Asia, and have affected near a billion people (in 1999–2008) causing an estimated 20–25% of all deaths associated with natural disasters (Kaltenbom et al. 2010). In 2009, more than 56 million people were severely impacted by floods and over one million people by the smaller, but often dangerous

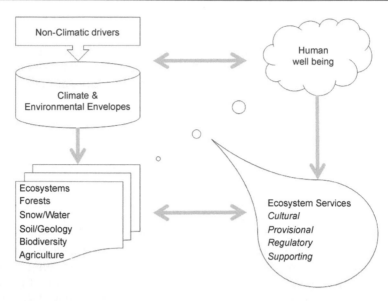

Figure 6.1. Linkage between climate change drivers, ecosystems, and human well being (adapted from MEA 2003).

flash floods (Fig. 6.2). Climate change is having direct impact on human life and livelihoods, affecting different sectors, including water, agriculture and food security and forest. Such affects would exacerbate all environmental and developmental problems in Himalaya.

Response Measures for the Management of Climate Change in the Himalayan Region

Review of current adaptation strategies in mountain region suggests several important lessons; the foremost is that all adaptation and development initiatives must be considered a continuum. In fact the adaptation and mitigation is a continuous process and should be dynamic in identification, adaptation and implementation phases. At one end are developments activities such as providing basic amenities like drinking water, energy and food security, while at the other end are the efforts required to reduce vulnerability to climate change and disaster risk and to build socio-environmental resilience. In the mountainous regions, livelihood which is the most important and key factor defining the interaction with the natural environment needs diversification and this one of a key adaptation strategies requires policy and institutional support. Likewise social networks, self-help groups and local institutions play a vital role in developing and realizing the adaptive capacity. Furthermore, socio-cultural norms and environmental

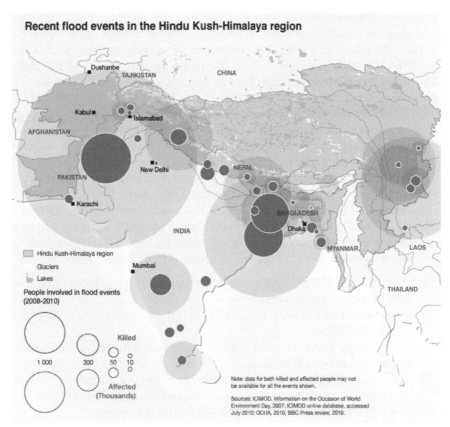

Figure 6.2 Recent flood events in Hindu Kush-Himalaya (Source: Kaltenborn et al. 2010).

Color image of this figure appears in the color plate section at the end of the book.

settings affect adaptive behavior of the community, but can also shift over time in response to variety and order of stresses including natural and climate induced disasters. If governance and planning including the community based initiatives takes climate change and resulting stresses into the account, infrastructure development can contribute to enhancing resource security and disaster risk reduction. Also policy at different levels and impletion at different scales plays a paramount role in understanding, adapting and mitigating the effects of climate change induced risk and disasters. Policy at international and national levels often has strong implications for local conditions, but higher level policies are often out of touch with local concerns and realities (ICIMOD 2009a,b).

Planned adaptation includes the angle of human systems hence adaptation decisions becomes the central idea of the approach, where adaptation is planned for reducing the impact on humans. This approach is very well suited for the people living in high altitude. The day to day hardships in these regions are very high, any adaptive measure which can reduce the effort of their work and simultaneously making then adapt for the climate variability can be the right approach of the planning. The framework for adaptation for the Himalayan region cannot be very different from other approaches but at the time of planning this above mentioned aspect has to be incorporated. The preliminary social surveys are very important to determine important parameters which need to be incorporated at the time of successful adaptation framework. The elements of efficiency, effectiveness, equity and legitimacy are important parameters for assessing the adaptation needs and successful implementation in any given region (Adger et al. 2005). The final and most important facet of climate risk adaptation is that it has to be continuous process. As rate of climate change is not constant and additional adaptive measure has to be planned in the framework. Taking all these factors in consideration we propose 5 steps pre-framework for effective adaptation:

1. Assesses the climate risk in the given region (on the basis of climatic and non-climatic drivers).
2. Assessment of intensity of risks occurred due to climate change and its prioritization.
3. Knowledge enhancement for the management of particular risk (prioritize in the above step).
4. Assessment of future climate- related risk and combined reactive and proactive steps of adaptation to optimize the cost of adaptation.
5. Continue the assessment of risks for future planning.

Adaptation Frameworks for Climate Change

The Hindu Kush region is considered to be most vulnerable to climate change impacts and it is extended from Afghanistan to Myanmar (IPCC 2007). This region has great pressure from non-climate drivers because of high population density and due to the presence of the poorest countries in this region. The climate variability in this region force people of this area to adapt with the available resources but rapid and frequent changes needs continuous evolution of adaption measure. One type adaptation framework cannot fit in the dynamic Hindu Kush region with a lot of climate and non-climatic drivers. In this chapter we discuss major frameworks developed by different groups to tackle this problem. The important frameworks are listed in Table 6.1 given below with all the essential input parameters. Essentially all the frameworks are classified in two broad categories: Impact and adaptation frameworks (Fig. 6.3). Different groups have used these

Table 6.1. Impact and adaptation centered framework.

Assumptions	Variables and parameters
Impact centered Frameworks (ICFs)	
Models measure impacts of Climate change; modeling impacts parameterizing adaptation	Impacts net of adaptation, not gross impacts Normative
Adaptation is incorporated in an unchanging equation assumed to take adaptation into account	This is not modeling adaptation
Adaptation cannot be varied in the model	Parameter set or assumed at static level
The amount of adaptation (or net of achievement) is assumed and is not verified or nor does it have an empirical basis	Output is based on the inclusion of or exclusion of adaptation, but adaptation itself is not being modeled
Adaptation centered Frameworks (ACFs)	
Allow for the variation of adaptation options or different levels of adaptation	Allows for how much can be accomplished through adaptation; given the impacts, how much adaptation could or would occur
Adaptation can be manipulated, assessed and evaluated	Potential to demonstrate the strengths/ weakness of adapting to climate change
ACFs demonstrate the ability of adaptation to reduce climate change impacts	ACFs are much more satisfactory than ICMs and they represent a more promising direction for future development

Example, in Rosenzweig and Parry (1994), the assumptions included no adaptation, 50% adaptation and full adaptation.

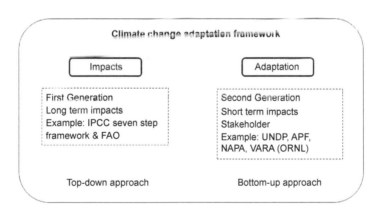

Figure 6.3. Introduction of climate change adaptation frameworks.

frameworks according to study areas under concern. For the Himalayan region both the frameworks can be applied separately or collectively.

Impact framework, adopted mainly for livelihood management, involves learning of managing present risks and developing resilience capacity for future risks. FAO has used this approach for food security (FAO 2007), has the following elements:

  i.   Legal or institutional
  ii. Policy and planning
  iii. Ecosystem
  iv. Livelihood
  v.   Integrated farming system
  vi. Low carbon intensive technologies

FAO has prescribed steps for designing livelihood adaptation towards climate change based on the elements given above (Fig. 6.4). This is an inclusive approach and tries to involve all the sectors and scales while designing the adaptation options for any particular region. FAO has adopted this approach in Nepal (LSP working paper 7).

The adaptation framework was a widely adopted framework and has application for a wide range of problems occurring due to climate problems in hilly regions. The case study conducted by ICIMOD in India and Nepal is the classic study case of adaptation framework; this examines the governance of embankments and its implications on adapting capacity of local people for floods at two sites in Assam and in basin of Koshi River in Nepal (see Box 6.1.). The adaptation framework can include the following approaches to propose adaption measures for a particular region, like participatory rural appraisal method, transect walks, focus group discussions, resource mapping, etc. These methods are important to determine the communities' response towards the change and willingness towards adaptation measures. The study reveals that local communities have developed indigenous coping strategies (see Box 6.2) for floods and other hazards; hence importance of traditional and local knowledge in the

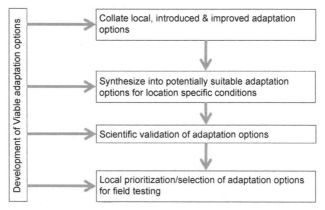

Figure 6.4. FAO framework of designing adaptation options for livelihood management in the climate change scenario.

BOX 6.1. Details of Two River Basins.

The Brahmaputra originating from the great glacier mass of Chema-Yung-Dung in the Kailash range of southern Tibet (elevation of 5,300 m), it traverses China (1,625 km) and India (918 km) before passing 337 km through Bangladesh and emptying into the Bay of Bengal. The Koshi River originates in the Himalayan range at an altitude of about 7,000 m in the Tibet Autonomous Region of China and 8,848 m (Mount Everest) in Nepal. The Koshi is a complex river system causing huge flooding, acute bank erosion, and migration of the channel every year. The sediment load carried by the river is extremely high; hence, its tendency to braid and shift course frequently. It shifted course by about 160 km between 1723 and 1948 (Mishra 2008).

BOX 6.2. Local knowledge of indigenous communities.

Examples: First—Some people living near the bank of river Jiadhal can predict flow of the river and then accordingly keep away from vulnerable places. Second—In Auniati and Na-Kalita villages on the Jiadhal River, families have moved near parts of the embankments that they see as strong or that have been recently renovated. Some of them live on the riverside of embankments to be able to cultivate fertile lands along the riverbanks. Although such a decision exposes them more to floods, they get a good harvest when a major flood does not happen. This is an example of risk trade-off where the risk of damage to households and property can be mitigated to some extent by the benefit of productive farming and by using the embankment as shelter if needed (source: ICIMOD).

---

field of adaption has clearly emerged in this case. The findings of this study validates the need of adaptation framework as it indicates that structural measures needs to be balance with other measures such as flood forecasting, disaster preparedness, watershed management and flood plain regulation.

The given classification between ICFs and ACFs is a preliminary categorization system. Such a categorization does not provide details about the applicability of these models. Therefore, the models can be further differentiated by subcategories. The most important among them are Integrated Assessment models (IAMs) which link social and economic aspects very beautifully with the scientific issues. IAMs haves specific use to assess policies for interventions and controls towards climate change and mitigation (Weyant et al. 1996). This leads to creation of interdisciplinary frameworks addressingissues and problems related climate change. These also determine influential forces that make different sectors sensitive to such challenges and to quantify the resultant environmental, non-environmental, social and economic problems by ranking climate change control benefits and detriments in developed and developing countries (IPCC 2001). One of the recent framework of UNEP is based on this ICM approach has wide application in Himalayan region. The Ecosystem-Based Adaptation Decision support Framework (EBA-DSF) of is described as:

This frame work is based on the principles to practice approach. This has four iterative steps which have feedback links from each other. The steps are setting of adaptive context, selection of appropriate option, design for change and implementation. The goal of framework is to feed in the national plans of the country and other adaptation actions. In the program it is expected that ecosystem management practices can help vulnerable communities. It has emphasis on least developed countries to help them to move towards building resilience for, restoration of, degraded or vulnerable ecosystems.

Looking Forward

Adopting a secure pathway towards adaptation is a challenging and complex question. This could be solved in light of available options, collateral datasets, investments and effects, interest of stakeholders and similar indicators. Such requisites make adaptation a science to identify key and fundamental indicators in the process of decision making and policy formulation. Adaptation is the process where people and communities come together to manage impacts of climate change. Adaptation is a rapid evolving process and development of one specific framework is difficult. The adaption framework can take direction from traditional approaches and other available solutions but it has to be region specific, community specific and above all time specific. The appropriate adaptation policy framework need to take care of marginalizing and has to suggest measures to lead to security. After studying different cases studies based on the above mentioned frameworks various priority areas has been identified for the hill regions (Fig. 6.5), the priority areas have some important subjects such as traditional knowledge of the region and ecosystem based approaches, these two subjects are important to have appropriate region specific adaption policy.

Adaptation, mitigation and development in light of any change and disaster should be seen in continuous band for any population and community. These activities should contribute to the communities' adaptive capacity and resilience. Livelihood diversification for poverty alleviation would require significant institutional and policy support. Such initiatives and strategies must be sensitive to the cultural contexts, norms and differences. The climate change adaptations approaches are now widely discussed even in the international accords emphasis on cross sectorial, cross boundary and trans-disciplinary approaches. The hilly region where most appropriate approaches are ecosystem based need to have better inventory of impacts on biodiversity and human well-being including health and its relationship between poverty and other development parameters.

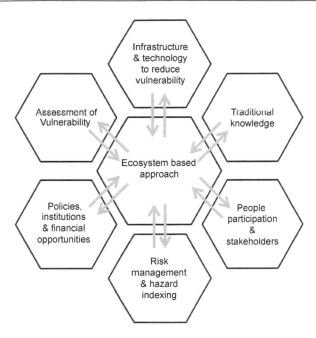

Figure 6.5. Priority areas of climate change adaption in hilly region.

References

Neil Adger, W.N., N.W. Arnell and E.L. Tompkins. 2005. Successful adaptation to climate change across scales. Global Environmental Change 15(2): 77–86.

Barnett, T.P., J.C. Adam and D.P. Lettenmaier. 2005. Potential impacts of a warming climate on water availability in snow dominated regions. Nature 438: 303–309.

Cruz, R.V., H. Harasawa, M. Lal and S. Wu. 2007. 'Asia.' pp. 470–506. *In*: M.L. Parry, O.F. Canziani, J.P. Palutikof, P.J. Van Der Linden and C.E. Hanson (eds.). Climate change 2007: Impacts, Adaptation, and Vulnerability—Contribution of Working Group II to the Fourth Assessment Report of the Intergovernmental Panel on Climate Change. Cambridge University Press, Cambridge, UK.

Dortch, J.M., L.A. Owen, W.C. Haneberg, M.W. Caffee, C. Dietsch and U. Kamp. 2009. Nature and timing of large landslides in the Himalaya and Transhimalaya of northern India. Quaternary Science Reviews 28: 1037–1054.

FAO. 2007. Adaptation to climate change in agriculture, forestry and fisheries: Perspective, framework and priorities. Report of the FAO Interdepartmental Working Group on Climate Change. Rome.

Grabherr, G., M. Gottfried and H. Pauli. 1994. Climate effects on mountain plants. Nature 369: 448.

Grabherr, G., M. Gottfried, A. Gruber and H. Pauli. 1995. Patterns and current changes in alpine plant diversity. pp. 167–181. *In*: F.S. Chapin III and C. Körner (eds.). Arctic and Alpine Biodiversity: Patterns, Causes and Ecosystem Consequences, Ecological Studies, 112. Springer Verlag, Berlin.

GTOS. 2008. Why focus on mountain regions? www.fao.org/gtos/tems/mod_mou.jsp accessed on 03.09.2009 (accessed 27 September 2011).

ICIMOD. 2009a. Local responses to too much and too little water in the greater Himalayan region. Kathmandu, Nepal: ICIMOD.

ICIMOD. 2009b. The changing Himalayas: ICIMOD.

Immerzeel, W.W., L.P.H. van Beck and M.F.P. Bierkens. 2010. Climate change will affect the Asian water towers. Science 328: 1382–1385.

IPCC. 2001. The Third Assessment Report: Climate Change 2001, Synthesis Report. Cambridge University Press, Cambridge, UK.

IPCC. 2007. The Fourth Assessment Report: Climate Change 2007, Synthesis Report. Cambridge University Press, Cambridge, UK.

Ives, J.D. 1987. The theory of Himalayan environmental degradation: its validity and application challenged by recent research. Mountain Res. Dev. 7: 189–199.

Kaltenborn, B.P., C. Nellemann and I.I. Vistnes. 2010. High mountain glaciers and climate change – Challenges to human livelihoods and adaptation. United Nations Environment Programme, GRID-Arendal, www.grida.no.

Kazakis G., Ghosn D., Vogiatzakis I.N., Papanastasis V.P. 2006. Vascular plant diversity and climate change in the alpine zone of Lefka Ori, Crete. Biodiversity Conservation 16, 1603-1615.

Khanal, N.R. and T. Watanabe. 2006. Abandonment of agricultural land and its consequences: a case study in the Sikles area, Gandaki basin, Nepal Himalaya. Mountain Research and Development 26(1): 32–40.

Körner, C. 1999. Alpine Plant Life. Springer, Berlin.

Millennium Ecosystem Assessment (MEA). 2005. Ecosystems and Human Well-being: Synthesis, Millennium Ecosystem Assessment Report. World Resources Institute, Island Press, Washington, DC.

Millennium Ecosystem Assessment (MEA). 2003. People and Ecosystems: A Framework for Assessment and Action. Island Press, Washington, DC.

Mishra, D.K. 2008. Bihar floods: the inevitable has happened. Economic Political Weekly 43: 8–12.

Nandargi, S. and O.N. Dhar. 2012. Extreme Rainstorm Events over the Northwest Himalayas during 1875–2010. J. Hydrometeor 13: 1383–1388.

Nogués-Bravo, D., M. Araújo, M.P. Errea and J.P. Martinez-Rica. 2006. Exposure of global mountain systems to climate warming during the 21st century. Global Environmental Change – Human and Policy Dimensions 17: 420–428.

Pauli, H., M. Gottfried, T. Dirnböck, S. Dullinger and G. Grabherr. 2003. Assessing the long-term dynamics of endemic plants at summit habitats. In: L. Nagy, G. Grabherr, C. Körner and D.B.A. Thompson (eds.). Alpine Biodiversity in Europe - A Europe-wide Assessment of Biological Richness and Change. Ecological Studies 195–207.

Weyant, J., O. Davidson, H. Dowlatabadi, J. Edmonds, M. Grubb, R. Richels, J. Rotmans, P. Shukla, W. Cline, S. Fankhauser and R. Tol. 1996. Integrated assessment of climate change: an overview and comparison of approaches and results. In: J. Bruce et al. (eds.). Climate Change 1995—Economic and Social Dimensions of Climate Change. Contribution of Working Group III to the Second Assessment Report of the Intergovernmental Panel on Climate Change (IPCC). Cambridge University Press, Cambridge.

Zobel, D.B. and S.P. Singh. 1997. Forests of the Himalaya: their contribution to ecological generalizations. Bio. Science 47: 735–745.

SECTION 4

GOVERNANCE AND LEGAL ISSUES IN MOUNTAINS

7

Governance and Laws of Mountains

Flavia Witkowski Frangetto#

INTRODUCTION

To regulate hills and mountains is probably more difficult than to climb a high mountain. This chapter has four contributions related to national and international laws with some examples from Brazil.

Both developing and developed nations face the challenge of deteriorating ecosystems in the mountains. In case of developing countries such as Brazil, although there are no mountains per se as compared to the Himalayas or other high mountains - there are just hills (with a height lower than the mountains) and these hills are also facing similar issues as mountains such as growing population and impacts and pressure of the communities to use the limited land.

It would be better to prohibit the occupation of such mountains by people in fragile areas where an equilibrium between nature and human beings deserves to be a priority. In the case of Brazil, Brazilian Forestry Code (the new Law n° 12.651, which came into force on the May 25, 2012) does not permit any construction in mountains at places higher than 100 meters. It is very important to implement this rule, to avoid the advance of risks typically associated with indiscriminate use and inhabitation of mountains.

Brazilian Environmental Law Institute, Environmental Superior Council of the Federation of Industries of the State of São Paulo (FIESP).
Corresponding author: flavia@frangetto.com

It is useful to analyze situations in which the use of the mountains is in risk and trying to use these lessons learnt as a warning to prevent further damage to mountains and its ecosystem, or to solve existing conflicts in the particular hill scenarios.

The focus of this chapter is on governance and laws in mountains—to deal with the real situation of mountains particularly in Brazil, as an example on how to obey limits of nature in mountains, instead of destroying its ecosystems. The chapter has four sections contributed independently by various authors.

The first section describes the contribution of International and National Law in the field of mountains. It shows how regulation is proportional to recognized environmental functions and services of the hills, mounts, mountains and ranges. Although the law has been passed in Brazil regarding Forest Code but it still neglects many requirements proposed by the environmentalists. A very interesting suggestion is to have a global international accord designed specially to protect mountains. However in the absence of such a global accord or treaty, at least some national rules and laws are important, such as the Brazilian provision of 'permanent protection areas' which also include mountains.

The second section of this chapter examines the Alpine Convention as a model for protecting transboundary areas of mountains. The efforts of protection are also analyzed by looking at what international organizations and research networks have been working on in relation to mountains. After this analysis, a socio-environmental perspective is presented which describes the ecological risks and disasters in hillsides and mountainous areas in Brazil as problems faced by some section of population living under vulnerable conditions in the mountain regions.

To illustrate that it is feasible to avoid damage in mountain areas, 'The Precautionary Principle' in Practice in Brazil demonstrates that it is possible to protect ecosystems and population, from both natural incidents (such as landslides) as well as accidents caused by changes in mountain regions.

Contribution of International and National Law

Marcia Dieguez Leuzinger[1] and
Solange Teles da Silva[2]

Mountains, hills, mounts and ranges have always played a major role in the cultures of different human societies. They perform key functions in keeping diverse ecosystems, climate systems and the production and distribution of water in balance. Just like water bodies, mountains also divide the planet to form different ecosystems and climatic conditions (Price et al. 2011). Diverse human cultures also consider many mountains, hills, mountains and ranges as sacred places. In addition to all those reasons, many mountain ranges have also been placed under protection due to their importance as landscapes or the need to limit human settlements and activities.

The legal protection of hills, mounts, mountains and ranges, however, is a challenge. Can legal tools from environmental law and land use regulations provide broad forms of protection for these geological formations, or do we need special tools to grant them specific protection? An analysis of legal protection for hills, mounts, mountains and ranges must consider both the general context for their protection and specific norms from a variety of regulatory fields required to achieve that protection. We will look first at emerging international expressions of protection for mountains, and then

---
[1] Professor of Environmental Law at Uniceub (Brasilia – Brazil), Programa de Mestrado e Doutorado em Direito, SEPN 707/907 Bloco 3 térreo, 70790-075 Brasília – DF/Brazil. Email: marcia.leuzinger@uol.com.br
[2] Professor of Environmental Law at Mackenzie Presbyterian University (São Paulo – Brazil), Programa de Mestrado e Doutorado em Direito Político e Econômico, Rua da Consolação n. 896, prédio 3 subsolo, 01302-000 São Paulo – SP/Brazil.
National Council for Scientific and Technological Development (CNPq).
Email: solange.teles@pq.cnpq.br
[#] Corresponding author: flavia@frangetto.com

examine Brazilian legislation and prospects for protecting or destroying hills, mounts, mountains and ranges arising from the bill of law to amend the Forest Code, soon to be approved by the National Congress.

While there is no cogent (global) international legal instrument to protect mountains, there is growing concern about ecological issues that affect them. That concern intensified in the 1990s, when mountains gained prominence as a global theme (Rudaz 2011).

The emergence of international law on mountains is a relatively recent phenomenon. In the Agenda 21 'soft law' document adopted in Rio de Janeiro at the UN Conference on the Environment and Development (Rio 1992), sustainable development for fragile mountain ecosystems received special attention and management suggestions in Chapter 13. 'Soft law' refers to non-binding norms which are therefore of limited value, since no sanctions can be applied if they are disobeyed. Even so, they open the way to future agreements on policy actions and are also recommendations for the kind of behavior states should adopt in domestic policies.

In Europe, the Convention on the Protection of the Alps and its Protocols (protection of nature, agriculture, forests, soil protection, energy, tourism, transportation and others) make up an international system with its own bodies (Alpine Conference, Permanent Committee, Compliance Committee and Permanent Secretariat) and official observers including the International Commission for the Protection of the Alps, created in 1952 within the International Union for Conservation of Nature (IUCN). Despite the work done and tools developed to protect mountains—for example the Alpine network of protected areas—this international system has no enforceable mechanisms, and instruments for the protection of mountains must still be harmonized domestically.

Cross-border cooperation to protect mountains has brought nature reserves into being in border areas between a few countries, such as the Tumucumaque Mountains National Park (Amapá, Brazil) and the Guiana Amazon Park (French Guiana, a French overseas territory) which, together, make up the world's largest protected area in a tropical forest.

Meanwhile, inside Brazil, mountains can be protected by creating special protection areas, combining conservation units with other sorts of environmental areas. Conservation units are one of the land-management categories envisaged by Law 9985/2000, which created the National System of Nature Conservation Units (SNUC). The units are defined as "territory and its environmental resources, including territorial waters with significant natural features, legally instituted by public authority with clearly defined limits and conservation objectives, under special administrative rules, to which appropriate assurances of protection shall be applied." [Art. 2(I) of Law 9985/2000]. There are two major groupings of conservation units: a) wholly protected units, which allow no direct use of natural resources and whose purpose is to preserve the natural environment with the least possible human intervention, such as national parks, ecological stations, biological reserves, natural monuments and wildlife refuges, and b) sustainable-use units, intended to make the conservation of nature compatible with the sustainable use of some of its natural resources, such as extractive reserves, sustainable-development reserves, environmental protection areas, areas of significant ecological interest, fauna reserves, natural forests and private natural-heritage reserves. Yet it is impossible to turn all areas containing hills, mounts, mountains and ranges into conservation units or some other category of protected areas. What other legal solutions might be possible?

The intrinsic link between protection of mountains and property rights refers us to the social function of property, which provides substance to those rights along with broad and binding administrative limits on their exercise. Law 4771/1965, known as the Forest Code, provides for 'permanent protection areas' (APPs), which place administrative constraints on property rights by setting aside plant cover that cannot be removed or suffer any type of intervention and whose environmental function is to preserve natural resources and to ensure population's well-being. The original purpose of the APPs was to protect the soil and water (Silva et al. 2010), but more recently it was realized that APPs also perform other environmental services, particularly for the conservation of biodiversity. To promote the welfare of residents and protect the soil—which, depending on the slope, has a greater tendency to slide down the sides and tops of hills, mounts, mountains and ranges—the Forest Code provides that hillsides steeper than 45° and the tops of these geological formations are all APPs, as well as the slopes coming down plateaus and mesas. In areas with slopes between 25° and 45°, the only possible activity is extraction of logs.

The importance of these APPs is readily perceived: conservation of biodiversity, enhancement of soil fertility, the maintenance of micro-climates and water production, as well as the safety of human communities threatened by landslides, most of which are caused by the removal of plant cover in APPs on the sides and tops of mounts, hills, mountains and ranges. This danger rises during the summer rainy season, often to catastrophic proportions. Climate change caused by global warming has in fact made rainstorms more intense and frequent. Protecting mountains is thus fundamental as one of the strategies to prevent disasters and help adapt to the impacts of climate change.

The Forest Code, however, has come under the attack from powerful interest groups, particularly representatives of agribusiness. Their heavy campaign to alter the standard of excellence for protection of forests and other forms of plant cover, including ecosystems on hill and mountain tops and on the sides of plateaus and mesas, culminated in the approval in May 2011 by the House of Representatives of Bill 1876/1999, which significantly cuts back APPs and legal reserve areas. Although the Senate improved the wording of the bill (PLC 30/2011) somewhat, the version it approved in December 2011 included various provisions that run against the interests of Brazilian society, which will no longer have access to remedies for environmental damage caused by a variety of players. Since the Senate did introduce some changes to the bill, it was sent back to the House, but then signed by the President to become a law in May 2012.

The specific provisions on hills, mounts, mountains and ranges have kept APPs for slopes steeper than 45°. But for hills and mountain sides between 25° and 45°, also subject to landslides, the law will now allow sustainable forest management, agro-forestry pastures and the infrastructure required for these activities. Plant cover will only be included in an APP on mountain and hill tops with altitudes over 100 meters and slopes averaging more than 25°, two requirements introduced by the new Forest Code.

These changes exemplify major losses for permanent preservation areas, since lower-altitude hilltops will no longer require plant cover, just as more moderate slopes will allow native vegetation to give way to a plethora of previously illegal activities. Nor will there be any further protection for rolling-hill landforms with gentle ridges and rounded tops. One therefore might wonder what the future will hold for Brazil's hills, mounts, mountains and ranges, 'water towers' of the world and global reservoirs of biodiversity, essential for the well-being of the population and for the capacity to respond to climate change.

References

Price, M.F., G. Gratzer, L.A. Duguma, T. Kohler, D. Maselli and R. Romeo. 2011. Mountain Forests in a Changing World—Realizing Values, addressing challenges. FAO/MPS and SDC, Rome.
Rudaz, G. 2011. The cause of mountains: the politics of promoting a global agenda. Global Environmental Politics 11(4): 43–65.
Silva, S.T., S. Cureau and M.D. Leuzinger. 2010. Código Florestal: desafios e perspectivas. Fiuza, São Paulo.

7B

The Alpine Convention—a Model?

Ricardo Rosario

In the words of Esty and Ivanova (2002), "The discrepancy between a globalized world and a set of inescapable transboundary problems on the one hand, and a dominant structure of national policymaking units on the other, has led to a gap in issue coverage National legislatures often do not see their role in addressing worldwide trans-boundary harms, while global bodies often do not have the capacity or the authority to address them".

Contiguous states may collaborate to sustain shared ecosystems and solve common problems. Indeed, many trans-boundary issues appear first at the regional level, affecting several neighboring countries" (Lian and Robinson 2002).

In this way The Convention on the Protection of the Alps, namely 'The Alpine Convention' is a treaty under international law to protect the Alps at a regional scale. The Convention was signed on November 7, 1991 in Salzburg (Austria) by Austria, France, Germany, Italy, Switzerland, Principality of Liechtenstein and the EU (Slovenia signed the convention on March 29, 1993 and Principality of Monaco became a party on the basis of a separate additional protocol). The Convention entered into force on March 6, 1995.

The scope of the Alpine Convention covers the entire alpine region, with some 190,000 square kilometers and 13.6 million people living in that region and it is visited by millions of tourists. Therefore many common challenges and questions of development have to be discussed through a responsible international coordination of spatial planning, transport, energy, tourism policy and other measures.

Brazilian Environmental Law Institute, Instituto de Estudos de Direito e Cidadania (IEDC).
   Email: rpgrosario@yahoo.com.br
Corresponding author: flavia@frangetto.com

It aims at providing conservation and restoration of unique, rare and typical natural complexes and objects of recreational and other importance situated in the heart of Europe, preventing them from negative anthropogenic influences through the promotion of joint policies for sustainable development among the seven countries of the region.

The Alpine Convention is the first convention for the protection of a mountain region worldwide that is binding under international law: for the first time a transnational mountain area has been considered in its geographical continuity, a common territory facing common challenges. This is how the Alpine Convention evolved, and has been followed by the Carpathian Convention. Today several other areas (Caucasus, Central Asia, Andes) look with interest at the experience of the Alpine Convention (MAP 2011).

For Lian and Robinson (2002), regional systems of environmental management are an essential component of global environmental governance improving efforts at the national and global levels, and for Dua and Esty (1997), the regional scale represents a critical middle ground between the global and national levels.

To understand the importance of The Alpine Convention we have to ask: Why protect the mountains in a same common way if they belong to different countries? Mountain areas cover 24% of the Earth's land surface and host 12% of its people. Mountains provide vital resources for both mountain and lowland people, including freshwater for at least half of humanity, critical reserves of biodiversity, food, forests and minerals. They are culturally rich and provide places for the physical and spiritual recreation of the inhabitants of our increasingly urbanized planet (MAP 2011).

The disconnection between environmental needs and environmental performance in the current international system is striking. New institutional mechanisms for better global governance are urgently needed (Esty and Ivanova 2002), the Alpine Convention can be one of those new mechanisms.

The Alpine Convention on one hand is a model because it is the first mountain convention to protect an entire region, on the other hand, the Alpine Convention needs to be really effective and implement indicators, measurements and accountability to verify the effectiveness of the convention.

References

Becker, Alfred and Harald Bugmann (eds.). 2001. Global Change and Mountain Regions. The Mountain Research Initiative.

Esty, Daniel C. and Maria H. Ivanova. 2002. Revitalizing global environmental a governance: function-driven approach. pp. 181–204. *In*: Daniel C. Esty and Maria H. Ivanova (eds.).

Global Environmental Governance: Options & Opportunities. Yale School of Forestry & Environmental Studies. New Haven. USA.

Lian, Koh Kheng and Nicholas A. Robinson. 2002. Regional environmental governance: examining the Association of Southeast Asian Nations (ASEAN) Model. pp. *In*: Daniel C. Esty and Maria H. Ivanova (eds.). Global Environmental Governance: Options & Opportunities. Yale School of Forestry & Environmental Studies.

MAP—The Multi-Annual Work Programme of the Alpine Conference. 2005–2010. Available at http://www.uibk.ac.at/alpinerraum/meetings/water_2006/map_c_gesamt.pdf. Last access on 01/12/2011.

Mountain Partnership. Organization, Membership and Governance. Available at http://www.mountainpartnership.org/files/pdf/governance_e_final.pdf Last access on 01/12/2011.

Muenger, François. Launch of the World Mountain Forum. Available at: http://klewel.com/verbiergps2011 Last access in 28/12/2011.

The Finish Long-Term Socio-Ecological Research Network. (FINLTSER). Available at: http://www.uibk.ac.at/alpinerraum/documents/3_psenner-lackner.pdf. Last access on 28/12/2011.

7C

International Organizations and Research Networks*

Ricardo Rosario

The Man and the Biosphere (MAB) Programme is a UNESCO interdisciplinary undertaking of environmental research. Shortly after the program was launched a panel of experts discussed the "Impact of human activities on mountain and tundra ecosystems" and recommended studies about: human settlements at high altitudes; the effects of land-use alternatives, impact of large-scale technology and effects of tourism and recreation on mountain ecosystems (FAO). As a consequence, a large number of case studies were carried out worldwide within the framework of the MAB Programme, in particular in the Andes and the Alps. Besides, several Biosphere Reserves were created in areas of mountains for its conservation by MAB Programme.

The International Partnership for Sustainable Development in Mountain Regions (hereinafter referred to as the 'Mountain Partnership') is an involving alliance of mountain stakeholders, to promote, inter alia, joint initiatives based on paragraph 42 of the Johannesburg Plan of implementation and other related instruments. The Mountain Partnership is a voluntary, broad-based alliance whose membership includes 50 countries, 16 intergovernmental organizations and 118 major groups and NGOs around the world. Members are addressing the challenges facing mountain regions by tapping the wealth and diversity of resources, knowledge, information and expertise, from and between one another, in order to stimulate concrete initiatives at all levels that will ensure improved quality of life and environments in the world's mountain regions (Mountain Partnership 2011).

Brazilian Environmental Law Institute.
  Email: rpgrosario@yahoo.com.br
* Corresponding author: flavia@frangetto.com

* This section will present some of the actions implemented by international organizations to protect mountains around the world.

Besides the action of organizations, researchers of the world are connected via some research institutions and think tanks such as MRI, Mountain Forum, IEA and LTSER. The Mountain Research Initiative (MRI) is a multidisciplinary scientific organization that addresses global change issues in mountain regions around the world. MRI promotes and supports: MRI regional networks, MRI events and MRI publications. The objectives of MRI includes a vision of a global change scientific program that: detects signals of global environmental change in mountain environments; defines the consequences of global environmental change for mountain regions as well as lowland systems dependent on mountain resources; informs sustainable land, water, and resource management for mountain regions at local to regional scales (Becker and Bugmann 2001). The MRI program has three elements: communication, framing the research agenda and funding.

The World Mountain Forum (WMF) provides a permanent innovative platform for concerned mountain actors to exchange information, establish a regular dialogue and connect different levels of actors. The vision of the World Mountain Forum (WMF) is to conserve, construct and celebrate mountain regions as vital ecosystems, by engaging their inhabitants and all those who benefit from mountains to jointly promote their conservation and sustainable development. Besides the WMF will generate innovative public private partnerships (PPPs) to promote and implement relevant activities for sustainable Mountain Development and will increase the awareness and recognition regarding the relevance and challenges of mountain people worldwide (Muenger 2011).

The Institute for Alpine Environment of European Academy (EURAC), a problem-oriented research between the conflicting priorities of ecology-economy was founded in 1995. In a range of international, national and regional projects and co-operations a multi-disciplinary team of young scientists is investigating the following fields of research by experimental field research, computer simulations as well as participative trans-disciplinary approaches: Ecosystem research and landscape ecology of mountain environments with a special focus on global change; biogeochemical cycles; functional biodiversity; ecosystem services; sustainable development (IAE 2011).

Besides those actions there is the project of implanting a net of researchers in the molds of a LTER—Long Term Ecological Research Network. The key feature of LTER is its long-term approach. Many of the central ecological processes and problems, such as biodiversity changes and effects of climate change on biogeochemical cycles, take place over the long time scales and gradually. Short term projects are not enough to elucidate trends in global change and the effects of policies on the environment. The permanent infrastructure is required for monitoring these changes.

LTER approach emphasizes long-term information management (From the information management point of view, long-term science is concerned with the research need to collect and keep records of the same measurements over long periods of time). Consistent data collection and meticulous data stewardship aim to maintain continuity, integrity and availability of the long-term data series. Information infrastructure supports the use and reuse of data throughout their protracted lifecycles by provisioning access to data and data sharing that are integral for interdisciplinary research collaboration (FINLTSER).

References

Becker, Alfred and Harald Bugmann (eds.). 2001. Global Change and Mountain Regions. The Mountain Research Initiative.

FAO. UNESCO's Man and the Biosphere Programme in mountain areas. Available at http://www.fao.org/docrep/x0963e/x0963e08.htm Last access in 28/12/2011.

Institute for Alpine Environment (IAE). Available at http://www.eurac.edu/en/research/institutes/alpineenvironment/Partners.html Last access in 01/12/2011.

Mountain Partnership. Organization, Membership and Governance. Available at http://www.mountainpartnership.org/files/pdf/governance_e_final.pdf Last access on 01/12/2011.

Muenger, François. Launch of the World Mountain Forum. Available at: http://klewel.com/verbiergps2011 Last access in 28/12/2011.

The Finish Long-Term Socio-Ecological Research Network. (FINLTSER). Available at: http://www.uibk.ac.at/alpinerraum/documents/3_psenner-lackner.pdf. Last access on 28/12/2011.

The Ecological Risks and Disasters in Hillsides and Mountainous Areas in Brazil: A Socio-environmental Perspective

Ricardo Stanziola Vieira,[1] *Nicolau Cardoso Neto*[2] and *Fernando José Passarelli Neme*[3]

"Disorder in nature is reflection of disorder in society"
Nonuya Leader interviewed in the
Columbian Amazon

Context—Introduction

The ecological disasters, either of natural or technological origin, can be understood as some of the important subjects of the current environmental law, for its aggravation before the climatic changes, intensification of the risk generation due to the fast technological development, but especially as a consequence of the environmental vulnerability generated by poverty,

---

[1] Brazilian Environmental Law Institute, Instituto de Estudos de Direito e Cidadania (IEDC).
Email: ricardo@ambientallegal.com.br

[2] Professor of Environmental Law and Public Health at Fundação Universidade de Blumenau (FURB) and Serviço Nacional de Aprendizagem Industrial (SENAI), Blumenau/SC - Brazil.
Email: nicolau@scambiental.com.br

[3] Secretaria Nacional de Ações com a Sociedade e o Governo (SASG - Meio Ambiente) da Comunidade Bahá'i do Brasil.
Email: fjpneme@gmail.com

[#] Corresponding author: flavia@frangetto.com

which contributes for a bigger exposition to the human rights violation, especially to the right of life. The effect of environmental disasters is felt in a differentiated way by different groups, individuals and communities because of their environmental vulnerabilities. It should be considered, as emphasized by the environmental justice movement, that environmental risks are not distributed in an equitable way and that factors such as poverty, ethnic or racial composition can cause these risks and environmental costs distribution.

When recognizing this difficult and not always desired connection, the systems of protection of human rights (national and international) could play an important role for protection of these individuals and vulnerable groups in situations of ecological disasters. In this direction, it is distinguished European Cut of Human Rights (Cut EDH) that has an innovative and consolidated jurisprudence in ambient substance, it recognizes that the right to the life is impinged upon by damages caused by natural or manmade disasters. An example of recurrent risks of ecological catastrophes, includes the cases of landslides and burials in mountainous areas. This chapter uses a case of landslide in Brazil as a basis for analysis for the recent disasters in Brazil (Santa Catarina 2008 and Rio De Janeiro 2010 and 2011).

Disasters and Climatic Changes in Brazil—Populations in Socio-environmental Vulnerable Situation

To begin with, it is important to understand the meaning of the term 'ecological disaster' and it is necessary to establish its relation with the ambient vulnerability before its effect, especially the one generated by poverty. Disaster is a complex concept that can be understood from different perspectives: social, ambient and economic. As described by Lienhard (1995, p. 91), the notion of disaster refers to an event of great magnitude, occurring in a particular location, at a precise moment and involving instantly many individuals. Similarly, the ecological disaster be caused strictly by human actions, such as developmental and technological advancements that involve a certain level of risk; or it can be because of natural phenomena aggravated by human interventions such as global warming. In many documents of international organizations it is clear that a lot of natural disasters can also be related to human activities (where human actions have aggravated the situation for natural calamities as well). In the Operational Guide on Human Rights and Natural Disasters, elaborated for the IASC—Inter-Agency Standing Committee (2008) natural disaster are exemplified as "the consequences of events triggered by natural hazards that overwhelm local response capacity and seriously affect the social and economic development of a region, are traditionally seen as situations creating challenges and problems mainly of a humanitarian nature. Less attention has been devoted to human rights protection which also needs to be provided in this particular context." (In Inter-Agency Standing Committee – IASC. June 2006. PROTECTING PERSONS AFFECTED BY NATURAL DISASTERS: IASC Operational Guidelines on Human Rights and Natural Disasters. June 2006, page 8/32. Acess: http://www.humanitarianinfo.org/iasc/downloadDoc.aspx?docID=4463&ref=4). UNEP (2008, p. 6) defines disaster as: "a serious disturbance in the functioning of a community or society causing generally losses human beings, materials, economic or ambient that exceed the affected capacity of the communities or societies to face using it its proper resources. A disaster is a function of the risk process. It results of the combination of perigos, insufficient conditions of vulnerability and capacity or ways to reduce potential the negative consequências of the risk" (In. Program of United Nation for the Environment. Environment and Disaster Risks—Emerging Perspectives 2008, p. 6).

We have defined disaster from technique-scientific perspectives to better contextualize the perspective of the communities and groups in condition of socio-environmental vulnerability. It is ironic that the same model of development in modern scientific thinking causes catastrophes (directly or indirectly) and an effort is needed to carry out a thorough diagnosis of the present crisis of socio-environmental risks to consider solutions. On other hand there are sustainable ways of development which will cause less impact on the environment. The irony is that the socio-economic vulnerable society (generally poor) contributes least to the environmental crisis leading to the disasters but suffer the most from the results of these disasters (such as floods or droughts or spills) because they are located at the areas that are most affected and do not have the means to relocate.

Research carried through for the Foundation João Pine in 2005 pointed a housing deficit of 7.902.699 in Brazil. This reflects the current picture of social exclusion of a right to own a house, probably caused by historical disorderly process of urbanization in the country (Brazil 2007). The absence of a place to live (at times due to urban development) forces people to occupy ambient fragile areas, especially at the edges of rivers and hillsides (or mountains/Brazilian mounts). In regions where heavy rains occur or it is projected to have more rain, such occupations on fragile lands (which lack any infrastructure), are more vulnerable to the extreme events of floods and landslides of hills involving material damages and losses in human lives. Since these areas have a higher risk of disaster but also provide important ecological functions, the Brazilian right guardianship has marked these areas for permanent preservation (APPs) to preserve the environment and the communities. The APP refers to the fragile areas for mandatory preservation since the good functioning of the water cycle and other natural cycles depend on these areas. According to Brazilian Forest code (Law 12,651/12 areas of permanent preservation, "covered or not by native vegetation, with the environmental function of preserving water resources, the landscape, the geological stability, biodiversity, the fauna and flora gene flow, to protect the soil and to ensure the well-being of human population". It essentially includes amongst others, the forests and forms of natural vegetation, situated throughout any water course; water springs; top of mounts, mounts, mountains; the hillsides with a slope of 45° degrees, equivalent to the 100% in the line of bigger declivity. The law of Ambient Crimes (Law 9,605/98) stipulates the destruction, to cause damages, to remove, to hinder or to make it difficult the regeneration of considered forest of permanent preservation, as a "crime". Brazilian 'Serra do mar' (seaside mountain range) and socio-environmental vulnerable areas are a dramatic example of disrespect to the legislation, the scientific knowledge and common-sense. Although there are scientific and historical evidences of natural landslides that occurred in the mounts/mountains of bioma of Mata Atlantic, especially in the regions of South and Southeastern Brazil, there is also a discussion that there is a high concentration of population that are socio-economically vulnerable living on this environmentally fragile land. There is also a discussion that even public power has to own some responsibility because it has big infrastructure projects but it did not provide housing to socio-economic vulnerable people and did not prevent people involved in relocation for large infrastructure projects from inhabiting fragile protected by the ambient legislation of Areas of Permanent Preservation. State government was aware of the situation of people living in the area and also the risk of landslides in the serrana region of Rio de Janeiro (Petrópolis, Teresópolis and Nova Friburgo) as well as in the hillsides of the region of the valley of the river Itajaí in Santa Catarina but still did not take any action. In all the cases, many deaths could have been avoided with the preventive measures taken by the State. The irresponsibility of the State led to an harmful impact on the society, perhaps it is time to consider making illegal settlements as legal to give them better housing and infrastructure. The analysis of the disasters characterizes the populations of the affected areas, as explicit of the 'Society of Risk' (BECK 1998), therefore coexisted, of conscientious form, with the possibility of higher risks. The acceptance of the social, economic, ambient risk as dynamic demonstrates the lack of the urbanization planning leading to re-occurring natural catastrophes.

Since different income groups have different capacities to deal with the disaster (because of difference in disposal income), the impact of damage is far more reaching for poor people. Therefore, in contrast of what is suggested, the social inequality also multiplies the risk and intensity with which vulnerable people are affected (ASCELRAD 2006).

Brief Conclusions

As discussed above socio-economic vulnerable people are settling in fragile ecosystems prone to disasters such as landslides. It is important to have a strategy or policy to prevent people from inhabiting such fragile land to minimize the impact of natural disasters. The recent changes to the Forest Code in Brazil shows disrespect to the reason why the statute came into being. The legal aberration voted by the House of representatives, on April 25, 2012 reflects a law which is blind to the scientific findings to protect fragile ecosystem and protecting socio-economic vulnerable people.

References

Acselrad, Henri. 2006. Mapa dos conflitos ambientais no estado do Rio de Janeiro. Rio de Janeiro: CD ROOM.
Beck, Ulrich. La sociedad del riesgo. Hacia una nueva modernidad. Tradução de Jorge Navarro, Daniel Jiménez e Maria Rosa Borrás. 1998. Madri: Ediciones Paidós Ibérica.
Brasil. Ministério das Cidades/Instituto de Pesquisas Tecnológicas—IPT Mapeamento de Riscos em Encostas e Margem de Rios/Celso Santos Carvalho, Eduardo Soares de Macedo e Agostinho Tadashi Ogura, organizadores—Brasília: Ministério das Cidades; Instituto de Pesquisas Tecnológicas—IPT, 2007.
Inter-Agency Standing Committee – IASC. June 2006. PROTECTING PERSONS AFFECTED BY NATURAL DISASTERS: IASC Operational Guidelines on Human Rights and Natural Disasters. June 2006, page 8/32. Acess: http://www.humanitarianinfo.org/iasc/downloadDoc.aspx?docID=4463&ref=4.
Lienhard, Claude. Pour un droit des catastrophes. 1995. Recueil Le Dalloz.

SECTION 5
CASE STUDIES

8

Global Change and Sustainable Mountain Tourism: The Case of Mount Kenya

Irandu Evaristus M.

INTRODUCTION

The significance of mountains as tourist destinations is illustrated by Ives (1992) who observes that mountains beckon as oases of spiritual calm and peace, a place in which to commune with nature and to rediscover the simple pleasures of life. Greenwood (2008) on the other hand highlights another important aspect of mountains by arguing that mountains are zones where the signals of global change are particularly clear. This is attributed to their verticality. The vertical zonation of habitats in mountains represents a gradual change of habitats from the Equator to the Poles within a short distance between them. Mountains, therefore, represent unique areas for the early detection of climate change and the assessment of climate-related impacts. This is possible because as climate changes rapidly with height over relatively short horizontal distances, so does vegetation and hydrology (Whiteman 2000). As a result of their vertical zonation of habitats, mountains exhibit high biodiversity, often with sharp transitions in vegetation sequences, and equally rapid changes from vegetation and soil to snow and ice (Fig. 8.1).

Professor, Department of Geography and Environmental Studies, University of Nairobi, P.O. Box 5207-00200, Nairobi, Kenya.
Email: iranduevaristus@yahoo.com

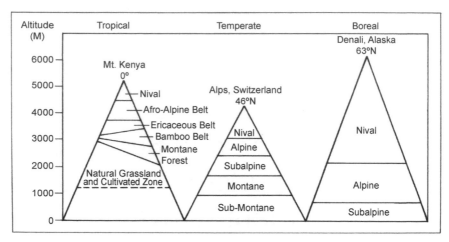

Figure 8.1. Vegetation zonations in tropical, temperate and boreal mountains (Source: Beniston, M. and Fox, D.A. (1996), pg 196).

The watershed in the development of mountain areas was the United Nations Conference on Environment and Development (UNCED) held in Rio de Janeiro (Brazil) which recognized that mountains are a major component of the global environment. The significance of mountains was highlighted in Chapter 13 of Agenda 21 part of which reads:

> *"Mountains* are important sources of water, energy, minerals, forest and agricultural products and areas of recreation. They are storehouses of biological diversity, home to endangered species and an essential part of the global ecosystem. From the Andes to the *Himalayas*, and from Southeast Asia to East and Central Africa, there is serious ecological deterioration. Most mountain areas are experiencing environmental degradation".

Some mountains such as those found in East Africa (e.g., Mt. Kenya, Mt. Kilimanjaro), appear like 'islands' rising above the surrounding plains (Hedberg 1964). In socio-economic terms, mountain landscapes are attractive destinations for exploration, journey and outdoor recreation as they offer a wide range of activity options such as trekking, mountaineering, water rafting, mountain biking, fishing (Maroudas et al. 2004). The environmental stress imposed by growing numbers of tourists in mountain areas places an increasingly heavy burden on mountain resources and on local communities especially in the developing world (Godde et al. 2000).

Mount Kenya like other mountains of the world is a zone where the signals of global change are quite apparent (Fig. 8.1). Already there are indications that the glaciers on Mt. Kenya are receding and snow on its summits is melting (Hastennath and Kruss 1992). The likely impacts of climate change on the mountain's tourism are unclear. The main issue is: Could climate change affect Mount Kenya's ecosystem so irreversibly that mountain tourism could become unsustainable? This is the issue addressed in this chapter.

Tourism Development in Kenya

Tourism in Kenya developed before 1930 when international tourists began arriving in the country in small numbers. Most of these early overseas visitors to the country were wealthy Europeans and Americans who could afford the time and resources for leisure recreation (Sindiga 1999, Kamau 1999). The rich wildlife resource was the base on which Kenya's tourism industry was founded. During this early period, several national parks were established in different parts of the country such as Nairobi National Park (1946), Tsavo (1948) and Mt. Kenya National Park (1949). At present, there are 25 terrestrial national parks, four marine national parks, 22 national reserves, five marine reserves, one national orphanage and one national sanctuary. In all, the protected areas cover about 8% of the country's total land surface and accommodate a wide variety of wildlife species such as the rare Bongo, Roan Antelope and Giant Forest Hog among others (KWS 2011).

Parallel to the development of wildlife tourism was beach tourism at Kenya's coast. Almost immediately after the establishment of British colonial rule in Kenya, the coast began to attract resident Europeans from upcountry especially European settler farmers. This was the beginning of domestic tourism in the country. The major attractions for domestic tourists at Kenya's coast were the three Ss (sea, sun and sand). The three Ss also form the base for international tourism at the coast even today (Irandu 2004a).

The tourism industry in Kenya has been on upward trend making it one of the leading foreign exchange earners in the economy besides tea and horticulture. For instance in 2006, tourism industry recorded upward growth despite world economic recession. The number of visitor arrivals reached an all time high of 1.6 million. Foreign exchange earnings from tourism rose from KES (Kenyan Shillings). 48.9 billion (US$ 489 million) in 2005 to KES. 56.2 billion (US$ 562 million) in 2006, representing a 15% increase in earnings (UNWTO 2007, RoK 2007). However, the number of visitor arrivals declined drastically in 2007 to below one million following eruption of violence due to election disputes in the country. The tourism sector has shown a lot of resilience and has recovered quickly after the election violence. Kenya received close to 1.1 million visitor arrivals in 2010 (RoK 2010). During the same year, foreign exchange earnings from tourism reached KES 73.68 billion (US$ 736.8 million).

The country continued to rely on her traditional tourism markets of Europe and North America with the United Kingdom leading with 174,051 visitor arrivals followed by United States with 107,842 arrivals. Italy and Germany took third and fourth positions at 87,694 and 63,011. France took the fifth position with 50,039 visitors. Asian markets are also growing in importance due to aggressive marketing by the government. For example, about 47,611 arrivals came from India followed by China (28,480) and UAE (14,874) in 2010 (RoK 2011).

Despite the remarkable growth recorded by Kenya's tourism industry as a whole, the country's Vision 2030 tourism goals are far from being achieved.

The country's Vision 2030 tourism goals include:

1. To quadruple tourism's GDP contribution to more than KShs. 200 billion.
2. To raise international visitor arrivals from 1.6 million in 2006 to 3 million in 2012.
3. To increase the number of beds from the current 40,000 to about 65,000.

One way of achieving the Vision 2030 Tourism Goals is by developing new forms of tourism such as mountain tourism especially in the era of climate change. This would help in diversifying the tourism sector which, for a long time has been dominated by wildlife and beach tourism (Irandu 2004a).

Development of Mountain Tourism in Kenya

Today, mountain tourism represents an important market segment of global tourism industry (Maroudas et al. 2004). Mountain tourism involves climbing mountains, mainly as a form of recreational activity. Mountain climbing is a popular tourist activity worldwide especially where hills are high enough to provide a challenge to participants. Jenik (1997) contends that climbers' rewards include physical exertion, the satisfaction of overcoming challenges by working as a team, the thrill of reaching a summit and the magnificent view from the mountain top. Other forms of mountain tourism include cultural tourism and ecotourism (Maroudas et al. 2004).

In Kenya mountain (alpine) tourism is only partly developed and a sustainable approach to its development is required (Kenya Wild Service 1990). This would ensure that visitor needs are satisfied while at the same time maintaining the capacity of the natural resources to provide long term benefits and meeting goals of social equity and environmental quality (Jenik 1997).

Mount Kenya the second highest peak in Africa (5,199 m) after Mt. Kilimanjaro is an extinct volcano which last erupted about three million years ago (KWS 2011). In April 1978 the area was designated a UNESCO Biosphere Reserve (UNEP 1998), making Mt. Kenya one of the 140 such

Biosphere Reserves in the world. The national park and the forest reserve were declared a UNESCO World Heritage Site in 1997 (UN 2008). Mount Kenya National Park and Reserve cover approximately 2800 sq km. Mt. Kenya National Park was gazetted in 1949 as a national park and covers about 715 km sq. It is estimated to have over 800 recorded plant species of which 81 are endemic to the area. Initially Mt. Kenya National Park was a forest reserve. At present the national park is within the forest reserve which encircles it (KWS 2011). The national reserve was gazetted in 2000 and covers 2124 sq. km.

The natural endowments of Mt. Kenya comprise animal species such as elephants, buffaloes and the rare bongo, indigenous plants and the mountain's scenic beauty. That is why it has been gazetted as a protected area, both as a national park and forest reserve because of its conservation value. The National Park covers the entire area above 3200 m and the protected area has now been extended to include the Mount Kenya Forest Reserve, in order to protect the catchment area and wildlife. The Mountain's biodiversity is one of the most valuable national and international tourist destinations in the whole country. Biodiversity is outstanding, owing to the succession of different bio-ecological zones at close range, extending from nival and afro alpine to forest and savannah. The summit area is a major destination for mountain tourism in Kenya, including trekking, mountaineering and game watching (Makunyi 2010).

Its picturesque landscape and its snow cap at the summit make it appealing mainly to nature loving tourists from all over the world. On average, Mt. Kenya attracts 30,000 tourists every year. However, more and more visitation by tourists especially by uncaring visitors is threatening the very fragile landscape by way of over use of climbing routes and littering along the trails thus causing more and more stress on the routes, on animals and on the vegetation (Kariuki 2005). Mt. Kenya is a major water tower in the country and is the only free standing feature on the Equator with glaciers. There are at present 12 remnant glaciers, all receding rapidly (Fig. 8.2). With its glacier-clad rocky summits, Mount Kenya is one of the most impressive and imposing topographical features in East Africa (Plate 8.1).

Despite Mt. Kenya's scenic beauty it is not yet developed as a recreational and tourist destination. Nevertheless, it continues to attract a large number of both domestic and international tourists such as mountaineers, walkers, bird-watchers and fishermen (Makunyi 2010). The tourism potential of the mountain's ecosystem and the adjacent areas has also drawn interest from various actors such as private sectors (hotels and lodges), NGOs and international development organizations as well as local communities. If this becomes a reality, it will greatly boost the economy and employment in the region. The development of new niche-based tourism products such

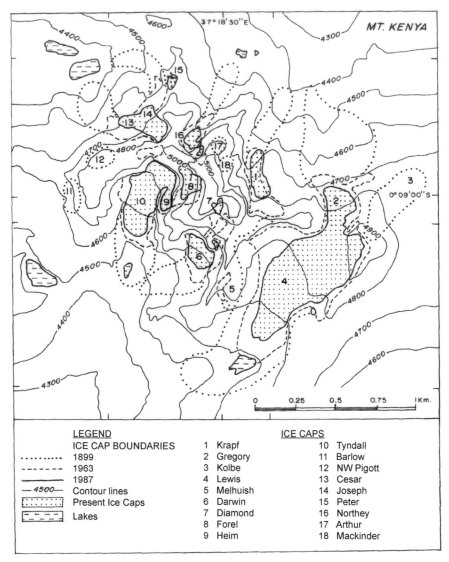

Figure 8.2. Map of Mount Kenya's glaciers (Source: Hastenrath, S. and Kruss, P.D. (1992), pg 128).

as adventure tourism and community-based ecotourism will continue to draw importance in the Mount Kenya area and its biodiversity and cultural diversity.

As Kloiber (2007) observes mountain areas all over the world have for long fascinated and attracted people of varying interests such as pilgrims, ascetics, naturalists, explorers, and, in recent years, mountaineers, trekkers

Plate 8.1 Mt. Kenya showing remaining Glaciers.

Color image of this figure appears in the color plate section at the end of the book.

and cultural tourists from near and far-off. Features that attract tourists are the pure, original nature, natural monuments and the healthy climate. Mt. Kenya like other mountains of the world has many tourist attractions such as mountaineering, hiking and trekking, sport fishing, cave tourism, religious tourism, adventure tourism and ferratta (metal way), These attractions can form the base for development of adventure tourism, ecotourism and alpine tourism (Makunyi 2010, Mburu 2011). To establish the potential of mountain tourism on Mt. Kenya, a SWOT analysis was carried out. Its results are summarized in Table 8.1.

Strengths

The snow-clad summits of Mt. Kenya lying on the Equator offer opportunities for outdoor adventure tours. Mount Kenya is a world class destination for mountaineers. It attracts many mountain climbers from all over the world every year (Makunyi 2010). Mountaineering is the mountain's flagship activity which motivates camping, hiking and trekking. Camps are used as bases for mountain climbing (Fig. 8.2). The picturesque

Table 8.1. SWOT Analysis.

Strengths	Weaknesses
Unique mountainous setting for mountain climbing Abundant and diverse wildlife populations for sight seeing UNESCO Heritage Site Status Highly experienced campsite operators, tour operators and guides Numerous rivers for sport fishing Eco-friendly experience Proximity to capital city of Nairobi attracts business travellers	Stiff competition among operators Operators not taking full advantage of unique attractions such as landscape No branding of products Poor transport and other infrastructure Poor marketing including limited web marketing strategy
Opportunities Untapped adventure tourism potential Internet technologies and online marketing Environmental awareness of tourists Interest in cultural tourism Increasing demand for adventure tourism products worldwide Increasing number of women campers Special events such as Lewa Marathon	**Threats** Aging population in source markets High cost of operating campsites Melting of snow and drying of rivers due to global warming and climate change Insecurity especially in remote campsites Illegal/unlicensed campsite and tour operators Illegal logging destroying wildlife habitats

Source: Compiled by author 2011.

and rugged mountain peaks are the major tourist attraction, drawing close to 20,000 climbers annually (KWS 2011). Several routes for climbing the mountain have been developed (Fig. 8.4). Mt. Kenya National Park located on the mountain slopes is a high altitude Park with abundant wildlife and rare fauna and flora species. Its unique geological structure and ecology earned the mountain the UNESCO Heritage Site status in 1997 giving it a competitive edge over many other national parks and game reserves in the country.

Weaknesses

The Mt. Kenya Porters and Guide Safari Club is a community-run tour business. Most of these businesses lack basic facilities to run their marketing activities (Kariuki 2005). With the marketing of tourism moving into the information age, community groups are clearly disadvantaged. Most of them only have e-mail addresses as their contact to the world outside their villages and even then, these are not often used due to limited internet services which can only be accessed in major nearby towns such as Nyeri, Nanyuki, Meru or Embu. Therefore, the greatest challenge facing the community tour business is marketing themselves to both local and international tourists, as they do not have the required infrastructure (Kariuki 2005).

Access to financing is a challenge and concern for tour operators. Some tour operators have difficulty obtaining financing to support key activities such as marketing, staff retention, professional development, or facility/equipment upgrades. When institutional financing is not available, operators must search out alternate sources (e.g., family and friends), but this can create difficult and stressful situations. Operators feel that they lack access to government grants or assistance.

Opportunities

The Mt. Kenya's outdoor adventure segment has massive untapped potential given the vast, undeveloped natural environment of the area. The diversity of the natural environment provides ideal locations for both hard adventure such as mountaineering and soft adventure such as nature trails and bird watching. Operators increasingly recognize that outdoor adventure tourism attracts a global market. Internet technologies and online marketing provide new ways to advertise and reach potential clients. Operators are taking advantage of websites to promote their business (e.g., posting videos about their products), and participate in networking and collective marketing activities through social media (e.g., Facebook, MySpace, blogs, forums, etc.). A key benefit to having an Internet presence is that it enables clients to research their own trips and find operators. Global Internet use is expected to increase, so it will be increasingly important for operators to develop Web capabilities in their business. Many are already taking advantage of these technologies.

Tourists are becoming ever more sensitive to the environmental impact of their travel choices. Environmental tourism is a trend that Mt. Kenya operators can capitalize on, as demand intensifies for activities with minimal or no impact on the natural environment, such as canoeing, kayaking and hiking. Operators believe tourists will increasingly choose to visit—and be willing to pay more for—camps that are self-sufficient and green (i.e., equipped with solar heating/lighting, etc.). An opportunity operators on Mt. Kenya are not currently taking advantage of is soft adventure offerings such as bird watching and trout fishing aimed at business travellers.

Threats

The high cost of operation for camps, high fuel and transportation prices make it difficult for operators to realize high profits. Rising fuel prices affect all aspects of business. The effect is particularly noticeable in the cost of aircraft charters and driving costs, but also skews food and supply costs. Increased insecurity in the study area has affected some operators and some visitors have had equipment and other valuables taken away by robbers while visiting some camps.

The unknown consequences of global warming and climate change could harm adventure travel or prevent its growth in the study area. The melting and receding of glacier on top of Mt. Kenya could spell disaster to the adventure travel and tourism industry in the area. Drying up of rivers flowing from the mountain could reduce the number of tourists attracted by trout fishing in the area. Wildlife habitats such as the mountain forest would disappear together with the associated animal communities.

Impacts of mountain tourism in Kenya

As Kruk et al. (2007) argue, tourism if well planned can generate employment and income opportunities for local people in mountain areas such as Mt. Kenya. Local jobs are created through opportunities to work as guides and porters, in lodges, general supplies to tourists, renting camp sites and performing cultural shows. A major concern though has been the negative impacts of tourism on the environment (e.g., deforestation, pollution of water sources, littering). Some scholars also argue that tourism has had negative impacts on local culture and values (Lama and Sherpa 1994).

Socio-economic impacts

Today, Mt. Kenya provides livelihood for more than six million people living adjacent to the forest including 1500 porters and guides who provide services such as porterage, guiding and cooking to the tourists (KWS 2011). The local communities have also benefited from various development projects funded by the KWS with funds accruing from tourism. Some of the projects include a Maternity Wing at Mutindwa Dispensary in Mara District (Tharaka Nithi County), a computer room and one classroom at Gatwe Secondary School in Kirinyaga County.

Environmental impacts

Tourism development also creates environmental problems and poses new challenges to mountain areas (Mburu 2010). Increasing amount of litter often found at high altitudes in the mountain areas of the world constitutes a major negative environmental impact of mountain tourism. If this trend continues unchecked, mountains may be converted into junkyards (Lama and Sherpa 1994). On Mt. Kenya, the major negative environmental impacts arising from growing mountain tourism include trail creation, littering and trampling of vegetation (Mburu 2010, Wilson 2011).

Trail creation is highly destructive to the fragile afro-alpine flora which takes very long to recover. Trail creation occurs because of some tourists avoiding well established trails on the mountain. On Mt. Kenya, littering causes pollution which is endangering wild animals and the surrounding natural forest, one of the few natural forests left in the world. Within four days only, a group of four mountain climbers can on average generate 20 kilogrammes of rubbish most of which is scattered on the mountainside (Wilson 2011).

Human encroachment activities in Mt. Kenya National Park have also adversely affected tourist resources of the mountain. In the Lower Forest Reserve, human interference has been very serious in the past. Human pressure from poor but rapidly growing population has been increasing. Poaching, illegal firewood collection, destructive honey harvesting and illegal hunting for bush meat are major threats to the mountain ecosystem and its rich biodiversity (Bussmann 1996). The effect of this wanton destruction of the mountain's natural resources has been the disruption of wildlife habitats, diminishing of biodiversity, and impairment of water catchment and retarded vegetation growth which has slowed the development of mountain tourism (Bussmann 1996, Mburu 2010).

Sustainable mountain tourism in Kenya

In order for the development of mountain or alpine tourism to succeed, the idea of sustainable development in mountain areas must be considered. The concept of sustainable development grew out of the limits to growth debate of the early 1970s. This debate was anchored on the issue of whether or not continuing economic growth would ultimately lead to severe degradation and societal collapse on a global scale. By the late 70s, it was obvious that economic development could not be sustained without the conservation of the environment. As Hunter and Green (1995) have observed, sustainable development constitutes a key element for the management of tourism that integrates concern for natural, built and cultural environment. Sustainable development is a multidimensional concept embracing essentially three dimensions or pillars: economic, social, and environmental sustainability (Ngunyi 2009).

The official use of the term sustainable development can be traced to 1987 when it received international recognition. In that year, the Brundtland Commission on Environment and Development defined sustainable development as "development that meets the needs of the present without compromising the ability of future generations to meet their own needs" (World Commission on Environment and Development, 1987). Sustainable tourism is defined as " tourism developed and maintained in an area in such a manner and such a scale that it remains viable over an indefinite period and does not degrade or alter the environment in any way that might prohibit the successful development and wellbeing of other activities and processes" (UNESCAP 2001). Sustainability principles refer to the environmental, economic, and sociocultural aspects of tourism development, sustainable mountain tourism should:

1. Optimize use of natural resources while at the same time maintaining ecological processes and helping conservation of biodiversity,
2. Promote respect for the socio-cultural authenticity of host mountain communities, conserve cultural heritage and traditional values, and
3. Contribute to poverty alleviation.

Sterly (1997) noted that one of the major challenges faced in the development of mountain tourism was to exploit it in such a way that it benefits mountain people and their environments while at the same time satisfying needs of tourists. Mountain tourism needs to be sustainable (Mburu 2010). As Lane (2005) observes, people and local ownership of tourism are at the core of sustainable tourism. Mountain people are the stewards of mountain ecosystems and any decision to develop mountain tourism must be made with their involvement and agreement.

Sustainable mountain tourism can take many forms such as ecotourism, community-based tourism, adventure tourism and nature based tourism among others (Maraudos et al. 2006). Mt. Kenya has great potential for the development of cultural tourism. This is mainly due to cultural values attributed to the mountain by all the various groups of people living around the forest. The mountain has been considered as the abode of the Kikuyu, Meru and Embu god named 'Ngai' or 'Murungu'. Mt. Kenya forest therefore, provides an important location for religion and other rituals for the local people. Prayers and rituals are carried out in several sacred areas in the forests in time of need, such as to bring rain and bless the community (Gathaara 1999). In recent times, the mountain has acquired new historical significance because the 'Mau Mau' war of independence was fought in its forest. The recognition has been done through the gazettement of 'Mau Mau' caves as historical sites under the National Museums of Kenya Act (Kariuki 2005).

It should also be borne in mind that many national parks and game reserves created in different parts of the country including Mt. Kenya, are surrounded by resentful people because their own land was expropriated for wildlife conservation. This left them without sufficient land for cultivation or grazing their livestock. To bridge the gap between local communities and wildlife and thereby minimize human-wildlife conflict, the govenment has encouraged the development of ecotourism projects throughout the country (Ngunyi 2009). Ecotourism was defined by the International Ecotourism Society in 1991 as: "responsible travel to natural areas that conserves the environment and sustains the well-being of local people" (The International Ecotourism Society 1991).

An example of an ecotourism project managed by the local community on Mt. Kenya is the Lake Nkunga Ecotourism Project. This consists of a crater lake sitting on 40 ha of land surrounded by a natural forest whose trees are in danger of depletion through illegal logging. The lake is situated near Meru Town in Meru County, and is considered sacred because in the traditional past the Ameru used it as a sacrificial area to their god (Kariuki 2005). The geological formation of the crater Lake offers tourism potential in the area. There are plans to develop water sports, picnic sites and nature trails. The project commenced in 2001 as a joint project of the United Nation Development Programme-Global Environmental Facility/Small Grants Programme (UNDP-GEF/SGP), the Lions Club of Meru and the local community through their elected representatives. The UNDP-GEF/SGP and Lions Club of Meru are the main sponsors of the project. The contribution of the local community is mainly providing labour. The conservation of the lake was designed to meet the conservation objectives of the area as well as the needs of the local people who are dependent on the lake's water for their household use.

Climate change and mountain tourism in Kenya

Mountain ecosystems are sensitive to climate change. Observable effects of climate change in mountain areas include less snow, receding glaciers, melting permafrost and more extreme events such as landslides and rock falls. Climate change is also likely to shift mountain flora and fauna, thereby affecting future tourism (Burki et al. 2003, Greenwood 2008, World Bank 2008). The receding of the glaciers of Mount Kenya is symbolic of the impact of climate change in Kenya. It also illustrates how the relationship between cause and effect is less than straightforward. Although much accentuated in the last 20 years, the glacier retreat seems likely to have commenced early in the 20th century. Recent human deforestation of the lower slopes has also played a part by altering the local ecosystem and micro-climate (Bussmann 1996, KWS 2011).

As already discussed above, tourism is a major source of foreign exchange for Kenya. Climate-change-related impact on natural resources therefore has direct effects on the country's tourism industry. Climate change could thus place tourism at risk, particularly in coastal zones and mountain regions (IPCC 2007). For example, the 1997/1998 coral bleaching episode observed in the Indian Ocean and Red Sea was coupled with a strong El Niño-Southern Oscillation (ENSO). In the western Indian Ocean region, a 30% loss of corals resulted in reduced tourism in Mombasa, and caused financial losses of about US$ 12–18 million (Payet and Obura 2004, IPCC

2007). Rising sea levels will gradually increase the risk of serious flooding in the city of Mombasa and the surrounding coastal region which will obviously impact negatively on beach tourism (McLeman and Smit 2004, IPCC 2007).

Glaciers on Mt. Kenya are melting at an alarming rate and if this continues, the fun of climbing this mountain may vanish forever (Fig. 8.2). The quantities of ice have reduced significantly on the mountain having lost about 92% of the original ice in the past 100 years.

About a century ago, there were over 18 glaciers on the mountain which stands at 5199 metres above sea level (RoK 2007, KWS 2011). But today there are only 12 glaciers left and four of them are a pale shadow of what they were 30 years ago, thanks to the effect of global warming and climate change. There are about 1500 guides who earn a living by guiding tourists on Mt. Kenya and whose livelihoods now hang on the balance. About 20,000 mountaineers attempt to conquer the various peaks of Mt. Kenya every year (KWS 2011). Most of these tourists are disappointed because of melting snow. A local Daily described the mood of tourists climbing Mt. Kenya today aptly by stating, "to a mountain climber, there is no joy when one has not reached the highest point, Batian, which has to be accessed by climbing through masses of ice" (Daily Nation 2009).

Global warming increases melting of permafrost and makes many mountain areas vulnerable to landslides. The warming in mountain areas also makes hiking and climbing more dangerous due to increasing rock fall. Increasing risks of rock fall have been reported on Mt. Kenya (Mburu 2011). Global warming is a challenge for the tourism industry in mountain areas. Over all, climate change is a threat for mountain tourism due to less snow, less glaciers, and more extreme events (e.g., landslides).

Climate is a principal resource for tourism, as it codetermines the suitability of locations for a wide range of tourist activities, is a principal driver of global seasonality in tourism demand, and has an important influence on operating costs, such as heating/cooling, snowmaking, irrigation, food and water supply, and insurance costs (UNEP 2008). Thus, changes in the length and quality of climate dependent tourism seasons (i.e., sun and sea or winter sports holidays) could have considerable implications for competitive relationships between destinations and therefore the profitability of tourism enterprises. Studies indicate that a shift of attractive climatic conditions for tourism towards higher latitudes and altitudes is very likely (Fig. 8.3).

Due to climate change coastal areas in Kenya are likely to be much hotter and therefore less appealing to tourists during the peak season in October–March (Fig. 8.3). Bourdeau (2009) has observed that there could be a shift in seasonal flow of tourists in Europe away from the sea in the summer months towards cooler mountain areas and vice versa in winter. He attributes this

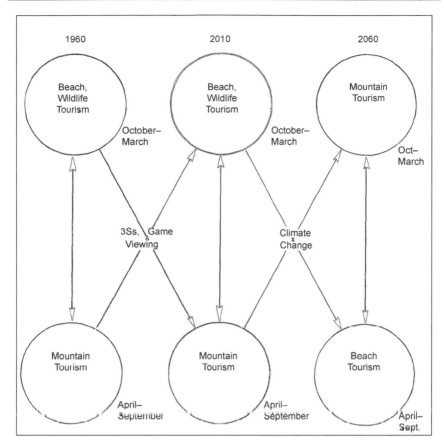

Figure 8.3. Possible seasonal shift in Tourist patterns in Kenya (Source: Modified from Bourdeau, P. (2009).

to the fact that a mountain acts as natural 'air-conditioned zone'. Although there are no distinct seasons on the Equator, there is a likelihood of a shift of tourism patterns in Kenya too, with mountain areas especially Mt. Kenya becoming more appealing during the peak season (October–March). During northern winter (October–March), Kenya's coast is usually very appealing to tourists from Europe and North America due to its sunshine. This may change as beaches may become hotter and unsuitable for tourism during the peak season in the next 50 years (Fig. 8.3).

Generally, mountain ecosystems possess higher levels of biodiversity some of which are endemic. Endemic species such as those found on Mt. Kenya, are expected to be sensitive to climate change. Climate change may alter the fitness of individual species and cause either species migration or extinction. The degree of the impact will vary by species and by region. For

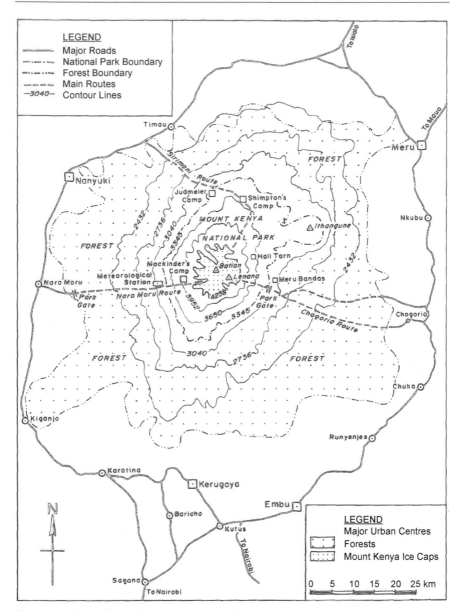

Figure 8.4. Mount Kenya Tourist Route map (Source: Compiled by the Author).

example, on Mt. Kenya many species will likely be able to migrate up slope as temperature increases and as rainfall patterns shift. Such changes may have serious repercussions because in Kenya, mountains are protected for conservation purposes and exploited for tourism. Therefore, any changes in the composition or loss of some of their wildlife that usually attracts tourists would lead to loss of foreign exchange and revenue to the country (Ogara and Awuor 1997).

Adaptation strategies for mountain tourism in changing climate in Kenya

There are many challenges faced in climate change adaptation by the communities living around mountains mainly because of their inadequate capacities (Bagoora 2010). Such communities need early warning systems to be able to respond in time and avoid any possible climate change triggered disasters. The high levels of poverty make the affected communities even more vulnerable and take longer to adapt to climate change impacts such as land slides and crop failure due to prolonged droughts.

In developing response strategies to address climate change and its impacts in mountain areas such as Mt. Kenya, there is need to heed the admonition of Vellinga and Stewart (1991) who stated that "the race to prevent global warming is like the marathon race that is held annually in New York by amateurs. Suddenly, we are aware that we have to run this marathon. We do not know what to expect but with the outcome of the IPCC report, the starting short has been fired. And this marathon is not a race of forty kilometres but a race for at least forty years" (Vellinga and Stewart 1991).

Climate change poses a serious challenge for the tourism industry in mountain areas (Burki et al. 2003). There are a wide array of climate change responses and adaptation strategies that mountain destinations such as Mt. Kenya would adapt to cope with reduced mountain snow and increased natural hazards such as rock fall and land slides (UNEP 2008, UNWTO and UNEP 2008). The responses and adaptation strategies should be implemented as soon as possible by the affected mountain destinations in order to reverse the adverse effects of global change. The responses and adaptation strategies developed in the mountain destination should be all inclusive by taking into consideration the diverse interests of all stakeholders. The key strategies adopted to minimize impacts of climate change on Mt. Kenya include conservation education and development of carbon sinks. These strategies are discussed below.

Conservation education

The local communities living around Mt. Kenya are sensitized by the Kenya Wildlife Service (KWS) on the value of conserving forests and other forms of biodiversity. The local people are made aware that forests act as carbon sinks and help in reducing global warming. They are also educated on the role of forests in protecting water catchment so that rivers and streams do not dry up even in the face of climate change. In this way, human encroachment into the forest and Mount Kenya National Park is reduced (Bussmann 1996).

Development of carbon sinks

Forest ecosystems have natural mitigation processes such as carbon sequestration. Worldwide, there is a large and growing interest in carbon sequestration as a method of combating climate change. This involves the establishment of new forests or the management of existing ones as carbon sinks (Earth Watch Institute 2011). On the slopes of Mt. Kenya, a reforestation project to sequester carbon with community development activities has been established. This is known as the Meru and Nanyuki Community Reforestation Project. It combines hundreds of individual tree planting activities and enables local communities to improve access to food and create additional sources of income beyond subsistence farming. The project has over eight thousand members. It enables members who are small scale farmers, to plant trees on their land. The farmers receive annual payments for each planted tree and additionally will, in the future, collect revenues as the trees grow and sequester carbon. The project also aims at reducing soil erosion, protecting water catchment and enhancing biodiversity (Carbon Neutral Company 2011).

Conclusion

Mt. Kenya like other mountains of the world represents a unique area for the early detection of climate change and the assessment of climate-related impacts. Yet, mountains did not attract a lot of international attention until after the 1992 Rio Summit. In the face of global change mountains around the world have become a focal point of international agenda. One major issue of concern is the sustainability of mountain tourism due to climate change. Tourism on Mt. Kenya is no exception. It can be developed as a way of diversifying the tourism sector which, for a long time has been dominated by wildlife and beach tourism. But, beach and wildlife tourism may become less significant in the face of global change.

Tourism on Mt. Kenya if well planned and managed, can generate local employment and confer other benefits to the local communities. On Mt. Kenya, some of the major environmental impacts identified include trail creation, littering and trampling of vegetation. Trail creation is highly destructive to the fragile afro-alpine biodiversity. In adapting to climate change the local communities living on the slopes of Mt. Kenya like those in other mountain areas of the world, face daunting challenges due to their poverty. Their capacity to respond and adapt to climate change is thus impaired.

The local communities living around Mt. Kenya are sensitized by the Kenya Wildlife Service (KWS) on the value of conserving forests and other forms of biodiversity. In this way, human encroachment into the forest and Mount Kenya National Park can be reduced. On the slopes of Mt. Kenya, a reforestation project to sequester carbon with community development activities has been established. This project enables members who are small scale farmers, to plant trees on their land to act as carbon sink and also to reduce soil erosion, protecting water catchment and enhancing biodiversity.

Acknowledgements

Some of the material presented in the chapter relate directly or indirectly to my own research experience on tourism in Kenya. However, the compilation of a book chapter such as this one requires wide reading of bibliographic material on the topic at issue. Therefore, quite naturally, the chapter draws extensively on the work of many scholars and only a small proportion of them are mentioned in the main text. I am especially grateful to Dr. Martin Marani a colleague in the Department of Geography and Environmental Studies of the University of Nairobi for his tireless effort and time to read through the manuscript and make very valuable comments. The author and Publishers are grateful to Chris 73/Wikimedia Commons (for Plate 8.1 under GNU Free Documentation Licence) and Mrs. Esther W. Makunyi (Plate 8.2). The authors of Figs. 8.1–8.3 are also duly acknowledged. They are: Beniston M. and Fox, D.A. 1996. Figure 8.1; Hastenrath S. and Kruss, P.D. 1992. Figure 8.2; Bourdeau, P. 2009. Figure 8.3.

Plate 8.2. Shipton's Camp used as Base for Climbing Mt. Kenya (Source: Makunyi 2010, Plate 2, pg. 20.

References

Bagoora, F.K. 2010. Challenges of Climate Change in Mountain Ecosystems in Africa. An Overview Presentation at the Side Event Organized Mountain Alliance Initiative, United Nations Climate Change Conference. Cancun, Mexico, 29 November–10 December.

Beniston, M. and D.G. Fox. 1996. Impacts of climate change on mountain regions. pp. 191–213. *In*: R.T. Watson, M.C. Zinyowera and R.H. Moss (eds.). Climate Change 1995: Impacts, Adaptations and Mitigation of Climate Change. Contribution of Working Group II to the Second Assessment Report of the IPCC. Cambridge University Press, New York, USA.

Bourdeau, P. 2009. Mountain Tourism in a Climate of Change. Institute of Alpine Geography (University of Grenoble), UMR Territoires-PACTE.

Bürki, R., H. Elsasser and A. Bruno. 2003. Climate Change—Impacts on the Tourism Industry in Mountain Areas, 1st International Conference on Climate Change and Tourism, Djerba, 9–11 April.

Bussmann, R.W. 1996. Destruction and management of Mount Kenya's forests. Ambio 25(5): 314–317.

Day-Wilson, Victoria. 2011. Cleaning Tourist Litter on Mt. Kenya. http://www.safariweb.com/Safarimate/litter.html. Retrieved on 2011-11-23.

Earth Watch Institute. 2011. Climate Change: Mitigation—Carbon Capture and Storage. London, U.K.

Gathara, G. 1999. Aerial Survey of the Destruction of Mt. Kenya, Imenti and Ngare Ndare Forest Reserves. Forest Conservation Programme, Kenya Wildlife Service, Nairobi.

Gichuki, Francis Ndegwa. 1999. Threats and opportunities for mountain area development in Kenya. Ambio 28(5): 430–435.

Godde, P.M., M.F. Price and F.M. Zimmermann (eds.). 2000. Tourism and Development in Mountain Regions. CABI Publishing, Wallingford.

Greenwood, G. 2008. Why mountains matter. *In*: Mountainous Regions: Laboratories for Adaptation. Magazine of the International Human Dimensions Programme on Global Environmental Change 2: 4–6.

Hastenrath, S. and P.D. Kruss. 1992. The dramatic retreat of Mount Kenya's glaciers between 1963 and 1987: greenhouse forcing. Journal of Glaciology 16: 127–133.

Hedberg, O. 1961. The Phytogeographical Position of the Afro-alpine Flora. Recent Advances in Botany 1: 914–919.

Hunter, C. and H. Green. 1995. Tourism and the Environment: a Sustainable relationship. Routledge. London

[IPCC] Intergovernmental Panel on Climate Change. 2007. Climate Change. The Physical Science Basis. Summary for policymakers. International Panel on Climate Change, Geneva.

Irandu, E.M. 2004a. The role of tourism in the conservation of cultural heritage in Kenya. Asia Pacific Journal of Tourism Research 9(2): 133–150.

Ives, J.D. 1992. Preface. *In*: P.B. Stone (ed.). The State of the World's Mountains: A Global Report. Zedi Books, London.

Jenik, J. 1997. The diversity of mountain life. pp. 199–235. *In*: B. Messerli and J.D. Ives (eds). Mountains of the World: A Global Priority. Parthenon Press, London and New York.

Kamau, M.A.N. 1999. The role of tourism in regional development: a case study of village tourist centres of the Coast Province, Kenya. M.A. Thesis, University of Nairobi.

Kariuki, J. 2005. Embracing community-based ecotourism. [IFRA] Les Cahiers d'Afrique de l'Est. Kenyan Studies. No. 28.

Kloiber, J. 2007. Mountain ecotourism. A case study of the High Pamir Mountain, Kyrgyzstan/ Tajikistan. M.A. Sustainable Tourism Management. Eberswalde University, Eberswalde.

Kruk, E. 2009. Tourism in the Himalaya—Mountains of Opportunities in a Changing Climate ICIMOD position paper.

KWS. 2011. Mount Kenya National Park. Come Touch the Sky, KWS, Nairobi.

Lama, W. and A. Sherpa. 1994. Tourism Development Plan for the Makalu Base Cam Trek and Upper Barun Valley Revised Draft Report. Makalu–Barun Conservation Project.

Lane, B. 2005. Sustainable rural tourism strategies: a tool for development and conservation. International American Journal of Environment and Tourism 1(1): 12–18.

Makunyi, E.W. 2010. A survey of methods used by the Kenya tourist board in marketing adventure tourism in the Mount Kenya region. M.Sc. Thesis. Hospitality and Tourism Management Department. Kenyatta University.

Maroudas, L., A. Kyriakaki and D. Gouvis. 2004. A community approach to mountain adventure tourism development. Anatolia: An International Journal of Tourism and Hospitality Research 15(1): 5–18.

Mburu, S.W. 2011. Challenges faced in the development of alpine tourism in Kenya. A case study of Mt. Kenya National Park. Project Paper. Bachelor of Tourism Management.

McLeman, R. and B. Smit. 2005. Assessing the security implications of climate change related migration. Preprint, Human Security and Climate Change: An International Workshop, Oslo, 20 pp.

Ngunyi, R.N. 2009. Ecotourism and sustainable development in Kenya. Doctorate Paper. School of Business, Department of Hotel and Tourism Management, Sun Yat University.

Ogara, W.O. and V.O. Awuor. 1997. Effects of climate change on wildlife and tourism. pp. 73–92. In: J.S. Ogola, M.A. Abira and V.O. Awuor (eds.). Potential Impacts of Climate Change in Kenya. Climate Network Africa. Nairobi.

Payet, R. and D. Obura. 2004. The negative impacts of human activities in the Eastern African region: an international waters perspective. Ambio 33: 24–33.

[RoK] Republic of Kenya. 2007. Economic Survey. Government Printer, Nairobi.

[RoK] Republic of Kenya. 2010. Tourism Performance Overview. Ministry of Tourism, Nairobi.

Sindiga, I. 1999. Tourism and African Development: Change and Challenge of Tourism in Kenya. African Studies Centre, Leiden, Netherlands.

Sterly, J. 1997. Simbu Plant-Lore Plants Used by the People in the Central Highlands of New Papua Guinea. Dietrich Reimer Verlag, Berlin.

The Daily Nation. 2009. Kenya: The vanishing snow of Mount Kenya. Thursday, 17 December.

Carbon Neutral Company. 2011. Meru and Nanyuki Community Reforestation Project, London.

[UN] United Nations. 2008. Mount Kenya National Park/Natural Forest. Archived from the original on 2006-12-30. http://web.archive.org/web/20061230202343/http.//whc.unesco.org/pg.cfm?cid=31&id_site=800. Retrieved 2008-02-23.

[UNEP] United Nations Environment Programme. 1998. Protected Areas and World Heritage. Archived from the original on 2007-02-12. http://web.archive.org/web/20070212211303/http://www.unep-wcmc.org/sites/wh/mt_kenya.html. Retrieved 2008-02-23.

[UNEP] United Nations Environment Programme. 2008. Climate Change Adaptation and Mitigation in the Tourism Sector: Frameworks, Tools and Practices. Oxford University Press.

[UNESCAP] United Nations Economic and Social Commission for Asia Pacific. 2001. Managing Sustainable Tourism Development. ESCAP Tourism Review, No. 22

[UNWTO] United Nations World Tourism Organisation. 2007. World Tourism Barometer. Madrid.

[UNWTO] United Nations World Tourism Organisation and [UNEP] United Nations Environment Programme. 2008. Climate Change and Tourism. Responding to Global Challenges. Madrid.

Vellinga, P. and R. Stewart. 1991. The greenhouse marathon: proposal for global strategy. pp. 129–134. J. Jäger and H.L. Ferguson (eds.). In: Climate Change: Science, Impacts and Policy. Cambridge University Press, Cambridge.

[WCED] World Commission on Environment and Development. 1987. Our Common Future, UN, New York.

Whiteman, D. 2000. Mountain Meteorology, Oxford University Press.

9

Managing Ecosystem Services for Enhancing Climate Change Adaptation in the Hindu Kush Himalayas

Nakul Chettri, * *Golam Rasul* and *Eklabya Sharma*

INTRODUCTION

Ecosystem services are benefits people derive from ecosystems, which include provisioning services (food and water); regulating services (regulation of floods, drought, land degradation, and disease); supporting services (soil formation and nutrient cycling); cultural services (recreational, spiritual or religious); and other non-material benefits (MA 2005). These services are critical to the functioning of the Earth's life support system that contributes to human welfare, both directly and indirectly, representing part of the total economic value of the planet (Costanza et al. 1997). Dynamic ecosystems and their services are intricately linked to human wellbeing and are linked with various drivers of change, such as excessive demands from a growing population, land use and cover change, and climate change to name a few, leading to biodiversity loss (SCBD 2010). The Millennium Ecosystem Assessment (MA) documented the importance of ecosystem services to human wellbeing and showed that continued supply of these services is threatened by unsustainable anthropogenic activities (MA 2005).

International Centre for Integrated Mountain Development, Kathmandu, Nepal.
* Corresponding author

Approximately 60% of the ecosystem services examined during the MA are being degraded or used unsustainably including food, freshwater, air and water purification, and the regulation of regional and local climate and natural hazards. As a result, the focus has been gradually widening from biodiversity conservation to management and sustenance of the ecosystems for adaptation to change on which are dependent for goods and services (TEEB 2010).

The Hindu Kush-Himalayas (HKH) is one of the most dynamic ecosystems in the world with a rich and remarkable biodiversity (Pei 1995, Guangwei 2002, Chettri et al. 2008a). Stretched over more than four million square kilometers, the HKH includes Bhutan and Nepal in their entirety and parts of six other countries: Afghanistan, Bangladesh, China, India, Myanmar and Pakistan. The region is endowed with a high level of endemism, diverse gene pools, species and ecosystems of global importance (Mittermeier et al. 2004). Numerous critical ecoregions of global importance can be found in this region (Olson and Dinerstein 2002). As a result, the HKH have been highlighted in many global conservation prioritization strategies (see Brooks et al. 2006). In terms of species diversity, the region is equally rich in flora and fauna (Chettri et al. 2008b, Chettri et al. 2010). It is a home to all four big cats of Asia: the snow leopard (*Uncia uncia*), tiger (*Panthera tigris*), common leopard (*Panthera pardus*), and clouded leopard (*Neofelis nebulosa*). Ungulates, a number of which are endemic, such as the Tibetan wild ass (*Equus kiang*), wild yak (*Bos grunniens*), Chiru (*Pantholops hodgsoni*), and Tibetan gazelle (*Procapra picticaudata*) are of special significance (Chettri et al. 2011).

This complex and fragile ecosystem with extreme heterogeneity in micro environments helps to stabilize headwaters, preventing flooding, and maintain steady year-round flows of ecosystem goods and services (Xu et al. 2009). These functions contribute to one third of the humanity for their wellbeing far beyond the immediate vicinity, benefiting entire river basins. As a result, the HKH have often been referred to as 'natural water towers' because they contain the headwaters of rivers, which are vital for maintaining human life in the densely populated areas downstream (Schild 2008). In addition, the HKH also represents a unique source of freshwater for agricultural, industrial and domestic use, and are an important economic component of tourism and hydro-electric power production and maintains water quality, regulates water flow (floods and droughts), and supports biodiversity (Trisal and Kumar 2008, Eriksson et al. 2009, Xu et al. 2009, Rasul et al. 2011). The region also plays an important role in mitigating the impacts of climate change by acting as carbon sinks (ICIMOD 2009, Trisal and Kumar 2008).

However, this diverse ecosystem of the HKH is facing overarching threats from various drivers of changes including climate change (Myers et al. 2000, Pandit et al. 2007, Xu et al. 2009, Singh et al. 2011). The ecosystems in the HKH are degrading mainly due to lack of incentive provisions for maintaining ecosystems and the goods and services provided by them. This is leading to development that is unsustainable including loss of biodiversity. Even the protected areas such as national parks, nature reserves and wildlife sanctuaries face tremendous pressures from external driving forces and communities living inside and outside (Sharma and Yonzon 2005). It is a paradox that in spite of being rich in biodiversity the region is also home to poorest of the world and the most vulnerable in the face of climate change (Chettri et al. 2010, Singh et al. 2011). So, there is a mounting challenge to balance conservation with development in the region. This chapter is an attempt to document the reconciling initiatives on maintaining ecosystem resilience through integrated conservation and development initiatives to address prevailing climate change challenges faced by the region with some evolving regional experiences.

Ecosystem Diversity, Their Services and Human Wellbeing

The ecosystems of the HKH are inherent component of the culture, landscape and environment of this high mountain area of Asia. Elevation zones across the HKH extend from tropical (>500 meters above sea level) to alpine ice snow (>6,000 meters above sea level), with a principal vertical vegetation regime composed of tropical and subtropical rainforest, temperate broadleaf deciduous or mixed forest, and temperate coniferous forest, including high altitude cold shrub or steppe and cold desert (Guangwei 2002, Pei 1995). The dominant vegetation types such as forests and rangelands have multiple functions: they harbor biodiversity, anchor soil and water, provide carbon sinks, regulate climate, and temper stream flow. Likewise the wetland ecosystems, which are the transitional ecosystems between terrestrial and aquatic habitats encompassing water, soil and organisms, supports rich agricultural and wild biodiversity as well as provides environmental services such as food, flood regulation, nutrient and sediment retention, maintenance of groundwater table and so on, that in turn, become valuable benefits to the society. They are also habitat for a wide variety of birds, mammals, reptiles, amphibians, fish and invertebrate species. They are sometimes described as 'the kidneys of the landscapes' because of the functions they perform in hydrologic and chemical cycles and as a store house of carbon as a result in the form of peat lands.

The HKH is characterized by five agro-climatic zones and farming practices viz. (i) specialized pastoralism (purely livestock based, high altitude transhuman subsistence livelihoods); (ii) mixed mountain agro-pastoralism (livestock, agriculture, and agroforestry livelihoods based in the mid hills); (iii) cereal based hill farming systems (agriculture based livelihoods in the low and mid hill areas); (iv) shifting cultivation (livelihoods based on rotational agroforestry with slash and burn practices); and (v) specialized commercial systems (livelihoods based on monoculture and other commercial crops). In each of these specialized zones there is a variation in crops and cropping patterns that supports a wide range of agro-biodiversity, which is the source of food, nutrients, and economic prosperity for the region (Sharma and Kerkhoff 2004). Among these farming systems, specialized pastoralism is one of the oldest and the most predominant systems in the HKH. Some of the well known pastoral communities found in the Himalayas are *Bakrawals, Gujjars, Gaddis, Kanets, Bhotias, Kaulis,* and *Kinnauras* of the North Indian Himalayas; *Bhotias* and *Sherpas* of the Khumbu valley of Nepal; the *Kirats* of eastern Nepal; *Lachungpas* and *Lachenpas* of Sikkim; *Changpas* of Ladakh; *Brokpas* of Bhutan; *Tibetans* of China; and the *Shimshal* of Pakistan. These people's age-old dependence on the high pastures and livestock products is embedded in their culture and practices, and governed by traditional knowledge and natural resources governance mechanisms (Chettri et al. 2012).

Ecosystems are capital assets of the HKH that provide a wide range of services. These include supporting services that maintain the conditions for life; provisioning services that provide direct inputs to livelihoods and the economy; regulating services such as those that provide flood and disease control; cultural services that provide opportunities for recreation, spiritual or historical sites; and supporting services that sustain and fulfill human life (MA 2005). The region supports 10 major river basins: the Indus, Ganges, Amu Darya, Brahmaputra (Yarlungtsanpo), Irrawaddy, Salween (Nu), Mekong (Lancang), Tarim, Yangtse (Jinsha), and Yellow River (Huanghe) basins (Table 9.1). Glacial melt makes an important contribution to river flow, varying from the lowest rate of 1.3% for the Yellow River to the highest rates of 40.2% for the Tarim and 44.8% for the Indus River Basin. It has been estimated that about 30% of the water resources in the eastern Himalayas are directly derived from the melt of snow and ice; this proportion increases to about 50% in the central and western Himalayas and becomes as high as 80% in the Karakoram (Xu et al. 2009). The region is also important for high altitude wetlands, source of freshwater resources, and plays an important role in water storage and regulating water regimes (Trisal and Kumar 2008). These regions also play an important role in mitigating the impacts of climate change by acting as carbon sinks. The peatlands on the Tibetan Plateau are one of the most important stores of carbon in the mountain region, storing 1500 to 4000 t ha^{-1} (Trisal and Kumar 2008).

During summer, the region further provides an anomalous mid-tropospheric heat source for southwestern Asia, and, thus, plays a prominent role in the Asian monsoon system (Yanai et al. 1992). Seasonal blocking episodes with associated anomalies in temperature and precipitation are

Table 9.1. The 10 major river basins in the Himalayan region.

River basins	Basin area (sq.km)	Countries	Population (x 1000)	Population density (per sq. km)
Amu Darya	534,739	Afghanistan, Tajikistan, Turkmenistan, Uzbekistan	20,855	39
Brahmaputra	651,335	China, India, Bhutan, Bangladesh	118,543	182
Ganges	1,016,124	India, Nepal, China, Bangladesh	407,466	401
Indus	1,081,718	China, India, Pakistan	178,483	165
Irrawaddy	413,710	Myanmar	32,683	79
Mekong	805,604	China, Myanmar, Laos, Thailand, Cambodia, Vietnam	57,198	71
Salween	271,914	China, Myanmar, Thailand	5,982	22
Tarim	1,152,448	Kyrgyzstan, China	8,067	7
Yangtze	1,722,193	China	368,549	214
Yellow	944,970	China	147,415	156
Total	8,594,755		1,345,241	

Source: Adapted from Eriksson et al. (2009).

also closely linked to the presence of mountains, which act as orographic barriers to the flow of moisture-bearing winds and control precipitation in neighboring regions (Manabe and Terpstra 1974, Kutzbach et al. 1993). For example, the Himalayas play an important role as a trigger mechanism for cyclogenesis through their perturbation of large-scale atmospheric flow patterns; they also act as a barrier to atmospheric circulation for both the summer monsoon and the winter westerlies. The summer monsoon dominates the climate of the region, but is longest in the eastern Himalayas, lasting five months (June–October) in Assam, four months (June–September) in the Central Himalayas, and two months (July–August) in the western Himalayas (Chalise and Khanal 2001). However, the Himalayas display a great variability in hydrometeorological conditions: the western Himalayas and north facing slopes are generally arid and dry, while the eastern Himalayas and south-facing slopes are generally humid and wet.

The livelihoods of people in the HKH are mainly based on forests, agro-forestry, wetland and rangeland resources (Chettri and Sharma 2006, Sharma et al. 2006). The long history of human presence in this ecosystem and maintenance of its fragility is a strong indicator of compatibility of satisfaction of community needs through traditional practices, with biodiversity conservation. The functioning of these systems, in which the human beings were also part of, greatly depended on surrounding ecosystem base for various goods and services. These natural resources were managed traditionally, with practices having been evolved by the local communities through trial and error over long period of time (Rai et al. 1994). Belief in Buddhism, Hinduism, and a varying blend with animistic beliefs cuts across all mountain people imparting a sense of compassion and awareness for all forms of life and the surrounding natural environment. Buddhist beliefs in 'hidden lands' or 'beyuls' (Sherpa 2003), and 'hidden treasures' or *ters* (Ramakhrishnan 1996) are often linked to the idea of conservation areas for human and nature and provide a strong organizing principle in how people relate to vast natural spaces and the biodiversity therein. Such beyuls are rich in biodiversity and often are named for dominant flora and fauna, such as the 'Beyul Khenpalung' or the 'Hidden Land of Artemisia' in Makalu Barun National Park in Nepal, and Demajong, or 'valley of rice' in Sikkim. These areas were usually marked by strict observances such as bans on hunting, mining, polluting rivers and streams, and harvesting of timber and plant resources. Transgressors were often punished through fines and other disciplinary actions. Likewise, the traditional natural resources management systems such as Dzumsa by the Pipons (village head) among the Lanchungpas in Sikkim (Rai et al. 1994), wise knowledge and sustainable natural resources use practiced among the Lepchas (Jha 2002), strong ethics for landscape level conservation among Sikkimese Buddhists (Ramakrishnan 1996), effective rangeland management by Kiratis and Limbuwans (Oli 2008) and Bhutias (Nautiyal et al. 2003) are some of the effective traditional conservation measures that address 'sustainability'. Thus, conservation was culturally enforced within many of these indigenous groups. This reveals that in the past, there was strong resilience between biological resources and the human needs. However, these practices, which are participatory and inclusive, have not been recognized by many of the national policies of the Himalayan region as effective conservation measures and are overshadowed by modern statuary strict and stringent obligations.

Potential Impact of Climate Change on Ecosystem Services

The HKH region is facing enormous pressures from an array of drivers of change including climate change (Erikson et al. 2009, Xu et al. 2009, Tse-ring et al. 2010). The region has shown consistent trends in overall warming during the past 100 years (Yao et al. 2007). Various studies suggest that warming in the HKH has been much greater than the global average of 0.74°C over the last 100 years (IPCC 2007, Du et al. 2004). For example, warming in Nepal was 0.6°C per between decades 1977 and 2000 (Shrestha et al. 1999). While the Fourth Assessment Report (4AR) of Intergovernmental Panel on Climate Change (IPCC) made a strong science-based rationale for the need for actions countering the potential ill effects of climate change globally (IPCC 2007), it also pointed out the lack of reliable data and limited data collection efforts in the HKH. It is evident that climate change in the HKH will affect all aspects of the climate, making rainfall less predictable, changing the character of seasons, and increase the risk to biodiversity (Xu et al. 2009, Chettri et al. 2010, Chettri et al. 2012). However, the impacts of climate change are not evenly distributed within the region, nor among different communities and sectors of society.

Potential impact ecosystems

Although there is no strict compartmentalization of vegetation along altitudinal gradients in the HKH region, elevation has important implications for its ecology, evolution, physiology and conservation and is highly relevant to species' composition and phenology patterns (Chettri et al. 2001, Carpenter 2005). As a result of microclimatic variations, most organisms found in the HKH are confined to specific ecosystems such as highland pastures, forests and so on. This is a special risk factor for highland species that are sensitive to climate change (Pounds et al. 2006) and more likely to be at risk of extinction. Globally, there is evidence of the shift of species towards the north in latitude (Hickling et al. 2006) or higher elevations (Wilson et al. 2007), especially for species in the transition zone between subalpine and alpine which are more vulnerable to climate change as they have limited scope for movement. Analyses for the Himalayas are few and limited to certain pockets of areas (Carpenter 2005). Observations have been made about the change in events related to plant and animal phenology and also to shifting of treelines and encroachment of woody vegetation into alpine meadows. Phenological changes, such as early budding or flowering and ripening of fruits in plants and hibernation, migration, and breeding in animals, could have adverse impacts on pollination patterns. Consequently, this may have an impact on the population of pollinators, leading to change in ecosystem productivity and species' composition in high-altitude habitats (Thuiller et al. 2008).

Potential impact species

Climate change increases the risk of extinction of species that use narrow geographic and climatic ranges but important for the ecosystem integrity (Hannah et al. 2007). According to the prevailing extinction theory, the larger and more specialized species are likely to be lost due to habitat destruction (Sodhi et al. 2004). This might be significant for the Himalayan region as habitat and forest destruction were seen to have increased although the quality of increasing and decreasing forests has not been assessed (Pandit et al. 2007). Since ecosystems in the Himalayas are layered as narrow bands along a longitudinal axis of the mountain range, they are greatly influenced and easily impacted by climatic variations. For example, the sub-tropical and temperate forests (broadleaved, coniferous, and mixed), include the tiger (*Panthera tigris*) and other members of the cat family (Felidae) which would be extremely vulnerable to climate change as would narrowly endemic taxa, such as Mishmi takin (*Budorcas taxicolor taxicolor*) and Hoolock gibbon (*Hoolock hoolock*), which are likely to face challenges to their conservation in the forests. The brow-antlered deer (*Cervus eldi*), locally known as Sangai, is endemic to the Manipur wetlands, especially Loktak Lake, and is the rarest and most localized subspecies of deer in the world.

Potential impact people's livelihood

The increasing risk for human livelihoods and wellbeing include increasing frequency and severity of extreme events such as cyclones, landslides and floods. Within the Himalayan ecosystems, the impact of these changes is often aggravated by existing environmental and socio-economic problems, such as poverty, water scarcity or food deficiency (Mertz et al. 2009). These in turn contribute to a downward-spiralling cycle with adversely impacting the livelihoods and driving people to desperate measures that decimate ecosystem services. Observational evidence indicates that the impacts related to climate warming are well underway on the Himalayas (Singh et al. 2011) and increasing threats to biodiversity and derived ecosystem goods and services (Chettri et al. 2010, Chettri et al. 2011). The poorer, more marginalized people of the high mountains are likely to suffer the earliest and the most. Given the evidence that many risks already threaten women disproportionately; and also the elderly, disabled, and indigenous groups, especially their poorer members; identifying changes in the cryosphere and alpine ecosystem most likely to affect them is of utmost importance. In addition, there are broader regional questions of which the more severe highland-to-lowland dangers relate to rapid melting events, floods caused by natural dam bursts, increased sedimentation, and droughts caused by reduced or changed flow patterns.

Potential and Evolving Adaptation Strategies

In the HKH, the classical approach of biodiversity conservation started with emphasis on the flagship species conservation. The assumption was that if the flagship species, which usually occupied the tip of the pyramid in the food web in an ecosystem, flourished, then the ecosystem was considered healthy. However, the approach changed significantly from species focused conservation to landscape level within last three decades (Sharma et al. 2010). As of 2007, there were 488 protected areas (IUCN category I–VI) within the region, covering more than 1.6 million km^2, representing about 39% of the region's terrestrial area (Chettri et al. 2008a). Interestingly, the proportion of terrestrial area covered by the protected areas in the region is much higher (39%) than in Central America (26%) (Chape et al. 2005). Such growth in the number and areas of protected areas is a significant achievement on the part of the region countries towards fulfilling their global commitment to conservation. Interestingly, the analysis showed that the protected areas in the HKH have adopted a shift away from strictly managed protected area systems to community based, as also observed by Zimmerer et al. (2004).

Several recent initiatives in the region offer significant opportunities for advancing and piloting innovative and regionally appropriate conservation and adaptation approaches. In particular, the importance of biodiversity conservation in protected areas, corridors and transboundary landscapes focussing on climate resilience by maintaining ecosystem integrity for enhancing flow of environmental goods and services have been the thrust for ICIMOD since last one decade. Here are some examples on the reconciling initiatives piloted by ICIMOD in the HKH.

Landscape/ecosystem approach in biodiversity conservation

Landscape/Ecosystem approach in biodiversity conservation is an evolving concept (Worboys et al. 2010). The concept has emerged primarily out of recognition that strict protection through a network of protected areas (e.g., national parks, sanctuaries, wildlife reserves) is an essential but insufficient biodiversity conservation strategy (Naughton-Treves et al. 2005, Ervin 2011). These researchers and others argue that protected areas are essential as these are the places where biodiversity conservation is the primary objective. However, many of the existing protected areas are too small to meet the ecological needs of viable population of wide ranging species in the changing climate (Ibisch et al. 2010). Thus, more than preserving isolated patches of sustained wilderness in the form of protected areas, the focus is now more on the necessity of maintaining landscape integrity, of viewing and conserving ecosystems as part of larger agro-ecological and socio-cultural landscapes to withstand the challenges posed by climate change (Worboys et al. 2010).

Application of landscape or ecosystem approach, as advocated by the Convention on Biological Diversity (CBD), recognizes the need of increased regional cooperation, in part due to the biophysical nature of these mountainous areas, the extreme heterogeneity of the region, inter-linkages between biomes, habitats and sectors, and the strong upstream –downstream linkages related to the provisioning of ecosystem services.

Seven critical 'Transboundary Landscapes' have been identified by ICIMOD (Fig. 9.1), highlighting the crucial role of improved cooperation amongst the countries of the region to enhance understanding the value of biodiversity and the potential impacts of environmental change on ecosystem goods and services. In these initiatives, ICIMOD is promoting transboundary cooperation for research and long term monitoring on both ecological and environmental aspects at a landscape level and piloted in a number of transboundary landscapes since late 1990s (see Sherpa et al. 2003, Sharma and Chettri 2005, Sharma et al. 2007, Zomer et al. 2010).

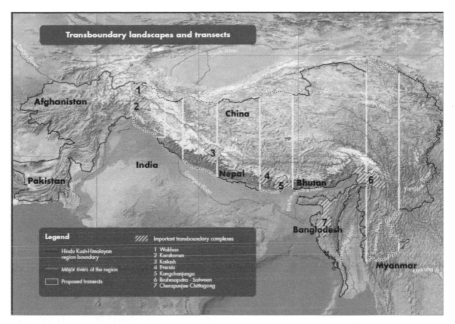

Figure 9.1. Map showing transbounday landscapes and transects in the HKH.

Color image of this figure appears in the color plate section at the end of the book.

Trans-Himalayan Transect approach

Climatic, environmental and other change processes across the HKH have both regional and global concerns (Messerli 2009). Nevertheless, the HKH is one of least scientifically studied or monitored areas in the world, and a 'data-deficit' region (IPCC 2007). Basic hydro-meteorological data are lacking, sparse or not readily available. This is true for other environmental data, e.g., biodiversity, land-use and land cover change, climate change impacts on various ecosystem goods and services, and carbon cycles. An improved understanding of these regional change processes is essential to provide the basis for informed decision making, risk and vulnerability mapping, adaptation and mitigation strategies and effective biodiversity conservation and management. ICIMOD, being an intergovernmental regional center, is working in the eight countries of the HKH and has been active in facilitating its regional member countries through various conservation and development approaches. The 'HKH Trans-Himalayan Transect', an approach to address the information gaps across the HKH, was conceptualized and discussed among global and regional stakeholders in 2008 (Chettri et al. 2009). Four 'transects' have been proposed considering representation from west to east, dry to wet and the south to north latitudinal expanse of the HKH region (Fig. 9.1). As indicated above, additionally, seven Transboundary Landscapes provide an initial opportunity for piloting of the concept and activities including a range of environmental monitoring and the initiation of long-term ecological and environmental research. The geographically defined 'transects' allow for co-locating research, monitoring and sampling sites, in-depth studies, and action research projects across the region, and for both comparative research and synergistic efficiencies. ICIMOD envisaged playing a facilitating role amongst the regional, national and local partners, and the global research community and other stakeholders through participatory and consultative processes encouraging regional cooperation and national ownership. On an experimental basis ICIMOD has initiated a number of pilot programs (Box 9.1).

Box 9.1. The Kailash Sacred Landscape Conservation Initiative (KSLCI).

Piloting the concept, ICIMOD has been engaged in partnership with UNEP, GTZ, and member countries in the Kailash Sacred Landscape. This transboundary landscape includes an area of the remote southwestern portion of the Tibetan Autonomous Region of China, and adjacent parts of northwestern Nepal, and northern India, and is comprised of a broad array of bioclimatic zones, rich natural and cultural resources, and a wide range of forest types. The initiative engages regional, national and local stakeholders in a consultative process for facilitation of transboundary, integrated approaches to sustainable development and conservation. Ecosystem management is promoted through the Regional Cooperation Framework development process, based upon a Conservation Strategy, supported by a Comprehensive Environmental Monitoring Plan, to address threats to the environmental and cultural integrity of this area, analyze change processes, and to develop a knowledge base upon which to build regional cooperation.

Zomer et al. 2010.

Valuation and rewarding ecosystem service providers

Humans benefit from biodiversity rich areas with the provision of ecological services such as climate regulation, soil formation and nutrient cycling; and from the direct harvest of biodiversity for food, fuel, fibers and pharmaceuticals. In the face of increasing human pressures on the environmental change, these benefits could act as powerful incentives to conserve nature (MA 2005), yet evaluating them has proved difficult because they are mostly not captured by conventional, market-based economic activity and analysis (Rasul et al. 2011). In the recent years, a new generation of conservation approaches with economic dimensions is rapidly emerging. They differ from traditional approaches in three critical and interrelated ways: a) they emphasize human-dominated landscapes; b) focus on ecosystem services, and c) utilize innovative finance mechanisms. Such concerns have moved beyond the science community to the global stakeholder and policy makers with the publication of the Millennium Assessment (MA 2005). The analysis acknowledges that biodiversity plays a significant role in directly providing goods and services as well as regulating and modulating ecosystem properties that underpin the delivery of ecosystem services. To rationalize the conservation value of biodiversity and the derived goods and services in the landscapes, ICIMOD developed an assessment framework paper (Rasul et al. 2011) and also identify and even quantify the ecosystems goods and services provided by conservation corridors (Pant et al. 2012). The economic benefits generated by the flow of selective forest ecosystem services in the three districts was around NPR 8.9 billion per annum (approximately US$ 125 million) equivalent to NPR 30,000 per ha/per year. Almost 80% of the total benefits (NPR 7.01 billion per annum or approximately US$ 98 million) was from provisioning services, i.e., goods from the forests used directly or indirectly. The average benefit per household from ecosystem services was estimated to be NPR 60,144 per year. The value of carbon sequestration services was also considerable at NPR 1.65 billion annually, close to 18% of the total value of the ecosystem services (see Pant et al. 2012). The study was important in terms of reconciling conservation and climate change as it rationalizes the need for enhancing ecological resilience through conservation intervention. Such studies have been replicated in other critical ecosystems such as Koshi Tappu Wildlife Reserve in Nepal (ICIMOD and MoFSC/GoN 2014) and Phobjikha Conservation Area of Bhutan (ICIMOD and RSPN 2014).

Ecosystem Resilience and Community Adaptation Opportunities

While acknowledging the significant diversity of ecosystems in the HKH region and the existence of a fair understanding of the important drivers of change, it is recognized that concerted efforts are needed to monitor and research the impacts of climate change on biodiversity. During the course of ICIMOD's learning, four priority thematic areas were identified to strengthen the reconciling process.

Long-term consistent monitoring of both climate change and biodiversity

The importance and need for establishing long-term, consistent monitoring of climate change and its impact on ecosystems and people's livelihoods have been clearly realized. Permanent plots and/or units need to be established on an altitudinal transect spanning the tropics to the alpine regions in order to monitor diverse ecosystems. An institutionalized monitoring system, however, requires standardization of monitoring parameters. In this respect a consistent, uniform methodology and a network of collaborative efforts to collect and analyze data and information regularly was made a prerequisite. Realizing the need for a facilitating institution at a regional level, ICIMOD has taken on the role by consensus. Academic and research institutions need to be engaged to establish and maintain the permanent research plots, carry out the regular monitoring and generate and analyze the data. The involvement of communities in the respective areas was seen to be critical in maintaining the plots, in participatory action research and in carrying out observations and sharing perceptions.

Focussed research on impacts, coping mechanisms and adaptation to climate change

Documentation on impacts is, as yet, anecdotal for the most part—there is a need, therefore, to document impacts as well as coping mechanisms of communities to change systematically. The most promising indicators seem to be agro-biodiversity, followed by other forms of biodiversity (both flora and fauna). Documentation of changes in crops and their performance and coping mechanisms of communities, focussing on changes in cropping patterns, crop shifts and cropping system management should be carried out on a priority basis. One important aspect requiring documentation and monitoring is the changes in nutritive value of crops as a result of the impacts of climate change. Systematic documentation and monitoring, however, will need a framework of institutional support and re-orientation of existing government research programs and institutions in regard to adaptive research.

Assessment of critical habitat linkages, protected area effectiveness, ecological and social vulnerabilities

In addition to the functional responses of existing protected areas to climate change (which could provide critical information about responses of natural systems to change and hence provide benchmark parameters), new critical habitats and the necessity for linkages of such habitats to existing ones need identifying. Existing protected areas will require constant monitoring to document changes in vegetation, identification, and census of indicative species (to monitor population dynamics as a function of changing climate impacts). The effectiveness of protected area management will have to be central to all research and feed into evolving responsive management approaches and technologies. Findings and conclusions from the above should provide insight into adaptive responses and into resilience of natural systems; and these will become critical elements in evolving decision-support systems and hence require priority. Research on institutional frameworks and their effectiveness in governance and assessments of good practices with examples of community-led conservation have to be central to formulation of an effective and responsive governance system. Strong emphasis has to be placed on indigenous knowledge systems, particularly in regard to natural resource management approaches and institutional frameworks, drawing upon traditional practices of management and governance especially in regard to sacred landscapes.

Policy analysis on climate change, adaptation and coping mechanisms; and relevant adjustments to existing policies

In order to support and strengthen community efforts to cope with change, an enabling policy environment is essential. Documentation and assessments indicate the need for policy dialogues focussing on areas identified. Policy dialogues would need to focus on areas where adjustments in existing policies are required, particularly in regard to economic benefits, governance frameworks, and local-level policy adjustments. A clear concern was the multiplicity of policy actors governing natural resource management and livelihood support and the need for convergence of different (often conflicting) policies under one forum for ease of implementation. Dialogues need to focus on this required convergence before moving on to sectoral details. There is a critical role for scientific institutions in regard to policy formulation concerning natural resource management, livelihood support and climate change. Policy makers require authentic data inputs and, more often than not, these are not available or not in a comprehensible form. Scientific institutions need to fill this gap so that policy making can be based on scientific findings.

Conclusion

Ecosystem services are benefits people derive from ecosystems. These services are critical to the functioning of the Earth's life support system and are intricately linked to human wellbeing. However, they are witnessing growing pressure increasing human population, land use and cover change, and climate change to name a few. The Hindu Kush Himalayas, with a complex and fragile ecosystem maintain steady year-round flows of ecosystem goods and services to one third of the humanity for their wellbeing far beyond the immediate vicinity, benefiting entire river basins. However, this diverse ecosystem of the HKH is facing overarching threats from various drivers of changes including climate change. Even the protected areas such as national parks, nature reserves and wildlife sanctuaries face tremendous pressures from external driving forces and communities living inside and outside. The most pressuring challenge faced by these protected areas is from the prevailing climate change.

The member countries sharing the Hindu Kush Himalayas are progressing towards adaptive measures through innovative and regionally appropriate conservation and adaptation approaches. In particular, the importance biodiversity conservation in protected areas, corridors and transboundary landscapes focussing on climate resilience by maintaining ecosystem integrity for enhancing flow of environmental goods and services have been the thrust for the region. As a regional knowledge development and learning center, ICIMOD is committed to strengthen the existing research network in the HKH and also facilitate data sharing mechanisms. In 2008, a framework on 'HKH Transboundary Landscapes and Trans-Himalayan Transect' was developed with the technical inputs from the recognized mountain experts with an objective to promote transboundary collaboration among the countries and filling the data deficit issue through seven representative landscapes across the gradients of precipitation, altitude and latitude. Since 2008, ICIMOD and its regional member counties have made significant progress in mainstreaming this Framework within ICIMOD through implementation of Transboundary Landscape Programme.

The need for biodiversity conservation has been rationalized by valuation of ecosystems services provided by various ecosystems, both inside and outside the protected areas. One of the studies showed that around NPR 8.9 billion per annum (approximately US$ 125 million) equivalent to NPR 30,000 per ha/per year worth of services are used by local communities from the forested ecosystem of which 80% of the total benefits (NPR 7.01 billion per annum or approximately US$ 98 million) was from provisioning services, i.e., goods from the forests used directly or indirectly. The average benefit per household from ecosystem services was estimated to be NPR 60,144 per year. The study was important in terms of reconciling conservation and climate change as it rationalizes the need for enhancing ecological resilience through conservation intervention.

The importance and need for establishing long-term, consistent monitoring of climate change and its impact on ecosystems and people's livelihoods have been clearly realized and the functional responses of existing protected areas to climate change and the effectiveness of protected area through policy analysis considering the contemporary challenges from climatic change and provision for adaptation and coping strategies were felt necessary.

Acknowledgements

We express our sincere gratitude to the Director General of ICIMOD, Dr. David Molden, for his inspiration and for providing the required facilities for the preparation of this chapter. We are also thankful to many ICIMOD colleagues for their support and contributions.

References

Brooks, T.M., R.A. Mittermeier, G.A.B. da Fonseca, J. Gerlach, M. Hoffman, J.F. Lamoreux, C.G. Mittermeier, J.D. Pilgrim and A.S.L. Rodrigues. 2006. Global biodiversity conservation priorities. Science 313: 58–61.

Carpenter, C. 2005. The environmental control of plant species density on a Himalayan elevation gradient. Journal of Biogeography 32: 999–1018.

Chalise, S.R. and N.R. Khanal. 2001. An introduction to climate, hydrology and landslide hazards in the Hindu Kush-Himalaya region. pp. 51–62. *In*: L. Tianchi, S.R. Chalise and B.N. Upreti (eds.). Landslide Hazard Mitigation in the Hindu Kush-Himalayas. ICIMOD, Kathmandu.

Chape, S., M. Harrison, M. Spalding and I. Lysenko. 2005. Measuring the extent and effectiveness of protected areas as an indicator for meeting global biodiversity targets. Phil. Trans. R. Soc. B 360: 443–455.

Chettri, N., A.B. Shrestha, Y. Zhaoli, B. Bajracharya, E. Sharma and Q. Hua. 2012. Real world protection from the 'third pole' and its people. pp. 113–133. *In*: F. Huettmann (ed.). Protection of Three Poles. Springer, Japan.

Chettri, N., B. Shakya and E. Sharma. 2011. Enhancing ecological and people's resilience: an approach for climate change adaptation in the Hindu Kush-Himalayan region. pp. 29–31. *In*: Contribution of Ecosystem Restoration to the Objectives of the CBD and a Healthy Planet for All People. Abstracts of Posters Presented at the 15th Meeting of the Subsidiary Body on Scientific, Technical and Technological Advice of the Convention on Biological Diversity, 7–11 November 2011, Montreal, Canada. Technical Series No. 62. Montreal, SCBD.

Chettri, N., B. Shakya, R. Thapa and E. Sharma. 2008a. Status of protected area system in the Hindu Kush Himalaya: an analysis of PA coverage. International Journal of Biodiversity Science and Management 4(3): 164–178.

Chettri, N., E. Sharma, B. Shakya, R. Thapa, B. Bajracharya, K. Uddin, K.P. Oli and D. Choudhury. 2010. Biodiversity in the Eastern Himalayas: status, trends and vulnerability to climate change. Climate change impact and vulnerability in the Eastern Himalayas—Technical Report 2. ICIMOD, Kathmandu, Nepal.

Chettri, N., E. Sharma and R. Thapa. 2009. Long-term monitoring using transect and landscape approaches within the Hindu Kush-Himalayas. pp. 201–208. *In*: E. Sharma (ed.). Proceedings of the International Mountain Biodiversity Conference Kathmandu, 16–18 November 2008. ICIMOD, Kathmandu, Nepal.

Chettri, N., E. Sharma and D.C. Deb. 2001. Bird community structure along a trekking corridor of Sikkim Himalaya: a conservation perspective. Biological Conservation 102(1): 1–16.

Chettri, N., B. Shakya and E. Sharma. 2008b. Biodiversity conservation in the Kangchenjunga landscape. ICIMOD, Kathmandu, Nepal.

Chettri, N. and E Sharma. 2006. Prospective for developing a transboundary conservation landscape in the Eastern Himalayas. pp. 21–44. *In*: J.A. McNeely, T.M. McCarthy, A. Smith, O.L. Whittaker and E.D. Wikramanayake (eds.). Conservation Biology in Asia: Society for Conservation Biology, Asia Section and Resources Himalaya Foundation, Kathmandu, Nepal.

Costanza, R., R. d'Arge, R.S. de Groot, S. Farber, M. Grasso, B. Hannon, K. Limburg, S Naeem, R.V. O'Neil, J. Paruelo, R.G. Raksin, P. Sutton and M. Van den Bel. 1997. The value of world's ecosystem services and natural capital. Nature 387: 253–260.

Duan, K., T. Yao and L.G. Thompson. 2004. Low-frequency of southern Asian monsoon variability using a 295-year record from the Dasuopu ice core in the central Himalayas. Geophys. Res. Lett. 31: L16206.

Eriksson, E., J. Xu, A.B. Shestha, R.A. Vaidya, S. Nepal and K. Sandström. 2009. The changing Himalayas: Impact of climate change on water resources and livelihoods in the greater Himalayas. ICIMOD, Kathmandu, Nepal.

Ervin, J. 2011. Integrating protected areas into climate planning. Biodiversity 12(1): 2–10.

Guangwei, C. (eds.). 2002. Biodiversity in the Eastern Himalayas. Conservation through Dialogue. Summary Reports of the Workshops on Biodiversity Conservation in the Hindu Hush- Himalayan Ecoregion. ICIMOD, Kathmandu, Nepal.

Hannah, L., G. Midgley, S. Andelman, M.I. Araújo, G. Hughes, E. Martinez-Meyer, R. Pearson and P. Williams. 2007. Protected area needs in a changing climate. Front. Ecol. Environ. 5(3): 131–138.

Hickling, R., D.B. Roy, J.K. Hill, R. Fox and C.D. Thomas. 2006. The distributions of a wide range of taxonomic groups are expanding polewards. Global Change Biol. 12: 450–455.

Ibisch, P.L., A.E. Vega and T.M. Herrmann. 2010. Interdependence of biodiversity and development under global change. Technical Series No. 54. Secretariat of the Convention on Biological Diversity, Montreal.

International Centre for Integrated Mountain Development (ICIMOD) and Ministry of Forests and Soil Conservation, Government of Nepal (MoFSC/GoN). 2014. An Integrated Assessment on Effects of Natural and Human Disturbance on a Wetland Ecosystem: A retrospective from the Koshi Tappu wildlife reserve, Nepal. ICIMOD, Kathmandu, Nepal (in press).

International Centre for Integrated Mountain Development (ICIMOD) and Ministry Royal Society for Protection of Nature (RSPN). 2014. An Integrated Assessment on Effects of Natural and Human Disturbance on a Wetland Ecosystem: A Retrospective from Phobjikha Landscape Conservation Area, Bhutan. ICIMOD, Kathmandu, Nepal (in press).

ICIMOD. 2009. Potential carbon finance in the land use sector of the Hindu Kush Himalayas: A Preliminary Scoping Studies. ICIMOD, Kathmandu, Nepal.

IPCC. 2007. IPCC Summary for Policymakers: Climate Change 2007: Climate Change Impacts, Adaptation and Vulnerability. IPCC WGII Fourth Assessment Report.

Jha, A. 2002. Ecological prudence for the Lepchas of Sikkim. Tigerpaper 29(1): 27–28.

Kutzbach, J.E., W.L. Prell and W.F. Ruddiman. 1993. Sensitivity of Eurasian climate to surface uplift of the Tibetan Plateau. J. Geolo. 101: 177–190.

Liu, X. and B. Chen. 2000. Climatic warming in the Tibetan Plateau during recent decades. Intern. J. Climat. 20: 1729–1742.

MA. 2005. Ecosystems and Human Well-being: Synthesis. Millennium Ecosystem Assessment Report. Published for World Resources Institute, Island Press, Washington, DC.

Manabe, S. and T.B. Terpstra. 1974. The effects of mountains on the general circulation of the atmosphere as identified by numerical experiments. J. Atmosph. Sci. 31: 3–42.

Mertz, O., C. Padoch, J. Fox, R.A. Cramb, S.J. Leisz, N.T. Lam and T.D. Vien. 2009. Swidden change in Southeast Asia: understanding causes and consequences. Human Ecology 37: 259–264.

Messerli, B. 2009. Biodiversity, environmental change and regional cooperation in the Hindu Kush-Himalayas. pp. 13–20. *In*: E. Sharma (ed.). Proceedings of the International Mountain

Biodiversity Conference Kathmandu, 16–18 November 2008. ICIMOD, Kathmandu, Nepal.

Mittermeier, R.A., P.R. Gils, M. Hoffman , J. Pilgrim, T. Brooks, C.G. Mittermeier, J. Lamoreaux and G.A.B. da Fonseca. 2004. Hotspots revisited: Earth's biologically richest and most endangered terrestrial ecoregions. Mexico City: CEMEX/Agrupación Sierra Madre.

Myers, N., R.A. Mittermeier, C.G. Mittermeier, G.A.B. da Fonseca and J. Kent. 2000. Biodiversity hotspots for conservation priorities. Nature 403: 853–858.

Naughton-Treves, L., M.B. Holland and K. Brandon. 2005. The role of protected areas in conserving Biodiversity and sustaining local livelihoods. Annual Review of Environment and Resources 30: 219–252.

Nautiyal, S., K.S. Rao, R.K. Maikhuri and K.G. Saxena. 2003. Transhumant pastoralism in the Nanda Devi Biosphere Reserve, India. Mountain Research and Development 22(3): 255–262.

Oli, K.P. 2008. Pasture, livestock, and conservation: challenges in the transborder areas of Eastern Nepal. pp. 91–96. *In*: N. Chettri, B. Shakya and E. Sharma (eds.). Biodiversity Conservation in the Kangchenjunga Landscape. ICIMOD, Kathmandu, Nepal.

Olson, D.M. and E. Dinerstein. 2002. The Global 200: priority ecoregions for global conservation. Annals of Missouri Botanical Garden 89: 199–224.

Pandit, M.K., N.S. Sodhi, L.P. Koh, A. Bhaskar and B.W. Brook. 2007. Unreported yet massive deforestation driving loss of endemic biodiversity in Indian Himalaya. Biodiversity and Conservation 16: 153–163.

Pant, K.P., G. Rasul, N. Chettri, K.R. Rai and E. Sharma. 2012. Value of forest ecosystem services: A quantitative estimation from the Kangchenjunga landscape in eastern Nepal. ICIMOD Working Paper 2012/5. ICIMOD, Kathmandu, Nepal.

Pei, S. 1995. Banking on biodiversity: report on the regional consultations on biodiversity assessment in the Hindu Kush Himalaya. ICIMOD, Kathmandu, Nepal.

Pounds, A.J., M.R. Bustamante, L.A. Coloma, J.A. Consuegra, M.P.L. Fogden, P.N. Foster, E.L. Marca, K.L. Masters, A. Merino-Viteri, R. Puschendorf, S.R. Ron, G.A. Sanchez-Azofeifa, C.J. Still and B.E. Young. 2006. Widespread amphibian extinctions from epidemic disease driven by global warming. Nature 439: 161–167.

Rai, S.C., E. Sharma and R.C. Sundriyal. 1994. Conservation in the Sikkim Himalaya: traditional knowledge and landuse of the Mamlay Watershed. Environmental Conservation. 15: 30 35.

Ramakrishnan, P.S. 1996. Conserving the Sacred: From Species to Landscape. Nature and Natural Resources. UNESCO, Paris.

Rasul, G., N. Chettri and E. Sharma. 2011. Framework for Valuing Ecosystem Services in the Himalayas. ICIMOD, Kathmandu, Nepal.

Schild, A. 2008. The case of the Hindu Kush-Himalayas: ICIMOD's position on climate change and mountain systems. Mountain Research and Development 28(3/4): 328–331.

Secretariat of the Convention on Biological Diversity (SCBD). 2010. Global Biodiversity Outlook 3. SCBD, Montreal, Canada.

Sharma, E., N. Chettri, J. Gurung and B. Shakya. 2007. Landscape approach in biodiversity conservation: A regional cooperation framework for implementation of the Convention on Biological Diversity in the Kangchenjunga Landscape. ICIMOD, Kathmandu, Nepal.

Sharma, E., N. Chettri and K.P. Oli. 2010. Mountain biodiversity conservation and management: a paradigm shift in policies and practices in the Hindu Kush-Himalayas. Ecological Research 25: 909–923.

Sharma, E., N. Chettri, K. Tse-ring, A.B. Shrestha, J. Fang, P. Mool and M. Eriksson. 2009. Climate change impacts and vulnerability in the Eastern Himalayas. ICIMOD. Kathmandu, Nepal.

Sharma, E. and N. Chettri. 2005. ICIMOD's Transboundary biodiversity management initiative in the Hindu Kush-Himalayas. Mountain Research and Development 25(3): 280–283.

Sharma, U.R. and P.B. Yonzon. 2005. People and Protected Areas in South Asia. Resources Himalaya Foundation, Kathmandu and IUCN World Commission on Protected Areas, South Asia, Bangkok.

Sharma, E. and E. Kerkhoff. 2004. Farming systems in the Hindu Kush-Himalayan region. pp. 10–15. *In*: R. Adhikari and K. Adhikari (eds.). Evolving Sui Generis Options for the Hindu Kush-Himalayas. South Asian Watch on Trade, Economics and Environment, Modern Printing Press, Kathmandu, Nepal.

Sherpa, L.N., B. Peniston, W. Lama and C. Richard. 2003. Hands around Everest: Transboundary Cooperation for Conservation and Sustainable Livelihoods. ICIMOD, Kathmandu, Nepal.

Sherpa, L.N. 2003. Sacred Beyuls and Biological Diversity Conservation in the Himalayas. Proceedings of the International Workshop on The Importance of Sacred Natural Sites for Biodiversity Conservation held in Kunming and Xishuangbanna Biosphere Reserve, People's Republic of China, 17–20 February 2003. UNESCO.

Shrestha, A.B., C.P. Wake, P.A. Mayewski and J.E. Dibb. 1999. Maximum temperature trend in the Himalaya and its vicinity: an analysis based on temperature records from Nepal for the period 1971–'94'. Journal of Climate 12: 2775–2786.

Singh, S.P., I. Bassignana-Khadka, B.S. Karky and E. Sharma. 2011.Climate change in the Hindu Kush-Himalayas: The state of current knowledge. ICIMOD, Kathmandu, Nepal.

Sodhi, N.S., L.P. Koh, B.W. Brook and P.K.L. Ng. 2004. Southeast Asian biodiversity: the impending disaster. Trends Ecol. Evol. 19: 654–660.

TEEB. 2010. The economics of ecosystems and biodiversity: Mainstreaming the economics of nature—A synthesis of the approach, conclusions and recommendations of TEEB. Geneva, Switzerland: TEEB Consortium (c/o UNEP) www.teebweb.org/Portals/25/TEEB%20Synthesis/TEEB_SynthReport_09_2010_online.pdf (accessed 18 August 2012).

Thuiller, W., C. Albert, M.B. Araujo, P.M. Berry, M. Cabezad, A. Guisan, T. Hickler, G.F. Midgley, J. Paterson, F.M. Schurr, M.T. Sykes and N.E. Zimmerman. 2008. Predicting global change impacts on plant species' distributions: future challenges. Perspectives in Plant Ecology, Evolution and Systematics 9: 137–152.

Trisal, C. and R. Kumar. 2008. Integration of high altitude wetlands into river basin management in the Hindu Kush Himalayas: Capacity building need assessment for policy and technical support. Wetlands International-South Asia, New Delhi.

Tse-ring, K., E. Sharma and N. Chettri. 2010. Climate change vulnerability of mountain ecosystems in the Eastern Himalayas; Climate change impact and vulnerability in the Eastern Himalayas—Synthesis report. ICIMOD, Kathmandu, Nepal.

Wilson, R.J., D. Gutierrez, J. Gutierrez and V.J. Monserrat. 2007. An elevational shift in butterfly species richness and composition accompanying recent climate change. Global Change Biol. 13: 1873–1887.

Worboys, G.L., W. Francis and M. Lockwood. 2010. Connectivity Conservation Management: A Global Guide. Earthscan, London, UK.

Xu, J., E.R. Grumbine, A. Shrestha, M. Eriksson, X. Yang, Y. Wang and A. Wilkes. 2009. The melting Himalayas: cascading effects of climate change on water, biodiversity, and livelihoods. Conserv. Biol. 23(3): 520–530.

Yanai, M., C. Li and Z. Song. 1992. Seasonal heating of the Tibetan Plateau and its effects on the evolution of the summer monsoon. J. Meteorol. Socit. Japan 70: 319–351.

Yao, T., J. Pu, A. Lu, Y. Wang and W. Yu. 2007. Recent glacial retreat and its impact on hydrological processes on the Tibetan Plateau, China, and surrounding regions. Arctic, Antarctic, and Alpine Research 39(4): 642–650.

Zimmerer, K.S., R.E. Galt and M.V. Buck. 2004. Globalization and multi-spatial trends in the coverage of protected-area conservation (1980–2000). Ambio 33(8): 520–529.

Zomer, R., E. Sharma, K.P. Oli and N. Chettri. 2010. Linking biodiversity conservation and climate change perspectives in bio-culturally rich transboundary areas in the Kailash sacred landscape region of China, India, and Nepal. pp. 142–144. *In*: Biodiversity and climate change: Achieving the 2020 targets, Abstracts of Posters Presented at the 14th Meeting of the SBSTTA of the CBD, May 2010, Nairobi, Kenya, CBD Technical Series No. 51. Montreal: Secretariat of the CBD.

10

Future Sustainability Challenges in Hindu Kush-Himalaya

Adaptation Challenges to Global Changes in Hindu Kush-Himalaya

Janak Pathak

INTRODUCTION: Why Hindu-Kush Himalaya is Important?

The Hindu-Kush Himalayan (HKH) mountain range functions as a complex interaction of "atmospheric, cryospheric, hydrological, geological and environmental processes that bear special significance for the Earth's biodiversity, climate and water cycles" (Yao et al. 2011). This region also plays a prominent role in global climate and function to generate the Asian monsoon system that sustains one of the largest populations on Earth (ICIMOD 2011). The ecosystems occupy about 3.6% of the world's area but contain 16 and 12% of global floral and faunal species respectively (UNEP RRC.AP 2001), including four global biodiversity hotspots, 488 protected areas, and 60 global eco-regions. This region provides ecosystem services including water, food, energy, biodiversity and hydrological regulating functions to support livelihoods of people living upstream and downstream (ICIMOD 2012b). Meltwater from the Himalayan glaciers is one of the main

Department of Early Warning and Assessment Branch, United Nations Environment Programme, P.O. Box, Kenya, 30552, Nairobi 00100, Nairobi, Kenya.

sources of freshwater reserves providing headwater to the 10 major river basins (Fig. 10.1), which are the lifeline of the regional economy. These basins directly sustain livelihoods of millions of people living in the region, especially people living in arid and semi-arid areas. Ecosystem services from these basins provide the basis for livelihoods contributing to a substantial part of the region's GDP (Eriksson et al. 2009). At varying degrees and times, more than 1.4 billion people living in the Himalayan river basins rely on both meltwater and monsoon waters to sustain their livelihoods, mainly for irrigation, drinking, sanitation and industrial uses (Xu et al. 2009, UNEP 2012a). Part of the water flow in these river basins depends on snow and glacial melt to perennial rivers, such as the Ganges, Indus, Brahmaputra, Mekong and Yangtze (Eriksson et al. 2009). In turn, the amount of snow- and ice melt influences runoff into lowland rivers and the amounts of water recharging river-fed aquifers. The greatest dependence is in arid and semi-arid areas, such as western China, northeastern Afghanistan, Uzbekistan and parts of Pakistan (Immerzeel et al. 2010, Thayyen et al. 2007). In western China, 25% of the population directly depends on meltwater in the dry season (Xu et al. 2007); there is less dependence in monsoon-dominated regions (Dyurgerov and Meier 2005).

However the region faces extreme vulnerability and risk due to climate change, warming at higher altitudes has raised concern of glacial retreat, and increased frequency of extreme events while persistent challenges of poverty, rapid urbanization, increased production and consumption, population growth and environmental degradation have exacerbated, threatening future sustainability of the region (ICIMOD 2011).

Is the Future Sustainability of Hindu Kush-Himalaya at Risk?

A globally dominating production and consumption

The region embraces globally dominating production and consumption, and the fastest growing economies (UNEP 2008). Figure 10.2 GDP per capita in 1995 and 2025 and the annual rate of growth between 1995 and 2025 are projected to grow in the HKH counties. Growth rate will be highest in Asia, ranging from 2.1 to 5.2% per year (Rosegrant et al. 2002). As a result the region has emerged as a significant contributor of greenhouse gas emissions although per capita emissions are still extremely low by international standards (World Bank 2009a). The emerging economies of China and India are expected to maintain 8 to 9% rate of economic growth over the next several years that will largely determine the future perspective of the region's environment (Bawa et al. 2010). The HKH region also faces the greatest population pressure on the land, with a three-fold increase in human population since 1950 (GWP 2011). Net irrigation-water

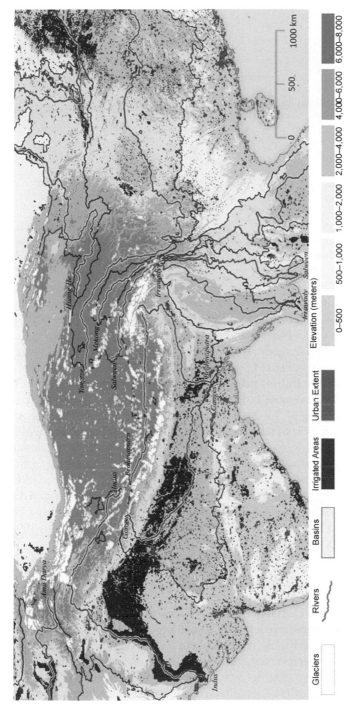

Figure 10.1. An overview of Hindu Kush-Himalayan region with the major river basins and human influences (Source: www.careclimatechange.org).

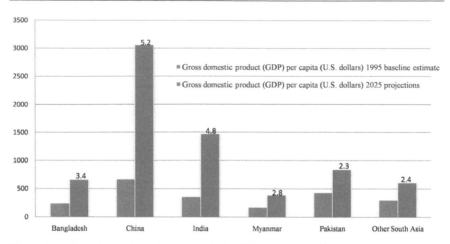

Figure 10.2. Gross domestic product per capita in 1995 and projected for 2025 in HKH countries. Numbers on the top of the bars graph is projected annual growth rate between 1995 and 2025 (Rosegrant et al. 2002).

demand is high in this region (Table 10.1), but per capita water availability is very low—around 2,000 to 3,000 m³/capita/year—which is far less than the world average of 8,549 m³/capita/year (UNEP 2008) and projected declining water availability per capita. Moreover, total non-irrigated water consumption increases in HKH (Table 10.2), 75% in the Yellow River basin and by over 100% in the Indus and Ganges basins compared to 1995 levels (Rosegrant et al. 2002). Arable land has declined from over 1 hectare per person at the beginning of the 20th century to less than 0.1 ha at present (GWP 2011). In this region of the world's fastest growing economies, water is becoming an increasing security priority of the Himalayan countries to sustain a growing economy, population growth and a bourgeoning of the middle class. With increasing urbanization, almost 40% of the population in the region now lives in the cities. As the result of a burgeoning middle class, more than half of Asia's population will be living in urban area by 2025 (Fig. 10.3), and increasing energy consumption by 50% between 1996 and 2006 (UNEP 2008, IPCC 2007b).

Is climate change a challenge for the future sustainability of HKH?

The ecologically fragile Himalayan region is one of the most economically underdeveloped and most densely populated mountain ecosystems on the planet making the region highly vulnerable to the impact of climate change. Climate change is projected to have negative effects and will impact the Himalayan rivers in two distinct ways; i) rising temperature is likely to affect the Himalayan glaciers then altering the basins hydrology, ii) global

Table 10.1. Projected water availability and demand. Based on a best guess of glacier area in 2050, this projection shows the number of people in the Himalayan area who could be threatened by food insecurity due to changes in the Himalayan glaciers. Increased mean upstream rainfall partly compensates for upstream water losses, although net irrigation demand may put more stress on the region's food security. The Yellow River basin, where there is an increase in upstream water yield, is an exception. The Yellow River basin only marginally depends on glacial melt; thus there is a notable 9.5 per cent increase in upstream water yield in the basin with an estimate of an increase of 3.0±0.6 million people that can be fed in the Yellow River basin. Upstream refers to the area above 2,000 m altitude (Immerzeel et al. 2010). *Upper Indus.

Parameter	Indus	Ganges	Brahmaputra	Yangtze	Yellow
Total population (millions)	209.619	477.937	62.421	586.01	152.72
Net irrigation water demand (mm/yr)	908	716	480	331	525
People threatened by food insecurity (millions)	26.3 ± 3.0	2.4 ± 0.2	34.5 ± 6.5	7.1 ± 1.3	
Percentage decrease in mean upstream water supply	8.4*	17.6	19.6	5.2	
Percentage increase in mean upstream rainfall	25	8	25	5	14

Table 10.2. Water consumption by non-irrigation sectors and future projection (Rosegrant et al. 2002).

Region/country	Domestic (km³)			Industry (km³)			Livestock (km³)			Total non-irrigation (km³)		
	1995	2010	2025	1995	2010	2025	1995	2010	2025	1995	2010	2025
China total	30	48	59.4	13.1	24.4	31.1	3.4	5.3	7.4	46.5	77.7	97.9
Brahmaputra	0.6	0.8	1	0.4	0.5	0.6	0.8	1.3	2.1	1.9	2.7	3.6
Indus	0.9	1.4	1.8	0.4	0.6	0.7	0.6	1	1.6	1.9	3	4.1
Ganges	9.6	15.6	20.1	3	6	6.5	0.4	0.6	1	13	22.2	27.6
Pakistan	3.1	5.1	7	1.6	2.7	3.9	0.9	1.4	2.1	5.7	9.2	13
Bangladesh	2.3	3.6	4.7	0.2	0.3	0.5	0.3	0.5	0.8	2.8	4.4	5.9
Other South Asia	1.6	2.8	4.6	0.1	0.2	0.3	0.5	0.7	1	2.2	3.7	5.8
Myanmar	0.7	1.4	2.1	0.1	0.1	0.2	0.3	0.5	0.7	1.1	2	3

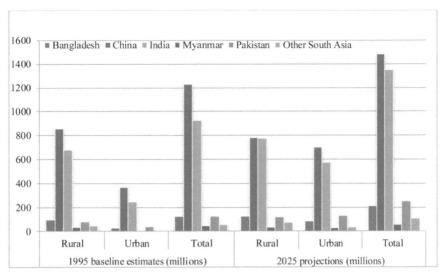

Figure 10.3. Rural and urban population and the future projection in HKH region (Rosegrant et al. 2002).

warming is likely to impact monsoon patterns, and any minor change in the monsoon pattern has an immediate and discernible effect on the Himalayan rivers (SFG 2011). Especially, shift in the location, intensity and variability of rain and snow due to climate change will likely have a greater impact on regional water supply (NAS 2012). An anticipated 40% exceeding water demand by 2030 and nearly 80% of the region's water is used in agricultural production, are likely to threaten the food security (GWP 2011).

The Himalayan countries with the largest population (more than 2.2 billion people are expected by 2050) (World Bank 2009a and Fig. 10.4) and the highest incidence of poverty on Earth, are also vulnerable to the impact of climate change from i) threats to water supply and food security; ii) urbanization; and iii) vulnerability to natural disasters (World Bank 2012b, GWP 2011). The rising food prices are already worsening poverty in Asia, where about 600 million people are living on less than US\$ 1.2 a day (Hidellage 2003). In addition, 70% of the 210 million people living in the HKH are poor, threatened by food insecurity, environmental problems and disasters (Hoermann and Kollmair 2009). A recent trend shows increasing numbers of natural disasters in the HKH region (Fig. 10.5). UNISDR reported seven of the top 10 global natural disasters by number of deaths occurred in four of the Himalayan countries (Bangladesh, China, India and Pakistan), and accounted 82% of the total natural disaster-related deaths worldwide in 2007 (UNEP 2008).

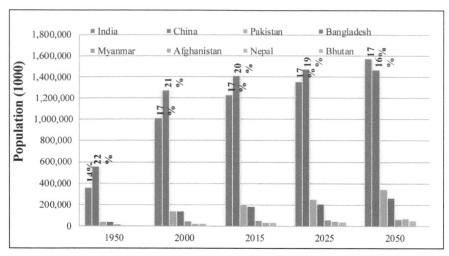

Figure 10.4. A projected population growth in Himalayan countries (United Nations Department of Economic and Social Affairs. Population Division).

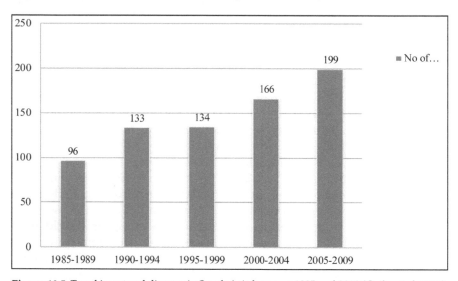

Figure 10.5. Trend in natural disaster in South Asia between 1985 and 2009 (Gaiha et al. 2010).

Monsoon bigger than glaciers for the basins' hydrology

Recent studies noted that meltwater from the Himalayan glaciers to the basins' hydrology vary temporally and spatially, and may not be substantial to all the Himalayan basins compared to water contributed by the Asian monsoon (Table 10.1) (Fig. 10.6). For example:

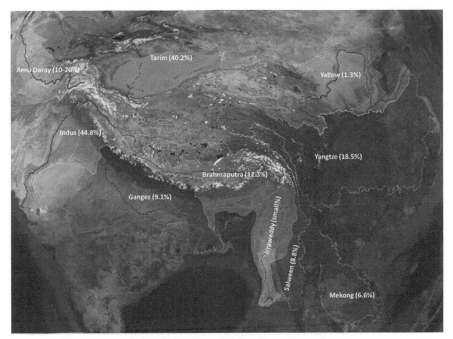

Figure 10.6. Predicted percentage of glacial melts contributing to basin flows in the Himalayan basins (Data from Xu et al. 2008. Shape files superimposed on background image from ESRI ArcGlobe 10.0 from UNEP 2012a).

Color image of this figure appears in the color plate section at the end of the book.

- The climate of the HKH is not uniform and strongly influenced by the South Asian monsoon and the mid-latitude westerlies. Glacial meltwater has relative importance to the basins' hydrology (NAS 2012).
- If all the Himalayan glaciers were to disappear, there would be about a 33% reduction in annual flow in the west, whereas the decline in the east would be only about 4–18% compared to the 1990 levels (Rees and Collins 2006).
- Precipitation has direct influence on Himalayan glacier catchment (the catchments where the glacier melt water contributes to the river flow during the period of annual high flows produced by monsoon) rather than mass balance changes of the glacier (Thayyen and Gergan 2010).
- Glacial melt contribution is minor in monsoon dominated catchments of the Ganges and Brahmaputra (Kaser et al. 2010).
- Precipitation dominates inter-annual runoff variation in a Himalayan glacier catchment (Thayyen and Gergan 2010).

- In northern India the millennial-scale, patterns of glacier advance and retreat correspond to patterns of monsoon variability (Scherler et al. 2010).
- Groundwater significantly influences the Himalayan river discharge cycle for the Ganges River basin. About two-thirds of its annual discharge comes from water travelling through groundwater reservoirs in the eastern HKH (Andermann et al. 2012).

The greater concern is impact of climate change on the Asian monsoon (bigger than glaciers), yet highly seasonal, with about 75% of the annual rainfall occurring during the monsoon months (June–September); the pressing question is how the monsoon will respond to global warming, especially in monsoon-dominated regions, where there is a decreasing trend in precipitation and runoff from the east to west due to the weakening of the summer monsoon as it moves westward along the Himalayan range (UNEP 2008, Armstrong 2010). The largest changes to the hydrological system in future will mostly likely be because of changes in the timing, location, and intensity of the monsoon (NAS 2012). These changes will lead to construction of dams and water transfer systems; India and China are planning or already implementing large interbrain schemes to transfer water to water-scare regions (Kohler and Maselli 2009). Construction of dams for future water security will potentially lead to conflict due to displacement of an estimated 50–70 million people in India, China, Bangladesh and Nepal by 2050 (SFG 2010).

Climate change and Himalayan glaciers

One of the main concerns in relation to climate change in the HKH region is the reduction of snow and ice. Many assessments link the receding of the Himalaya glaciers to global warming, and state that the melting of glaciers is a clear indicator of climate change (Xu et al. 2009). Also note that glacier change is the most visible and obvious indicator of changing temperatures (Armstrong 2010, Winkler 2010). Temperatures at some locations in the Himalayan region have risen faster than the global average. From 1982 to 2006, the average annual mean temperature in the region increased by 1.5°C with an average increase of .06°C per year, although the rate of warming varies across seasons and ecoregions (Shrestha et al. 2012). It stands to reason that the rising temperature in the Himalayas would affect glacier melt (Barnett et al. 2005). Also, climate change is predicted to lead to major changes in the strength and timing of the Asian monsoon and winter westerlies. However, uncertainties exist about the current state of the Himalayan glaciers (Bamber 2012, Kargel et al. 2010, UNEP 2009), because studies suggest that HKH glaciers have not responded uniformly to observed changes in climate; latest climate studies provide no meaningful insight into the significance of recently observed changes in climate compared to the natural variability of the region's climate (NAS 2012); and incomplete understanding of the processes affecting Himalayan glaciers under the current climate, make any projection on the Himalayan glaciers unclear.

The rate of retreat and growth of individual glaciers is highly dependent on glacier characteristics and location; small glaciers at low elevation and with little debris cover are the most vulnerable (NAS 2012). There is evidence of glacier retreat in the eastern and central Himalayas with rates accelerating over the past century, while glaciers in the western Himalayas appear to be more stable overall, glaciers around the Karakoram region show complex behavior and some have advanced and thickened (Gardelle et al. 2012, Scherler et al. 2011) indicating an apparently atypical behavior that has prevented drawing conclusion about the future state of Karakoram glaciers (Perkins 2012) (Fig. 10.7). Gangotri Glacier has a long record of terminus fluctuation with multidecadal terminus retreat rate variations since the 19th century. The retreat has stopped in the last decade (Fig. 10.8) (Kargel et al. 2011).

Black carbon affecting glaciers and Asian monsoon

Atmospheric loading of aerosol in the form of a brownish haze known as Atmospheric Brown Clouds (Ramanathan et al. 2005), from anthropogenic activities has significantly increased, making the Indo-Gangetic Plain one of the most polluted region in the world (Fig. 10.9). Black Carbon (BC) from the incomplete combustion of fossil fuels, biofuels, and biomass is a key component of ABC that strongly absorbs solar radiation contributing to increase in average global temperature (Fig. 10.10), and widely known to have an effect on snow/ice albedo that lead to surface dimming, especially strong in sensitive regions like the HKH (USEPA 2011). Recent studies have shown that black carbon because of its ability to heat the atmosphere, can affect the regional and global water cycles by altering the radiation balance of the Earth's atmosphere and surface, and modulating cloud and rain formation processes (Lau et al. 2010, Ramanathan et al. 2005, Rosenfield et al. 2008).

At a regional level, changes in monsoon such as weakening of summer monsoons over the past 50 years, a decreasing trend in the length of the monsoon (Dash et al. 2009, Ramesh and Goswami 2007), a model output indicate reduced rainfall over India, with a small increase over the Tibetan Plateau (Meehl et al. 2008). A consistent finding shows that aerosols can substantially influence the timing and intensity of the South Asian monsoon (NAS 2012) by the mixture of dust (desert dust) and BC in the deep aerosol

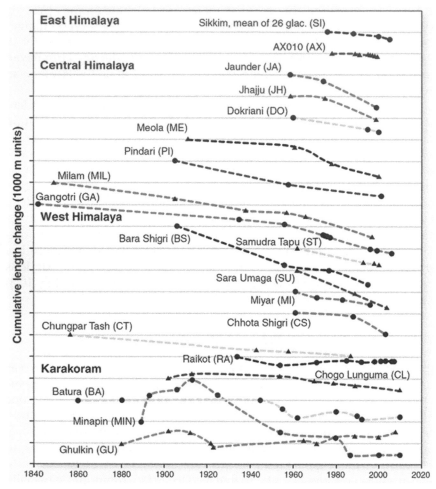

Figure 10.7. An overview of Himalayan glaciers retreat. Most of the glaciers in the eastern and central Himalayas show retreat since the mid-19th century. Glaciers in the western Himalayas appear to be more stable overall, and some may be advancing. For example glaciers at Nanga Parbat in the northwest (RA, CL) and glaciers in the Karakoram, which show a complex behavior (Bolch et al. 2012).

layer provides efficient heating of the atmosphere then interacts with the warm moist monsoon air and maximizes the atmospheric water cycle feedback that may significantly modulate the summer monsoon rainfall (Lau et al. 2006, Lau et al. 2008).

At the local level, snow and ice have greater energy-albedo feedbacks, small changes in the amount and timing of snow, and in the overall energy balance can have large effects on a glacier's mass balance. Fresh snow has an albedo range of about 0.75 to 0.95 while glacier ice has an albedo

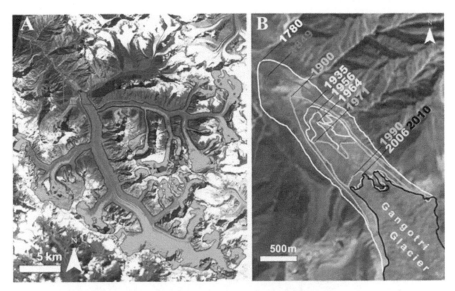

Figure 10.8. A retreating trend of Gangotri glacier since 1780. It is one of the largest glaciers in the Himalayas feeding meltwater to the Ganges river. It has been receding since 1780. Outlines for 1780–1971 are from the work by Vohra (4), the outline for 1990 is from our analysis of a Landsat Thematic Mapper (TM) image (November 15, 1990), and the outlines for 2006 (October 9) and 2010 (November 29) are from ASTER image analysis (Source: Kargel et al. 2011).

range of about 0.3 to 0.4. Keeping other factors constant, removing snow from a glacier means a 200 to 300% energy increase to the surface of the glacier, causing exposed glacier ice to heat and then melt. This also melts nearby snow, resulting in more energy delivery, and more glacier wastage (NAS 2012). Recent studies implicate snow darkening by black carbon and dust as a possible cause of glacier retreat. Menon et al. (2010) noted the increased aerosol in the Indian subcontinent, with a large contribution from BC emission from coal and biofuel is responsible for a 0.9% reduction in snow/ice cover over the Himalayas region between 1990 and 2000 (NAS 2012). In addition, particles associated with the ABC are likely to have significant public health impacts, such as, cardiovascular and pulmonary effects leading to chronic respiratory problems. Due to inhalation of ABCs outdoor in China and India, it is inferred that 337,000 deaths per year, with a 95% confidence interval of 181,000–492,000 (UNEP 2008).

Impacts and Risks in Hindu Kush-Himalayan

The effect of climate change on glaciers, monsoon behavior, and flood and drought intensity are already impacting the livelihoods of millions. Most of the mountain people in the HKH countries depend upon agriculture for

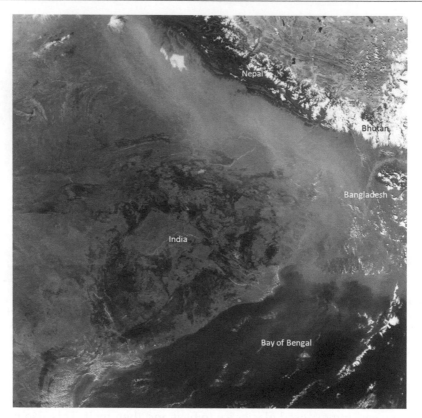

Figure 10.9. Atmospheric brown cloud seen over the Himalayan region appears to cause warming. The haze seen south of the Himalayan come from human activities, such as, agricultural fires, cooking sources that rely on wood, kerosene or dung; industrial and vehicle emissions. Melting of Himalayan glaciers, more extreme weather system, and darkening city skies is due to ABC. ABC resulting from fossil fuel and biomass burning—primarily affected air quality in Asia, and likely worsened the effects of climate change in the region (United Nations Environment Programme report ABC). In the figure, a river of haze (ABC) travels along the Ganges River flowing before spreading over the Bay of Bangla in the Indian Ocean. NASA image courtesy MODIS Rapid Response, NASA Goddard Space Flight Center (Image from: NASA Earth Observatory, Jesse Allen).

their livelihood. A majority of 150 million people from the region are farmers (ICIMOD 2003). About 80% of total water use goes towards agricultural need (GWP 2011) (Fig. 10.11). In 2030, percentage water availability on per capita cubic meter is estimated to decline in China, India, Bangladesh and Nepal by 13.5, 28, 22.13 and 35.29% respectively and reducing yield significantly (SFG 2010).

The cascading effects of anthropogenic activities and climate change have already stressed mountain ecosystems, and put more pressure from

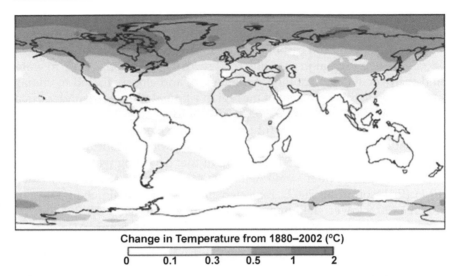

Change in Temperature from 1880–2002 (°C)

Figure 10.10. Result from computer models show the impact of soot on global temperature between 1880 and 2002. More than 25 percent of the increase in average global temperature between 1880 and 2002 is likely to be due to soot contamination of snow and ice worldwide. Snow and ice contaminated with black carbon absorbs incoming solar radiation unlikely pure snow and ice that reflect large amount of incoming radiation back into space. According to the estimate by the climate scientists from NASA and Columbia University, a soot content of only a few parts per billion (ppb) can reduce snow's ability to reflect incoming radiation by 1 percent. Image from: NASA Goddard Institute for Space Studies (NASA Earth Observatory).

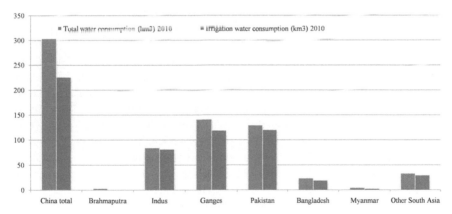

Figure 10.11. Water consumption by irrigation sector compared to total water consumption in 2010 in HKH region. Under Business As Usual (BAU) scenario, 17 percent projected increase in potential irrigation water demand due to growth in irrigated area from 1995 level. A much smaller increase of 4 percent in China. China and India faces more severe increase in water scarcity than in the developing countries as a whole. Data are IMPACT-WATER projections, 2002 in (Rosegrant et al. 2002).

population growth, economies, land-use change and urbanization. Climate change is also likely to alter ecosystem composition, structure and function; biodiversity, loss of soil carbon and, groundwater recharge (Xu et al. 2008). Biodiversity is crucial in ensuring food security, nutrition, access to water and the overall sustainability of the region, and closely linked to climate change. Ten to thirty percent of the region's faunal species are under threat of extinction (World Bank 2009b, UNEP 2012b).

Some projections on the future Himalayan glaciers may have serious implication on water resources and food security. Based on a projected estimate glacier area in 2050, it is thought that declining water availability will eventually threaten some 70 million people with food security (Table 10.1) (Eriksson et al. 2009). Expected impacts of climate change are; as much as 20% lower yield of major crops, an anticipated a 15–30% decline in the productivity of most cereals and rice across South Asia. Due to cumulative effect of water scarcity, glacial retreat, change in monsoon, flooding, desertification, pollution and soil erosion, a massive reduction in the production of rice, wheat, maize and fish will be felt. With the increasing water demand due to population growth and economic development, the agricultural sector will continue to be the major consumer of water in the HKH region. China's 65% of water for agricultural consumption will decline to 55% by 2030. In India, almost 90% agricultural uses of water will decline to 70–75% by 2050. Moreover, depletion of water due to pollution and losses caused by inefficient management is also considerable. In the Yellow River, 34% of the river water is unfit for drinking, aquaculture, and agriculture and pollution of the tributaries of Yangtze River extent to 30%. In India, the Yamuna River, the main tributary of the Ganges is polluted up to of 50% (SFG 2010).

About 90% of rice is grown and consumed in Asia (Gujja and Thiyagarajan 2009). Per capita water availability in India was 1730 cubic meter per person per year in 2006, close to World Bank's 'water stressed' level which is 1700, and expected to decline to 1240 by 2030 which is extremely close to becoming the World Bank's 'water scare' level (1000 cubic meters). India's water utilization rate is 59%. This is already crossing the threshold standard of 40% (i.e., without providing natural capacity to recharge adequately) (SFG 2010). Most of the harvest comes from irrigated agricultural land around the Ganges (two-third of Delhi's water comes from the Ganges), and the Yamuna. The Yangtze irrigates more than half of China's rice (Larmer 2010), and the Yellow River basins, have experienced slowdown in the growth rates of rice yields since the mid-1980s. The Yellow and Ganges Rivers are expected to lose between 15 to 30% water due to glacier depletion and will turn into seasonal rivers by the second half of the century. Similarly, the Yangtze and Brahmaputra will lose about 7 to 14% of the annual flow due to the same reason (SFG 2010). China and India will face 30–50% drop in wheat and rice yield by 2050 while demand for food grains will go up by at least 20%. To accommodate this increased demand, an estimated 200–300 million tons of wheat and rice is required in China and India (SFG 2010).

Hindu Kush-Himalaya is prone to more natural disasters

The ecologically fragile HKH is highly susceptible to natural disasters and impacts are already evident, with an increasing trend in natural disasters (Fig. 10.12). Over the past two decades there have been over 230,000 deaths and about US$ 45 billion in damage (World Bank 2008). Frequency of landslides, flash floods, debris flows, etc., are projected to increase in the uplands (300–3000 m), and fluvial and coastal flooding below 300 m (Xu and Rana 2005). The HKH region is outpacing others for flood and earthquakes disasters costing billions of dollars (Fig. 10.13). Past events show that the monsoon dominates erosion and sediment transport is high in the HKH region. For example, significantly higher number of extreme events leads to higher erosion volumes and greater fluvial mass transport rates (Bookhagen 2010) (Fig. 10.14). The most dramatic negative impacts are expected such as economic losses and damage of high-value infrastructure, high incident of vector and water-borne diseases (World Bank 2009a).

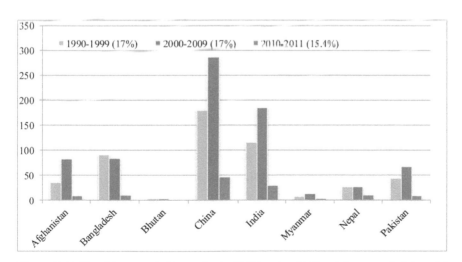

Figure 10.12. Trend in natural disaster in the HKH region. Taking into account of drought; earthquake (seismic activity); epidemic; extreme temperature; flood; insect infestation; mass movement dry; mass movement wet; storm; volcano; wildfire. Flood and earthquake are frequent in HKH region (http://www.emdat.be/about).

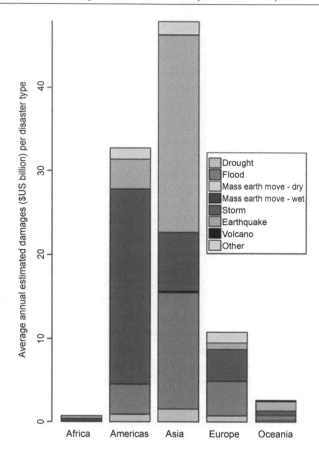

Figure 10.13. Average annual damages (US$ Billion) caused by reported natural disaster 1990–2011. EM–DAT: The OFDA/CRED International Disaster Database—www.emdat. be—Université Catholique de Louvain, Brussels—Belgium (http://www.emdat.be/about).

Color image of this figure appears in the color plate section at the end of the book.

Case study 1: Glacier lake outburst flood (Fig. 10.15)

Several glacial lakes that have developed in the HKH are dammed by unstable moraines (ICIMOD 2011). These lakes burst releasing enormous amount of lake water causing serious floods downstream, generally known as Glacial Lake Outburst Flood (GLOF). It occurs when water dammed by a glacier or a moraine is rapidly released by failure of the dam (Bajracharya et al. 2007). In the recent past, there have been several occurrences of GLOF event in different parts of the HKH region. Growing number and size of glacial lakes have resulted from thinning and retreat of glaciers in the HKH

Figure 10.14. A catastrophic flood and debris filled the channel of Nepal's Seti River. Flood contained a mixture of mud, sand, gravel and rocks, saturated with water was triggered by a rock fall on May 5, 2012 (Source: NASA image by Jesse Allen and Robert Simmon using EO-1 ALI data (NASA Earth Observatory)).

region (Fig. 10.15). Many glacial lakes in Nepal and Bhutan are growing considerably increasing a threat of GLOF. In Nepal, 24 GLOF events have occurred in the recent past, causing considerable loss of life and property. However, there are projects in Nepal and Bhutan now underway to lower water levels of potentially dangerous lakes to minimize the risk of GLOF (NAS 2012).

Case study 2: Indus Flood (Figs. 10.16–10.19)

About 20 million people in Pakistan has been affected by the intense monsoon rains in 2010, a fifth of the country was submerged, and extensive damage to agricultural lands, about 10% of the country's cotton crop was flooded and one-fifth of the rice crop. The rain was 70% and 102% above normal in July and August. Dam failure in Sindh caused flooding to the west of the valley. Flood covered at least 37,280 sq km of Pakistan, caused 1,985 deaths, and damaged 1.7 million houses (NASA Earth Observatory).

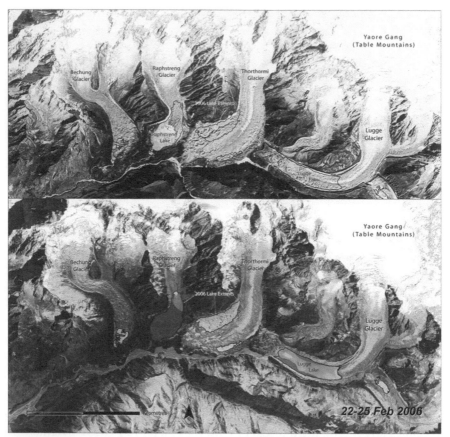

Figure 10.15. Formation of glacial lakes and glacier lake outburst flood in Bhutan. Rocky debris pile up forming moraines, acting as natural dams for lakes filled with melt water at the terminus of glacier. Rising water levels and failing of moraines dams of glacial lakes can create catastrophic glacial outburst floods. A partial collapse of the moraine at the southwest corner along the edge of the Luggye Lake in 1994, released a glacial outburst flood killing 21 people and swept away livestock, crops and homes. These lakes were not exist in 1969, now officials are concern about the Thorthormi Glacier Lake and the unstable moraine separating it from Raphstreng Lake. An outburst flood from Thorthormi into Raphstreng could easily cause the lower lake to overflow (Michott Scott NASA Earth Observatory, modified version adopted from UNEP 2012a).

What are the Implications?

Is Melting expected in decades to come?

A significant time lag (several decades or longer for most mountain glaciers) occurs between changing climatic conditions and the resulting change in glaciers (Armstrong 2010). This could be because of temperature in the

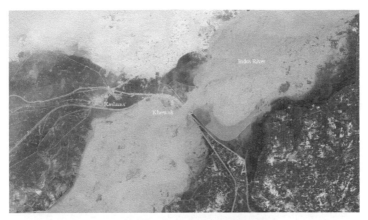

Acquired August 12, 2010

Acquired August 9, 2009

Figure 10.16. The extreme monsoon floods in Pakistan inundated much of the Khewali city and the surrounding farmland in August 2010. Top image acquired by the Landsat 5 satellite on August 12, 2010 showing flooding near Kashmor. The lower image show normal situation in August 2009. Stream gauges at the Guddu Barrage recorded extremely high levels of water (more than 910,000 cubic feet per second) flowing down the Indus. The flow rate increased on August 13 as the second wave of flooding reached the barrage. The extreme monsoon floods in Pakistan inundated much of the Khewali city and the surrounding farmland in August 2010. The lower image of August 2009, provided for context, to show the magnitude of flood in August 2010. Stream gauges at the Guddu Barrage recorded extremely high levels of water (more than 910,000 cubic feet per second) flowing down the Indus. The flow rate increased on August 13 as the second wave of flooding reached the barrage. On August 18, the Pakistan government stated that 15.4 million people had been directly affected by the flood with nearly a million homes damaged or destroyed. The land along the Indus River is prime farmland, and nearly 80 percent of the flood victims are farmers who have lost crops, animals, and equipment. At least 3.2 million hectares of crops had been destroyed as of August 18, reported the United Nations Office for the Coordination of Humanitarian Affairs (Holli Riebeek, NASA Earth Observation). NASA image by Robert Simmon, based on Landsat 5 data from the USGS Global Visualization Viewer (NASA Earth Observatory).

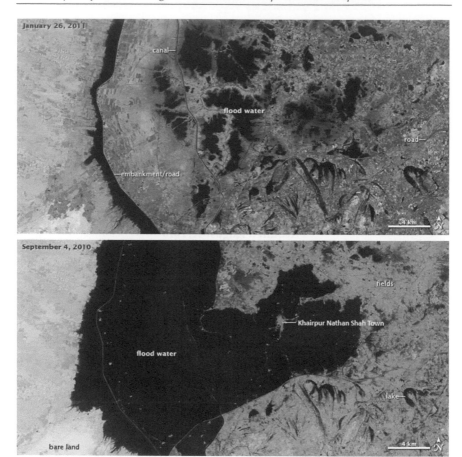

Figure 10.17. Indus river after flood in 2010. Many areas remained under water for months after the rains subsided from 2010's extreme monsoon flood in northwestern Pakistan. The effects were lasting, flood water in Sindh Province did not fully receded that provided the perfect breeding ground for malaria carrying mosquitoes and waterborne diseases such as cholera. In early 2011, flood water receded considerably, but some areas remained submerged preventing many people returning to their homes in parts of Sindh and Balochistan Provinces. These false-color satellite images show flood water in western Sindh province in September 2010, November 2010, and January 2011. It is apparent that roads and other infrastructure constrained the flow of flood water. NASA images by Robert Simmon, using data from Landsat 5 (NASA Earth Observatory).

HKH are far below freezing point and slowly rising, and may take several decades to reach above 0°C (between 50–150 years). For example, AX010 glacier takes 8 years to exhibit its terminus response to changing climate, and takes between 29 and 56 year (volume response time) and between 37 and 70 year (length response time) following the changes in mass balance condition (Adhikari et al. 2009). Response time of the majority of glaciers

Figure 10.18. An aerial view taken from a helicopter shows area submerged due to monsoon caused flash flood. Many houses submerged in the flood-hit Nowshera district. This photo released by Khyber Pakhtunkhwa Information Department (KPID) on July 30, 2010. **AFP/** Getty Images. The Sacramento bee.com (July 30, 2010).

Figure 10.19. Pakistani villagers move to a safer place from a flood hit village near Nowshera, Pakistan on Thursday, July 29, 2010. One of the worst floods hit northwestern region of Pakistan that cost the lives of more 400 people and thousands were made homeless AP/Mohammad **Sajjad**. The Sacramento bee.com (July 30, 2010).

in HKH region is most likely to be decades to centuries (Armstrong 2010). This has the potential to have serious implications in the future on water availability and the situation could worsen if the melting is expected in decades to come.

Filling the knowledge gaps: an urgent need

Studies on glacial melt contributing to the basin's hydrology lack direct evidence and sometimes appear to be inconsistent (Kaser et al. 2010). This has raised doubts that melting glaciers provide a key source of water in downstream areas across the entire Himalayan region (UNEP 2012a). Due to uncertainty about the future state of the climate, as well as a lack of long-term comprehensive *in situ* monitoring of glacial melt contributing to the Himalayan basins, some projections of their future have serious implication on water resources (NAS 2012, Bolch et al. 2012). Understanding the future implications of climate change to the Himalayan basins is urgently required because more than 1.4 billion people depend on water from the Indus, Ganges, Brahmaputra, Yangtze and Yellow Rivers (Immerzeel et al. 2010). Their dependency on these water resources underpins environmental sustainability, food security and livelihoods of the Himalayan community.

Indus Case study 3: Future implication Indus Basin—Fig. 10.20

The Indus River Basin covers an area of about 1 million sq. km and touches four countries (China, India, Afghanistan, Pakistan). It has one of the world's largest irrigation networks and the most depleted basins in the world (Sharma et al. 2010), is Pakistan's primary source of freshwater and can been seen as its lifeline. The Indus River irrigates 80% of Pakistan's 21.5 million ha of agricultural land and is critical for Pakistan's 160 million people (Rizvi 2001, CIA 2006). About 90% of Pakistan's agriculture depends on the river and much of the world's cotton comes from the Indus River Valley. On average, about 737 billion gallons of water are being withdrawn from the Indus River annually to grow cotton (enough water to supply Delhi residents for more than two years). In addition, the river is used for hydropower generation in Pakistan and India. In India, about 60 million live in the basin. The drainage 321,289 sq km of the Indus River basin encompasses nearly 10% of the total geographical area of India. Five% of 9.6 million hectares of cropland of India, of which 30% is irrigated (Rosegrant et al. 2002). Glacial melt contributes as much as half of the region's flow (Wheeler 2011, Winiger et al. 2005). Also, meltwater is crucial for upstream reservoirs to store and release water to downstream areas when most needed. The Indus Basin Irrigation System gets its water supply from the Tarbela dam on the Indus River and the Mangla dam on the Jhelum River; both are located in the upper Indus basin and are fed largely by glacier

Figure 10.20. Seasonal changes in the Indus River. Indus that irrigates an estimated 18 million hectares of farmland, experiences substantial fluctuation every year. Indus water is highly sensitive to variation in weather and climate and faces threat from climate change because of high dependency on glacier water. The water levels are lower in June 2010 than in June 2009. The river is fed by glacial meltwater in the Himalaya and Karakoram mountain ranges. Highest flow mostly between mid-July to mid-August, as rain and snow melt spike around the same time. The barrage (a type of dam) is designed to control flow of water on the river and reduce the risk of flooding, supporting irrigation throughout the year. The river and the irrigation infrastructure has been sustaining a population of millions. The Thematic Mapper on the Landsat 5 satellite observed these seasonal changes in the Indus River 1st picture taken June 6, 2009 and the 2nd June 9, 2010 (NASA Earth Observatory).

meltwater. This shows that any change in the discharge from melting glaciers will have a considerable effect on the millions of people living downstream (UNEP 2012a).

Yellow Case study 4: Future implication Yellow River Basin—**Fig. 10.21**

The Yellow River is the second largest river in China, crisscrossing nine provinces stretching 5,464 km, covering 7% of China's land area. The basin is of utmost importance for food production, natural resources, and

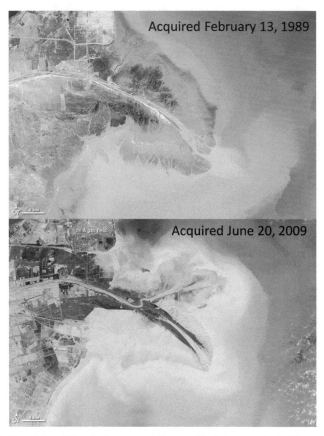

Figure 10.21. The Yellow River is the most sediment-filled river on Earth. The sediment have reshaped the cost. Compared to earlier image, the 2009 photo shows an increase in developed land. The shape of the coastline has been significantly changed. The lower reaches of the river and the delta have been extensively engineered to control flooding and to protect coastal development. Water and sediment flows to the delta have declined dramatically since the 1970s, due to reduced rainfall and intensive urban and agricultural demand for water upstream. In the 1990s, the river ran dry well before reaching the delta. Levees, jetties, and seawalls allow to slow erosion and direct the flow of the river however challenge to protecting the delta's natural wetlands, and its agricultural and industrial development remains (NASA images by Robert Simmon. NASA Earth Observatory).

socioeconomic development, for example 12.9 million hectares crop area in the basin, of which 31% is irrigated and supports 136 million people (11% of the country population). The basin contains 13% of the total cultivated area in China and holds 3% of the country's water resources. Interruption of flow in the lower Yellow River is declining groundwater levels, disappearing lakes, and silting up of river beds causing increased water scarcity in the basin. Income and population growth are driving rapid increases in water consumption in the domestic, industrial, and livestock sectors increasing water stress level from 0.72 to 0.90. Although total irrigated area has seen increases by 23 percent, irrigation water consumption declines or barely increases. It is estimated that by 2025, 5% decline in irrigation water compared to 1995 will slow agricultural production (Rosegrant et al. 2002), this decline would have serious future implication to the people who directly depend their livelihoods from Yellow river basin.

A Broad Adaptive Strategies is Needed in Hindu Kush-Himalayan Region

Since the HKH is proving to be the engine of economic growth in the 21st century, and also taking into account the global economic downturn and increasing environmental challenges, the rethinking of growth and development strategy is required. More importantly, improving security in the region is of foremost importance and a pivotal component for sustainable development. Building peace and security using water as an instrument for cooperation between countries can foster understanding among the Himalayan countries on the shared river basins. Raising awareness of security risks associated with the water crisis in the Himalayan region to explore confident building and conflict prevention measures that requires a paradigm shift from the mindset of conflict to a mindset of cooperation and security (SFG 2011).

Adaptation and mitigation are not adequate

Broadly, two approaches to address climate change in HKH. Adaptation focusing mainly on the rural, agriculture and water sectors while mitigation strategies focusing on the energy sector and alternative technologies. It is imperative to continue our efforts to address climate change that are broadly based on these two approaches. Mitigation measures for emission reduction to stabilize greenhouse gas concentration in the atmosphere. For example technological and socio-economic transition, improving energy efficiency and scaling up of renewable energy. Implementation of adaptation measures for adjusting to actual or expected climate change. Adaptation measures could even include relocations of vulnerable populations, such as those in the way of anticipated sea level rise (WB and UN 2010). However, these approaches may not be enough to adequately lower the chance of catastrophes and may not avail with the ongoing pace of climate change and the slow responses from the international community. In addition, the frequency of immediate threat and consequences of climate-related extreme events is expanding and bringing a range of newly emerging risks. The global costs of natural disasters have risen 15-fold since 1950, from US $38 billion (at 1998 values) in 1950–1959 to $652 billion in 1990–1999 (World Bank 2006). The impact is one of large-scale human suffering, loss of lives and an exponential rise in financial costs (Gaiha et al. 2010). Therefore, a third option is becoming increasing important: there is an urgent need to develop a robust and integrated early warning system for climate-related extreme events to provide actionable warnings to the most vulnerable parts of the region.

Food and water security in HKH

Another challenge to the Himalayan community is to ensure water and food security from the changing climate. Water is facing unsustainable demand from users in the region and is affecting its availability for food production. Environmental uses of water—a key to ensuring the sustainability in the long run, often get neglected in water resource management. A sustainable water scenario allows water for environmental uses maintaining food production under a business as usual scenario. In addition, broader strategies are needed in infrastructural development to promote water supply for irrigation, domestic and industrial purposes, and reform in water management and policy to conserve and improve the efficiency of water use (Rosegrant et al. 2002). This report shows that if current situation continue, farmers will find it difficult to meet the growing demand for food. Appropriate policies and investments is one of the options suggested for sustainable use of water that will ensure water security (Bawa et al. 2010).

Regional cooperation is vital to save HKH environment

Environmental problems are increasingly a challenging issue faced by the HKH countries and one that will require careful cooperation and advocacy using scientific collaboration to streamline political barriers between countries. Many HKH countries have almost 100% of their territory and population within international basins. Almost, 39% of the region's area is within national protected areas, and 20% of this crosses national boundaries. Receding glaciers, increased floods, incident of GLOF and food and water security are the common challenges to the region. In addition, conflicting demands for these international waters and existing tensions both within and between countries indicate cooperation is imperative for peace and security in the region (World Bank 2012a). The HKH countries have intensified cooperation for managing transboundary landscapes. For example, at the 11th meeting of the Conference of the Parties to the Convention on Biological Diversity (CBD), in Hyderabad, India, representatives from India, Bhutan, China, India, Nepal and Pakistan showed their interest to intensify cooperation for the management of the Brahmaputra-Salween, Kailash, Kanchenjunga, and Karakoram-Pamir transboundary landscapes (ICIMOD 2012a).

Case study 5: Need for a regional cooperation for monitoring and research:
Fig. 10.22

The Hindu Kush-Himalayan region has one of the least available data on glaciers and has very limited monitoring or understanding of the thresholds of climate change on the cryosphere, hydrosphere, biosphere, and on human society (Kohler and Maselli 2009, Xu 2007). Records of minimum of ten or more year are relevant for climate and hydrological variability and trend studies, but only two glaciers barely meet this requirement

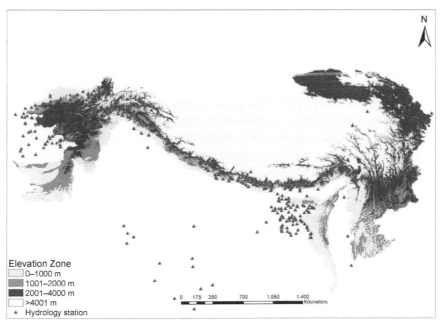

Figure 10.22. Monitoring stations in the Hindu-Kush Himalayan (HKH) region. The World Glacier Monitoring Service record shows only 97 monitoring stations in the HKH region in May 2011 adopted from UNEP 2012a.

(Armstrong 2010). Data and information is sparse and lack consistency, multi-temporal recording and field validation (Cogley 2011, Kargel et al. 2010) (Fig. 22). The region has been known as a "white spot", a term used in the IPCC 2007 report to refer to an area for which there is "little or no data". This has also constrained in-depth investigation of vulnerability and adaptation to climate change, which has become central to climate science and policy (Stigter and Winarto 2012). Additionally, available data is not always accessible, especially on transboundary water sharing, often for the reason "that concern politics and diplomacy rather than science" (Nature 2011). Hence, there is a need for greatly increased long-term monitoring for reliable and consistent time-series data for integrated research to understand highland complexities through cross-boundary scientific collaboration promoting regional cooperation, and present policy options based on best scientific understanding (UNEP 2012a).

China-India Cooperation for Sake of Regional Sustainability

China and India's economic, demographic development, urbanization and industrialization will play a dominant role in shaping the environmental outcomes in the region. While much of the world is facing an economic downturn, China and India in 2009 were projected to achieve high Gross Domestic Product growth (China, 8.4%; India, 6.2%) (Bawa et al. 2010). Import of raw materials from other countries, consumption of natural resources will significantly determine future environmental, social and economic outcomes. Cooperation between two countries can significantly and positively influence environmental problems in the region, such as, biodiversity loss, climate change and deforestation. For example, the creation of peace parks in alpine zones where armies have been deployed is an ideal for conservation of transboundary protected areas (Bawa et al. 2010). Their cooperation would influence their neighboring countries in addressing their environmental problems.

Promoting green economy in HKH region

A green economy concept may prevent the water insecurity situation in HKH region by reducing water and food waste through increased efficiency. For example: providing alternatives to cereal in animal feed, recycling waste, boosting of small-scale famer productivity and generating small-scale business opportunities, investments in green, small-scale technology and development (ICIMOD 2009). In the HKH, micro hydro, biogas, community-based natural resources management (community forestry, integrated conservation and development projects), eco-tourism, watershed management are some good practices and success stories that can be reported as a promising foundation for a green economy and good environmental governance, and recommended investments in green projects and policy reforms to provide incentives to the sectors such as agriculture, natural resources and industrial development (ICIMOD 2011).

One of the viable options is exploring low carbon growth strategies in energy, transport and urban sectors to reduce carbon emissions, such as introducing projects aiming to increase efficiency of energy production, promoting of alternative technologies like hydropower and small renewable power installations. Promoting climate-friendly interventions such as, promoting low-carbon growth, building climate-resilient rural economies, there is also a need to build awareness, ownership and capacity (World Bank 2009a).

Climate change into development agenda

Integrating climate change efforts into broader development and poverty reduction agendas can foster climate-resilient livelihoods with particular focus on highly vulnerable regions with the poorest and marginalized communities in the HKH region (IPCC 2007a). Development without consideration of climate risk and opportunities may not be a viable option. Although a range of development activities contribute to reducing vulnerability to many climate change impacts, in some cases, may increase vulnerability to climate change. For example, planning coastal development without taking into account projected sea level rise will put projects at risk and prove unsustainable in the long term. Integrating the consideration of long term climate risks into national planning processes, as well as in budgets, and building capacity of local governments and project planners to better assess and implement mitigation and adaptation measures (OECD 2009). Appropriate policy reform is urgently required to build up and enhance natural capital, such as, forests, water and soil (UNEP 2011).

Water storage for adaptation to climate change

With water scarcity looming, water storage is and will be a key strategy for climate change adaptation in the Himalayan region. The HKH is the major source of stored water in the cryosphere and the biosphere. The HKH region has a total area of more than 100,000 sq km—the largest bodies of ice outside the polar caps, providing important intra- and inter-annual water storage facilities. Developing a water storage system and management strategies in the context of climate change and making it available when it is needed the most is recognized as an appropriate strategy (ICIMOD 2009).

Glacier lakes may offer storage potential and can be accessed with appropriate technology and infrastructure to mitigate risk of outburst in the HKH region. The lakes that are not potentially dangerous in the region could offer an enormous capacity for water storage, for example, the Tibetan Plateau alone has more than 1000 lakes, with a total area of approximately 45,000 sq km (ICIMOD 2009).

Around 665 sq km of wetland, about 16% of the total area of the HKH comprises wetlands that play an important role in water storage and regulating water regimes. For example, the wetlands in the Ruoergai Marshes on the Qinghai-Tibetan Plateau in Southwest China, located at 11155–12795 feet above sea level, play an important role to maintain a natural system of water storage. Moreover, these wetlands play a prominent role in mitigating the impact of climate change by acting as carbon sinks. The peatlands in the Tibetan Plateau stores 1500–4000 tonnes per ha of carbon (Trisal and Kumar 2008).

Groundwater aquifers in the Himalayan region are important for water storage however there is little information available. Ways have been identified to store monsoon flows underground by recharging aquifers in the Ganges basin, for example (i) water spreading in the piedmont deposit (Bhabar zone) north of the Terai belt of springs and marshes; ii) slowdown runoff and increase infiltration using bunds at right angles to the flow lines; iii) increasing seepage from irrigation canals during monsoon (Revelle and Laksminarayana 1975). The unconsolidated Bhabar zone and the Terai plains provide a very large groundwater reservoir in the Himalayan region. In Nepal, 2800 million cu. m of groundwater recharge takes place in the Bhabar zone every year and 8800 million cu. m in the Terai belt (ICIMOD 2009).

Better use of green water (water stored in the soil profile), especially in the Ganges and Indus Rivers basins that have the lowest availability of green and blue water (water from surface bodies and aquifers) per capita. The role of improving transboundary water management is a potentially powerful adaptation strategy since large populations share rivers and aquifers—operationalizing shared management of transboundary water, such as, availability, access and conflict of use (GWP 2011).

Local participation for mitigation and adaptation

Projects exist that are being implemented by HKH countries that adheres to the notions adopted by programmes like Reducing Emissions from Deforestation and Forest Degradation (REDD) and green economy, and contributes to addressing climate change adaptation activities. For example, the community forestry programme in Nepal has been the largest and longest participatory green initiative wherein 40% of Nepal's population is involved in managing 25% of the country's forest area (ICIMOD 2011).

Case study 6: Community Forestry: Local participation for climate change adaptation and mitigation

Community forestry is a system in which communities are involved in the conservation of forests and regulation of forest resources. Local users organized as Community Forest User Group (CFUGs) take the responsibility of managing forests while the government facilitate in promoting community forestry. Community forestry is a central place in forest management in Nepal. About one-fourth of the country's forest is managed by more than 35% of the total population. This has provided employment, income generation from forest protection. More importantly, community forestry also plays important role in mitigation and adaptation to climate change. For example: nearly 65 million t C per year stored by the Indian Himalayan forests, equal to 15–20% CO_2 emissions from fossil fuels combustion from India around the year 2000. A study led by Dr. Margaret under the project "Kyoto: Think Global Act Local" in several African countries, Nepal and India, noted that a well-managed community forest in Uttarakhand can store a considerable amount of carbon, while meeting their day-to-day needs from the forest. The study also suggests that at least twice the size of agricultural area should be forest cover by providing incentive schemes to local people (Singh 2008).

Conservation and development through local participation

ICDP is an innovative approach to the conservation of biodiversity and ecological systems, at the same time promotion of human development by linking conservation and development, and complimenting each other For example, Annapurna Conservation Area Project (ACAP) launched in 1986, to achieve sustainable balance between nature conservation and socio-economic improvement, is the first conservation of the largest protected area in Nepal, covering 7,629 sq km that accommodates 1,226 species of flowering plants, 102 mammals, 474 birds, 39 reptiles and 22 amphibians, living in harmony with local communities. Addressing the environmental challenge of multifaceted problems, an integrated, community based conservation and development approach—an experimental model to promote the concepts of 'Conservation Area' through an 'Integrated Conservation and Development Programme' was first tested as a pilot programme in the Ghandruk Village Development Committee (VDC) in 1986, and covering entire ACAP area in 1992. It is the first protected area that allowed local resident to live within their boundaries, own their private property and maintain their traditional rights and access to the use of natural resources. All the revenue collected from tourism in the area is then used to implement conservation and development activities and projects in the area (ACAP).

Case study 7: Adaptive agricultural practice

According to Paustian et al. (2006) increasing productive, erosion control measures, reduced tillage and improving cropping practices of agricultural soil can capture CO_2 (Singh 2008). The fundamental approach that needs to be incorporated in agricultural development programmes, for example strengthening the ecological foundation of food security through sustainable food system without undercutting the basic natural conditions needed to produce food (e.g., water, soil formation, biodiversity). Broadly, these are: i) resource base agriculture (e.g., water and land) and, ii) supporting ecosystem services (e.g., soil formation and nutrient recycling; on-farm and off-farm biodiversity; climate condition and processes) (UNEP 2012b).

Conclusion

The dynamic and complex HKH mountain system is vulnerable to global changes. Globally dominating production and consumption, population growth and urbanization are the main anthropogenic activities changing the HKH environment. Since anthropogenic activities are the main engine of changing the HKH region, it is imperative for decision makers to revisit and redesign their growth and development strategy, along with new approaches apart from the ongoing adaptation and mitigation measures. All this however is not possible without bundling peace and security in the region. Since, the HKH's environmental issues are transboundary, regional cooperation is imperative for peace and security and for sustainable development.

In addition, uncertainties about the rate and magnitude of climate change and potential impacts, and inadequate scientific studies, are hampering the development of future projections on likely impacts on human and ecosystems. This in turn, hinders the effective action to adapt to anticipated changes in the HKH region. The region needs transboundary scientific cooperation to establish an effective communication between the scientific community and policymakers to identify knowledge gaps for better understanding the complexities of the HKH region, and allow policy options based on appropriate scientific evidence.

References

Adhikari, S., S.J. Marshall and P. Huybrechts. 2009. A comparison of different methods of evaluating glacier response characteristics; application to glacier AX010, Nepal Himalaya. The Cryosphere Discuss. 3: 765–804.

Andermann, C., L. Longuevergne, S. Bonnet, A. Crave, P. Davy and R. Gloaguen. 2012. Impact of transient groundwater storage on the discharge of Himalayan rivers. Nature Geosciences 5(2): 127–132.

Annapurna Conservation Area Project (ACAP). http://www.ntnc.org.np/project/annapurna-conservation-area-project.

Armstrong, R.L. 2010. The Glaciers of the Hindu Kush-Himalayan Region. A summary of the science regarding glacier melt/retreat in the Himalayan, Hindu Kush, Karakoram, Pamir, and Tien Shan mountain ranges. Technical Paper. Kathmandu. ICIMOD and USAID.

Asian Development Bank and World Bank (ADB and WB). 2010. Pakistan Floods 2010: Preliminary Damage and Needs Assessment. Islamabad, Pakistan.

Bajracharya, S., P.K. Mool and B. Shrestha. 2007. Impact of Climate Change on Himalayan Glaciers and Glacial Lakes: Case Studies on GLOF and Associated Hazards in Nepal and Bhutan. Kathmandu, Nepal: International Centre for Integrated Mountain Development (ICIMOD), Kathmandu, Nepal.

Bamber, J. 2012. Climate change: shrinking glaciers under scrutiny. Nature 482: 482–483.

Barnett, T.P., J.C. Adam and D.P. Lettenmaier. 2005. Potential impacts of a warming climate on water availability in a snow-dominated region. Nature 438: 303–309.

Bawa, K.S., L.P. Koh, T.M. Lee, J. Liu, P.S. Ramakrishnan, D.W. Yu, Y. Zhang and P.H. Raven. 2010. China, India and the Environment. Science 327(5972): 1457–1459. DOI: 10.1126/science.1185164.

Bolch, T., A. Kulkarni, A. Kääb, C. Huggel, F. Paul, J.G. Cogley, H. Frey, J.S. Kargel, K. Fujita, M. Scheel, S. Bajracharya and M. Stoffel. 2012. The state and fate of Himalayan glaciers Science 20 April 2012: 336(6079): 310–314.

Bookhagen, B. 2010. Appearance of extreme monsoonal rainfall events and their impact on erosion in the Himalaya. Geomatics, Natural Hazards and Risk 1(1): 37–50.

Boos, W.R. and Z. Kuang. 2010. Dominant control of South Asian monsoon by orographic insulation versus plateau heating. Nature 463: 218–222.

Cogley, G. 2011. Present and future states of Himalaya and Karakoram glaciers. Annals of Glaciology 52(59): 68–73.

Dash, S.K., M.A. Kulkarni, U.C. Mohanty and K. Prasad. 2009. Changes in the characteristics of rain events in India. Journal of Geophysical Research 114(D10): D10109.

Dyurgerov, M.B. and M.F. Meier. 2005. Glaciers and the Changing Earth System: A 2004 Snapshot. Institute of Artic and Alpine Research, University of Colorado, Boulder. http://instaar.colorado.edu/other/download/OP58_dyurgerov_meier.pdf.

Eriksson, M., J. Xu, A.B. Shrestha, R.A. Vaidya, S. Nepal and K. Sandstörm. 2009. The Changing Himalayas: Impact of climate change on water resources and livelihoods in the greater Himalayas. Perspectives on water and climate change adaptation. ICIMOD Kathmandu.

Gaiha, R., K.Hill and G. Thapa. 2010. Natural Disasters in South Asia. ASARC Working Paper 2010/06 http://www.crawford.anu.edu.au/acde/asarc/pdf/papers/2010/WP2010_06.pdf.

Gardelle, J., E. Berthier and Y. Arnaud. 2012. Slight mass gain of Karakoram glacier in the early twenty-first century. Nature Geoscience 5: 322–325. doi: 10.1038/ngeo1450.

Global Water Partnership (GWP). 2011. Climate Change, Food and Water Security in South Asia: Critical Issues and Cooperative Strategies in an Age of Increased Risk and Uncertainty. Synthesis of Workshop Discussions. A Global Water Partnership (GWP) and International Water Management Institute (IWMI) Workshop, 23–25 February 2011, Colombo, Sri Lanka.

Gujja, B. and T.M. Thiyagarajan. 2009. New Hope for Indian Food Security? The System of Rice Intensification. Gatekeeper 143. International Institute for environment and development, London.

Hidellage, V. 2003. Is There a Need for a South Asian Response on Technology for Poverty Reduction? South Asia Conference on technologies for Poverty Reduction, New Delhi, 10–11 October 2003. http://practicalaction.org/docs/region_south_asia/vishaka_hidellage.pdf.

Hoermann, B. and M. Kollmair. 2009. Labour migration in the Hindu Kush-Himalayas—A core livelihood strategy. ICIMOD, Kathmandu, Nepal.

Hua, O. 2009. The Himalayas-water storage under threat. Sustainable Mountain Development 56: ICIMOD, Kathmandu, Nepal.

Immerzeel, W.W., L.P.H. Beek and M.F.P. Bierkens. 2010. Climate change will affect the Asian water towers. Science 328: 1382.

Intergovernmental Panel on Climate Change (IPCC). 2007a. Climate Change 2007: Synthesis Report: An Assessment of the Intergovernmental Panel on Climate Change, Valencia, Spain.

Intergovernmental Panel on Climate Change (IPCC). 2007b. Contribution of Working Group II to the Forth Assessment Report of the Intergovernmental Panel on Climate Change, 2007. http://www.ipcc.ch/publications_and_data/ar4/wg2/en/contents.html.

International Centre for Integrated Mountain Development (ICIMOD). 2003. Mountain Agriculture in the Hindu Kush-Himalayan Region: Proceedings of an International Symposium held May 21–24, 2001 in Kathmandu, Nepal. ICIMOD, Kathmandu, Nepal.

International Centre for Integrated Mountain Development (ICIMOD). 2012a. Countries of the Hindu Kush Himalayas to Intensify Co-operation in Managing Transboundary Landscapes http://www.icimod.org/?q=8818.

International Centre for Integrated Mountain Development (ICIMOD). 2012b. Sustainable Mountain Development RIO 2012 and beyond. ICIMOD, Kathmandu, Nepal.

International Centre for Integrated Mountain Development (ICIMOD). 2011. Regional Assessment Report for RIO+20: Hindu Kush Himalaya and SE Asia Pacific Mountains. ICIMOD, Kathmandu, Nepal.

International Centre for Integrated Mountain Development (ICIMOD). 2001. Inventory of Glaciers, Glacial Lakes and Glacial Lake Outburst Floods: Monitoring and Early Warning Systems in the Hindu Kush-Himalayan Region Nepal. ICIMOD, Kathmandu, Nepal.

International Centre for Integrated Mountain Development (ICIMOD). 2009. Water Storage: A Strategy for Climate Change Adaptation in the Himalayas. Sustainable Mountain Development #56. ICIMOD, Kathmandu, Nepal.

Kargel, J.S., J.G. Cogley, G.J. Leonard, U. Haritashya and A. Byers. 2011. Himalayan glaciers: the big picture is a montage. Proceedings of the National Academy of Sciences of the United States of America (PNAS) 108(36): 14709–14710.

Kargel, J.S., R. Armstrong, Y. Arnaud, S. Bajracharya, E. Berthier, M.P. Bishop, T. Bolch, A. Bush, G. Cogley, K. Fujita, R. Furfaro, A. Gillespie, U. Haritashya, G. Kaser, S.J. Singh Khalsa, G. Leonard, B. Molnia, A. Racoviteanu, B. Raup, B. Shrestha, J. Shroder and C. Van der Veen. 2010. Satellite-era glacier changes in High Asia. http://www.glims.org/Publications/2009Dec-FallAGU-Soot-PressConference-Backgrounder-Kargel.pdf.

Kaser, G., M. Großhauser and B. Marzeion. 2010. Contribution potential of glaciers to water availability in different climate regimes. Proceedings of the National Academy of Sciences of the United States of America (PNAS) 107: 20223–20227.

Kohler, T. and D. Maselli. 2009. Mountains and Climate Change—From Understanding to Action. Published by Geographica Bernensia with the support of the Swiss Agency for Development and Cooperation (SDC), and an international team of contributors. Bern.

Lamer, B. 2010. The big melt. National Geographic Magazine. April 2010. http://ngm.nationalgeographic.com/print/2010/04/tibetan-plateau/larmer-text.

Lau, K.M., M.K. Kim and K.M. Kim. 2006. Asian summer monsoon anomalies induced by aerosol direct forcing: the role of the Tibetan Plateau. Climate Dynamics 26(7-8): 855–864.

Lau, K.M., M.K. Kim, K.M. Kim and W.S. Lee. 2010. Enhanced surface warming and accelerated snow melt in the Himalayas and Tibetan Plateau induced by absorbing aerosols. Environmental Research Letters 5(2).

Lau, K.M., V. Ramanathan, G.-X. Wu, Z. Li, S.-C. Tsay, C. Hsu, R. Sikka, B. Holben, D. Lu, G. Tartari, M. Chin, R. Koudelova, H. Chen, Y. Ma, J. Huang, K. Taniguchi and R. Zhang. 2008. The joint aerosol-monsoon experiment: a new challenge for monsoon climate research. Bulletin of the American Meteorological Society 89(3): 369–383.

Meehl, G.A., J.M. Arblaster and W.D. Collins. 2008. Effects of black carbon aerosols on the Indian monsoon. Journal of Climate 21(12): 2869–2882.

Menon, S., D. Koch, G. Beig, S. Sahu, J. Fasullo and D. Orlikowski. 2010. Black carbon aerosols and the third polar ice cap. Atmospheric Chemistry and Physics 10: 4559–4571.

National Academy of Science. 2012. Committee on Himalayan Glaciers, Hydrology, Climate Change, and Implications for Water Security, Board on Atmospheric Studies and Climate, Division on Earth and Life Studies, National Research Council. Himalayan Glaciers: Climate Change, Water Resources, and Water Security 2012. The National Academies Press, Washington D.C.

Nature. 2011. Climate action a 'moral responsibility'. Nature spoke with Qin Dahe. Nature. Published online 20 October 2011. doi:10.1038/news.2011.604.

Organization for Economic Co-operation and Development (OECD). 2009. Integrating Climate Change Adaptation into Development Co-operation: Policy Guidance. Paris, France.

Perkins, S. 2012. Renegade glaciers gain ice. Nature. 15 April 2012. doi:10.1038/nature.2012.10448.

Ramanathan, V., C. Chung, D. Kim, T. Bettge, L. Buja, J.T. Kiehl, W.M. Washington, Q. Fu, D.R. Sikka and M. Wild. 2005. Atmospheric brown clouds: impacts on South Asian climate and hydrological cycle. Proceedings of the National Academy of Sciences of the United States of America (PNAS) 102(15): 5326–5333.

Ramesh, K.V. and P. Goswami. 2007. Reduction in temporal and spatial extent of the Indian summer monsoon. Geophysical Research Letters 34(23): L23704.

Rees, H.G. and D.N. Collins. 2006. Regional differences in response of flow in glacier-fed Himalayan Rivers to climate warning. Hydrological Processes 20(10): 2157–2169.

Revelle, R. and V. Laksminarayana. 1975. The Ganges water machine. Science 188: 611–616.

Rizvi, Muddassir. 2001. Forecasting Water Flows in Pakistan's Indus River. International Development Research Council. Ottawa, Canada. http://web.idrc.ca/en/ev-5441-201-1-DO_TOPIC.html.

Rosegrant, M.W., X. Cai and S.A. Cline. 2002. World Water and Food to 2025: Dealing with Scarcity. International Food Policy Research Institute, Washington D.C.

Rosenfeld, D., U. Lohmann, G.B. Raga, C.D. O'Dowd, M. Kulmala, S. Fuzzi, A. Reissell and M.O. Andreae. 2008. Flood or drought: how do aerosols affect precipitation? Science 321(5894): 1309–1313.

Scherler, D., B. Bookhagen and M.R. Strecker. 2011. Spatially variable response of Himalayan glaciers to climate change affected by debris cover. Nature Geoscience 4: 157–159.

Scherler, D., B. Bookhagen, M.R. Strecker, F. von Blanckenburg and D. Rood. 2010. Timing and extent of late Quaternary glaciation in the western Himalaya constrained by 10Be moraine dating in Garhwal, India. Quaternary Science Reviews 29(7–8): 815–831.

Sharma, E., S. Bhuchar, M.A. Xing and B.P. Kothyari. 2007. Land use change and its impact on hydro-ecological linkages in Himalayan watersheds. Tropical Ecology 48(2): 151–161.

Shrestha, U.B., S. Gautam and K.S. Bawa. 2012. Widespread climate change in the Himalayas and associated changes in local ecosystems. PLoS One 7(5): e36741. Doi:10.1371/journal.pone.0036741.

Singh, S.P. 2008. Climate change in relation to the Himalayas. Climate Leaders Initiative—India. http://www.climate-leaders.org/wp-content/uploads/climtechange-spsingh.pdf.

Stigter, C.K. and Y.T. Winarto. 2012. What Climate Change Means for Farmers in Asia. Earthzine. 4 April 2012. http://www.earthzine.org/2012/04/04/what-climate-change-means-for-farmers-in-asia/.

Strategic Foresight Group (SFG). 2010. Himalayan Challenges: Water Security in Emerging Asia, 2010. Strategic Foresight Group. Mumbai, India.

Strategic Foresight Group (SFG). 2011. Himalayan Solutions: Co-operation and Security in River Basins. Strategic Foresight Group. Mumbai, India.

Thayyen, R.J. and J.T. Gergan. 2010. Role of glaciers in watershed hydrology: Himalayan Catchment. The Cryosphere 4: 115–128.

Thayyen, R.J., J.T. Gergan and D.P. Dobhal. 2007. Role of glaciers and snow cover on headwater river hydrology in monsoon regime—Micro-scale study of Din Gad catchment, Garhwal Himalaya, India. Current Science 92(3).

The World Bank (WB). 2006. Hazards of Nature, Risks to Development, Independent Evaluation Group, Washington D.C.

The World Bank (WB). 2008. Climate change: Why is South Asia Vulnerable? UN Climate Change Conference in Poznan. November 25, 2008. http://go.worldbank.org/CVAV2OB8N0.

The World Bank (WB). 2009. Climate Change Strategy for the South Asia Region (Draft). http://go.worldbank.org/DEOKW48F50.

The World Bank (WB). 2009. Climate Change: Are South Asia Ecosystem at the brink of Extinction? http://go.worldbank.org/HGRLJFBNV0.

The World Bank (WB). 2012a. Kathmandu to Copenhagen: A Regional Climate Change Conference. http://go.worldbank.org/7ZP5PBIOB0.

The World Bank (WB). 2012b. Regional Cooperation & Integration Water: Challenges and Benefits. http://go.worldbank.org/PF5DLQH980.

The World Bank and The United Nations (WB and UN). 2010. Natural Hazards, UnNatural Disasters: The Economics of Effective Prevention. Washington D.C.

Trisal, C.L. and R. Kumar. 2008. Integration of high altitude wetlands into river basin management in the Hindu Kush-Himalayas. Wetlands International—South Asia, New Delhi.

U.S. Environmental Protection Agency (USEPA). 2011. Report to Congress on Black Carbon: External Peer Review Draft. Washington D.C.

United Nations Environment Programme (UNEP). 2008. Atmospheric Brown Clouds. Regional Assessment Report with Focus on Asia. UNEP, Nairobi, Kenya.

United Nations Environment Programme (UNEP). 2009. Recent Trends in Melting Glaciers, Tropospheric Temperatures over the Himalayas and Summer Monsoon Rainfall over India.UNEP, Nairobi. Kenya.

United Nations Environment Programme (UNEP). 2012a. Measuring Glacier Change in the Himalayas. UNEP. Nairobi, Kenya.

United Nations Environment Programme (UNEP). 2012b. Avoiding Future Famines: Strengthening the Ecological Foundation of Food Security through Sustainable Food Systems. UNEP, Nairobi, Kenya.

United Nations Environment Programme (UNEP). 2011. Towards a Green Economy: Pathways to Sustainable Development and Poverty Eradication. UNEP, Nairobi, Kenya. www.unep.org/greeneconomy.

United Nations Environment Programme Regional Resource Centre for Asia and the Pacific (UNEP RRC.AP). 2001. South Asia: State of the Environment 2001, United Nations Environment Programme Regional Resource Centre for Asia and the Pacific (UNEP RRC.AP), Thailand.

United Nations Framework Convention on Climate Change (UNFCCC). 2010. Framework Convention on Climate Change. 9th Plenary Meeting 10-11 December, 2010. Cancun, Mexico.

United States Central Intelligence Agency (CIA). 2006. Washington D.C. https://www.cia.gov/cia/publications/factbook/geos/in.html.

Wheeler, W. 2011. India and Pakistan at odds over shrinking Indus River. Irrigation and hydroelectric projects are draining the river's flow, while glaciers are melting in Kashmir. For National Geographic News. Published October 12, 2011. http://news.nationalgeographic.com/news/2011/10/111012-india-pakistan-indus-river-water/.

Winiger, M., M. Gumpert and H. Yamout. 2005. Karakoram-Hindu Kush-Western Himalaya: assessing high-altitude water resources. Hydrological Processes 19(12): 2329–2338.

Winkler, S., T. Chinn, I. Gartner-Roer, S.U. Nussbaumer, M. Zemp and H.J. Zumbuhl. 2010. An introduction to mountain glaciers as climate indicators with spatial and temporal diversity. Erdkunde 64(2): 97–118.

Xu, J. 2007. The Highlands: A Shared Water Tower in a Changing Climate and Changing Asia. ICRAF Working Paper. ICRAF, China.

Xu, J.C. and G.M. Rana. 2005. Living in the mountains. pp. 196–199. In: T. Jeggle (ed.). Know Risk, U. N. Inter-Agency Secretariat of the International Strategy for Disaster Reduction, Geneva.

Xu, J., A. Shrestha and M. Eriksson. 2009. Climate change and its impacts on glaciers and water resource management in the Himalayan Region. Assessment of Snow, Glacier and Water Resources in Asia. International Hydrological Programme of UNESCO and Hydrology and Water Resources Programme of WMO.

Xu, J., A. Shrestha, R. Vaidya, M. Eriksson and K. Hewitt. 2007. The Melting Himalayas. Regional Challenges and Local Impacts of Climate Change on Mountain Ecosystems and Livelihoods ICIMOD Technical Paper, Kathmandu, Nepal.

Xu, J., R.E. Grumbine, A. Shrestha, M. Eriksson, X. Yang, Y. Wang and A. Wilkes. 2008. The Melting Himalayas: Cascading Effects of Climate Change on Water, Biodiversity, and Livelihoods. Conservation Biology 23(3): 520–530.

Yao, T.D., L.G. Thompson, V. Musbrugger, Y.M. Ma, F. Zhang, X.X. Yang and D. Joswiak. 2011. UNESCO-SCOPE-UNEP Policy Briefs Series. Third Pole Environment. UNESCO-SCOPE-UNEP, Paris.

11

Climate Change and its Impacts on Community Food and Livelihood in Kumaun Himalaya: A Case Study of Dabka Catchment

Pradeep K. Rawat* and Prakash C. Tiwari

INTRODUCTION

Intergovernmental Panel for Climate Change-IPCC (2001, 2007a, 2007b) warns that the progress in human development achieved over the last decade may be slowed down or even reversed by climate change, as new threats emerge to food and livelihood security, agricultural production and access, and nutrition and public health. Agriculture constitutes the backbone of most developing economies throughout the world and in turn, food and fiber production is essential for sustaining and enhancing human welfare (Rosenzweig and Parry 1994, Fischer et al. 1996). Consequently, agriculture has been a major concern in discussions on climate change. Agronomic and economic impacts from climate change depend primarily on two factors, i.e., (i) the rate and magnitude of change in climate attributes and the agricultural effects of these changes, and (ii) the ability of agricultural production to adapt to changing environmental conditions. Temperature,

Department of Geography, Kumaun University, Nainital, India.
  Email: geopradeeprawat@hotmail.com
* Corresponding author

precipitation, atmospheric carbon dioxide content, the incidence of extreme events and sea level rise are the main climate change-related drivers which impact agricultural production (Adams 1998). Climate change, however, is considered as posing the greatest threat to agriculture and food security in the 21st century, particularly in many of the poor, agriculture-based countries with their low capacity to cope effectively (Darwin and Kennedy 2000, Adams et al. 1995). Mountain agriculture is already under stress as a result of population increase, industrialization and urbanization, competition over resource use, degradation of resources and insufficient public spending for rural infrastructure and services. The impact of climate change is likely to exacerbate these stresses even further. The outlook for the coming decades is that agricultural productivity needs to continue to increase and will require more water to meet the demands of growing populations. Ensuring equitable access to water and its benefits now and for future generations is a major challenge as scarcity and competition increase. The amount of water allocated to agriculture and water management choices will determine, to a large extent, whether societies achieve economic and social development and environmental sustainability (Iglesias et al. 2000, Easterling et al. 1993).

Food security and rural livelihoods are intrinsically linked to water availability and use. Food security is determined by the options people have to secure access to own agricultural production and exchange opportunities (Parry et al. 1999). These opportunities are influenced by access to water. Making these water-livelihoods linkages is important for a more complete understanding of the nature of vulnerability of households to climate-related hazards such as drought, and the multi-faceted impacts that water security has on food and livelihood security (Rawat et al. 2011). In order to highlight such linkages, there has been a move in recent years towards looking at water issues through sustainable livelihood frameworks (Rosenzweig et al. 1999). One main feature of climate change adaptation at the local level is its attempt to increase the resilience of populations to climate-related hazards. This means assessing the populations at risk of water and food insecurity. First, lack of access to adequate water supply, both in quality and quantity, for domestic uses can be a major cause of declining nutritional status and of disease and morbidity. Second, domestic water is often a production input. Such production is essential for direct household consumption and/or income generation. Third, the amount of time used to collect water, and related health hazards, can be immense, especially for women and girls, and has been well documented (Iglesias 2007b).

This chapter reviews current knowledge about the relationships between climate change, community food security and livelihood in the Himalayan region. Climate change and land-use degradation affects community food and livelihood through accelerating several hydrological hazards, i.e., flash flood, river-line flood, erosion, non-seismic land slide in the monsoon period and drought hazard as drying up of natural water springs and decreasing streams discharge in the non-monsoon period. An attempt has been made in the present study to assess the trends of land-use pattern in a fragile watershed located near a seismic and tectonically active region of the Himalaya. The watershed lies between the latitude 29°24'09"–29°30'19"E and longitude 79°17'53"–79°25'38"N in the North west of Nainital Township (Fig. 11.1). The region encompasses a geographical area of 69.06 km² between 700 m and 2623 m altitude above mean sea level. The total population of the watershed is 9250 people, which includes 16 villages. The population density is 76.02 person/km². About 95% population of the total population depends on agriculture and forest resources but the forest cover is decreasing 0.36 km² per year and the agricultural production is also decreasing due to climate change and drought hazard (Rawat 2013). The Geographical Information System (GIS) and Remote Sensing (RS) techniques have recently been widely applied to study land-use/land cover changes

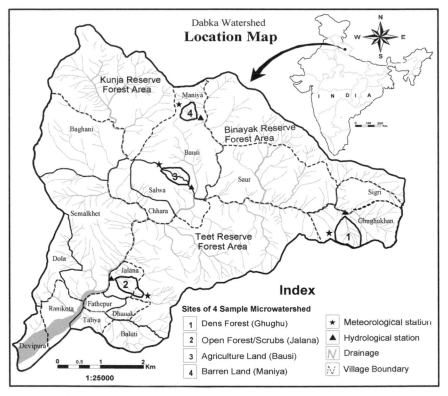

Figure 11.1. Location map showing selected sample micro-watersheds with 16 villages and location of hydrological and meteorological stations installed in the study area.

(Mohanty 1994, Minakshi et al. 1999, Brahmabhatt et al. 2000, Chauhan et al. 2003). In the Himalaya, a variety of changes have emerged in the traditional resource utilization structure mainly in response to population growth and resultant increased demand of natural resources, ineffective technology transfer, market forces, inappropriate land tenure policies, faulty environmental conservation programs, irrational rural developmental schemes, and increasing economic and political marginalization, during the recent years (Hamilton 1987, Tiwari and Joshi 1997). These emerging negative trends in the socio-economic profile have resulted in rapid exploitation and transformation of land resources and largescale land-use changes in the region (Tiwari 1995, Joshi and Gairola 2004). Under the impact of various land-use systems, the land and whole environment of a geographical region changes positively or negatively. The impact of some land-use changes is limited to the area in which they are operated while that of others reaches far in the surrounding ecosystems (Kostrowicki 1983, Sharma et al. 2001, Sharma et al. 2003, Scheling 1988). The extensive land-use changes in the Himalaya have not only disrupted the fragile ecological balance of the watersheds in the region through deforestation, erosion, landslides, hydrological disruptions, depletion of genetic resources, but have also threatened the livelihood security and community sustainability in mountains as well as in adjoining plains ecosystem (Tiwari 2000, Tiwari and Joshi 2002, 2005). Land-use degradation due to climate change affecting water resources as drying up of natural water springs and decreasing trends of streams discharge and as a whole triggering other hydrological hazards such as high runoff, flash floods, river-line floods and non-seismic landslide, etc. which are mainly responsible for several socio-economic consequences in mountainous terrain (Ives 1989, Valdiya and Bartarya 1989, Cruz 1992, Jain et al. 1994, Sing 2006, Rawat et al. 2012).

Methodology

The study comprises mainly two components, (a) laboratory/desk study and (b) field investigations. The procedure adopted has been outlined in Fig. 11.2 depicting that the study was carried out through GIS database management system (DMS). GIS-DMS is a set of computer programs for managing an integrated spatial and attribute database for such a task as map and data input storage, search, retrieval, manipulation and output. Existing DMS is constituted of four different GIS modules consisting of spatial map layers with their attribute data. These four GIS modules are: climate informatics, land-use informatics, hydro-informatics and agro-informatics as described below:

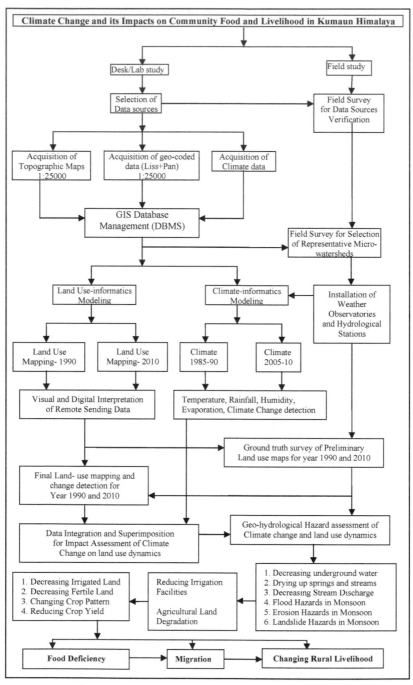

Figure 11.2. Procedure Adopted for Study.

Climate informatics module

Climate informatics module consists of spatial distribution of climate and its change detection through daily, monthly and annual weather data (temperature, rainfall, humidity and evaporation) of two study periods, i.e., during 1985–1990 and 2005–2010. To assess the climate change during last 25 years, a comprehensive meteorological study carried out for period 2005–2010 and compared the results of this study with the previous one carried out during 1985–1990 (Bight 1991) from the same study area (i.e., Dabka watershed). Consequentially the spatial distribution of climate throughout the study area has been carried out as subtropical climate, temperate climate and moist temperate climate in respect to meteorological data of both study periods. These meteorological data recorded at five meteorological observatories. Four observatories located in sample micro-watershed established at different elevation and running by geology and geography department of the Kumaun University and one located at a lower elevation of the study area established and run by Irrigation Department of Uttarakhand state government. The four meteorological station run by the Kumaun University funded by Govt. of India under different agencies as per their requirement, i.e., Department of Environment (1985–1990), Department of Science and Technology (2005–2010) government of India.

Land use informatics module

Land use informatics module consists of decadal and annual change detection in spatial distribution of the land-use pattern. Indian Remote Sensing Satellite (IRS-1C) LISS III and PAN merged data of 1990 and 2010 was used for the analysis and mapping of land cover/land use for the respective years (Fig. 11.2). Supplementary data and information required for the study have been generated from various primary as well as secondary sources. The primary information was generated through field surveys, mapping, interviews, etc., and the relevant secondary data was collected from various sources, such as, Census of India—2001, Government, Land Records, forests maps, etc. radiometric corrections were done employing dark pixel subtraction technique. The satellite images of the study area were registered geometrically using SOI Topographical Sheets (56 O/7NE and 56 O/7NW) of the area at scale 1:25000. For carrying out this important exercise uniformly distributed common Ground Control Points (GCPs) were selected and marked with root mean square (rms) error of one pixel and the images used were resampled by cubic convolution method. Both the data sets were then co-registered for further analysis initially, the LISS and PAN data were co-registered with root mean square (rms) error of 0.3 pixel and the output FCC was transformed into Intensity, Hue and Saturation (IHS) color space images. The reverse transformation from IHS to RBG was performed substituting the original high-resolution image for the intensity component, along with the hue and saturation components from the original RBG images. This merge data product obtained through the fusion of IRS–1C LISS–III and PAN was used for the generation of land cover/land-use map of the study area for the year 2001, and digital image processing techniques supported by intensive ground truth surveys were used for the interpretation of the remote sensing data (Fig. 11.2). In order to enhance the interpretability of the remote sensing data for digital analysis several image enhancement techniques, such as, PCA, NDVI, etc. were employed. In the Himalayan mountain terrain the interpretability of the remote sensing data to a large extent is affected by the complexity of the terrain as due to the effect of elevation and slope and its aspect, the spectral signature of same objects are often different or vice versa. In order to overcome these constraints and also to attain the best possible level of accuracy in the interpretation, intensive ground truth surveys were carried out in the study region and a visual interpretation key was evolved for primary land cover/land-use classification (Table 11.1). This was followed by the digital classification of land cover/land-use through on screen visual recording and rectification. To monitor the dynamics of land utilization pattern in the study area the land-use maps generated for the years 1990 and 2010 were overlaid using Geographic Information System (GIS) and land-use changes were detected and mapped (Fig. 11.2).

Table 11.1. Land Use Interpretation Key.

Land Cover/Land Use Classes	Enhanced LISS – III + PAN	
	Tone	**Texture**
Oak	Red, bright red in south aspects or gentle slopes, dull red and dull tar green in north and steep slopes.	Fine
Pine	Green in the north aspects, red in the south aspects or on the ridges.	Coarse
Mixed Forests	Red interspersing with reddish and dull green.	Coarse
Cultivated Land	Reddish and yellowish.	Medium/Coarse
Scrub/Barren land	Light cyan interspersing with reddish or yellowish in both the aspects.	Medium
River Beds & Water bodies	Black and blackish.	Coarse

Hydro-informatics module

The hydro-informatics module consists of daily, monthly and annual record of spring hydrology and stream hydrology of study area. To assess the geo-hydrological impacts of climate change and land-use degradation on mountain ecology a comprehensive study carried out for under groundwater table, spring discharge and stream discharge throughout the study area during 2005–2010 and then compared the results with the previous study carried out during 1985–1990 (Bisht 1991, Rawat et al. 2011).

Agro-informatics module

The agro-informatics module appraises the integrated impacts of climate change, land-use degradation and hydrological hazards (drought, floods, soil erosion and landslides) on community food and livelihood through negative impacts on agricultural ecosystem, i.e., reduced irrigation facilities, changing crop pattern, decreasing production rate, reducing livestock and milk production, etc. This module carried out through extensive household surveys and then compared the results with village level census data of India for 1981 and 2011.

Results and Discussions

Trends of climate change in lesser Himalaya

The climate-informatics module consists of the comparative study of climatic parameters for a 20 year period during 1990 to 2010 to appraise the impacts of climate change on land-use pattern because it reversely influences the ecology of the watershed and accelerated several environmental and socio-economic risks through high runoff, erosion and sediment delivery during the rainy season. Monthly and annual climate data records of two study periods is used to climate spatial distribution mapping for respective year (Table 11.2 and Fig. 11.3). A brief account follows:

Temperature: The five-station and five-year average mean annual temperature of the Dabka watershed was 19°C in 2005–2010 and about 17°C in 1985–1990 (Table 11.2) although it varies by elevation throughout the study area. May is the driest and hottest month and has the highest temperature especially on south-facing barren land, whereas January is the coldest month and has the lowest temperature in the year. The average maximum annual temperature was 37°C in 2005–2010, whereas it was only

Table 11.2. Monthly and Annual Climate Change (Carried out by comparative study of two study periods, i.e., 1985–1990 and 2005–2010).

Study Periods	Meteorological Parameters	Monthly Results												Annual Ave.
		Jan.	Feb.	Mar.	Apr.	May	Jun.	Jul.	Aug.	Sep.	Oct.	Nov.	Dec.	
Existing (2005–2010)	Temperature (°C)	4.00	10.00	16.00	24.00	37.00	27.00	24.00	23.00	21.00	20.00	14.00	12.00	19.00
	Rainfall (mm)	12.00	69.00	76.00	12.00	76.00	262.00	486.00	566.00	261.00	31.00	13.00	9.00	156.00
	Humidity (%)	52.00	65.00	64.00	57.00	51.00	58.00	82.00	79.00	65.00	57.00	49.00	49.00	60.00
	Evaporation Rate (mm)	400.00	620.00	745.00	812.00	950.00	700.00	651.00	550.00	500.00	500.00	400.00	401.00	602.00
Previous (1985–1990)	Temperature (°C)	7.00	9.00	14.00	22.00	32.00	24.00	21.00	20.00	18.00	17.00	12.00	11.00	17.00
	Rainfall (mm)	24.00	84.00	99.00	33.00	97.00	328.00	497.00	589.00	284.00	48.00	21.00	15.00	177.00
	Humidity (%)	58.00	69.00	68.00	61.00	59.00	64.00	92.00	84.00	78.00	64.00	56.00	56.00	67.00
	Evaporation Rate (mm)	370.00	580.00	634.00	745.00	812.00	620.00	570.00	460.00	430.00	430.00	380.00	395.00	536.00
Changes (1985–2010)	Temperature (°C)	–3.00	1.00	2.00	2.00	5.00	3.00	3.00	3.00	3.00	3.00	2.00	1.00	2.00
	Rainfall (mm)	–12.00	–16.00	–23.00	–20.00	–21.00	–67.00	–11.00	–23.00	–22.00	–17.00	–9.00	–6.00	–21.00
	Humidity (%)	–6.00	–5.00	–4.00	–5.00	–8.00	–6.00	–11.00	–5.00	–13.00	–8.00	–8.00	–8.00	–7.00
	Evaporation Rate (mm)	30.00	40.00	111.00	67.00	138.00	80.00	81.00	90.00	70.00	70.00	20.00	6.00	67.00
Annual Changes	Temperature (°C)	–0.15	0.05	0.10	0.10	0.25	0.15	0.15	0.15	0.15	0.15	0.10	0.05	0.10
	Rainfall (mm)	–0.64	–0.78	–1.14	–1.02	–1.05	–3.33	–0.53	–1.16	–1.11	–0.84	–0.43	–0.30	1.03
	Humidity (%)	–0.30	–0.23	–0.20	–0.23	–0.40	–0.30	–0.53	–0.25	0.65	0.38	0.38	0.38	0.35
	Evaporation Rate (mm)	1.50	2.02	5.57	3.35	6.90	4.00	4.03	4.50	3.49	3.49	1.00	0.28	3.34

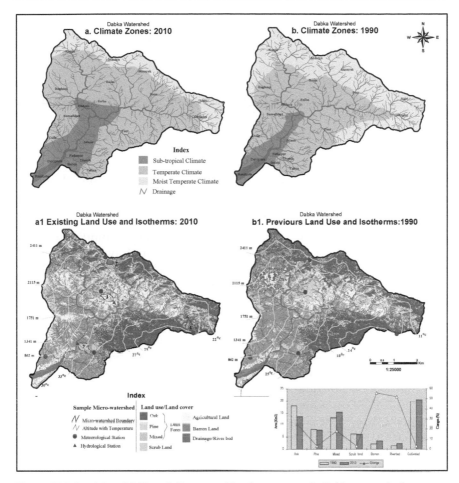

Figure 11.3. Spatial variability of climate and land use pattern in Dabka watershed.

32°C in 1985–1990. The average minimum annual temperature was 7°C in 1985–1990, whereas it had gone down to 4°C by 2005–2010. The climate data indicate increasing trends of extreme cold and hot weather in the months of January and May. Consequently the average annual temperature increased about 2°C in 1985–2010 at a rate of 0.10°C a^{-1}.

Rainfall: In the 2005–2010 period, average annual rainfall within the watershed varied between 1623 mm at Maniya (on an environmentally stressed barren hill slope) and 2187 mm at Ghughu (on dense forest hill slope), while it was 1969 mm at Bausi (on agricultural land), and 2086 mm at Jalna (on fairly dense forest/shrub land). Throughout the watershed the five-station average annual rainfall was 1874 mm (156 mm per month) in 2005–2010, whereas during the 1985–1990 period it was 2120 mm (177 mm per month) (Table 11.2). The difference in temporal distribution of monthly rainfall between these two time periods suggests that the annual rainfall pattern has changed with a higher concentration of extreme rainfall in July and August, which increases the risk of hydrological hazards such as erosion, sediment transport, floods, landslides and denudation. The rainfall records from 20 years indicate that the rainy season covers four months (June to September). In 1985–1990 average rainfall was higher in all months than in 2005–2010. The latter period was characterized by increasing drought events during non-monsoon months and heavier rainfall within short time periods during the monsoon months. Average annual rainfall decreased at a rate of 1.03 mm a^{-1} (Table 11.2).

Humidity: Due to climate change impacts the average annual humidity decreased at 0.35% a year (Table 11.2). The annual average humidity decreased about 7% during 1985–2010—it was 67% during 1985–1990 but decreased to 60% by 2005–2010 (Table 11.2). The two study periods' data analysis suggests that humidity varied between 86% in dense oak forest and 64% in the south-facing barren land.

Evapotranspiration Loss: The increasing average temperature (0.10°C a^{-1}) accelerated the evaporation loss at a rate of 3.34 mm a^{-1} (Table 11.2). During 1985–1990 the average annual evaporation loss was estimated at 536 mm whereas during 2005–2010 it increased by about 66 mm (Table 11.2). In 2005–2010 it approached 1072.3 mm on the south-facing barren land and dropped to 559.4 mm in the north-facing oak forest areas of the watershed. It was 786.7 mm and 811.4 mm on fairly dense forest/shrub land and agricultural land.

Spatial Distribution of Climate Change: The climate zone maps of the two study periods (1985–1990 and 2005–2010) (Figs. 11.3a and 11.3b) derived from climatic parameter distribution indicate that all the existing climatic zones have spread towards higher altitudes due to degrading climatic conditions of the mixed forest, pine forest, and oak forest in the subtropical, temperate and moist temperate climate zones. The results also show that the rates of climate change are increasing in higher elevations (Fig. 11.4), which dominantly affects the natural vegetation cover throughout the Himalaya up to the snow line (6000 m). Above the snow line it also affects the snow cover, which could be the major reason for the melting of glaciers. In order to monitor the impacts of climate change on land-use patterns, IRS LISS-III and PAN merged data for 1990 and 2010 were analyzed with the land-use informatics module.

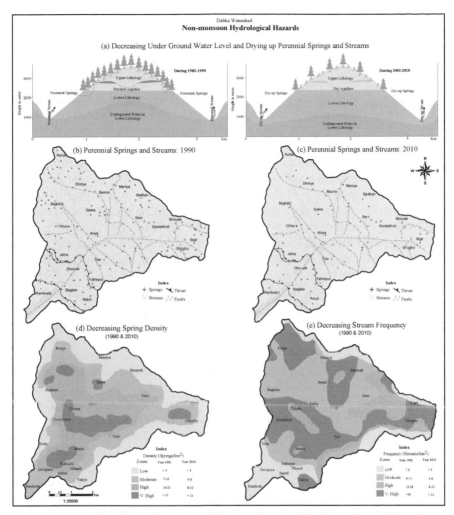

Figure 11.4. Geo-hydrological impacts of climate change and land-use degradation during 1990–2010 in Himalaya, i.e., decreasing underground water level and dry up of perennial springs and streams depicting by schematic diagram (a), decreasing perennial springs and streams during 1990 and 2010 (respectively b and c), decreasing perennial springs density (d) and decreasing perennial streams frequency (e).

Accelerated land use degradation through climate change

As mentioned in the methodological section that the land use-informatics module consists of comparative land-use land-cover mapping for year 1990 and 2010 to assess the changes and identified the accelerating factors for these changes as discussed in below. In order to monitor the dynamics of land transformation process land-use interpretation was carried out for the years 1990 and 2010 using IRS LISS–III and PAN merged data for the respective years. The exercise revealed that oak and pine forests have decreased respectively by 25 % (4.48 km²) and 3% (.28 km²) thus bringing a decline of 4.76 km² forest in the watershed during 1990 to 2010. But, due to climate change the mixed forest taking place of oak forest in certain pockets and consequently the mixed forest in the catchment increased by 18 % (2.3 km²) during the same period which reduced the overall loss of forests in the region but it is not eco-friendly as the oak forest as the broad leaved and wide spread roots of oak trees helps in controlling the several hydrological hazards such as accelerated runoff, erosion, landslides, flash flood and river-line flood during monsoon period and drought during non-monsoon period. As a result, the watershed recorded a total decline of 2.46 km² or 6% forest area during 1990 to 2010 (Table 11.3 and Fig. 11.3). The non-forest area has increased dramatically due to lopping and cutting of trees, accelerated runoff, soil erosion, and growing agricultural activities. The non-forest area has mainly been confined to barren land, riverbed and cultivated land. Barren land increased 1.21 km² (56%), riverbed increased 0.78 km² (52%) and cultivated land increased about 0.63 km² (3%) during the period of 1990 to 2010 (Table 11.3 and Fig. 11.3). The results of land-use dynamics presented on Table 11.4, advocate that the overall accelerating factor of land-use dynamics in the study area broadly categoriezed as a dominant and supporting factor. Out of the total seven classes of the land use-land cover, five classes (i.e., oak, pine, mixed, barren and riverbed) are changing dominantly due to the climate change factor and anthropogenic factors play a supporting role whereas only two classes (scrub land and agricultural land) are changing dominantly by anthropogenic factors and climate change factors play a supporting role. Expansion of mixed forest land brought out due to upslope shifting of existing forest species because of the climate change factor, only because upslope areas are getting warmer than in the past with the rate of 9°C–12°C/two decades (Table 11.3 and Fig. 11.3).

Geo-hydrological impacts of accelerated land use degradation

Increasing rates of land-use degradation due to climate change accelerating several geo-hydrological hazards in the Himalaya. The major geo-environmental impact is deforestation and land degradation which trigger several hydrological hazards during the monsoon and non-monsoon seasons. The major non-monsoon hydro-hazards are decreasing the underground water table, drying up of natural water springs and decreasing trends of streams discharge whereas monsoon hydro-hazards found high monsoon runoff, flash floods, river-line floods, soil erosion and non-seismic

Table 11.3. Minimum and Maximum Annual Average Temperature Obtained from five Meteorological Stations in the Watershed for 1990 and 2010 have been used to assess Climate Change (Section A) and its Impact on Land Use Dynamics (Section B).

(A) Climate Change

Meteorological Observatories with their Location and Established Years		Elevation (in m)	Minimum Average Temperature (in °C)		Maximum Average Temperature (in °C)		Annual Average Temperature (in °C)		Twenty Years Climate Changes (Average in °C)	Established and Running Authorities of Meteorological Observatories
No.	Location		1990	2010	1990	2010	1990	2010	1990-2010	1985-ongoing
M¹	Maniya (1985)	2411	4	2	21	30	13	16	3.0	Kumaun University
M²	Ghughu (1985)	2115	5	2	24	32	15	17	2.0	Kumaun University
M³	Bausi (1985)	1751	7	4	29	35	18	20	2.0	Kumaun University
M⁴	Jalna (1985)	1341	9	5	31	37	20	21	2.0	Kumaun University
M⁵	Devipura (1978)	862	10	7	34	40	22	24	2.0	Irrigation Department
	Study Area (Average)	1696	12	17	17	37	17	19	2.0	All Above

(B) Land-Use Dynamics

Land Use and Land Cover Classes	Year 1990		Year 2010		Twenty Year Dynamics		Accelerating Factors of Land Use Dynamics	
	km²	%	km²	%	km²	%	Dominant Factor	Supporting Factor
1. Oak	17.99	26.05	13.51	19.56	-4.48	25	Climate Change	Anthropogenic
2. Pine	8.21	11.89	7.93	11.48	-0.28	3	Climate Change	Anthropogenic
3. Mixed Forest	13.03	18.87	15.33	22.20	2.3	18	Climate Change	Nil
Total Forest	39.23	56.81	36.77	53.24	-2.46	6	Climate Change	Anthropogenic
4. Scrub land	6.38	9.24	6.22	9.01	-0.16	3	Anthropogenic	Climate Change
5. Barren Land	2.18	3.16	3.39	4.91	1.21	56	Climate Change	Anthropogenic
6. Riverbed	1.50	2.17	2.28	3.30	0.78	52	Climate Change	Anthropogenic
7. Cultivated Land	19.77	28.63	20.40	29.54	0.63	3	Anthropogenic	Climate Change
Total Non forest	29.83	43.19	32.29	46.76	2.46	8	Climate Change	Anthropogenic
Study Area (Total)	69.06	100.00	69.06	100.00		100.00	Climate Change and Anthropogenic Factors	

Table 11.4. Changes in spring and stream hydrology of the Dabaka watershed due to climate change and Land use Degradation (discharge is average of existing springs or streams in respective study periods).

Study Periods	No. of Perennial Springs and Streams	Average Discharge of Total Perennial Springs and Streams	Monthly												Annual
			Jan.	Feb.	Mar.	Apr.	May	Jun.	Jul.	Aug.	Sep.	Oct.	Nov.	Dec.	Average
Existing (2005–2010)	56 Springs	Spring Discharge (l/s)	2.00	2.00	1.50	1.00	0.50	1.00	2.00	4.00	4.00	2.00	2.00	1.50	1.96
	248 Streams	Stream Discharge (l/s/km²)	1.70	2.43	2.32	0.30	18.50	24.76	37.56	41.89	21.56	2.64	2.76	1.59	13.17
Previous (1985–2010)	116 Springs	Spring Discharge (l/s)	3.00	3.00	2.00	3.00	1.50	2.00	3.00	6.00	6.00	4.00	3.00	3.00	3.29
	162 Streams	Stream Discharge (l/s/km²)	3.00	4.85	6.56	0.78	26.90	30.61	42.61	48.08	27.44	4.92	3.27	2.90	16.82
Change (1985–2010)	–60 Springs	Spring Discharge (l/s)	1.00	1.00	0.50	2.00	1.00	1.00	1.00	2.00	2.00	2.00	1.00	1.50	1.33
	–80 Streams	Stream Discharge (l/s/km²)	1.30	2.42	4.24	0.48	8.40	5.85	5.05	6.19	5.88	2.28	0.51	1.31	3.65

landslide, etc. (Fig. 11.4 and Fig. 11.5). These hydrological hazards are mainly responsible for several socio-economic consequences in the mountains. A brief discussion is given below:

Decreasing under Groundwater Level: Land-use degradation and deforestation reduced the protective vegetal cover as a result that the significant proportion of rainfall goes waste as flood water without replenishing the groundwater reserve. It has been found that groundwater level throughout the watershed is gradually going down due to deforestation and high flood runoff. The results advocate that 20 years back during 1985–1990 the underground water was easily approachable up to 2000 m

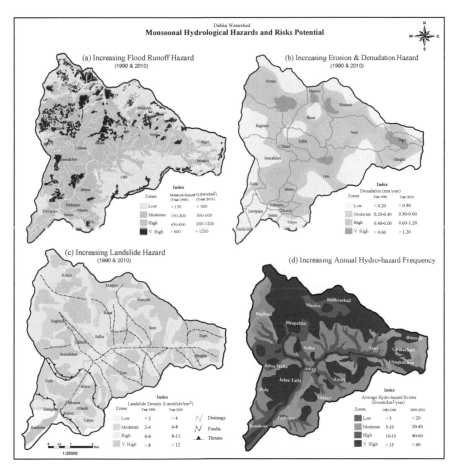

Figure 11.5. Monsoonal hydrological hazards due to climate change and land-use degradation during 1990–2010 in the Himalaya, i.e., increasing flood runoff hazard (a), increasing erosion and denudation hazard (b), increasing landslide hazard (c), integrated annual hydrological hazards frequency (d).

altitude whereas now it is quite difficult due to deficiency of underground water because the water table has been gone down to just 1200 m altitude (Fig. 11.4a). The decreasing trends of underground water table affecting the spring hydrology and stream hydrology in the study area as depicted by a schematic diagram in Fig. 11.4a for year 1990 and 2010.

Drying up Natural Water Springs: It was observed that, the springs are drying-up or becoming seasonal due to reduced groundwater recharge in the catchment. This has serious implications on water resources and on the livelihood and food securities as natural springs constitute the main source of drinking water and irrigation in the region. The investigation carried out in the region revealed that there were a total of 116 perennial springs in the watershed in 1990 (Bisht 1991), out of which 24% (28) have gone dry, and 28% (32) springs have become seasonal since 2010 (Fig. 11.4b and 11.4c). The spatial distribution of perennial spring density has been carried out which depict four categories of density, i.e., low, moderate, high and very high. All these categories suggested the decreasing trend of spring density during 1990 to 2010 and also decreasing water discharge (Table 11.4 and Fig.11.4d).

Decreasing Trends of Stream Discharge: The natural springs are drying up and streams are depicting decreasing trends of annual stream discharge and becoming perennial to seasonal streams in the study area. The hydrological results suggest that the existing average annual discharge of Dabka watershed is 13 l/s/km² whereas during 1985–1990 it was quite high, i.e., 17 l/s/km² (Table 11.4 and Table 11.5). It has also been found that due to decreasing trends of stream discharge a number of perennial streams have dried up and as a result decreasing the perennial stream frequency throughout the sturdy area. Figures 11.4c and 11.4e depict four categories (low, moderate, high and very high) of the spatial distribution of perennial streams frequency and advocating that perennial stream frequency has decreased during 2005 to 2010 due to climate change through drying up of natural water springs and decreasing trends of stream discharge.

Increasing Flood Runoff: Flood runoff means the flowing off of precipitation of a watershed through a surface channel. The comparative hydrological evaluation of two study periods suggesting that the monsoon flood runoff (June to September) is increasing because of accelerated land-use degradation due to climate change. These data advocate that the dense forested land with broad leaved species of trees is very high and the deforested barren land has very low water retention capacity within their hydrological system. Although all monsoon months have high runoff, particularly August receives maximum monsoon runoff due to extreme rainfall (Table 11.5). Consequently the average monsoon flood runoff of

Table 11.5. Changes in Hydro-meteorological Parameters of he Dabaka Watershed due to Climate Change and Land use Degradation.

Study Periods	Hydro-meteorological Parameters	Monthly Results												Annual Results	Denudation rate in mm/year
		Jan.	Feb.	Mar.	Apr.	May	Jun.	Jul.	Aug.	Sep.	Oct.	Nov.	Dec.		
Existing (2005–2010)	Temperature (in °C)	10.00	12.00	16.00	19.00	31.00	27.00	24.00	23.00	21.00	20.00	14.00	12.00	27.00	0.68
	Rainfall (mm)	11.50	68.65	76.27	12.33	76.27	261.78	486.24	566.34	261.47	30.88	12.71	9.34	2000	
	Stream Discharge (l/s/km²)	3.00	4.85	6.56	0.78	26.90	30.61	42.61	48.08	27.44	4.92	3.27	2.90	16.82	
	Flood Runoff (l/s/km²)	101.79	147.90	225.89	181.48	568.44	806.44	1005.98	1235.05	818.13	67.36	39.53	25.71	435.31	
	Bed Load (t/km²)	0.00	0.00	1.87	2.49	1.87	18.73	24.95	34.85	24.90	7.48	6.03	0.00	124.50	
	Suspended Load (t/km²)	0.41	0.12	0.24	0.59	0.24	8.95	11.60	17.55	11.70	3.51	2.93	0.56	58.50	
	Dissolved Load (t/km²)	0.31	0.10	0.14	0.28	0.14	5.20	7.63	10.35	7.26	2.42	0.35	0.31	34.50	
	Total Load (t/km²)	**0.72**	**0.22**	**2.24**	**3.35**	**2.24**	**33.45**	**44.50**	**66.63**	**44.40**	**11.11**	**8.88**	**4.44**	**224.45**	
Previous (1985–1990)	Temperature	6.00	8.00	10.00	13.00	24.00	21.00	18.00	16.00	13.00	11.00	8.00	7.00	18.00	0.42
	Rainfall	24.32	84.24	99.12	32.75	97.27	328.46	496.85	589.44	283.71	47.66	21.33	15.24	2120	
	Stream Discharge (l/s/km²)	1.70	2.43	2.32	0.30	18.50	24.76	37.56	41.89	21.56	2.64	2.76	1.59	13.17	
	Flood Runoff (l/s/km²)	89.48	119.20	195.90	158.45	513.76	762.80	790.34	1076.40	741.54	43.80	21.30	11.12	377.01	
	Bed Load (t/km²)	0.00	0.00	1.10	1.37	1.76	13.76	17.43	21.87	16.67	5.48	4.98	0.00	84.92	
	Suspended Load (t/km²)	0.30	0.10	0.19	0.38	0.18	6.80	9.65	14.78	9.65	2.67	2.10	0.39	47.19	
	Dissolved Load (t/km²)	0.28	0.08	0.10	0.21	0.10	3.90	5.80	8.50	5.61	1.86	0.21	0.17	26.82	
	Total Load (t/km²)	**0.58**	**0.18**	**1.39**	**2.46**	**2.04**	**24.46**	**32.88**	**45.15**	**31.93**	**10.01**	**7.29**	**0.56**	**158.93**	

Table 11.5. contd....

Table 11.5. contd.

Study Periods	Hydro-meteorological Parameters	Monthly Results												Annual Results	Denudation rate in mm/year
		Jan.	Feb.	Mar.	Apr.	May	Jun.	Jul.	Aug.	Sep.	Oct.	Nov.	Dec.		
Changes (1985–2010)	Temperature	4.00	4.00	6.00	6.00	7.00	6.00	6.00	7.00	8.00	9.00	6.00	5.00	9.00	0.26
	Rainfall	9.82	15.59	22.85	15.42	11	16.68	10.61	23.1	22.24	16.78	8.62	5.9	48.49	
	Stream Discharge (l/s/km²)	1.30	2.42	4.24	0.48	8.40	5.85	5.05	6.19	5.88	2.28	0.51	1.31	3.65	
	Flood Runoff (l/s/km²)	12.31	28.70	29.99	23.03	54.68	43.64	215.64	158.65	76.59	23.56	18.23	14.59	58.30	
	Bed Load (t/km²)	0.00	0.00	0.77	0.62	0.11	4.97	7.52	12.98	8.23	2.00	1.05	0.00	39.58	
	Suspended Load (t/km²)	0.11	0.02	0.05	0.21	0.06	2.15	1.95	2.77	2.05	0.84	0.83	0.17	11.31	
	Dissolved Load (t/km²)	0.03	0.02	0.04	0.07	0.04	1.30	1.83	1.85	1.65	0.56	0.14	0.14	7.68	
	Total Load (t/km²)	0.14	0.04	0.85	0.89	0.20	8.99	11.62	21.48	12.47	1.10	1.59	3.88	65.52	

August during 2005–2010 found 1235 l/s/km² whereas in the same month during 1985–1990 it was quite low with the rate of 1076 l/s/km² (Table 11.5). The spatial distribution of flood runoff for year 1990 and 2010 has been demonstrated in Fig. 11.5a.

Accelerated Soil Erosion: Under natural circumstances the total sediment load of streams varied from 66 tons/km²/year in the dense forest to 398 tons/km²/year in the barren land in the watershed (Table 11.5). Anthropogenic activities have accelerated the rate of load generation by four times (302 tons/km²/year) in agricultural land and six times (398 tons/km²/year) in barren land. Presently the average rate of total load delivery from Dabka watershed stands at 224.45 tons/km²/year whereas 20 years back during 1985–1990 it was just 158.93 tons/km²/year (Table 11.5). The average annual rate of soil erosion of the Dabka watershed stands at 0.68 mm/year at the present time, however it was just 0.42 mm/year 20 years back during 1985–1990 (Table 11.5). The spatial distribution map of soil erosion and denudation depicting five zones and advocating that the soil erosion and denudation rates has been increased (Fig. 11.5b and Table 11.5) which accelerated several hydrological hazards such as river-line flood, flash flood, landslides and slope failure, etc.

Increasing Landslide Hazard: After delineating all the exiting landslides during field work a spatial distribution and density map of landslides have been carried out by GIS mapping following grid and the isopleth's technique (Fig. 11.5c). The map suggested that the maximum area of the watershed had very high to high spatial density of landslide which stands respectively 39% (26.93 km²) and 32% (22.10 km²) area of the watershed whereas minimum area of the watershed about 7% (4.83 km²) occupied by low spatial density of landslide. Moderate spatial distribution density of landslide covers about 22% (km²) area of the watershed (Fig. 11.5c). Low and moderate spatial landslide density zone have respectively less than 4 and 4–8 landslide/km² whereas high to very high landslide spatial density zone have 8–12 and above 12 landslide/km² (Fig. 11.5c). Comparatively Fig. 11.5c also depicts that 20 year back the landslide density was quite low for all landslide hazard zones.

Increasing Hydro-hazard Frequency: The integrated monsoon hydro-hazard frequency map generated after overlaying flood, erosion and landslide hazard maps, shows four hydro-hazard frequency zones in the watershed (Fig. 11.5d). On evaluating the area statistics of present hydro-hazard frequency it was found that out of the total area of the watershed 34% is under very high hydro-hazard frequency (above 60 events/year/km²) 21% under high hydro-hazard frequency (40–60 events/year/km²), 17% as moderate hydro-hazard frequency (20–40 events/year/km²) and

28% is under low hydro-hazard frequency (below 20 events/year/km²). Comparatively this frequency was quite low in all four zones during 1985–1990 which suggested an increasing trend of hydro-hazard frequency in the study area (Fig. 11.5d). On the basis of overall assessment of the watershed it can be said that very high frequency is mostly found on the steep slopes with the lithology of Lariakantha-Bhumiadhar Formation and along major streams of the watershed and high frequency was found in moderately steep areas and along major tributary streams whereas, moderate and low frequency were confined to dense forests with the lithology of the Krol Formation and human managed landscape, like agriculture and pasture land, etc. (Fig. 11.5d).

Impacts of hydrological hazards on community food and livelihood

Farming is the main source of community food and livelihood throughout the Himalaya but unfortunately it is the most susceptible sector of climate change as the meteorological data shows that the trend of agro-ecology is gearing toward hotter and a less humid environment. This is attributed to the fact that climate change affects the two most important direct agricultural production inputs, precipitation and temperature. Climate change also indirectly affects agriculture by influencing the emergence and distribution of crop pests and livestock diseases, exacerbating the frequency and distribution of adverse weather conditions, reducing water supplies and irrigation; and enhancing severity of soil erosion. The results suggested that climate change accelerated land-use dynamics and hydrological hazards, i.e., drought (decreasing underground water, drying up natural springs and decreasing trends of stream discharge) in the no-monsoon period and flash flood, river line flood, soil erosion and landslide in the monsoon period which ultimately affected community food and livelihood as given below:

Dwindling Irrigation Facilities during Non-monsoon: Check dams, spring ponds, canals, gullies, water tanks and pipe lines are the main measures for irrigation throughout the Himalaya as well as in the study area. The results suggested that these measures are declining due to climate change and its impacts on water resources, i.e., drying up of perennial springs and streams (Fig.11.4 and Table 11.4). Twenty five years back during 1985–1990 throughout the study area there were a total 86 check dams, 38 spring ponds to provide water towards all irrigation canals and their gullies (Fig. 11.5 and Table 11.6). Consequently 90% agricultural land was under irrigation due to very vigorous network of canal and their gullies (184 km). But at the present time only 45% land is under irrigation due to very poor network of canals (42 km), check dams (33) and spring ponds (12). The annual results of the analysis concluded that the spring ponds, perennial streams, water

Table 11.6. Decreasing irrigation facilities during 1985–2010.

Irrigation Measures	During 1985–1990	During 2005–2010	Changes	
			1985–2010	Annual
Spring Ponds (number of ponds)	38	12	–26	–1
Perennial Streams (number of Streams)	92	39	–53	–3
Water Tanks (number of tanks)	60	14	–46	–2
Check Dams (number of dams)	86	33	–53	–3
Canals (km)	184	42	–142	–6

tanks and check dams were decreasing respectively with the rate of 1, 3, 2 and 3 by each year whereas the irrigating canal decreasing 6 km/year (Table 11.6). During 1985–1990 out of the total agricultural land (1977 hectares) 90% land (1787 hectares) was irrigated land whereas during 2005–2010 it had decreased up to 45% (916 hectares) of total agricultural land (Fig. 11.6 and Table 11.7). Subsequently only 10% (190 hectares) land out of the total agricultural land was non-irrigated land during 1985–1990 and it increased up to 55% (1124 hectares) out of total agricultural land (2040 hectare) during 2005–2010 (Fig. 11.6 and Table 11.7). Figure 11.7 depicts that the 25 years study concluded that 2% irrigated land changed into non-irrigated land but the spatial distribution of agricultural land suggested that the non-irrigated land increased with extreme high rates (22% by each year) because of two reasons :

- Irrigated land converted into non-irrigated land due to dwindling water resources and irrigation facilities.
- Because of low yield rates of non-irrigated land people increased their agricultural land to produce the required annual family food.

Agricultural Land Degradation during the Monsoon: The annual trends of agricultural land degradation due to geo-hydrological hazards has been analyzed in the existing study period during 2005–2010 (Table 11.8 and Fig. 11.8). The results suggested that in average about 443 hectares of agricultural land degraded each year through geo-hydrological hazards which accounts about 22% of the total existing agricultural land (Table 11.8 and Fig. 11.8). The maximum about 7% (149.73 hectares) degraded by flash flood or high runoff hazard whereas the minimum about 3% (65.20 hectares) degraded by landslide hazard (Table 11.8 and Fig. 11.8). Soil erosion and river-line flood hazard respectively degraded about 6% (130.57 hectares) and 5% (97.73 hectares) of the total agricultural land each year (Table 11.8 and Fig. 11.8). The spatial distribution of the degraded agricultural land during 2005–2010 suggested that flash flood and river-line flood mostly degraded high fertility irrigated agricultural land because such type of land are found along all rivers and their streams which are highly vulnerable for both types of flood

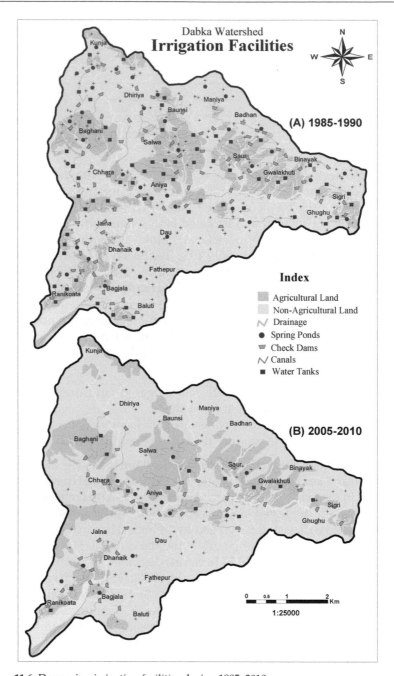

Figure 11.6. Decreasing irrigation facilities during 1985–2010.

Color image of this figure appears in the color plate section at the end of the book.

Table 11.7. Spatial distribution of irrigated and non-irrigated land under non-monsoon crops during 1985–2010.

Non-Monsoon Crop Pattern	Irrigated Agricultural Land				Non-Irrigated Agricultural Land				Total Agricultural Land			
	Covered Area (In Hectare)		Changes In %		Covered Area (In Hectare)		Changes In %		Covered Area (In Hectare)		Changes In %	
	1985–1990	2005–2010	1985–2010	Annual Change	1985–1990	2005–2010	1985–2010	Annual Change	1985–1990	2005–2010	1985–2010	Annual Change
Wheat	754	380	-50	-3	94	670	613	25	848	1050	24	0.95
Mustards	346	80	-77	-4	90	167	79	3	436	247	-45	-1.79
Gram	165	54	-67	-3	1	28	2800	112	166	82	-50	-2.00
Pulses	49	40	-18	-1	.5	22	4400	176	49.5	62	26	1.05
Radish	49	41	-16	-1	.5	21	4200	168	49.5	62	26	1.05
Pea	130	100	-23	-1	1	65	6500	260	131	165	27	1.07
Potato	147	110	-25	-1	1	75	7500	300	148	185	26	1.03
Onions	57	45	-21	-1	1	28	2800	112	58	73	28	1.10
Garlic	57	45	-21	-1	.5	28	5600	224	57.5	73	28	1.11
Chilly	33	21	-36	-2	.5	20	4000	160	33.5	41	24	0.96
Total	1787	916	-49	-2	190	1124	492	20	1977	2040	3	0.13
% of Total	90	45	-50	-3	10	55	550	22	100	100	3	0.13

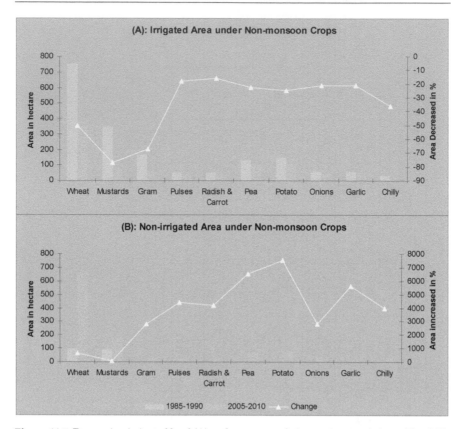

Figure 11.7. Decreasing irrigated land (A) and consequently increasing non-irrigated land (B).

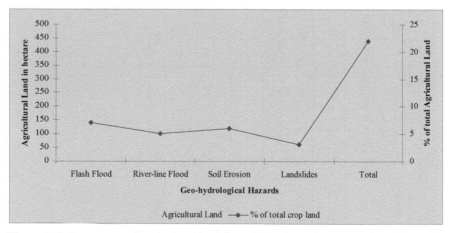

Figure 11.8. Average annual agricultural land degradation through geo-hydrological hazards.

Table 11.8. Agricultural land degradation through geo-hydrological azards during monsoon Period.

Years	Total Agricultural Land	Geo-hydrological Hazard and Agricultural land Degradation (in Hectare)					% of Total Agricultural land
		Flash Flood or High Runoff	River Line Flood	Soil Erosion	Landslide and Slumps	Total of All Hazards	
2005	2040	119.78	78.25	104.33	52.16	354.52	17
2006	2040	165.19	107.23	143.93	71.48	487.83	24
2007	2040	155.53	101.43	136.21	67.62	460.79	23
2008	2040	137.17	89.84	119.78	59.89	406.68	20
2009	2040	89.84	58.93	78.25	39.61	266.63	13
2010	2040	230.87	150.70	200.93	100.46	682.96	33
Total	**12240.00**	**898.38**	**586.38**	**783.43**	**391.22**	**2659.41**	**130**
Annual/Existing	**2040.00**	**149.73**	**97.73**	**130.57**	**65.20**	**443.24**	**22**
% of Existing Agricultural Land		7	5	6	3	22	-

hazard. Landslide hazard degraded mostly non-irrigated agricultural land because such type of land is mainly found in moderate to high slope areas of hills which are highly vulnerable to landslide hazard. The erosion hazard is also a major responsible factor for agricultural land degradation in the Himalaya mountains because it takes place with both types of hazard either as a landslide hazard or flood hazard. Consequently erosion hazard affects irrigated and non-irrigated land equally.

Decreasing Agricultural Productions: Climate change and accelerated hydrological hazards during the monsoon and non-monsoon periods affects the agricultural production. Drying up water resources during the non-monsoon period decreasing the irrigation land whereas as the floods, high runoff, soil erosion and landslide during monsoon period affects fertile agricultural land. As a result these hydrological hazards cause decreasing average yield rates of the monsoon and non-monsoon crops (Table 11.9 and Fig. 11.9). During two decades period (1990–2010) the average yield rate of the non-monsoon crops and monsoon crops have been decreasing about 26 and 28%, whereas the population has increased with the growth rate of about 37% during same time period (Table 11.9). Consequently the annual average rate of the yield decreasing accounts for about 1.35 and 1.40% respectively for the non-monsoon and monsoon crops (Table 11.9). It is also observed that climate change and its outputs (land-use degradation and hydrological hazards) not only affected agricultural production but also fruit production in the study area and poses a serious threat for community food and livelihood.

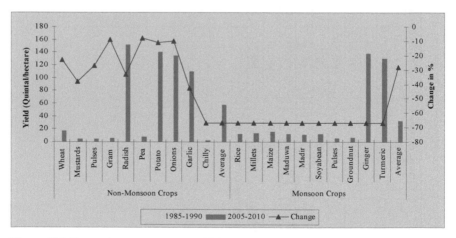

Figure 11.9. Decreasing average crop yield during 1985–2010.

Table 11.9. Decreasing average crop yield and increasing population growth during 1985–2010.

Annual Crop Pattern of Dabka Watershed		Crop Productions				Population			
		Yield (Quintal/hectare)		Yield Deficit Rate (In %)		Total Population		Growth Rate (In %)	
		1985–1990	2005–2010	1985–2010	Annual	1985–1990	2005–2010	1985–2010	Annual
Non-Monsoon Crops	Wheat	22	17	-23	-1.15	1075	1482	37	2
	Mustards	8	5	-38	-1.90				
	Pulses (Masur and Arhar)	7	4	-43	-1.35				
	Gram	9	6	-33	-0.45				
	Radish	168	152	-10	-1.65				
	Pea	11	8	-27	-0.40				
	Potato	154	140	-9	-0.55				
	Onions	147	135	-8	-0.50				
	Garlic	124	110	-11	-2.15				
	Chilly	6	2	-67	-3.35				
	Average	65.60	57.90	-26.90	-1.35				
Monsoon Crops	Rice	17	12	-29	-1.45				
	Millets	21	14	-33	-1.65				
	Maize	24	16	-33	-1.65				
	Maduwa	18	12	-33	-1.65				
	Madir	16	11	-31	-1.55				
	Soyabean	17	12	-29	-1.45				
	Pulses (Rajma, Urad, Gahat)	9	6	-33	-1.65				
	Groundnut	11	7	-36	-1.80				
	Ginger	152	138	-9	-0.45				
	Turmeric	146	130	-11	-0.55				
	Beans	13	9	-31	-1.55				
	Average	40.36	33.36	-28.00	-1.40				

Food Deficiency and Changing Livelihood: Results suggested that the average annual yield rate (total of non-monsoon and monsoon) has decreased 27% during the period of two decades (1990–2010) in the study area whereas the population increased with the growth rate of about 37% during same time period (Table 11.9). Consequently due to food deficiency the people seek other resources to manage and adopt other occupations which ultimately affect rural livelihood. The study brought out the facts that during 2005–2010 only 38% of total household families were completely involved in primary sectors occupations and 46 and 16% involved respectively in secondary and tertiary sectors whereas during 1985–1990 out of total household families 91% were involved in primary sectors occupation whereas only 7 and 2% involved in secondary and tertiary sectors (Table 11.10). The average annual results of the 25 years analysis concluded that on the one hand the household families increased at the rate of 4% every year due to high rate of population growth whereas on the other hand, 2% household families changed their livelihood occupations from the primary sector to secondary and tertiary sectors by the same year (Table 11.10).

Table 11.10. Changing livelihood pattern due to decreasing crop production and food deficiency in Dabka Watershed.

Occupational Structure of Livelihood	Household Families							
	1985–1990		2005–2010		Changes (1985–2010)		Annual Changes	
	No. of Families	In %	No. of Families	In %	No. of Families	In %	No. of Families	In %
Primary Sector	1154	91	905	38	−249	−53	−10	−2
Secondary Sector	89	7	1096	46	+1007	+39	+40	+2
Tertiary Sector	25	2	381	16	+356	+14	+14	+1
Total	1268	100	2382	100	+1114	+87	+45	+4

Conclusion

The study concluded that the climatic zones shifting towards higher altitudes due to global climate change and affecting the favorable conditions of the existing land-use pattern and decreased the oak and pine forests in upstream areas. Consequently the high rates of deforestation accelerated hydrological hazards in downstream areas during the monsoon and non-monsoon periods. The non-monsoon hydrological hazards (i.e., decreasing underground water level, drying up perennial springs and decreasing trends of stream water discharge) reducing irrigation facilities and decreasing the irrigated land by 2% each year whereas the monsoon hydrological hazards (i.e., flood, soil erosion and water induce landslide) degraded 22%

agricultural land each year. These climate changes outputs (hydrological hazards) resulted in a deficit agricultural production rate during 1985–2010. Consequently due to food deficiency the people sought other resources to manage and adopt other occupations which ultimately affected rural livelihood. The average annual results of the 25 years analysis concluded that the household families increased at the rate of 4% every year due to high rate of population growth whereas on the other hand, 2% household families changed their livelihood occupations from the primary sector to secondary and tertiary sectors each year.

References

Adams, R.M., R.A. Fleming, C.C. Chang, B.A. McCarl and C. Rosenzweig. 1995. A reassessment of the economic effects of global climate change on US agriculture. Climatic Change 30: 147–167.

Adams, R.M., B. Hurd, S. Lenhart and N. Leary. 1998. The effects of global warming on agriculture: an interpretative review. Journal of Climate Research 11: 19–30.

Bisht, M.K.S. 1991. Geohydrological and geomorphological investigations of the Dabka catchment district Nainital, with special reference to problem of erosion. Unpublished PhD. thesis, pp. 37–115.

Brahma Bhatt, V.S., G.B. Dawadi, S.B. Chhabra, S.S. Ray and V.K. Dadhwal. 2000. Landuse/land cover change mapping in Mahi Canal Command area, Gujarat, using multi-temporal satellite data. Journal of Indian Society of Remote Sensing 28(4): 221–232.

Chauhan, P.S., M.C. Porwal, L. Sharma and J.D.S. Negi. 2003. Change detection in Sal forest in Dehradun forest division using Remote Sensing and Geographic Information System. Journal of Indian Society of Remote Sensing 31(3): 211–218.

Cruz, R.A.D. 1992. The determination of suitable upland agricultural areas using GIS technology. Asian pacific Remote Sensing Journal 5: 123–132.

Darwin, R. and D. Kennedy. 2000. Economic effects of CO_2 fertilization of crops: transforming changes in yield into changes in supply. Environmental Modeling and Assessment 5(3): 157–168.

Easterling, W.E. III, P.R. Crosson, N.J. Rosenberg, M.S. McKenny, L.A. Katz and K.M. Lemon. 1993. Agricultural impacts of and responses to climate change in the Missouri-Iowa-Nebraska-Kansas (MINK) region. Climatic Change 24: 23–61.

Fischer, G., K. Frohberg, M.L. Parry and C. Rosenzweig. 1996. Impacts of potential climate change on global and regional food production and vulnerability. *In*: T.E. Downing (ed.). Climate Change and World Food Security, NATO ASI Series 137. Springer, Berlin.

Hamilton, L.S. 1987. What are the impacts of Himalayan deforestation on the Ganges-Brahmaputra lowland and delta? Assumptions and facts. Mountain Research and Development 7: 256–263.

IPCC. 2001. Climate change 2001. *In*: J.T. Houghton, Y. Ding, D.L. Griggs, M. Noguer, P.J. van der Linden, X. Dai, K. Maskell and C.A. Johnson (eds.). The Scientific Basis: Contribution of Working Group I to the Third Assessment Report of the Intergovernmental Panel on Climate Change. Cambridge University Press, Cambridge.

IPCC. 2007a. Climate change 2007. *In*: S. Solomon, D. Qin, M. Manning, Z. Chen, M. Marquis, K.B. Avery, M. Tignor and H.L. Miller (eds.). The Physical Science Basis: Contribution of

Working Group I to the Fourth Assessment Report of the Intergovernmental Panel on Climate Change. Cambridge University Press, Cambridge.

IPCC. 2007b. Summary for policymakers. pp. 7–22. *In*: M.L. Parry, O.F. Canziani, J.P. Palutikof, P.J. van der Linden and C.E. Hanson (eds.). Climate Change 2007: Impacts, Adaptation and Vulnerability. Contribution of Working Group II to the Fourth Assessment Report of the Intergovernmental Panel on Climate Change. Cambridge University Press, Cambridge.

Iglesias, A.L. Garote, F. Flores and M. Moneo. 2007b. Challenges to manage the risk of water scarcity and climate change in the Mediterranean. Water Resources Management 21: 227–288.

Iglesias, A., C. Rosenzweig and D. Pereira. 2000. Prediction spatial impacts of climate in agriculture in Spain. Global Environmental Change 10: 69–80.

Ives, J.D. 1989. Deforestation in the Himalaya: the cause of increased flooding in Bangladesh and Northern India. Land Use Policy 6: 187–193.

Jain, S.K., S. Kumar and J. Varghese. 1994. Estimation of soil erosion for a Himalayan watershed using GIS technique. Geol. Soc. London 151: 217–220.

Joshi, P.K. and S. Gairola. 2004. Land cover dynamics in Garhwal Himalayas—a case study of Balkhila sub-watershed. Indian Society of Remote Sensing 32(2): 199–208.

Kostrowicki, J. 1983. Land use systems and their impact on environment: an attempt at and classification. The Geographer 30(1): 6–14.

Minakshi, R. Chaurasiya and P.K. Sharma. 1999. Land use/land cover mapping and change detection using satellite data—A case study of Dehlon block, district Ludhiana, Punjab Journal of Indian Society of Remote Sensing 27(2): 225–235.

Mohanty, R.R. 1994. Analysis of urban land use change using sequential aerial photographs and SPOT data: An example of north Bhubnaswar, Orissa. Journal of Indian Society of Remote Sensing 22(5): 225–235.

Parry, M.L., C. Fischer, M. Livermore, C. Rosenzweig and A. Iglesias. 1999. Climate change and world food security: a new assessment. Global Environmental Change 9: S51–S67.

Rawat, Pradeep K., P.C. Tiwari and C.C. Pant. 2011. Modeling of stream runoff and sediment output for erosion hazard assessment in lesser Himalaya; need for sustainable land use plane using Remote Sensing and GIS: a case study. Natural Hazards 59(3): 1277–1297.

Rawat, Pradeep K., P.C. Tiwari, C.C. Pant, A.K. Sharma and P.D. Pant. 2012. Spatial variability assessment of river-line floods and flash floods in Himalaya: a case study using GIS. International Journal of Disaster Prevention and Management 21(2): 135–159.

Rawat, Pradeep K. 2013. GIS modeling on mountain geodiversity and its hydrological responses in view of climate change, Lambert Academic Publishing, Saarbrücken, Germany.

Rosenzweig, C., A. Iglesias, G. Fischer, Y. Liu, W. Baethgen and J.W. Jones. 1999. Wheat yield functions for analysis of land use change in China. Environmental Modeling and Assessment 4: 128–132.

Rosenzweig, C. and M.L. Parry. 1994. Potential impact of climate change on world food supply. Nature 367: 133–138.

Sharma, V.V.L.N., G.M. Krishna, B. Malini and K.N. Rao. 2001. Land use/land cover change detection through Remote Sensing and its climate implication in the Godavari delta region. Journal of Indian Society of Remote Sensing 29(1&2): 85–91.

Scheling, D. 1988. Flooding and road destruction in Eastern Nepal [Sun-Koshi]. Mountain Research and Development 8: 78–79.

Sharma, K.P., N.R. Adhikari, P.K. Ghimire and P.S. Chapagian. 2003. GIS based flood risk zoning of the Khando river basin in the tarrai region of East Nepal. Himalayan Journal of Science 1(2): 103–106.

Sing, S.K. 2006. Spatial variability in erosion in the Brahmaputra basin: causes and impacts. Curr. Sci. 90(9): 1272–1276.

Tiwari, P.C. 1995. Natural Resources and Sustainable Development in Himalaya. Shree Almora Book Depot, Almora.

Tiwari, P.C. and B. Joshi. 1997. Wildlife in the Himalayan Foothills: Conservation and Management. Indus Publishing Company, New Delhi.

Tiwari, P.C. 2000. Land use changes in Himalaya and their impact on the plains ecosystem: need for sustainable land use. Land Use Policy 17: 101–111.

Tiwari, P.C. and B. Joshi. 2002. Integrated resource management in forestry using Remote Sensing and GIS. pp. 242–258. *In*: A.S. Rawat (ed.). Forest History of the Mountain Regions of the World.

Tiwari, P.C. and B. Joshi. 2005. Environmental Changes and Status of Water Resources in Kumaon Himalaya. pp. 109–123. *In*: Jansky Libor et al. (eds.). Sustainable Management of Headwater Resources: Research from Africa and Asia

Valdiya, K.S. and S.K. Bartarya. 1989. Problem of mass-movement in part of Kmaun Himalaya. Current Science 58: 486–491.

12

Land-Use/Cover Changes in the Kewer Gadhera Sub-Watershed, Central Himalaya

Vishwambhar Prasad Sati

INTRODUCTION

Land is a complex and dynamic combination of factors—geology, topography, hydrology, soils, microclimate and communities of plants and animals. They are continually interacting under the influence of climate and of people's activities (Hudson 1995). Land-use/cover changes have recently become a major issue of research and a debate among academicians, researchers, policy makers and governments. Land use stands for the pattern of man's activity or economic function on a piece of land and land cover changes relate to the type of feature changes on the surface of the Earth (Lillesand and Kiefer 1994). It is the result of anthropogenic interaction with the natural environment. Besides affecting the quality of life of the people living in the area, land-cover changes also affects surface run-off as also erosion intensity (Piyoosh 2002). With increase in population and thus, in human settlements, urbanization, and industrialization, the change in all categories of land use seems to have led to tremendous changes in landscape. Similar to the global scenario, mountain regions have also been characterized by land-cover changes. In many of the mountain areas,

Professor of Geography, Mizoram University (Central), Aizawl - 796004, India.
  Email: sati.vp@gmail.com

expansion of human settlements and agricultural land over the marginal mountain niche is becoming very common which has resulted in shrinking of forestland. Population growth has forced the local people to bring more land under cultivation (Rawat et al. 1996). Depletion of forest from the fragile landscape of mountain has led to environmental hazards in the forms of landslides, debris flow, mass movement and flash flood. Musudiar (1993), Ukhimath [(Chamoli District), and Mapla (Pithoragarh District) (both in 1998)] landslides were the consequences of forest depletion which resulted in heavy damage to the property and loss of life. In the Central Himalaya, existing land-use pattern is a result of centuries old practices. Forestland occupies the highest geographical area (above 65%) while agricultural land is limited (12%). Cropping pattern is dominated by the practices of traditional subsistence cereal farming mainly millets. Currently, this pattern has changed as the populace of the region has largely migrated to the urban areas in search of better livelihood. Non-availability of agricultural land, unsuitability, and instability of terrain, harsh environmental conditions and inaccessibility of forestland are other aspects, which affect the land-use/cover changes (Sati and Kumar 2004). This study reveals that high population growth rate was observed in the Kewer Gadhera sub-watershed, increase in forestland and decrease in agricultural land was also observed during the last four decades. It was mainly due to the large-scale emigration on the one hand and land abandonment on the other. The main objective of this chapter is to examine land-use/cover changes during the last four decades in the Kewer Gadhera.

Description of Study Area

Kewer Gadhera, a perennial stream, is a sub-watershed of the Pindar River, Central Himalaya (Fig. 12.1). It originates from the root of Kanpurgarhi top (2,891 m), the highest point of the watershed, and confluences into the Pindar River at Narainbagar service centre. It extends between 30° 5' N and 30° 8' N and between 79° 20' E and 79° 23' E. The total area of the sub-watershed is 5,684 ha. It has four sub-tributaries. The whole watershed comprises steep to gentle slopes. During the monsoon season, it flows above the danger mark due to heavy downpour received by the watershed. Contrary to that, it flows below level during the winter season. Throughout its course, water of the stream is unused due to its rough, rugged, and precipitous landscape except in few agricultural patches, where land is irrigated. The whole watershed of the Kewer Gadhera is prone to soil erosion largely during the monsoon season, when it receives heavy downpour (mostly from the south-eastern wind). Consequently, landslides and debris flows are very common. There were several instances when the villages located along the stream and on the fragile slopes and associated

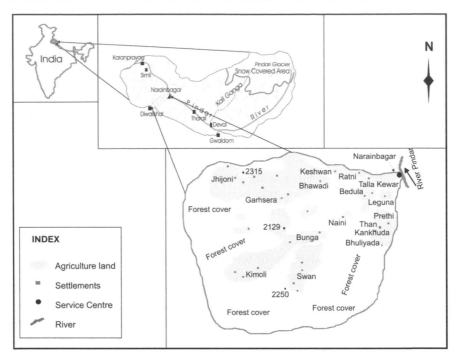

Figure 12.1. Location map of Kewer Gadhera sub-watershed showing recent land-use pattern. *Color image of this figure appears in the color plate section at the end of the book.*

agricultural land, were washed away with heavy loss of life and property, during the past decades. Infrastructural facilities such as transportation, banking, market, and communication are lagging behind in this region. Although, the state government has framed a plan to connect most of the villages by roads, yet the pace of implementation of this plan is very slow. Until recently (2010), only 20% villages were connected by roads. Also, the frequency and number of vehicles, running on the roads, is low and the condition of roads is bad. The people have to walk miles to do their daily work. A small service centre namely, Narainbagar is located on the bank of the Pindar River and on the roadsides connecting Karanprayag and Gwaldom service centres. This service centre fulfils the necessary needs of the villagers. Subsistence cereal farming is the main occupation of the people and livelihood of the populace largely depends upon the practices of cereal farming, livestock production, and forest-based non-timber products on a subsistence basis. On the other hand, the availability of natural resources—forest, water, soil, and manpower is enormous. Forests (mostly temperate) cover above 65% of the geographical area. Emigration to the urban areas the and Ganges valley is a very common phenomena.

From Kaub village alone, about 40% population emigrated during the last four decades. Land-use pattern is characterized by forestland (pine, oak and coniferous), agricultural land (including fallow land), and other land including cultivable waste, land under fruit trees and wasteland. Most of the land on the course of the Kewer Gadhera and its tributaries are unused because of unsuitable terrain. Similarly, the environmental services as water and non-timber forest products are also largely unused.

A case study of 17 villages of the Kewer Gadhera sub-watershed was carried out. The villages are Kewer Talla and Malla, Bhagoti, Ratni, Keshwan, Gadseer, Bunga, Jhijodi, Ali, Leguna, Bedula, Chirona, Kaub, Naini, Swan Malla and Talla and Kimoli. Village wise data on land use (1971–2010) were gathered from the primary and secondary sources. The collected data were analyzed and interpreted to know the actual changes in land use/cover. Rapid field visit of the case study villages was undertaken to observe land-use/cover changes on the ground and to facilitate further interpretation of data from time to time. Farmers, extension workers, and officials of agriculture, horticulture and forest departments were interviewed.

Land-Use Pattern (1971–2010)

Land-use pattern has been categorized as community forestland, agricultural land—irrigated and unirrigated land, and other land use comprises as land under fruit trees, barren land, community grassland and cultivable wasteland. Data of the last four decades from 1971 to 2010 were collected from the Census of India and the Patwari Circle Narainbagar Table 12.1 describes the land-use pattern (percentage of geographical area) between 1971 and 2010. All categories of data changed during these four decades. Total area under settlements, agriculture and community forest is 1421 ha. Changes in the community forestland (7.6%) were observed between 1971 and 2010, as it was 10.90 and 18.50% respectively. Irrigated land was 71.05% in 1971 and 16.98% in 2010 (Fig. 12.2). There was an increase in unirrigated land from 77.85 to 99.72%, during this period. Agricultural land, as a whole, decreased from 46.99 in 1971 to 41.10% in 2010. Other land use also decreased from 42.11 in 1971 to 40.38% in 2010.

Forestland

Total forestland is above 65%. Table 12.1 shows community forestland called *Van Panchayat* (VP). The community people owned the land and both the forest department and villagers manage it. In the lower altitude between 1000 m to 1800 m, pine is found extensively. Oak forests, including

Table 12.1. Land-Use Pattern between 1971* and 2010** (Percentage of Geographical Area).

Village name	Total area (ha) 2010	Community Forestland[1]		Agricultural land						Other land use[2]	
				Irrigated		Un-irrigated		Total			
		1971	2010	1971	2010	1971	2010	1971	2010	1971	2010
Kewer Talla	31.1	3.75	-	4.39	-	95.60	100	56.87	57.55	39.37	42.44
Kewer Malla	45.4	2.52	-	-	-	100	100	38.73	40.30	58.73	59.69
Bhagoti	104.5	1.4	2.48	-	-	100	100	40	62.00	58.6	35.509
Ratni	37.8	-	4.49	-	-	100	100	50	50.79	50	44.70
Keshwan	25.5	-	-	-	-	100	100	48	28.23	52	100
Gadseer	171.6	2.56	4.37	-	-	100	100	41.02	38.05	56.41	57.57
Bunga	129.4	10.71	53.01	-	-	100	100	35.71	34.15	53.57	12.82
Jhijodi	162	19.60	6.60	-	-	100	100	22.54	44.62	57.84	48.76
Ali	11.1	3.471	52.25	-	2.27	100	97.72	38.26	39.63	58.26	8.10
Leguna	19	14.281	36.84	-	11.94	100	88.40	57.14	36.31	28.57	26.84
Bedula	71	6.00	31.54	-	2.77	100	97.22	34.63	35.49	59.35	32.95
Chirona	27	48.50	72.96	-	-	100	100	18.65	18.88	32.83	8.14
Kaub	228.6	8.72	21.74	-	-	100	100	40.15	41.95	51.12	36.30
Naini	22.6	25.43	25.22	-	-	100	100	40.35	37.16	34.21	37.61
Swan Malla	28.2	50	18.432	-	-	100	100	50	34.39	-	47.16
Swan Talla	58.3	7.092	16.80	-	-	100	100	53.19	53.85	39.71	29.33
Kimoli	247.9	16.61	18.79	66.66	-	33.33	100	100	35.45	-8.34	45.74
Total	1421	10.90	18.50	71.05	16.98	77.85	99.72	46.99	41.10	42.11	40.38

Source: Data were gathered from the secondary sources; the Census of India 1971*, and the Patwari Circle 2010**, Narain Bagar, District Chamoli.
[1] Community forestland (*Van Panchayat*) is found around the villages.
[2] Other category of land use includes land under fruit plants, barren land, community grassland, and cultivable wasteland.

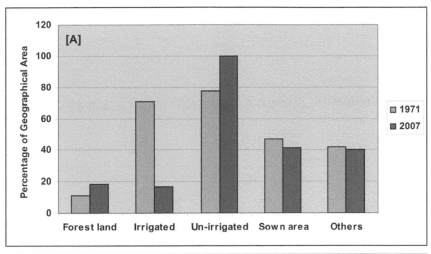

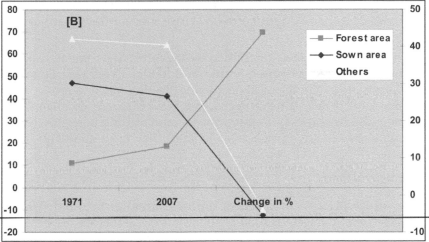

Figure 12.2. Land-use/cover changes in the Kewer Gadhera sub-watershed (percentage of Geographical area).

Black Mulberry is found between 1800 m and 2600 m and Deodar, Spruce, and Silver Fir are found above 2600 m. As indicated in Table 12.2, a large increase took place in areas under pine forest in the sub-watershed. Pine forest invaded temperate forest of oak at a large scale, mostly on the south facing slope (Sati 2010). The people of the surrounding villages opine that this situation has arisen due to climate change. Pine forests are grown comparatively in the warmer climate. As the temperature is increasing in the middle higher slopes, the pine forests are spreading towards higher elevation and invading oak forests. The other reason of increase in pine

Table 12.2. Tree Species and Trend in Changes over 40 Years in the Kewer Gadhera Sub-Watershed.

English Name	Local Name	Scientific Name	Changes	Area under tree species (%)
Preferred in construction				
Deodar*	Deyar/Kaol	*Cedrus deodara*	Increase	10
Pine***	Chid	*Pinus wallencia*	Large increase	40
Spruce*	Rai	*Picea Smithiona*	Increase	05
Silver fir*	Tos	*Abies pirdrow*	Increase	05
Preferred for winter fodder				
Oak**	Banj	*Quercus himalayana*	Decrease	20
Black Mulberry**	Tilonj	*Morus* spp.	Decrease	10
Preferred fuel				
Oak**	Banj	*Quercus himalayana*	Decrease	
Pine***	Chid	*Pinus wallencia*	Large increase	
Shrubs***	Kathi	*Indigofers* spp.	Constant	10

* High altitude trees grown over 2600 m, ** between 1800 to 2600 m, *** between 1100 to 1800 m.

forest area is large-scale afforestation by the forest department. Oak is an important species largely used for fuel and fodder. Every year, large-scale oak forests are cut by the local people for meeting fuel and fodder needs and as a result of this, its area is decreasing. Black mulberry a similar species of oak is also used widely. Deodar, Spruce, and Silver Fir are grown in the high altitudes and are very inaccessible. Thus, their use is almost negligible.

Agricultural land

Agriculture (41.10% of the geographical area) is rain-fed (99.72%) characterized by the cultivation of subsistence cereal crops (41.10%). The crops are nomenclature '*Barahnaja*' (twelve seeds), which include millets, pulses and oilseeds (Sati 2009). Farming is mostly practiced on the narrow patches of terraced fields in the middle and upper reaches of the watershed. Few patches in the valley regions are cultivable. The considerable loss of soil, due to large-scale erosion, has resulted in low fertility consequently low production and per ha yields (approx. 5 quintal/ha) of crops. Further, the mode of cultivation is traditional. Crops are grown in two different seasons. The first season is *rabi*, which starts from October-November and lasts upto April-May. In this season, wheat is the main crop followed by barley and mustard. Until the 1970's, barley was the main crop. The second season is *kharif*. Rice, millets, maize, pulses, and oilseeds are grown in this season. It begins from May and lasts till September. Land is kept fallow for six months from October to March. Changes in the farming systems were noticed during the last decades as in many areas, the cropped land that was characterized by cereals are converting into cultivation of paddy and wheat crops. As the population increased, the cereal crops were not enough to feed the people. Meanwhile, the production of paddy and wheat crops is considerably high. Keeping the high yield of paddy and wheat crops in view, large-scale changes in the cropping pattern was noticed during the 1980's. This happened mainly in the mid-altitudes and valley regions. In the highland villages, subsistence crops dominate in the cropping pattern.

Other land use

Another category of land includes land under fruit plants, barren land, community grassland, and cultivable wasteland. It covers 40.38% area. In and surrounding of the villages, extensive grassland slopes are found. These are community grassland on which livestock rearing is dependent. Fruits are also grown in the villages however; the area under fruit trees is considerably low.

Land-cover change

Village wise land-cover changes are given in Table 12.3. During the last four decades, community forest land increased by 69.72%. Unirrigated land also increased by 28.09% while irrigated land decreased –23.3%. There was a considerable decrease in agricultural land (–12.53%). Negative changes (–4.10) are also found in other categories of land use. Village wise changes in community forest land show that Ali village registered 1405.33% increase followed by Bedula village (425.66%). Bunga village had the third place with 394.95% increase in forest cover. Out of total 17 villages, five villages obtained considerable decrease in community forestland. These villages are located along the stream or near roadsides. Irrigated land is proportionately less. It is available only in the five villages with limited proportion of land. There are very little changes in agricultural land in reference to irrigation. Agricultural land also decreased during the past mainly due to land abandonment. In the villages, Kaub, Naini, Ali, Kewer Talla, and Kewer Malla, emigration took place at a large-scale. This has led to land abandonment on the one hand and increase in forestland, on the other. During the same period, Uttarakhand State as a whole, registered 1.3% increase in forestland. In Dehradun district alone, 10.6% forestland was increased. As a result of this, Dehradun district restored its natural beauty for which, it has been known for the centuries.

Table 12.3. Land-Cover Changes between 1971* and 2010** (Percentage of Geographical Area).

Village name	Total area	Community Forestland	Agricultural land			Others[1]
			Irrigated	Un-irrigated	Total	
Kewer Talla	−2.81	−100	−4.39	4.60	1.19	7.79
Kewer Malla	−4.42	−100	No Change	No Change	4.05	1.63
Bhagoti	4.5	77.14	No Change	No Change	55	−39.40
Ratni	0.53	100	No Change	No Change	1.58	−10.6
Keshwan	−49	No Change	No Change	No Change	−41.18	92.30
Gadseer	10	70.70	No Change	No Change	−7.24	2.05
Bunga	15.53	394.95	No Change	No Change	−4.36	−76.06
Jhijodi	−20.58	−66.32	No Change	No Change	97.95	−15.69
Ali	−3.47	1405.33	2.27	−2.28	3.58	−86.09
Leguna	35.71	157.96	11.94	−11.6	−36.45	−6.05
Bedula	−0.83	425.66	2.77	−2.78	2.48	−44.48
Chirona	0.74	50.43	No Change	No Change	1.23	−75.20
Kaub	−1.29	149.31	No Change	No Change	4.48	−28.99
Naini	−0.87	−0.82	No Change	No Change	−7.90	9.93
Swan Malla	135	−63.13	No Change	No Change	−31.22	47.16
Swan Talla	3.36	136.88	No Change	No Change	1.24	−26.13
Kimoli	2.43	13.12	−66.66	200.03	−64.55	−648.44
Total	**−0.47**	**69.72**	**−23.3**	**28.09**	**−12.53**	**−4.10**

Source: Data were gathered from the secondary sources; the Census of India 1971* and the Patwari Circle 2010**, Narain Bagar, District Chamoli and calculated by the author.
[1] Other category of land use includes land under fruit plants, barren land, community grassland, and cultivable wasteland.

Drivers of Land-Use/Cover Change

People's participation in forest management

Local people's participation in both decision-making processes and the implementation of management plans is crucial for sustainable forest management. In this region, the involvement of the local people was achieved in many areas. Extensive community grasslands and fodder trees are managed by the local people themselves and they also support the government forest department. During the 1980s, the popular "*Chipko* Movement*" (hugging trees against illegal felling) was launched by the local people in Uttarakhand and that movement spread throughout the Himalayan region. The people were able to protect the forest of the region from the hands of the contractors appointed by the Forest Department for mass cutting of trees. This movement had multiple impacts on the conservation of forest and on the local people. The forest cover increased and the people were able to run their livelihood from the forest and its products. In India, similar approaches were adopted to conserve forest resources. The forerunners of the *Chipko* movement were highly appreciated worldwide. Here, Gram Sabha (village assembly), the lowest unit of governance in the federal system of India, works for taking decision on the various developmental works including forest resource management. Gram Sabha has the control over community forestland and grassland and they manage it according to the needs of the people. The forest department and the Gram Sabha take initiatives to manage the forest resources, also fulfil the fuel wood, and fodder needs of the villagers.

Establishment of VP

VP literary means organization of villagers who look after the plantation of trees and conserve them in the community land located around the villages. The concept of VP in the Uttarakhand Himalaya is a century old but started in an organized way during the British rule. However, recently it got momentum and everywhere in the region, VPs were established. With establishment of VP forestland increased tremendously.

Government initiatives towards afforestation

The Soil Conservation Department (SCD) was established during the 1980s as a sister organization of the Forest Department. The forest department owns and manages the forest of the state except the community forest. This department is working mainly to look after forest related matters. Meanwhile, due to large forest cover and undulating terrain, this department could not look after forest depletion and soil erosion. To cope with these problems, SCD was established. SCD is largely working in the areas where forest depletion and soil erosion is excessive. It delineates the areas for wide afforestation and its conservation. Apart from that the department has the responsibility to check the illegal felling of trees. There are more than five sub-offices of the SCD in each development block with its one headquarter. Recently, most of the patches, where complete depletion of forests took place during the past, now reforestation has taken place. These patches are located along the courses of small perennial streams. Forest Bill (FB) of 1982 was successfully implemented and as a result of its rigorous implementation, regeneration of forests was noticed in the entire area.

Impact of Climate Change on Land-Use Pattern

To penetrate the reality of climate change an appraisal of climate data is required. Currently, the perception of all groups of society towards the impact of climate change in mountain regions is unanimous. The land, which was largely devoted for the cultivation of fruit crops earlier, is currently abandoned. Low production and productivity of fruit crops and consequently abandonment of productive land is considerably due to changes in the climatic conditions, as observed by the farmers. In discussions with the fruit growers, it was noticed that tremendous changes in the farming system is due to climate change. The areas where intensive cultivation of apple fruit was carried out during the past are now no better for its cultivation. This belt has been sifted greatly to the higher elevation, as observed. It is the same situation with the cultivation of other fruits—citrus, nut and stone fruits. Similarly, oak forests are being invaded by pine forests (between 1000 to 1800 m) in most of the areas particularly in the south-facing slopes (Fig. 12.3D). This impact can be noticed in Chakrauta area (Tons valley), Pauri district (Nayar valley), Rudraprayag district (Mandakini valley), and Chamoli district (Alaknanda, Nandakini, and Pindar valleys). Many sources of spring water have dried up due to the disappearance of oak trees. The impact of climate change can also be observed from the fact that musk deers have disappeared from the Nandadevi wild life sanctuary.

Figure 12.3. (A) Panoramic view of Kewer Gadhera sub-watershed, (B) Dense pine forest, (C) Land abandonment, (D) Invasion of Oak forest by pine trees.

Land abandonment

Changes in the cropping pattern can be noticed because of land abandonment. In the lower elevations and along the roads, land abandonment is due to large-scale transformation of cultivable land for commercial uses. In the highlands, land abandonment is basically due to mass emigration of the populace to the urban centres (Fig. 12.3C). It was noticed that households earning money through remittances, also abandon their agriculture land.

Conclusions

Land-use/cover changes over a period, is very common phenomenon in mountain regions. In the Kewer Gadhera sub-watershed, tremendous changes in land use/cover have occurred during the last decades. These changes have a tremendous impact on the agro-biodiversity of the region. Changes in the cropping pattern seem that farmers substantially benefit by replacing traditional food crops into cash crops but at the cost of increased vulnerability to climatic and market uncertainties. Abandonment of traditional crops means a loss of agro-biodiversity that remains 'lesser known' or 'unknown' to wider communities and associated indigenous technologies and knowledge. Expansion of agriculture on marginal land and declining crop yields are considered to be major unsustainable trends in the Himalaya (Eckholm 1979, Jodha 1990). Adaptive responses to stress factors by farmers played a significant role in evolution of traditional agriculture in the past when farming was the only option for securing livelihood. In the present circumstances, it often becomes more cost effective for the farmers to find employment than to spend his costly and scarce resources for rehabilitating the land of low productivity (Turner 1982). As at present, 40% of the total reported area is classified as pasture land, cultivable wasteland and fallow land, besides, a significant acreage of land classified as forestland (65%) in the study area. A cultivable wasteland may have potential for the development of vegetables (Singh 1991). Similarly, the abandoned land can be utilized for horticulture. With increasing needs as well as pressure of population, the traditional farming has become unsustainable both economically and ecologically (ICIMOD 1996).

Land-use pattern is typical in the Kewer Gadhera sub-watershed. Landscape is characterized by steep and precipitous slope. Therefore, agriculture is practiced on the narrow patches of terraced fields and agricultural land is limited (41.10%). The fertility of soil is considerably low because every year the upper layers of soil are eroded due to heavy rain and as well as by agricultural and other development activities. Over use of agricultural land resulted in low production and low per ha yield. The scope of expansion and modernization of agriculture is also limited. This led to a large-scale emigration of the populace towards the urban centres of the country. The trend of declining agricultural land (–4.34%) is also due to the said factor. Mass emigration can be checked through augmenting employment that can be achieved through increased production of cash crops and sustainable use of environmental services. Tremendous depletion in forestlands of India has been noticed after independence as the forest cover area decreased from about 21% in the 1940s to 19.7% in the 1990s. When, India as a whole registered tremendous growth in population including the Uttarakhand state, depletion of forestland was obvious. Meanwhile, in the Kewer Gadhera sub-watershed, 69.72% increase in the community forestlands was registered during 1971 to 2010. These are the various driving forces, which manifested in increase of forestlands. The increase under forest cover and decrease in net agricultural land shows that people from the rural hilly areas have emigrated to the other parts of the country for better living standards instead of extending their agricultural field. During the 1980s, the Government of Uttar Pradesh and Department of Forest set up 'Soil Conservation Department' aimed to conserve soil through launching plantation programs in the hilly areas. This program achieved tremendous success. India's Forest Act of 1982 has also contributed for the conservation of forest. Uttarakhand is a pioneer state, where the world famous '*Chipko* Movement' against illegal felling of trees was started during the 1980s. Reduced requirement of firewood and timber due to availability of LPG for fuel and cement and bricks for construction of houses also paved a way for increase in forest cover. Further, people's participation in the plantation activity and conservation of forests with the help of forest officials was significant.

References

Eckholm, E. 1979. Planting for the Future: Forestry for Human Need (World Watch Paper 26). World Watch Institute, Washington D.C., p. 73.

Hudson. 1995. Bridging the Gap between Communities and GIS Participatory 3-D Modeling, India.

ICIMOD. 1996. Background Note for Regional Meeting of Experts on Development of Micro Enterprises in Mountain Area 25–26 July, Unpublished Text.

Jodha, N.S. 1990. Mountain perspective and sustainability: a framework for development strategies. Paper presented at International Symposium on Strategies for Sustainable Mountain Agriculture, International Centre for Integrated Mountain Development, Kathmandu, Nepal.

Lillesand, T.M. and R.W. Kiefer. 1994. Remote Sensing and Image Interpretation. John Willey & Sons, Inc., New York, p. 170.

Piyoosh Rautela, Rahul Rakshit, V.K. Jha, Rajesh Kumar Gupta and Ashish Munshi. 2002. GIS and remote sensing-based study of the reservoir-induced land-use/land-cover changes in the catchment of Tehri dam in Garhwal Himalaya,Uttarakhand (India). Current Science 83(3): 308–311.

Rawat, D.S., N.A. Farooquee and R. Joshi. 1996. Towards sustainable land use in the hills of Central Himalaya, India. International Journal of Sustainable Development and World Ecology 3: 57–65.

Sati, Vishwambhar Prasad. 2010. Synthesis report of the global e-conference on Climate Change and the Himalayan Glaciers 7–30 May 2010. URL: http://www.freewebs.com/climatehimalaya.

Sati, Vishwambhar Prasad. 2009. Conservation of Agro-Biodiversity through Traditionally Cultivating '*Barahnaja*' in the Garhwal Himalaya. MF Bulletin Vol. IX Issue 2 July 2009. www.mtnforum.org.

Sati, V.P. and K. Kumar. 2004. Uttarakhand: Dilemma of Plenties and Scarcities. Mittal Publication, New Delhi, 278 p.

Turner, S.D. 1982. Soil conservation: administrative and extension approaches in Lesotho. Agricultural Administration 9: 147–162.

13

Hindu Kush, Karakorum, Himalaya and Tibetan Plateau Glaciers: Tipping Point

Iqbal Syed Hasnain

INTRODUCTION

The Hindu Kush, Karakorum, Himalaya and Tibetan Plateau mountain systems hold the largest ice mass outside the Polar Regions (Xu et al. 2009). The region is important as a source of freshwater to more than 1.5 billion people of two major emerging economies: India and China. Together, they contribute to large amounts of black carbon aerosols, released from incomplete combustion, and in process modifying the hydrology and radiative forcing over the entire region. The greenhouse gases and black carbon aerosols driven warming is visible on shrinking glaciers, shifting monsoons, declining crop yields, and unprecedented increase in the degradation of the mountain ecosystem.

Fifty five thousand glaciers on the Tibetan Plateau and surrounding the Hindu Kush, Karakorum and Himalaya mountains, in the tropical/sub-tropical region exert a direct influence on social and economic development in the surrounding regions such as China, India, Nepal, Tajikistan, Pakistan, Afghanistan, Bhutan, Bangladesh and Myanmar. It is subjected to influences by multiple climatic systems, complicated geomorphologies and various internal and external geological impacts. The area demonstrates

Distinguished Fellow, Stimson Center, Washington DC 20006, USA.
  Email: hasnainsyed09@gmail.com

considerable feedbacks to global environmental changes, while at the same time sensitive to current fluctuations. It is also influencing downstream hydrological systems, the alpine ecosystems and their services are affected by human activities and exacerbating the degradation.

Glaciers in the region are undergoing accelerated retreat, though the extent differs according to location. All glaciers are showing negative mass balance and are tipping which is reflected in increased discharges in melt water streams. Recent studies recognize the impact of black carbon or soot on glacial melt on the southern Tibetan Plateau, which receives black soot via the Indian monsoon during summer and from the west via winter westerly's. Black soot quantitative modeling on glacier dynamics is a current challenge. Nevertheless, some recent studies indicate that black soot is the important factor accelerating the HKH-Tibetan glacial melting (Ramanathan et al. 2007).

If irregular challenges caused by climate change go unchecked, they could lead to large scale regional conflicts as countries compete for dwindling water resources. Emerging global cooperation on the climate system and the Himalaya/Tibetan plateau adds complications as to how South Asian countries shape their own national strategies.

During the recent years, population dynamics, new economic growth and climate change in South Asia and China have occurred so intensely and so rapidly that traditional and balanced sustainable growth is fast losing their efficacy. The vast ice reserves of mountains are in peril. Accelerated glacial melting and environmental degradation threaten the traditional role that mountain regions have played as water reservoir for millions of its people.

The Tibetan plateau and surrounding mountain system of the Himalaya, Karakorum and Hindu Kush are often referred to as the 'Water towers of Asia' because they are the source of 10 of the largest rivers and contributes more than 40% of the world's freshwater. The major river basins of the region are the Amu Darya, Indus, Ganga, Brahmaputra, Irrawaddy, Salween, Mekong, Yangtze, Yellow and Tarim as shown in Fig. 13.1.

Unchecked global warming via greenhouse gases and enhanced heating by black carbon aerosols threaten to substantially reduce the Himalayan glaciers that nourish the region's major rivers. Nevertheless, what makes glacial melt so critical, even though in some it is a relatively small percentage of a river's annual flow, is the timing at which it occurs—melt drives flows during the dry spring and fall months, the so-called 'shoulder months' just before and after the monsoon rains (Orville 2010). As de-glaciations continue, however, melt water flows will wane. According to Immerzeel et al. (2010), receding glaciers could trim the water supply by 2050 in the Indus (–8.4%), Ganges (–17.6%), Brahmaputra (–19.6%), and Yangtze (–5.2%). Rivers, implying considerable shifts in stream flows that are expected to

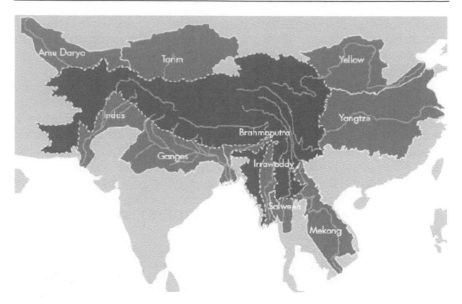

Figure 13.1 Hindu Kush Himalaya region with major river basins.

Color image of this figure appears in the color plate section at the end of the book.

balance ecosystems, irrigation schemes, and hydroelectric turbines. A study (Etienne et al. 2007) has suggested that 915 km² of Himalayan glaciers in the headwaters of the Chenab River, India, thinned by an annual average of 0.85 m between 1990 and 2004. A USAID report (Elizabeth 2010) succinctly raises the dangers of glacier lake outbursts floods (GLOF) caused by melting "Himalayan/Karakorum glaciers involving large impoundments by short-lived, unstable debris/ice dams that block tributaries of Indus river system….causing outburst floods." The tipping mass of glaciers and the unpredictability in the amount of runoff is causing great concern in water-dependent sectors, such as agriculture and power generation. Most Himalayan glaciers are in retreat for three reasons. First the overall warming tied to carbon emissions. Second, rain and snow patterns are changing, so that less new snow is added to replace what melts. Third, air pollution by cook stoves and trucks falling as black carbon or soot on surfaces became darker and less reflective—causing them to melt more quickly.

Hindu Kush, Karakorum and Himalaya Rivers System and Glaciers

The tallest and spectacular mountains of the world comprise the mountain ranges of Hindu Kush, Karakoram and Himalaya. The total area of valley glaciers in these ranges is approximately 2700 km², 16,600 km² and 33,050 km² respectively (Dyurgerov and Meir 2005). Climatically, the northern slope of Himalaya is cold and dry, and southern slope warm and moist, indicating a general decrease, from south to north, in precipitation and temperature. The topography of the Himalaya-Tibetan plateau, which acts as an elevated heat source, is crucial to the onset of the monsoon, and sets this region apart from some other tropical/sub-tropical regimes (Webster et al. 1998). Broadly, the major controls on climate in the region are latitude, altitude and position relative to the Indian monsoon flow. Summer monsoon or Southwestern monsoon is critical in controlling regional climate. Upon reaching the Himalaya by late June monsoon air rises and cools, the moisture condenses and heavy precipitation ensues. It also penetrates into Southeast Tibet by moving along the Brahmaputra valley. There are some 24,488 glaciers that spread across mountains of Southwestern China, which includes the Himalaya and the Nyainqntanglha, Tanggula and Hengduan mountains. Together they make up an area of almost 30,000 square kilometers. According to (Zongxing Li et al. 2011) climate change has had devastating affects on glaciers of mountains of Southwestern China. Their data shows that between 1970 and 2001, the Penggu basin's 999 glaciers lost a combined surface area of 131 square kilometers and 12 cubic kilometers of mass. The Gangrigabu Mountains also showed significant losses. There, 102 glaciers disappeared between 1915 and 1980, equaling a loss of more than 41 square kilometers in area and six cubic kilometers in mass. The Yalung glacier alone receded more than 1,500 meters between 1980 and 2001, resulting of swelling of nearby glacier lakes. The three major South Asian river basins provide water to more than 1.5 billion people vary considerably in characteristics (Table 13.1). The Indus and Brahmaputra have extensive upstream areas (>2000 m) and larger glaciated areas. All support large–scale irrigation system and thousands of ecosystem and play a crucial role in sustaining human populations. During last 30 years a large number of glaciers in warm and humid valleys of south of Himalayan arc

Table 13.1. Characteristics of the three major south Asian rivers.

All in %	Indus	Ganga	Brahmaputra
Snow & Glacier Melt in total Q	44.8	9.1	12.3
Upstream Area > 2000 m in %	40	14	68
Glaciated Area in %	2.2	1.0	3.1
Upstream ppt. > 2000 m in %	36	11	40
Downstream ppt.	64	89	60
NMI: Upstream Q /Downstream Q (NMI=index)	151	10	27

Source: *Science*, v. 328, 2010

reflect monsoon dynamics and are down wasting (i.e., stationary decaying) instead of just retreating in response to atmospheric warming (Hasnain 2010). There has been shrinkage of 5.5% in the volume of glaciers in China and similar rates are found in Nepal, India and Bhutan (Edwards et al. 2010).

Glaciers in the eastern and central part of Himalaya are expected to be especially sensitive to present atmospheric warming, due to their summer-accumulation type nature (Ageta and Higuchi 1984). An increase in summer air temperature not only enhances ice melt but also significantly reduces the accumulation by altering snowfall to rain. In contrast, winter-accumulation type glaciers receive their main accumulation at lower temperatures and are thus less sensitive to an increase in air temperatures. Rapid retreat in plateau glaciers is presumably driven by warming due to increasing greenhouse gases and by black carbon or soot (Baiqing et al. 2009). Quantitative modeling of the effect of black soot on glacier dynamics is a current challenge, but some indication in studies by (Yasunari et al. 2010) suggest that black soot is responsible for a secondary factor that drives substantial melting in glaciers.

Given the size and remoteness of glaciers in the Himalaya, satellite imagery is widely used to obtain a comprehensive regional estimate of glacier mass balance. Recently, in Bhutan, glacier retreat shows a north-south gradient with larger retreat rates in the south as the influence of the monsoon decreases towards the north, according to the studies conducted by Kaab (2005). He combined ASTER and SRTM (Shuttle Radar Topographic Mission) data to produce synthetic digital elevation model and then map the contrasted dynamical behavior of north and south facing glaciers in the northernmost section Bhutan Himalaya ridge (called 'Lunana'; 28°N, 90–91° E) separating the Tibetan plateau to the north from the central Himalaya to the south. The northbound glacier tongues show speeds of several tens to over 200 m year^{-1}. They are almost debris free as the ice flux drains the large accumulation areas through the northbound valleys, in order to keep the glacier in geometric equilibrium. In contrast the southern glaciers have high debris cover and velocity around 40 m y^{-1} near the tongues, which appear to be nearly stagnant. The response to atmospheric warming for these glacier tongues is down wasting—essentially decoupled from the dynamics of upper glacier parts. Most of the Himalaya glaciers south of ridge lie in a warm and humid climate system are down wasting, blocks dead ice having no or loose contact with the active part of glacier. As a consequence moraine dammed glacier lakes are commonly formed at the glacier tongues. Studies by (SAC 2011) on three basins from Ganga, Brahmaputra and Indus show upward movement of snowline and negative mass balance for large number of glaciers. The Kolahoi, Indian Kashmir the biggest glacier and principal source of water to Jhelum River has shrunk over 3 km^2 during the past three decades (Hasnain 2010). Similarly, the

Siachen glacier—site of an Indian and Pakistani military standoff for past 25 years—has shrunk to half its size.

Most of the studies attribute the retreat of the Himalayan glaciers to rising air temperatures. The warming is much more pronounced in elevated levels of the HKH (Hindu Kush-Himalaya) region (Tandong et al. 2009).

Impact of Climate Change on Water Supply

For energy-constrained economies in south Asia, the prospect of diminishing river flows in the future and the possibility that energy potential from hydropower may not be achieved, will have far reaching economic consequences. Another reason of immediate concern is the danger of Glacier Lake Outburst Flood (GLOF) which causes catastrophic discharges from the failure of temporary glacial lakes dammed by loose earth (moraines) materials formed by rapidly melting glacier ice. Observations indicate that the frequency of GLOFs in the eastern Himalaya including Nepal, Sikkim, Bhutan has increased during the last decade of 20th century and threatening the very existence of many hydropower plants constructed recently on Himalayan rivers.

The Indus and Sutlej Rivers originate in the ice fields of the western Tibetan plateau, an area experiencing exceptionally high warming—1.8°C or 0.3°C per decade, over the past 50 years, twice the rate of observed global warming. The retreat in the Tibetan Plateau is driven by warming due to the thickening of the global greenhouse gas blanket, but the rapidity of glacier retreat during the past 25 years suggests additional warming may be playing a critical role. Black soot in aerosol pollution can warm the troposphere, enhancing surface melt. According to a report by the United Nation Environment Program (www.unep.org), Integrated Assessment of Black carbon and Tropospheric Ozone, the Himalayan and Tibetan plateaus are regions where black carbon is likely to have profound impacts on the ice fields and glacial melting.

Any change in upstream water supply or hydrology with changing climate will have profound effect on millions of people in lower riparian countries. Projections show that countries of South Asia will suffer from water stress by 2050. Another area of concern in South Asia is extremely limited water storage capacity in countries like Pakistan and India, for example, less than 250 cubic meters of water per capita compared to more than 5000 cubic meters per capita in countries like Australia and USA. The lack of water storage capacities leaves the already vulnerable populations at great risk of fluctuations in water flows and changes in monsoon patterns. Investments can be increased in natural and constructed water storage systems.

Glacial melt is critical before and after the rainy season (shoulder months) in the region, when it supplies a greater portion of flow in every river from the Yangtze (which irrigates more than half of China paddy fields) to the Ganges and Indus (important to the agricultural heartland of India and Pakistan respectively). Chinese scientists have monitored more than 680 glaciers on the Tibetan Plateau and all of them are rapidly shrinking, with heaviest losses on its southern and eastern edges. These glaciers are not simply retreating, but they are losing mass from surface down. The ice cover in southern and eastern plateau has shrunk more than 6% since the 1970s and the damage is still greater in Tajikistan and northern India, with 35 and 20% declines respectively over the past five decades. If the current trend holds (Tandong et al. 2009), it is believed that 40% of the plateaus glaciers could disappear by 2050. Full scale glacier shrinkage is inevitable and it will lead to ecological catastrophe.

The Copenhagen and Cancun Accords (CHA) in December 2009 and 2010 were accords on framework not on binding targets for reductions in carbon dioxide emissions. The 2°C barrier agreed at COP 15 translates into a 2.5 Wm^{-2} barrier for energy additions. However, the barrier poses a huge dilemma for policy makers because the blanket of manmade GHGs that surround the planet as of 2005 has already trapped 3 Wm^{-2} and the barrier of 2°C has already been exceeded by 20% (Ramanathan and Xu 2010). The most crucial climate tipping point is the changes being felt on the plateau is the rise of temperature of up to 0.3°C per decade approximately three times of global warming rate. Reduction of snow albedo by black carbon deposition will reduce reflectivity by 2.0 to 5.2% and thus enhancing about 70–204 mm of water equivalent runoff from a typical Tibetan glacier (Yasunari et al. 2010). Another study by (Lau et al. 2006) indicate as a response to radiative forcing by dust and black carbon in the Indo-Gangetic plain and Himalayan foothills, the atmosphere over the plateau is anomalously heated and moistened via elevated heat pump (EHP) effect. The warm and moist atmosphere overlying the Tibetan Plateau land surface, causes a reduction in surface sensible and latent heat fluxes from land to atmosphere, i.e., net heat gain by the land surface. The net heat gained is used for melting more snow and ice over the Tibetan Plateau.

Impact on society and ecosystem

The strategic position of Tibet has became more obvious in recent years as climate change has the potential to reduce snow pack and glacier mass which cascade down and alter the hydrological system. Now, China has sovereign rights on the world's largest freshwater resources outside the Polar Regions. These water resources vulnerable to global and regional warming are very critical for sustaining the food and water security of

South Asia. Should China be the lone stakeholder to the fate of waters in Tibet? What happens in the lower riparian nations that depend heavily on these rivers? (Sinha 2010). China has a robust glaciological program and knows fairly well how long snow/ice resources will last. Therefore, they have embarked on integrated water resource management (IWRM) of all the rivers emanating from Tibetan Plateau. Rivers diversion projects, if implemented, will have enormous ecological issues for all lower riparian countries.

Brahmaputra (Yarlung Tsangpo) River

China aims to build 59 reservoirs on Tibetan Plateau to save glacier runoff. Construction is in full swing at Zangmu for a 540 MW run of the river power project and feasibility studies have been completed to construct five more similar projects further upstream on Yarlung Tsangpo. Tapping the power of the Tsangpo (Brahmaputra for Indians) River as it bends and plunges from the Himalayan roof of the world down towards Indian and Bangladeshi flood plains has long been a dream of Chinese politicians and hydro-engineers (Peoples Daily 2009). Metog will be site of a mega project at a huge bend inside Canyon which is about 3.1 mile deep and 198 miles in length. This will involve construction of a series of tunnels, pipes, reservoirs and turbines to generate 40,000 MW power and exploit the spectacular 2000 meters fall of the river as it curls down towards India. The water diversion project was an essential part of the 10th five year plan. The project will cost 62 billion US dollars and water will be diverted through three channels in the eastern, central and western regions respectively (Tiwari 2010). The entire staff which constructed the Lhasa–Beijing railway line has been assigned to this mega project to execute it fast. The project would have ominous consequences for millions of Indians and Bangladeshi population. Chinese conservationists have admitted that the canyon is home to more than 60% biological resources on Tibetan Plateau and many indigenous communities reside in the canyon.

The Brahmaputra has always been considered the very soul of the Indian state of Assam as poets and ordinary people alike consider the river as a part of their folklore and culture. China, in her own interests, could use the water for power generation as run of the river projects and allow lower-riparian countries to use water for agricultural purposes during fall and dry months. The entire NE region of India and Bangladesh would be starved of nutrient-rich sediments and water which is lifeline for lower riparian communities.

Chinese scientists hold the view that upstream reservoirs would alleviate floods and erosion in the Brahmaputra. Ironically, this makes little sense, since flooding could actually get worse due to relentless silting which, will be accelerated by the slowing down (reduced velocity) of the river flow. It may be noted that flooding normally happens not as much because of snow/glacier melt waters in the Yarlung Tsango section, but more from the monsoon rains from on the southern side of the Himalaya. Many tributaries join it in Arunachal Pradesh to make it into a huge water resource. At this place the pre-rainy season flow averages well above 120,000 ft^3 per second, rising to 1 million ft^3 per second during the monsoon season. All rainfall in the Assam hills discharge into this river, making it at places 10 kilometers wide. Recently, China and India (Hasnain 2010) agreed for a joint mechanism to share hydrological data on rivers Brahmaputra and Sutlej for the rainy season only. From water resources point of view, data sets during low flow periods are critical and interestingly this is not the part of joint agreement in spite of India's insistence.

For India and Bangladesh, water resources are already overstretched by increasing demands from growing populations, new economic growth and intensifying agriculture. The proposed dam on the Tsangpo canyon will likely to reduce flows in the Brahmaputra by 20 to 30% during shoulder months.

Indus River

The Indus River, which originates in western China, runs through Indian Kashmir and the Gilgit-Baltistan region of Pakistan. The 1947 partition gave India the headwaters of four tributaries of the Indus, namely Jhelum, Beas, Ravi, and Chenab, and gave Pakistan the main artery of the Indus and also the Sutlej in the Tibetan Plateau. The Indus water treaty (IWT) signed by India and Pakistan in 1960, was brokered by the World Bank to equitably allocate the rivers of Indus system to India and Pakistan. The treaty assigned three eastern rivers (Ravi, Beas, and Sutlej) to India for its exclusive use, which is about 20% of the total flow of the six rivers. The treaty gave the flow of the Indus proper and two of its tributaries (Jhelum and Chenab) to Pakistan for its use, which is about 80% of the total flow (with a provision that India could use western rivers for the development of energy through 'run of the river' projects without storing water).

The World Bank's facilitation of negotiations contributed hugely to the success of the IWT and the bank continues to be an important player in resolving disputes over the construction of power projects by India on western rivers. Over the years, the World Bank has remained fully involved by appointing neutral experts and arbitration panels to resolve disputes.

The IWT essentially does not have any provisions for joint management of the river basin. It only covers the division of water. However, it has provisions to furnish hydrological, hydraulic and engineering data to Pakistan, the lower riparian. India claims that it has scrupulously fulfilled its obligation to share data on all the projects undertaken on western rivers. Nonetheless, Pakistan denies that India shares such data. Both countries engage in endless debates about the equity of water allocation, both inside the commission meetings and in public forums. While India tries to use permissive provisions, Pakistan applies restrictive ones. This upper and lower riparian saga is going on and on.

Six rivers that belong to the Indus System originate in the Hindu Kush-Himalaya-Tibetan region and are fed by snow and glacier melt waters, which account for about 44.8% of the total river flow (Table 13.1). The proportion of glacial melt in the Indus, Jhelum, and Chenab is predicted to increase, as glaciers in Karakorum-Himalaya and Tibetan plateau recede. Broadly, the major controls on the climate are latitude, altitude, and position relative to the Indian monsoon and the westerly.

Traditionally, trans-boundary water conflicts center on infrastructure built by upper riparian country/countries to alter hydrological flows, affecting the quantity and timing of flows, especially during low flow periods. Currently, India and Pakistan are gridlocked over the interpretation of various provisions under the IWT. Pakistan is currently objecting to India's energy project on the Kishenganga, a lower tributary of the Jhelum in Indian Kashmir, and declared this issue to be a 'dispute' to be taken to the court of arbitration. Nevertheless, Pakistan has not recognized the emerging realities of climate change's role in rapidly shrinking glaciers in the Tibetan Plateau, Indian Kashmir and Gilgit-Baltistan region. Tipping mass of all glaciers and variability in precipitation are driving huge flows in the Indus tributaries. Therefore, effectively addressing water issues linked to emerging concerns relating to climate change through confidence-building measures (Track II) in South Asia will address the root causes of the present tensions. The climate change platform provides an opportunity to India and Pakistan to measure, monitor and model glacier systems across the Line of Control (LOC) and ensure trans-boundary water security.

Science-policy gap

Water as an instrument and tool of bargain and trade-off will assume predominance because the political stakes are high in South Asia. Water issues between India and China and India and Pakistan have the potential to become catalysts of conflicts. The opportunity in the least integrated region of the world is to build 'Knowledge Action Networks' to Foster Cross Border Cooperation on Trans Boundary Rivers. It is imperative for stakeholders like policy makers, scientists, and members of civil society in South Asia to better understand, assess and cooperate on the links between global and regional climate change processes impacting water resources, ecosystem stability, and as a fall-out of the growing ecological footprint of economic development. The problem, however, is that existing narratives and projections are fraught with uncertainties.

Unfortunately, in the security establishments of South Asia, there has been a lack of understanding the roles that snow and glaciers play in sustaining water flows in the Himalayan rivers. The knowledge deficit about high altitude snowfields and valley glaciers and their role in sustaining low flows during post and pre-monsoon periods (shoulder months) has hampered the ability of security experts as well as the water bureaucracy to gauge and grapple with the full ramifications of rising water stress in the Indus Basin.

If irregular challenges caused by climate change go unchecked, they could lead to large scale regional conflicts as countries compete for dwindling water resources. Each South Asian country should design and carry out its own assessment of glacier and monsoon changes, with international support, not direction. Both India and Pakistan have knowledgeable leaders who can forge relationships with decision makers, but there aren't enough decisionmakers in either country who are sufficiently versed in these issues. The critical mass sufficient to characterize the multiple impacts of climate change on glaciers and communicate them to decision-makers is significantly lacking. Capacity building, as suggested by Secretary of State Hillary Clinton as the 'first stream' of US water strategy, is therefore a critical issue in the region and needs to be addressed as a priority. We will help to support knowledge action networks which can create a two-way flow of information, knowledge and methods between local communities, opinion leaders, scientists and decision-makers, and their regional, national and global counterparts. The function of this networks includes acquiring and disseminating knowledge about climate change. Knowledge networks can appeal to the UN for assistance in measuring, monitoring, and modeling glacier changes.

Conclusion

The compounding impacts of regional climate warming, monsoonal rainfall variability and rising emissions of black carbon aerosols are accelerating the melt of ice and reducing the accumulations of snow on these glaciers, leading to a significant loss of ice mass over large portions of mountain regions. Continued widespread melting of glaciers during the coming years will lead to floods/water shortages, declining crop yields and habitats of local communities. Thus far, conclusions are being made on limited data coming from satellite imageries and field research on a few selected glaciers, but the extreme conditions found in the region make field research expensive, time-intensive, dangerous and physically challenging.

References

Ageta, Y. and K. Higuchi. 1984. Estimation of mass balance components of a summer-accumulation type glacier in the Nepal Himalaya, Geografiska Annaler. Series A, Physical Geography 63(30): 249–255.

Baiqing Xu, Junji Cao, James Hansen, Tandong Yao, Daniel R. Joswia, Ninglian Wang, W.U. Guangjian, Mao Wang, Huabiao Zhao, Wei Yang, Xianqin Liu and Jianqiao He. 2009. Black soot and the survival of Tibetan glaciers. PNAS 106(52): 22114–22118.

Dyurgerov, M.B. and M.F. Meir. 2005. Glaciers and the changing earth system: A 2004 snapshot. Institute of arctic and alpine research, Occasional paper, 58.

Edwards, S., L. Catherine and L. Stanbrough. 2010. The waters of Third Pole: sources of threat, sources of survival. UCL Hazard Research Center, Kings college, London, pp. 48.

Elizabeth, M.L. 2010. Changing glaciers and hydrology in Asia: addressing vulnerabilities to glacier melt impacts, USAID special Report, pp. 81.

Etienne, B., Y. Arnaud, R. Kumar, S. Ahmad, P. Wagnon and P. Chevallier. 2007. Remote sensing estimates of glacier mass balances in the Himachal Pradesh. Remote Sensing Environment 108: 327–338.

Hasnain, S.I. 2010. Studies of Benchmark glaciers in the Himalaya, technical report 1, TERI, New Delhi (unpublished).

Hasnain, S.I. 2010. River Runs Through it, Times of India, New Delhi, Edit Page, Aust 17, 2010.

Immerzeel, W.M., P.H. Ludovicus, van Beek and M.F.P. Bierkens. 2010. Climate change will affect the Asian water towers. Science V. 328.

Kaab, A. 2005. Combination of SRTM3 and repeat Aster data for deriving alpine glacier flow velocities in the Bhutan Himalaya, remote sensing Environment 94: 463–474.

Lau, W.K.M., M.K. Kim and K.-M. Kim. 2006. Asian summer monsoon anomalies induced by aerosol direct forcing: the role of Tibetan Plateau. Climate Dynamics 26: 855–864.

Orville, S. 2010. The Message from the Glaciers. The New York Review of Books, May 27, New York, USA, pp. 9.

Peoples'Daily (on line). 2009. China Launches key Construction along water diversion projects, June 5, 2009, Beijing.

Ramanathan, V., M.V. Ramana, G. Roberts, D. Kim, C. Corrigan, C. Chung and D. Winkev. 2007. Warming trends in Asia amplified by brown cloud solar absorption. Nature 448–578.

Ramanathan, V. and Y. Xu. 2010. The Copenhagen Accord for limiting global warming: criteria, constraints, and available avenues. PNAS 107(18): 8055–8062.

SAC (Space Application Center, Ahmedabad). 2011. Snow and glaciers of the Himalayas, project Report Space Application Center, pp. 253 (unpublished).

Tandong, Y., L.G. Thompson and V. Mosbrugger. 2009. Report of the 1st Third Pole environment Workshop, Institute of Tibetan Plateau Research (ITP), CAS, Beijing, China.

Tiwari, R. 2010. India, China Renew Brahmaputra Pact, Indian Express, New Delhi, online, Aprol 29, 2010.

Sinha, U.K. 2010. Tibets watershed challenge, The Washington Post, June 14, 2010.

UNEP. 2011. Integrated assessment of black carbon and tropospheric ozone; summary for decision makers.

Yasunari, T.J., P. Bonasoni, P. Laj, K. Fujita, E. Vuillermoz, A. Marinoni, P. Cristofanelli, R. Duchi, G. Tartari and K.M. Lau. 2010. Preliminary estimation of black carbon deposition from Nepal Climate observatory-pyramid data and its possible impact on snow albedo changes over Himalayan glaciers during the pre-monsoon season. Atmospheric Chemistry and Physics 10: 6603–6615.

Webster, P.J., V.O. Magana, T.N. Palmer, J. Shukla, R.A. Tomas, M. Yanai and T. Yasunari. 1998. Monsoons: processes, predictability, and the prospects for prediction. Journal of Geophysical Research 103: 14,451–14,510.

Zongxing, Li, Hi Yuanqing, A. Wenling, A. Linlin, Z. Wei, C. Norm, W. Yan, W. Shijin, L. Huancai, C. Weihong, H.T. Wilfred, W. Shuxin and D. Jiankuo. 2011. Climate and glacier change in southwestern china during the past several decades. Environ. Res. Lett. 6: 24.

14

Glacier Lake Outburst Floods (GLOFs)—Mapping the Hazard of a Threat to High Asia and Beyond

Manfred F. Buchroithner[1,]* and *Tobias Bolch*[2]

INTRODUCTION

Lakes are scenic assets for any high-mountain landscape. They can, however, also be dangerous. This applies particularly to lakes in the immediate neighbourhood of or even on glaciers. These lakes are known to be prone to outbursts due to various reasons, and these high-energy events can be significant threats to life, property and infrastructure and rather disastrous along the stretches further down the valleys (Ives 1986, Richardson and Reynolds 2000, Ives et al. 2010, Han et al. 2013). Globally, the glaciers are in a general state of retreat, most probably because of climatic warming (WGMS 2008). They often leave behind voids filled by melt water called glacial lakes which tend to burst because of internal instabilities in the natural moraine dams retaining the lakes (e.g., as a result of hydrostatic pressure, erosion from overtopping, or internal structural failure) or as a result of an external trigger such as a rock or ice avalanche, or even earthquake.

[1] Institute for Cartography, Dresden University of Technology, Germany.
[2] Department of Geography, University of Zurich, Switzerland, and Institute for Cartography, Dresden University of Technology, Germany.
* Corresponding author

Hence, classically such a Glacier Lake Outburst Flood (GLOF) occurs when water dammed by a glacier or a moraine is released. A water body that is dammed by the front of a glacier is called a marginal or proglacial lake, a water body that is on glacier ice is called supra-glacial lakes, and a water body capped by the glacier is called a subglacial or englacial lake (Ives 1986, Buchroithner 1996). Supra-glacial lakes often appear to merge with moraine-dammed lakes, or may develop contemporaneously as composite forms. When a marginal lake bursts, it may also be called a marginal lake drainage. When a sub-glacial lake bursts, it may be called a jökulhlaup.

Thus, a jökulhlaup describes a sub-glacial outburst flood. It is an Icelandic term that has been adopted into the English language, originally only referring to glacial outburst floods from Vatnajökul, which are triggered by volcanic eruptions, but later it has been accepted to describe any abrupt and large release of sub-glacial water (Ives 1986). Initially, this term has, however, frequently erroneously been used for GLOFs in general (cf. Buchroithner 1996).

Glacial lake volumes vary significantly, but may hold millions to hundreds of millions of cubic metres of water (cf. Buchroithner 1985, Cenderelli and Wohl 2001). Catastrophic failure of the containing ice or glacial sediment contained in the 'dam' can release this water over periods of minutes to days. Peak flows between 2000 m^3 up to 30.000 m^3 per second have been calculated for such events (Cenderelli and Wohl 2001, Richardson and Reynolds 2000) suggesting that a v-shaped canyon of a normally small mountain stream could suddenly develop into an extremely turbulent and fast-moving torrent some 50 metres high (cf. chock block in gorge near Pangpoche after the 1977 Nare Drangka GLOF; Buchroithner et al. 1982).

There are a number of imminent deadly GLOFs situations that have been identified worldwide. Imja Tsho located in the Everest region (Watenabe et al. 2009) and Tsho Rolpa glacier lakes gained international fame. The latter one is located in the Rolwaling Valley, about 110 kilometres northeast of Kathmandu, at an altitude of 4,580 m. It is dammed by a 150 m high unconsolidated terminal-moraine dam. The lake is growing larger every year due to the melting and retreat of the Trakarding Glacier (cf. Fig. 14.1), and has become the largest and one of the most dangerous glacier lake in Nepal, with approximately 90 to 100 million m^3 of water stored (ICIMOD 2011; http://www.dhm.gov.np/tsorol/background.htm; accessed December 2012). Remarkable mitigation measures, supported by the international community, have in this instance been materialized.

Figure 14.1. Development of Thso Rolpa 1978–2007 (Photos courtesy of K. Fujita).

Beginning of Scientific Interests

GLOFs of potentially dangerous glacial lakes (PDGLs) have been known for long and have been studied by means of remote sensing since the 1970s. K.J. Hsü (1975) mentions (probably as one of the first ones worldwide) catastrophic debris streams, which he calls 'Sulzstroms', generated by rockfalls. Louis Lliboutry (Spanish: *Luis Lliboutry*) was a French geographer, glaciologist and mountaineer who, since the 1950s, partly lived in Chile. In 1977 he published three papers in the renowned Journal of Glaciology about glaciological problems set by the control of dangerous glacial lakes in the Cordillera Blanca in the Peruvian Andes (Lliboutry 1977, Lliboutry et al. 1977a, 1977b).

Soon after the aforementioned paper by Buchroithner et al. (1982) Wilfried Haeberli reported about 'glacier floods' in the Swiss Alps (Haeberli 1983). In 1985 some Japanese scientists and one Nepali published a short note on catastrophic floods in Nepal originating from glacier lakes (Fushimi et al. 1985). In the following years Vuichard and Zimmermann (1986, 1987) reported in detail about the Langmoche flashflood in Khumbu Himal, Nepal caused by the sudden drainage of Dig Thso (cf. Fig. 14.2). In 1986 a first comprehensive monograph about GLOFS, initiated by the Kathmandu-based International Centre for Integrated Mountain Development (ICIMOD), was issued by Jack D. Ives (Ives 1986).

Figure 14.2. Recent photo of Dig Thso which drained catastrophically in 1985. The breached dam is still well visible (Photo courtesy of K. Fujita).

Color image of this figure appears in the color plate section at the end of the book.

In a volume dedicated to earth-observing space photogrammetry and its application to mountain cartography, Robert Kostka investigated the potential of spaceborne stereo-photography for glaciological purposes and demonstrated it with detailed mappings of glacial lakes and their surroundings in the Khumbu Himal and the South Tibetan Pekhu Tso area (Kostka 1987). In the same volume Manfred Buchroithner (Buchroithner 1987) issued so far unpublished additional details about catastrophic GLOFs in the Nepalese Himalaya described in previous papers (Buchroithner et al. 1982, Buchroithner 1984a, 1984b, 1985, Danninger and Posch 1986). In 1988 and 1991 the Soviet glaciologist N.V. Popov reported a study earlier already published in Russian, concerning the control of PDGLs in the Northern Tien Shan (Popov 1988, 1991).

Subsequently, in 1990—and maybe triggered by Ives' report—one of the first regional inventories of PDGLs in the form of a small-scale map was made for ICIMOD by Manfred Buchroithner on the initiative of Surendra Shrestha, then founder and head of MENRIS and later Director of UNEP-EAP. This 'Jokulhlaup Hazard Map' was, however submitted as an internal copy for ICIMOD and only published several years later in a diminished form (Buchroithner 1996). Based on December 1982 high-resolution stereoscopic colour-infrared Metric Camera imagery taken from the NASA Space Shuttle that covered the High Himalaya region south of the Lhasa-Kathmandu Friendship Highway between Kodari in the West and the eastern border of Sikkim and identified clusters of 162 risk sites with surficial PDGLs. This map-preparation was based on in-depth field- and office experiences of the author from his studies carried out in the 1980s (Buchroithner 1984a, 1984b, 1985).

It is remarkable that in all three cases of these described flash-flood traces, their places of origin were situated in areas where major geological lineaments intersect minor ones. This observation suggests that tectonic forces, i.e., earthquakes, could, in the very first instance, be responsible for these catastrophes (Buchroithner 1984b).

Calculation of the real, non-projective area eroded during the flash-flood event detected by Manfred Buchroithner in the upper Tamur River yielded an acreage of some 479000 m^2, a figure 92000 m^2 (24%) larger than the projective area (Buchroithner 1985). The possibility to calculate real areas and, consequently, also volumes by prior- and post-state comparisons, clearly demonstrates the potential of photogrammetric space imagery, in the above case of Metric Camera photographs. In particular for high-relief terrain it is essential to refer not only to the nadir-projective but—first of all—to the real acreage.

GIT-based GLOF Hazard Research in Times of Recent Gobal Climate Change

Since the 1990s, climate change and the concomitant glacier recessions have been causing an increasing number of continuously extending glacial lakes in mountain areas round the world which are leading to an intensified risk of lake outbursts, both in High Asia and around the world. Consequently, the body of scientific literature is increasing. In High Asia the majority of the studies focused on the Himalaya (e.g., Richardson and Reynolds 2000, Iwata et al. 2002, Ma et al. 2004, Bajracharya et al. 2007, Bolch et al. 2008, Gardelle et al. 2011, Hewitt and Liu 2011) but a few studies are also available for the Pamir (Mergili and Schneider 2011), the Tibetan mountains (Wang et al. 2011, Ma et al. 2004), and the Tien Shan (Bolch et al. 2011a, Narama et al. 2010). The increasing availability of suitable high-resolution satellite imagery facilitated the investigation of the development of glacial lakes in the remote mountainous areas (cf. Fig. 14.3). Special mention may be given to the monograph prime-authored by Jack D. Ives about a quarter century after his initially mentioned 1986 monograph about the GLOF hazards in the Himalayan region (Ives et al. 2010), which actually served as the basis of the report about glacial lakes and glacial lake outburst floods in Nepal issued by ICIMOD (2011).

The more recent the literature, the more obvious became the interrelation between glacier recession and glacial lake generation. To properly evaluate hazards posed by alpine glacial lakes, systematic information must be collected on lake types, dam characteristics, outburst mechanisms, down-valley processes and possible cascades of processes (cf. Haeberli et al. 2010).

In a comprehensive article based on the initial work by Peters (2009) and Bolch et al. (2011a) reported about the identification of PDGLs and their possible impact by means of remote sensing techniques using the northern

Figure 14.3. Termini of Imja Glacier showing also the occurrence of the Imja Lake based on 1970 Corona and 2007 Cartosat-1 images (Bolch et al. 2011b).

Tien Shan as an example. According to this paper and in part adapted on the basis of ICIMOD (2011), in the following subsections the above mentioned determining parameters will be described.

Glacier lake mapping

Water bodies can be in principle be identified well using remote sensing imagery. For this purpose several methods like ratioing and the Normalised Differential Water Index (NDWI) with different band combinations (e.g., Blue, Green, NIR, SWIR) may be used (Huggel et al. 2002, Bolch et al. 2008, 2011a). One of the biggest challenges when mapping glacial lakes is to obtain a highly accurate delineation of lakes with (partial) ice coverage and of turbid lakes with minimum classification error. Optimum results can be obtained using the NDWI approach by employing NIR and Green bands (Green – NIR/Green + NIR). The water index using Blue mostly performs better in the shaded areas while the index with Green shows fewer problems with ice on the water bodies. The applied thresholds have, however, to be adapted for each individual scene (Table 14.1). One major

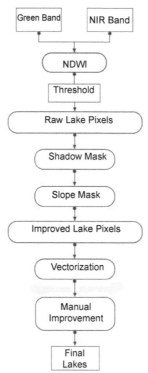

Figure 14.4. Proposed optimised procedure for lake mapping based on multi-spectral remote sensing data (Bolch et al. 2011a).

Table 14.1. Overview of the utilized NDWI thresholds for different sensors (Bolch et al. 2011).

Sensor	Landsat MSS	ASTER	Landsat ETM+
Threshold	0.45–0.9	0.3–0.7	0.3–0.9

drawback of the NDWI is that cast shadows are also included in the mask and must be excluded in a post-processing step. A shadow mask may be used for this purpose (cf. Huggel et al. 2002) or the classification needs to be done separately for shadowed and non-shadowed areas (Chen et al. 2012). In a last step, visual checks and manual improvement are necessary especially for lakes in shadow and turbid lakes.

Hazard assessment

It is obvious and has already been mentioned in the literature repeatedly (cf. i.a. Haeberli et al. 2010, ICIMOD 2011) that the hazardous processes can interact with each other and with less climate-sensitive parts of the involved geomorphic system. Disasters have mostly resulted from cascading processes rather than single phenomena (Haeberli et al. 2010). Therefore it is necessary to apply an integrated system approach to avoid missing important processes (Huggel et al. 2004, Bolch et al. 2011a, 2011b, Mergili and Schneider 2011). Uncertainties at all stages of the process cascade are considerable. Actually, uncertainties also cascade, i.e., they increase along the process chain. Hence, systematic and detailed observations are essential for objective and sound assessments (Kääb et al. 2005).

The design of appropriate observation systems is thereby greatly facilitated by modelling scenarios of possible developments including the following ones (after Haeberli et al. 2010): Landslides in bedrock (1a) and moraines (1b) due to debuttressing effects from glacier retreat since the Little Ice Age, ice avalanches from polythermal steep glaciers due to increasing firn and ice temperatures (2a), and from dissected temperate steep glacier parts (2b), rock fall and avalanches in relation with permafrost degradation (3a), regular rock fall from areas of warm bedrock permafrost (3b) fuelling debris flow (4) initiation zones, lake formation (5) due to glacier retreat or down-wasting, posing outburst floods hazards that are aggravated by potential impacts from multiple mass movement processes.

Several factors need to be taken into account when assessing the hazard potential of a glacial lake outburst. Bolch et al. (2011a) adapted a methodology developed by Huggel et al. (2002) and Bolch et al. (2008) introducing a higher number of variables for the hazard assessment. The latter can be summarized by four major parameter groups: (a) lake characteristics, (b) characteristics of the lake surroundings, (c) characteristics of the adjacent glaciers, (d) impact on downstream areas. Each of those groups consists of several variables. Table 14.2 gives an overview of these variables and their applicability using remote sensing.

Table 14.2. Key factors contributing to the hazard risk of a glacial lake and its investigation using remote sensing data (Bolch et al. 2011a).

Characteristics group	Factor	Remote sensing data source and applicable techniques	Suitable for automatization
Lake characteristics	Lake area and volume	Detection using multi-temporal multi-spectral (MS) satellite data	Yes
	Rate of lake formation and growth	Change detection using multi-temporal (MT) and MS satellite data	Yes
Glacier characteristics	Fluctuations of the glacier	Investigation of area and volume change of the glacier based on MS and MT satellite data, MT digital terrain models (DTMs)	Yes
	Activity of the glacier	Derive glacier velocity using feature tracking or DInSAR based on MT optical or radar data	Yes
	Geomorphometric characteristics of the glacier	Geomorphometric DTM analysis, slope classification	Yes
Characteristics of the lake surrounding	Freeboard between lake and crest of moraine ridge	Geomorphometric DTM analysis	Partly
	Width and height of the moraine dam	Geomorphometric DTM analysis	Partly
	Stability of the moraine dam/presence of dead ice in the moraine dam	Investigation of surface deformation based on MT DTM analysis, permafrost modeling	Partly
	Possibility of mass movements into the lakes	Mapping of ice cover and geology using MS data, Geomorphometric DTM analysis of the surrounding catchment areas	Yes
Impact of an GLOF todownstream areas	Affected Area	Flow modeling	Yes
	Infrastucture down-valley	Detection of human infrastructure based on MS satellite data analysis	Partly

Lake characteristics

One of the most important variables for analyzing the potential danger of a GLOF is the change of the glacial lake. Today, normally these changes are identified using multi temporal space imagery (cf. Table 14.2). The growth of a supraglacial or proglacial lake depends primarily on the glacier characteristics and retreat. However, it has to be noted that also strong and fast glacier advances (surges) from a side valley into the main valley can block the river and also lead to PDGL. This situation is especially known for the Karakoram (Hewit and Liu 2010) The volume of typical proglacial lakes may be addressed based on the empirical formula (Equation 1) suggested by Huggel et al. (2002) which is primarily based on 15 lakes with existing depth measurements.

$$V = 0.104 \, A^{1.42} \hspace{4cm} \text{Equation 1}$$

It is, however, important to emphasize that this scaling formula may only serve as a rough estimate, since the lake volume depends on several variables. For instance, by applying this formula to the lakes in the Kishi Almaty Valley of the northern Tien Shan with existing bathymetric measurements (Kasatkin and Kapista 2009, Tokmagambetov 2009) reveals an overestimation of the volume of up to 20% (Bolch et al. 2011a). Similar uncertainties were also mentioned by Huggel et al. (2002).

Characteristics of adjacent glaciers

In the first instance the recent outlines of glaciers have to be mapped based on multi-temporal imagery. These outlines allow to assess the changes of the glacier front in detail. Glacier velocity is usually estimated from multitemporal optical or SAR imagery based on feature tracking using cross-correlation techniques (cf. Kääb 2005, Berthier et al. 2005, Bolch et al. 2008). Here, the open source software 'Cosi–Corr' (Leprince et al. 2007) may serve the automated estimation of the velocity of mountain glaciers as well (Scherler et al. 2008). Due to the better contrast than in the higher resolution panchromatic band, the use of the near infrared band is recommendable. A glacier may be called stagnant in its distal part if the calculated velocity is below the uncertainty of one pixel.

The average slope of the glacier surface gives a hint where glacial lakes can develop or an existing lake can extend in the near future. A threshold of 2° for supraglacial lake formation on debris-covered glaciers in the Himalaya (Reynolds 2000, Quincey et al. 2007, cf. Bolch et al. 2008) or of 5° for the formation of proglacial lakes in overdeepening of debris-free glaciers in the Alps (Frey et al. 2010) have been reported.

Characteristics of lake surroundings

Mass movements like rock fall or ice avalanches into a lake are important triggering mechanisms for an outburst. Hence, an analysis of the surrounding topography is of high importance. A simple but robust model for this purpose is the Modified Single-Flow Model (MSF, Huggel et al. 2003). The model is a modified D8 flow direction algorithm and calculates the likelihood that a raster cell will be affected by such a mass movement. Similar methods to model rock and ice avalanches were applied by Allen et al. (2009) and Salzmann et al. (2004).

If no detailed ground information is available, it might be appropriate to model the probability of a rock fall based on Kaibori et al. (1988), who presented detailed statistics for the slope at the detachment zone and for the angle of friction. The angle of friction is defined by the average slope between the starting and endpoints of the mass movement (Hsü 1975). In contrast to Kaibori et al. (1988) who chose average values, one might, however, use minimum values so that 90% of all occurred events are included.

For ice avalanches, based on the empirical work of Alean (1985), thresholds of 45° for the slope of the detachment zone for cold glaciers and 25° resp. 17° for warm glaciers may be used (cf. Bolch et al. 2011a). Although these values are based on studies in the Alps and high mountains in Japan, these values seem to be reasonable for a first estimation as they represent a worst-case scenario. Van der Woerd et al. (2004) estimated a gradient of about 45° for the origin of ice avalanches for cold glaciers in the Central Asian Kunlun Range. An example for a modelling of ice avalanches is shown in Fig. 14.5.

The probability of a dam failure depends mainly on the characteristics of the lake dam itself. Most of the lake dams consist of moraine-ice material. The width and height of a dam as well as the freeboard between the lake level and the crest can be visually determined by means of high-resolution DTMs, nowadays possibly derived from satellite imagery. SRTM data can provide a hint but the resolution is too coarse for detailed investigations (Fujita et al. 2008). Since 2013 the TanDEM-X interferometric SAR mission of DLR acquires global DTM data with a planimetric resolution of 12 metres and an accuracy of better than two metres which seems to be well suited for a baseline dataset.

A dam can become unstable if it contains permafrost or buried ice which thaws or will thaw due to changing temperature conditions (Richardson and Reynolds 2000). Here, a comparison of multi-temporal high resolution DTMs can give a hint for the thawing of the ice content and the extent of the lowering of the dam can also be detected (Fujita et al. 2008).

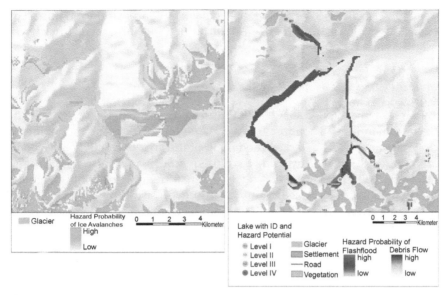

Figure 14.5. Examples from the flow modelling. Left: probability of an area affected by ice avalanches, right: probability of an area affected by flash floods and mudflows (Bolch et al. 2011a, Peters 2009).

Color image of this figure appears in the color plate section at the end of the book.

In order to obtain some indications whether a moraine dam is currently within the permafrost zone and could be affected by thawing, permafrost may be modelled using a simple empirical model, e.g., based on Permakart (Keller 1992). This model is based on empirical findings of the permafrost distribution as well as geomorphometric parameters, Mean Annual Air Temperature (MAAT), which can be computed using a DTM, and additional data. Bolch et al. (2011a) extended this model and included the solar radiation as additional information. The limits of the permafrost distribution are geographically strongly differing, and frequently small-scale variability (e.g., caused by the land cover) cannot not be captured. Climate change also has an impact on the permafrost distribution, and the global permafrost area diminished during the last 130 years (cf. i.a. Marchenko et al. 2007). It can be assumed that a dam may become unstable if it is located outside the continuous permafrost.

Although being a rough estimate, especially when taking into account that the blocky moraine material itself may retard thawing (Gorbunov et al. 2004) this approach provides a relatively quick evaluation of the possible current existence and condition of permafrost in a dam (Bolch et al. 2011a).

However, it has to be noted that the material composition and important processes which can initialize a dam failure like piping can hardly be assessed based on remote sensing and modelling.

Impact of glacial lake outburst floods

As already mentioned GLOFs can present a risk for infrastructures and even human lives. Therefore it is advisable to calculate the probability of downstream affects based on modelling of the area potentially affected by an outburst flood, e.g., by using the aforementioned MSF model of Huggel et al. (2003). Flash floods often lead to or transform into debris flows. Besides the availability of loose sediments, a certain velocity of the water is needed to transport the debris. The kinetics and mechanics of such an extreme case of flooding can be described by the principle of a shallow-water wave (non-linear differential calculus; Scheidegger 1975). This again depends mainly on the steepness of the presumptive water channel. Haeberli (1983) and Huggel et al. (2002) suggested that debris flows confine if an average inclination of 11° is reached. According to Allen et al. (2009) lower thresholds seem advisable, reaching down to an angle of friction of less than 3° (Table 14.3). These thresholds are, however, only rough estimates, and in reality there exist transitions with different flow types occurring within one event (cf. Table 14.4).

Further models which were used to model flash floods include RAMMS (Rapid Mass Movements, Christen et al. 2010) and FLO-2D (O'Brian 1993) (cf. Mergili et al. 2011).

Thus, the relative probabilities of an affected downstream area can be calculated (Fig. 14.5). A major decisive parameter, though, is the quality of the applied DTM.

Table 14.3. Parameters and their thresholds used for modelling of the probability of mass movements Parameter Rock avalanche (Bolch et al. 2011a).

Parameter	Rock avalanche (Kaibori 1988)	Ice avalanche (Alean 1985)	Debris flow (Heaberli 1983, Huggel et al. 2004)	Floodwave (Allen et al. 2009)
Slope at the detachment zone	30°	25°	n.c.	n.c.
Angle of friction (average incline)	20°	17°	11°	3°

debris flows. Notably, large debris flows can form during breaching of moraine dams (Clague and Evans 2000, Haeberli et al. 2001).

Table 14.4. Weighting factors of the selected variables (Bolch et al. 2011a). Further explanation see page 336.

Variable	Exemplary weighting factors	Key for initial determination of weighting factors
Lake area change	0.1661	0: Shrinkage or no significant growth 0.5: growth < 50% of the initial area 1: growth < 100% of the initial area 1.5: growth < 150% of the initial area 2: growth > 150% of the initial area
Risk of ice avalanche	0.1510	1: Modelled deposits hit lake 0: Modelled deposits do not hit lake
Risk of rock fall/ avalanche	0.1359	1: Modelled deposits hit lake 0: Modelled deposits do not hit lake
Instable dam	0.1208	1: Dam is within discontinuous permafrost 0: Dam is outside discontinuous permafrost
Debris flow	0.1057	1: Debris-flow would occur if an outburst would happen 0: Debris-flow would not occur if an outburst would happen
Flash flood	0.0906	1: Flash flood would occur if an outburst would happen 0: Flash flood would not occur if an outburst would happen
Contact to glacier	0.0755	1: Lake is in direct contact with glacier 0: Lake is not in direct contact with glacier
Lake area	0.0604	0.5: Small (size < 50,000 m²) 1.0: Medium (> 50,000 and <100'000 m²) 1.5: Large (> 100,000 m²)
Glacier shrinkage	0.0453	1: Significant glacier shrinkage 0: No significant glacier shrinkage
Glacier slope <5° at the terminus	0.0302	1: Glacier has slope angels below 5° adjacent to the lake 0: Glacier has slope angels above 5° adjacent to the lake
Stagnant ice at the terminus	0.0151	1: No significant glacier velocity was detected at terminus 0: Significant glacier velocity was detected at the terminus
Sum of the weights	1.000	

River sediments

Sediment flux in river systems is strongly influenced by exposure of loose moraine material due to glacier wastage and by deeper or even complete thaw of perennially frozen glacial debris and talus. New exposure of erosion-susceptible sediment sources can increase sediment loads in rivers or trigger

Identification of Potentially Dangerous Glacial Lakes (PDFLs) based on the interconnectivity of systems

An integrated assessment is required for the identification of potentially dangerous glacial lakes in an automated and objective way. Bolch et al. (2011a) suggest to combine the conditioning parameters based on a numerical approach on the basis of additive ratio scales similar to those used in business studies (Kahle 1998). The aim is to have an efficient tool for decision making. In this case, this approach helps to ascertain which lakes are potentially of high danger and should be further investigated. Each introduced variable has to be tested if it applies to the investigated lake. If so (e.g., if a potential ice avalanche would reach a lake or a lake is in direct contact to the glacier), a value of one is assigned to the lake (otherwise a zero). This approach is, however, not applicable for lake area and lake growth. Usually a larger lake area contains more water and can therefore cause higher damage. Bolch et al. (2011a) introduced three size classes (small, medium, large) and assigned the factors of 0.5, 1 and 1.5 to each lake according to its area. Lake growth was treated in a similar way.

In contrast to reality, for easy modelling the applied variables should be independent. As already mentioned above this is practically not the case: For example, the increase of lake area depends at least partly on the glacier retreat, if a lake is in direct contact to the glacier. Glacier-flow velocity and slope (below 5°) are also not independent. These two issues can be considered while assigning the weighting factors to each variable. A weighting scheme is also needed in order to account for the different impacts on the potential danger of the lakes investigated. However, the weighting is at least partly subjective, depends also on the special situation in the study are and should hence be done by an expert.

Bolch et al. (2011a) suggest a weighting scheme after a sequential order of the parameters, as this is most objective and each variable is treated

Figure 14.6. Highly dangerous lake No. 6 in Kishi Almaty (Malaya Almatinka) valley after the surface lowering by deepening of the outflow channel in 2010 (left view from dam to glacier, right view to dam, photos: V. Blagoveshchenskiy).

separately. The first and crucial step for the suggested scheme is the ordering of the variables after the estimated hazard potential from the highest to the lowest. Knowledge, e.g., from literature, about the general setting and past GLOF events (Table 14.1) can be considered for this step. Then, the weights are linearly distributed while the 2nd lowest weight is two times the lowest weight, the 3rd lowest is the sum of the 2nd lowest plus the lowest weight and so on. The sum of the weighting factor is by default set to 1 (Table 14.4).

The variables which are applied for each lake are then multiplied with the weighting factor and subsequently added up. Thereafter, a total of nine parameters are included and modelled in a GIS environment for the current situation, and two additional parameters (glacier slope at terminus <5° and stagnant terminus) are also included which indicate whether the glacial lake may continue to grow in the near future (Bolch et al. 2011a).

In most cases the characteristics of the moraine dams (width, height, freeboard) can only be addressed visually while other critical measures of dam stability like material composition or piping can hardly be addressed from remote sensing.

The final classification of PDGLs can be established by the definition of qualitative threshold values ranging from very low potential danger to high danger. A lake should have a very low hazard potential only if no or only one factor with low weight applies to the lake. Hence, 0.1 may be taken as the first threshold. Bolch et al. (2011a) considered that a lake can be of potentially high danger if the four most important factors apply to the lake or a combination of several factors reaching the sum of the weights of the four most important factors. The threshold between the low and medium potentially dangerous glacial lakes is suggested to be the mean value between class 1 and 4. The weighting and classification scheme may be evaluated based on visual interpretation of the morphometric variables in satellite imagery of selected case studies, previous GLOF events and possible knowledge from field visits.

Discussion

An approach which is based on remote sensing analysis and modelling can be successfully applied to identify potentially dangerous glacial lakes. It is certainly suitable for a first comprehensive assessment of PDGLs for a larger area and addresses levels 1 and 2 of the approach suggested by Huggel et al. (2002). Bolch et al. (2011a) combined the manifold conditioning parameters which had not been addressed previously in such a comprehensive way. Their approach is easily reproducible as it is based on well-developed methods such as the detection of water bodies using multispectral imagery (Huggel et al. 2002), the automated detection of glaciers (Paul et al.2002, Bolch and Kamp 2006), and their velocities (Bolch et al. 2008, Kääb 2005) and simple but robust models like the modelling of an outburst path or rock/ ice avalanches (Huggel et al. 2003), or a permafrost model (Keller 1992).

Frequently, historical imagery like Corona or Hexagon proves to be suitable to extend the analysis back in time which was already shown for some glaciers (Narama et al. 2010b, Bolch et al. 2010, 2011b). Using the described approach it is possible to map PDGLs for a larger area such as whole mountain ranges within a short period of time. The results need, however, to be carefully evaluated and the weighting scheme possibly to be adjusted for the special situation in the respective study region. Data from former outburst events may serve as valuable sources for calibration.

Drawbacks of this geomatics-based approach are that: (1) the dam characteristics and the probability of a dam failure can only be addressed marginally, (2) lakes in shaded areas and turbid lakes are difficult to be identified automatically, and (3) the modelling is frequently based on DTMs with an insufficient spatial resolution.

The result of the modelling of the outburst path highlights the most endangered areas. Comparisons with historical outbursts show that frequently path length and, hence, affected areas are underestimated. It is essential that the most accurate DTMs are used.

The major limitation of a remote sensing based study is, however, that characteristics and stability of the moraine/lake dam cannot or only roughly taken into account. The described permafrost model (Keller 1992) is coarse and does not consider the material composition which can strongly alter the thermal conditions within the materials and also retard the thawing. However, this model can be easily applied to different mountainous regions in order to provide an estimate about the existence of permafrost. For a first order assessment the global permafrost map modelled by Gruber (2012) may serve the needs. Fully physical based permafrost model (e.g., Marchenko 2001) will provide a better estimate about the permafrost existence but the required input data are often not available for the remote mountainous areas where the glacial lakes are situated. Other critical measures of the stability of the dam are the types of drainage. While the drainage over the dam could be detected, at least by using high-resolution remote sensing imagery, outflows under or through the dam or piping can, however, not be addressed (cf. Bolch et al. 2011a).

A further limitation is that the water volume of the lakes can only be roughly calculated. Field studies would be necessary to determine the lake depth and to address the grain size distribution of the moraine.

Despite the fact that the described methodology for the classification of the glacial lakes produces reasonable results, the applied weighting according to the importance of the variables should, however, be carefully chosen by an expert who knows the study region.

Conclusions and Recommendations

Since continuous climate warming and the resultant permafrost thawing and glacier recession will increase the potential danger of glacial lake outbursts (cf. also see earlier), the UNEP RRC Inventory of Glaciers, Glacial Lakes and Identification of Potential Glacial Lake Outburst Floods, Affected by Global Warming in the Mountains of Himalayan Region (http://www.rrcap.ait.asia/issues/glof/) has a special section on 'Glacial Lake Outburst Flood Monitoring and Early Warning System'. Within this initiative UNEP through its facilities at Environment Assessment Program for Asia-Pacific (EAP.AP) at the Asian Institute of Technology, Bangkok, is establishing an operational early warning system to monitor GLOF hazards in the Hindu Kush-Himalayan Region. EAP.AP will implement the project in collaboration with the International Center for Integrated Mountain Development (ICIMOD), Nepal, to mitigate the hazards of glacier lake outburst flood. The project will also help in assessing the environmental conditions of the high mountainous regions. The expected outputs of the proposed study are (1) an inventory of existing glacier lakes along the Hindu Kush Himalaya (2) monitoring of potential risk lakes for draining, and (3) an operational early warning mechanism for GLOF hazards. Similar activities are on their way in other parts of the world.

References

Alean, J. 1985. Ice avalanches: some empirical information about their formation and reach. Journal of Glaciology 31: 324–333.

Allen, S.K., D. Schneider and I.F. Owens. 2009. First approaches towards modelling glacial hazards in the Mount Cook region of New Zealand's Southern Alps. Natural Hazards and Earth System Sciences 9(2): 481–499.

Armstrong, R.L. 2010. The Glaciers of the Hindu Kush-Himalayan Region. A summary of the science regarding glacier melt/retreat in the Himalayan, Hindu Kush, Karakoram, Pamir, and Tien Shan mountain ranges. Technical Paper. Kathmandu. ICIMOD and USAID.

Bajracharya, S.R., P.K. Mool and B.R. Shrestha (Compilers for ICIMOD). 2007. Impact of Climate Change on Himalayan Glaciers and Glacial Lakes. Case Studies on GLOF and Associated Hazard in Nepal and Bhutan. ICIMOD in cooperation with UNEP Regional Office Asia and the Pacific (UNEP/ROAP).

Bamber, J. 2012. Climate change: Shrinking glaciers under scrutiny. Nature 482: 482–483 (23 February 2012). http://www.nature.com/nature/journal/v482/n7386/full/nature10948.html (accessed December 2012).

Berthier, E., H. Vadon, D. Baratoux, Y. Arnaud, C. Vincent, K.L. Feigl, F. Rémy and B. Legrésy. 2005. Surface motion of mountain glaciers derived from satellite optical imagery. Remote Sensing of Environment 95(1): 14–28.

Bolch, T. 2007. Climate change and glacier retreat in northern Tien Shan (Kazakhstan/Kyrgyzstan) using remote sensing data. Global and Planetary Change 56: 1–12.

Bolch, T., M.F. Buchroithner, S.R. Bajracharya, J. Peters and M. Baessler. 2008. Identification of glacier motion and potentially dangerous glacier lakes at Mt. Everest area/Nepal using spaceborne imagery. Natural Hazards and Earth System Sciences 8(6): 1329–1340.

Bolch, T., M.F. Buchroithner, A. Kunert and U. Kamp. 2007. Automated delineation of debris-covered glaciers based on ASTER data. pp. 403–410. *In*: M.A. Gomarasca (ed.). GeoInformation in Europe (= Proc. 27th EARSeLSymposium 2007, Bolzano, Italy). Millpress, Netherlands.

Bolch, T. and U. Kamp. 2006. Glacier mapping in high mountains using DEMs, Landsat and ASTER data. Grazer Schriften der Geographie und Raumforschung. Proceedings 8th Int. Symp. high mountain remote sensing cartography, Vol. 41, March 2005, La Paz, Bolivia, 13–24.

Bolch, T., A. Kulkarni, A. Kääb, C. Huggel, F. Paul, J.G. Cogley, H. Frey, J.S. Kargel, K. Fujita, M. Scheel, S. Bajracharya and M. Stoffel. 2012. The State and Fate of Himalayan Glaciers. Science 336: 310–314.

Bolch, T., J. Peters, A. Yegorov, B. Pradhan, M.F. Buchroithner and V. Blagoveshchensky. 2011a. Identification of potentially dangerous glacial lakes in the Northern Tian Shan. Natural Hazards 59: 1691–1714.

Bolch, T., T. Pieczonka and D.I. Benn. 2011b. Multi-decadal mass loss of glaciers in the Everest area (Nepal, Himalaya) derived from stereo imagery. Cryosphere 5: 349–358.

Bolch, T., T. Yao, S. Kang, M.F. Buchroithner, D. Scherer, F. Maussion, E. Huintjes and C. Schneider. 2010. A glacier inventory for the western Nyainqentanglha Range and Nam Co Basin, Tibet, and glacier changes 1976–2009. The Cryosphere 4: 419–433.

Buchroithner, M.F. 1984a. Identification of Extensive Flash Flood Erosions in Remote Areas Using Landsat MSS and Metric Camera Data.—Proc. 1984 World Conf. Remote Sensing, Bayreuth (FRG), October 1984, Bayreuth.

Buchroithner, M.F. 1984b. Geological mapping of remote mountainous regions using metric camera imagery. Initial Experiences with Photogrammetric Space Images. Mitt. Österr. Geol. Ges. 77, 1984, 115–149.

Buchroithner, M.F. 1985. Thematic Mapping and Erosion Monitoring with Metric Camera Imagery Using Computer-Aided Methods. SSS Cerma Internat. Conf. Series, Proceedings Workshop Remote Sensing & Geogr. Info. Systems, Washington D.C., pp. 3-1–3-10.

Buchroithner, M.F. 1987. Geologische Kartierung entlegener Gebirgsregionen mittels MetricCamera-Aufnahmen. *In*: R. Kosta (ed.). Die erderkundende Weltraumphotographie und ihre Anwendung in der Gebirgskartographie. Mitteilungen der Geodät. Institute der Technischen Universität Graz. 57: 117-156.

Buchroithner, M.F. 1996. Jökulhlaup mapping in the Himalaya by means of remote sensing. Kartographische Bausteine, Institute for Cartography, Dresden University of Technology, Germany 12: 75–86.

Buchroithner, M.F. and T. Bolch. 2007. An automated method to delineate the ice extension of the debris-covered glaciers at Mt. Everest based on ASTER imagery. *In*: Proceedings of the 9th International Symposium on High Mountain Remote Sensing Cartography, 14–22 September 2006, Graz, Austria, Grazer Schriften der Geographie und Raumforschung 43: 71–78.

Buchroithner, M.F., G. Jentsch and B.Wanivenhaus. 1982. Monitoring of recent geological Events in the Khumbu Area (Himalaya, Nepal) by Digital Processing of Landsat MSS Data. Rock Mechanics 15(4): 181–197.

Cenderelli, D. and E. Wohl. 2001. Peak discharge estimates of glacial-lake outburst floods and "normal" climatic floods in the Mount Everest region, Nepal. Geomorphology 40: 57–90.

Chen, W., H. Fukui, T. Doko and A. Gu. 2012. Improvement of glacial lakes detection under shadow environment using ASTER data in Himalayas, Nepal. Chinese Geographical Science 22: 1–11.

Clague, J. and S.G. Evans. 2000. A review of catastrophic drainage of moraine-dammed lakes in British Columbia. Quaternary Science Reviews 19: 1763–1783.

Christen, M., J. Kowalski and B. Bartelt. 2010. RAMMS: Numerical simulation of dense snow avalanches in three-dimensional terrain. Cold Regions Science and Technology 63: 1–14.

Danninger, H. and H. Posch. 1986. Gletscherseeausbrüche in der Khumbu-Region des Himalaya, Nepal. Österreichische Wasserwirtschaft, 38.

Frey, H., W. Haeberli, A. Linsbauer, C. Huggel and F. Paul. 2010. A multi-level strategy for anticipating future glacier lake formation and associated hazard potentials. Natural Hazards and Earth System Science 10: 339–352.

Fujita, K. and T. Nuimura. 2011. Spatially heterogeneous wastage of Himalayan glaciers. PNAS 108(34): 14011–14014.

Fujita, K., R. Suzuki., T. Nuimura and A. Sakai. 2008. Performance of ASTER and SRTM DEMs, and their potential for assessing glacial lakes in the Lunana region, Bhutan Himalaya. Journal of Glaciology 54(185): 220–228.

Fushimi, H., K. Ikegami, K. Higuchi and K. Shankar. 1985. Nepal case study: catastrophic floods. *In*: G.J. Young (ed.). Techniques for Prediction of Runoff from Glacierized Areas. IAHS Publ., 149: 125–130.

Gardelle, J., Y. Arnaud and E. Berthier. 2011. Contrasted evolution of glacial lakes along the Hindu Kush Himalaya mountain range between 1990 and 2009. Global and Planetary Change 75: 47–55.

Gorbunov, A.P., S.S. Marchenko and E.V. Severskiy. 2004. The thermal environment of blocky materials in the mountains of Central Asia. Permafrost Periglacial Processes 15(1): 95–98.

Gruber, S. 2012. Derivation and analysis of a high-resolution estimate of global permafrost zonation. The Cryosphere 6: 221–233.

Haeberli, W. 1983. Frequency and characteristics of glacier floods in the Swiss Alps. Annals of Glaciology 4: 85–90.

Haeberli, W., J.J. Clague, C. Huggel and A. Kääb. 2010. Hazards from lakes in high-mountain glacier and permafrost regions: climate change effects and process interactions. Proceedings XI Reunión Nacional de Geomorfología, Solsona 2010. Avances de la geomorfología en España 2008–2010.

Haeberli, W., A. Kääb, von der D. Mühl and Ph. Teysseire. 2001. Prevention of outburst floods from periglacial lakes at Grubengletscher, Valais, Swiss Alps. Journal of Glaciology 47/156: 111–122.

Han, Y., P. Huang, L. Li, A.B. Shrestha and S.R. Bajracharya. 2013. Xichang, China: Risk assessment and cartography of flash floods. pp. 56–62. *In*: A.B. Shrestha and S.R. Bajracharya (eds.). Case studies on Flash Flood Risk Management in the Himalayas: In support of specific flash flood policies. Kathmandu: (ICIMOD).

Hewitt, K. and J. Liu. 2010. Ice-dammed lakes and outburst floods, Karakoram Himalaya: historical perspectives on emerging threats Physical Geography 31. 528–551.

Hsu, K.J. 1975. Catastrophic debris streams (Sulzstroms) generated by rockfalls. Geological Society of America Bulletin 86: 129–140.

Huggel, C., A. Kääb, W. Haeberli, P. Teysseire and F. Paul. 2002. Remote sensing based assessment of hazards from glacier lake outbursts: a case study in the Swiss Alps. Canadian Geotechnical Journal 39: 316–330.

Huggel, C., A. Kääb, W. Haeberli and B. Krummenacher. 2003. Regional-scale GIS-models for assessment of hazards from glacier lake outbursts: evaluation and application in the Swiss Alps. Natural Hazards and Earth System Science 3: 647–662.

Huggel, C., W. Haeberli, A. Kääb, D. Bieri and S.D. Richardson. 2004. An assessment procedure for 448 glacial hazards in the Swiss Alps. Canadian Geotechnical Journal 41: 1068–1083.

ICIMOD/GFDRR (Global Facility for Disaster Reduction and Recover)/The World Bank. 2011. Glacial lakes and Glacial Outburst Floods in Nepal. ICIMOD, Kathmandu, 99 pp.

Ives, J.D. 1986. Glacial lake outburst floods and risk engineering in the Himalaya. ICIMOD Occasional Papers, No. 5. Kathmandu, 42 p.

Ives, J.D., R.B. Shrestha and P.K. Mool. 2010. Formation of glacial lakes in the Hindu Kush-Himalayas and GLOF risk assessment. ICIMOD, Kathmandu.

Iwata, S., Y. Ageta, N. Naito, A. Sakai, C. Narama and Karma. 2002. Glacial lakes and their outburst flood assessment in the Bhutan Himalaya. Global Journal of Environmental Research 6(1): 3–17.

Kääb, A. 2005. Combination of SRTM3 and repeat ASTER data for deriving alpine glacier flow velocities in the Bhutan Himalaya. Remote Sensing of Environment 94: 463–474.

Kääb, A., C. Huggel, L. Fischer, S. Guex, F. Paul, I. Roer, N. Salzmann, S. Schlaefli, K. Schmutz, D. Schneider, T. Strozzi and Y. Weidmann. 2005. Remote sensing of glacier- and permafrost-related hazards in high mountains: an overview. Natural Hazards and Earth System Sciences 5: 527–554.

Kahle, E. 1998. Betriebswirtschaftliche Entscheidungen: Lehrbuch zur Einführung betriebswirtschaftlicher Entscheidungstheorie. Munich.

Kaibori, M., K. Sassa and S. Tochiki. 1988. Betrachtung über die Bewegung von Absturzmaterialien. International Symposium Interprevent 1988/07, 2: 227–242.

Kasatkin, N.E. and V.P. Kapista. 2009. Themorainic lakes dynamics in the Ile Alatau. pp. 55–58. *In*: Materials of the Internat. Conf. 'Mitigation of natural hazards in mountain areas', September 2009, Bishkek.

Keller, F. 1992. Automated mapping of mountain permafrost using the program PERMAKART within the geographical information system ARC/INFO. Permafrost Periglacial Processes 3(2): 133–138.

Kostka, R. 1987. Die Aussagekraft für glaziologische Zwecke. *In*: R. Kosta (ed.). Die erderkundende Weltraumphotographie und ihre Anwendung in der Gebirgskartographie. Mitteilungen der Geodät. Institute der Technischen Universität Graz. 57: 157–169.

Leprince, S., S. Barbot and F. Ayoub. 2007. Automatic and precise orthorectification, coregistration, and subpixel correlation of satellite images, application to ground deformation measurements. IEEE Transactions on Geosciences and Remote Sensing 45(6): 1529–1558.

Lliboutry, L. 1977. Glaciological problems set by the control of dangerous lakes in Cordillera Blanca, Peru. II. Movement of a covered glacier embedded within a rock glacier. Journal of Glaciology 18(79): 255–274.

Lliboutry, L., B.M. Arnao, A. Pautre and B. Schneider. 1977a. Glaciological problems set by the control of dangerous lakes in Cordillera Blanca, Peru. I. Historical failures of morainic dams, their causes and prevention. Journal of Glaciology 18(79): 239–254.

Lliboutry, L., B.M. Arnao and B. Schneider. 1977b. Glaciological problems set by the control of dangerous lakes in Cordillera Blanca, Peru. III. Study of moraines and mass balances at Safuna. Journal of Glaciology 18(79): 275–290.

Ma, D., J. Tu, P. Cui and R. Lu. 2004. Approach to mountain hazards in Tibet, China. Journal of Mountain Science 1(2): 143–154.

Marchenko, S. 2001. A Model of Permafrost Formation and Occurrences in the Intracontinental Mountains. Norsk Geografisk Tidskrift - Norwegian Journal of Geography 55: 230–234.

Marchenko, S.S., A.P. Gorbunov and V.E. Romanovsky. 2007. Permafrost warming in the Tien Shan Mountains, Central Asia. Global and Planetary Change 56(3–4): 311–327.

Mergili, M. and J.F. Schneider. 2011. Regional-scale analysis of lake outburst hazards in the southwestern Pamir, Tajikistan, based on remote sensing and GIS. Natural Hazards and Earth System Science 11: 1447–1462.

Mergili, M., D. Schneider, R. Worni and J. Schneider. 2011. Glacial lake outburst floods in the Pamir of Tajikistan: challenges in prediction and modelling. *In*: 5th International Conference on Debris-Flow Hazards Mitigation: Mechanics, Prediction and Assessment. Padova, Italy, 14–17 June 2011.

Narama, C., M. Duishunakunov, A. Kääb, M. Daiyrov and K. Abdrakhmatov. 2010. The 24 July 2008 outburst flood at the western Zyndan glacier lake and recent regional changes in glacier lakes of the Teskey Ala-Too Range, Tien Shan, Kyrgyzstan. Natural Hazards and Earth System Sciences 10: 647–659.

O'Brien, J., P. Julien and W. Fullerton. 1993. Two-dimensional water flood and mudflow simulation. Journal of Hydraulic Engineering 119: 244–261.

Paul, F., A. Kääb, M. Maisch, T. Kellenberger and W. Haeberli. 2002. The new remote sensing derived Swiss glacier inventory: I. Methods. Annals of Glaciology 34: 355–361.

Peters, J. 2009. Identifizierung und Kartierung potenziell gefährlicher Gletscherseen im nördlichen Tien Shan auf Basis von multi-temporalen Fernerkundungsdaten. Unpubl. Diploma Thesis, Institute for Cartography, Dresden University of Technology.

Popov, N.V. 1988. Die Kontrolle gefährlicher Gletscherseen im nördlichen Tienschan. *In*: Proceedings of the International Symposium Interpraevent. 4: 29–41.

Popov, N.V. 1991. Assessment of glacial debris flow hazard in the North Tien-Shan. *In*: Proceedings of the Soviet-China-Japan Symposium and field workshop on natural disasters, 2–17 Sept. 1991, 384–391.

Quincey, D.J., S.D. Richardson, A. Luckman, R.M. Lucas, J.M. Reynolds, M.J. Hambrey and N.F. Glasser. 2007. Early recognition of glacial lake hazards in the Himalaya using remote sensing datasets. Global and Planetary Change 56(1–2): 137–152.

Reynolds, J.M. 2000. On the formation of supraglacial lakes on debris-covered glaciers. IAHS Publication, Vol. 264 (= Debris-covered Glaciers), 153–161.

Richardson, S.D. and J.M. Reynolds. 2000. An overview of glacial hazards in the Himalayas. Quaternary International 65/66(1): 31–47.

Salzmann, N., A. Kääb, C. Huggel, B. Allgöwer and W. Haeberli. 2004. Assessment of the hazard potential of ice avalanches using remote sensing and GIS-modelling. Norsk Geografisk Tidskrift 58: 74–84.

Scheidegger, A.E. 1975. Physical Aspects of Natural Catastrophes. Elsevier Sci. Publ., Amsterdam, Oxford, New York, 289 p.

Scherler, D., S. Leprince and M.R. Strecker. 2008. Glacier-surface velocities in alpine terrain from optical satellite imagery—accuracy improvement and quality assessment. Remote Sensing Environment 112(10): 3806–3819.

Shrestha, A.B. and S.R. Bajracharya (eds.). 2013. Case Studies on Flash Flood Risk Management in the Himalayas: In support of specific flash flood policies. Kathmandu: (ICIMOD), 64 p.

Tokmagambetov, T.G. 2009. The moraine-dammed glacial lakes current state in the Iliy Alatau. pp. 82–83. *In*: Materials of the International conference on "Mitigation of natural hazards in mountain areas", September 2009, Bishkek.

UNEP RRC (n.d.) Inventory of Glaciers, Glacial Lakes and Identification of Potential Glacial Lake Outburst Floods, Affected by Global Warming in the Mountains of Himalayan Region (http://www.rrcap.ait.asia/issues/glof/; accessed December 2012).

Vuichard, D. and M. Zimmermann. 1986. The Langmoche flashflood, Khumbu Himal, Nepal. Mountain Research and Development 6(1): 90–93.

Vuichard, D. and M. Zimmermann. 1987. The 1985 catastrophic drainage of a moraine-dammed lake, Khumbu Himal, Nepal: Cause and consequences. Mountain Research and Development 7(?): 91–110.

Wang, X., S. Liu, Y. Ding, W. Guo, Z. Jiang, J. Lin and Y. Han. 2012. An approach for estimating the breach probabilities of moraine-dammed lakes in the Chinese Himalayas using remote-sensing data. Natural Hazards and Earth System Science 12: 3109–3122.

Wang, W., T. Yao, Y. Gao, X. Yang and D.B. Kattel. 2011. A first-order method to identify potentially dangerous glacial lakes in a region of the Southeastern Tibetan Plateau. Mountain Research and Development 31: 122–130.

Watanabe, T., D. Lamsal and J.D. Ives. 2009. Evaluating the growth characteristics of a glacial lake and its degree of danger of outburst flooding: Imja Glacier, Khumbu Himal, Nepal. Norsk Geografisk Tidskrift—Norwegian Journal of Geography 63: 255–267.

WGMS [Zemp, M., I. Roer, A. Kääb, F. Paul, M. Hoelzle and W. Haeberli]. 2008. Global glacier changes: Facts and figures. Geneva, 88 pp.

Woerd, van der J., L.A. Owen, P. Tapponnier, X. Xu, F. Kervyn, R.C. Finke and P.L. Barnard. 2004. Giant, M8 earthquake-triggered ice avalanches in the eastern Kunlun Shan, northern Tibet: characteristics, nature and dynamics. Geological Society of America Bulletin 116(3): 394–406.

Yao, T.D., L.G. Thompson, V. Moosbrugger, Y.M. Ma, F. Zhang, X.X. Yang and D. Joswiak. 2011. UNESCO-SCOPE-UNEP Policy Briefs Series. Third Pole Environment. June 2011. UNESCO-SCOPE-UNEP, Paris.

15

Impact of Socioeconomic Factors and Environmental Changes on Sri Lanka's Central Highlands

Jürgen Breuste[1,]* and *Lalitha Dissanayake*[2]

INTRODUCTION

Central Highlands of Sri Lanka occupy a unique position among the main geographical zones of the country. It is an area elevated 300 m above the mean sea level occupying about 17% of the country's land area (Wickramagamage 1990). Though it is small in land extent, being located within the country, this zone has a diverse blend of most of the world's climatic features. The Central Highland area is also the watershed for 103 main rivers and more than 1,000 feeder streams joining the main rivers (Madduma Bandara 2000). This area is the heart of the entire country because of its important ecological conditions and as a driver of the economy for the whole country (Fig. 15.1). Although a small area of the Sri Lankan Highlands, it is of high importance for the country regarding both an ecological and an economic point of view. It has several unique qualities for ecosystem services related

---

[1] Department of Geography & Geology, University of Salzburg, Austria, Hellbrunner Straße 34: A 5020 Salzburg.
  Email: juergen.breuste@sbg.ac.at
[2] Department of Geography, University of Peradeniya, Sri Lanka.
  Email: dissanayakedml2011@gmail.com
* Corresponding author

Central Highlands of Sri Lanka

Figure 15.1. Location of the Central Highlands of Sri Lanka.

to national and international importance, which is one of the hotspots of biodiversity in the world. UNESCO has recognized three areas as UNESCO world heritage natural sites. The cultural importance of the Sri Lankan highlands is extremely high; before colonization it was the core area of the national kingdom for centuries and the tooth relic of Buddha placed in that palace is the most pretigious place for Buddhist people in the world. On the other hand, the management of the sensitive ecosystems of the Sri Lankan highlands is extremely weak and destructive. Due to climatic changes, deforestation has taken place since colonization, with population growth and urbanization, water mismanagement, etc. the present situation of the highland had been drastically changed. A strategy for sustainable utilization together with the preservation of natural potential must be developed and implemented in order to preserve the values of the unique tropical highlands.

Values of the Sri Lankan Highlands

Rich in biodiversity

Highlands of Sri Lanka have been recognized as one of the most beautiful tropical highland areas in the world. The World Heritage Committee, held its 34th session in Brasília, in 2010 proclaimed this area as one of the World Heritage Sites. The site comprises the Peak Wilderness Protected Area, the Horton Plains National Park and the Knuckles Conservation Forest. These are the rain forests at an elevation of 2,500 m (8,200 ft.) above sea level. They is rich in biodiversity: more than half of Sri Lanka's endemic vertebrates, half of the country's endemic flowering plants and more than 34% of its endemic trees, shrubs, and herbs are restricted to these diverse mountain rain forests and adjoining grassland areas (UNESCO 2012).

These Highlands are richly endowed with biological resources manifested in a wide range of ecosystems, such as montane forests, evergreen forests (in the lower parts of the highlands), inland wetlands, savanna grasslands and riparian ecosystems (Gunatillake et al. 2008, ME and NR 2007). Besides UNESCO, Conservation International (CI) environment activist group has identified this area as one of 25 biodiversity hotspots in the world. Indeed, the wet zone and highland consists a quarter of Sri Lanka's territory, containing, 88% of the flowering plants of the island and 95% of its angiosperm endemics. There are more than 450 known bird species in the hotspot, of which about 35 are endemic. More than 20 species are endemic to Sri Lanka, mostly are from the lowland rainforests and montane forests of the island's southwestern region. Both the Western Ghats in India and the island of Sri Lanka are considered as Endemic Bird Areas by Birdlife International ([CI] Conservation International 2012).

Kandy is the second largest city with in Central Highland, Sri Lanka and its history extends far beyond the colonial period. Most of the British governors preferred Kandy's cool climate for their local residences. The British governor Sir Edward Barnes (1824–1831) was known for encouraging human settlements in Sri Lanka's Central Highlands: its climate is close to the climate of Britain and it was considered as the second home for the British. Nuwara Eliya *Barnes Hall*, for example, which is an exclusive hotel today, was originally built as a recreation and hunting post. The British Governor William Gregory further developed the outpost Nuwara

Eliya town in Sri Lanka's Central Highlands by adding several attractions and landscaping from 1872 to 1877. To reach the highlands, several roads and road networks were built in the British colonial time: Colombo—Kandy road from Peradeniya via Ramboda; Colombo—Avissawella road passing Ginigathhena Haton, Thalawakale and via Nanuoya Badulla—Bandarawella road via Walimada to Hakgala, from Colombo—Kandy road passing, Hguranketha, Rikillagaskada, Padiyapalella, Ragala via Kadapola. Within the road network settlements were developed as urban centers in the Central Highlands of Sri Lanka (GoldenSriLanka.com 2011). As a result the highland became the most famous place for the tourism both for international and the locals. Tourism is one of the most important drivers of growth and development in the Sri Lankan economy and is a key focus in the governmental development strategy (Ranasinghe and Deshyapriya 2010).

Importance of agriculture

Agriculture in Sri Lanka can be categorized into two groups on the basis of its economic value: subsistence agriculture (for domestic consumption) and commercial agriculture. In recent history, subsistence agriculture started during the Kandy Kingdom. Since then, with the increasing population, cultivation of paddy and garden crops was increased. During the Portuguese period in the 16th century, migration from low land and settlement expansion processes, and the Chena cultivation these human influences had less impact on the highland. However, land-use patterns were environmentally friendly and the old terraced paddy fields and the Kandyan forest garden systems have prevailed up to the present. Forests were cleared for Chena farming (shifting cultivation), and the land was allowed to regenerate by allowing natural vegetation to grow in such locations without disturbance for a long period of time and the impact level was minimal (Wickramagamage 1990). During the Dutch period, cinnamon was popular in the lower lying areas of the highlands. Large scale commercial agriculture for coffee plantation was started by the British. Coffee from Sri Lanka enjoyed premium prices on the world market. By 1869, the numbers of coffee crop owners were approximately 1,700, but the production dropped rapidly due to a disease set in and every effort to revive coffee production failed. Coffee had to be replaced by tea plantations and the British had early access to vast extents of land while tea fetched high prices in the world market. The climatic conditions were very favorable, massive labor forces were needed and cheap labor was imported from southern India and tea production reached a maximum level. Tea production was started in 1873 and it already amounted to 81.3 tons in 1880. In 1890, it reached 22,899 tons and in 1927 it increased up to 10,000 tons (Holsinger 2002). From that period to up to date Sri Lankan tea occupies a special place in the world market today (Fig. 15.2).

In Sri Lanka, except for tea, the Central Highlands at an elevation over 600 m are exploited mainly for potato and vegetable production. The land area under potato and vegetable cultivation is around 60,000 ha, which is comparatively low compared to the area under tea cultivation (188,966.4 ha) (Perera and Jayasuriya 2008, Holsinger 2002). Therefore, potato and vegetables are cultivated in the region on an intensive and commercial scale. Potato, carrots, capsicum and other vegetables have become common in the highlands during the last two to three decades. The Sri Lankan Highlands also contribute more than 25% of the livestock sector. The livestock sector contributes around 1.2% to the national GDP and livestock primarily provides a crucial source of high quality protein. In addition, cattle and buffalo are the primary source of renewable and draught power for a variety of agricultural operations and transport (Perera and Jayasuriya 2008) (Fig. 15.3).

Figure15.2. Mackwood's Tea plantation Labokele (Breuste 2010).

Figure 15.3. Vegetable production near Nuwara Eliya (Breuste 2010).

Drivers of Change in Central Highlands of Sri Lanka

Climate change

According to current predictions for Sri Lanka, the effects of climate change by 2050 will be marginal, reaching only +0.50°C temperature increase and +5% evaporation/rainfall (wet season only) in the large scenario (Yogaratnam 2010). Central Highlands studies have also revealed that the amount of rainfall on the eastern and western slope of the area increases with the altitude to a maximum at a height of about 1,000 m; further up it decreases (Nianthi 2005). These trends show that within the average, the intensity of dry weather and rainfall may increase. Therefore, climate change could have increasingly significant effects in the scenario for 2070. Studies on weather patterns and crop yields for the past years (Ratnasiri et al. 2008, Yogaratnam 2010) have shown that drought affects tea by reducing the yields. Direct impacts will result from increased carbon dioxide levels, which affect photosynthesis and rising temperatures which, in turn, cause heat stress and increased evapotranspiration in crops. Indirect impacts will result from changes in moisture levels, an increased incidence of pests and growing spoilage of agro-products as a result of enhanced microbial activity. These effects could result in reduced yields and shifts in productivity. On the other side, irregular patterns of rainfall and high seasonal concentrations in Sri Lanka's Central Highlands, with attendant increases in run-off ratios, could result in soil erosion, land degradation and the loss of productivity of plantation crops. Hydropower generation in Sri Lanka in June 2012 dropped by 14.8% 323 GWH due to receding water levels of hydropower reservoirs due to less rainfall (Abeywicreama 2012). Therefore, these issues may need to be addressed more through laws and regulations, while at the same time taking into account developmental needs. In this context, balancing enforcement measures with awareness and training through assistance programs appears to be a big challenge for the relevant authorities (Ministry of Environment 2010).

Deforestation

Sri Lanka has a striking variety of forest types brought about by spatial variations in rainfall, altitude, and soil. The forests have been categorized; wet sub-montane forests (at elevations between 1,000–1,500 m in the wet zone); wet montane forests (at elevations of 1,500–2,500 m); with reverence vegetation along river banks; the wet lowland forests transform into sub-montane and montane forests (UNESCO 2012).

The present natural forest cover of Sri Lanka is a little less than 25% of the total land area (Ministry of Forestry and Environment 1999). In the 1880s, the forest cover was estimated to be still around 80% of the country's area. However, during the colonial period, mostly tropical montane forest and tropical moist evergreen forest cover rapidly decreased. By the time the British left the island in 1948, the forest cover was down to about 54 to 50%. In fact this trend has accelerated after independence; largely tropical lowland wet evergreen, tropical dry mixed evergreen and riverine forest have continually been reduced from about 44 to 24% from 1956 to 1992 due to resettlement programs. Considerable areas of the forest were also submerged by the hydropower reservoirs that were constructed during the past three decades under the Accelerated Mahaweli Project (Ministry of Forestry and Environment 1999). The average rate of deforestation between 1956 and 1992 was approximately one km^2 per day (Hewawasam 2010). After 1992, large parts of the forest patches in the Central Highlands were cleared to expand towns and villages and to develop infrastructure facilities (Fig. 15.4, Table 15.1).

Figure 15.4. Recent settlement extensions near Nuwara Eliya (Breuste 2010).

Table 15.1. Forest cover in Sri Lanka.

Year	Percentage
1881	84
1900	70
1956	44
1983	27
1992	24

Source: (Wicramagamage 1990, MF and E 1999, Ratnayake et al. 2011).

Water mismanagement and soil degradation

Water management schemes have changed the natural flow regimes of the major rivers, resulting in sometimes higher flood flows, lower dry-season flows, and the degradation of riparian and wetland habitats in the lower catchment. Downstream water quality in the lower catchment area is being degraded as a result of nutrient loading and pesticides from farm runoff. Sediments and agrochemicals in runoff water lead to eutrophication of the reservoirs. Chemical inputs to the agricultural systems in the upper and lower catchment areas are degrading local surface and groundwater quality.

The increase in the cultivation of annual crops in contrast to perennials in hilly regions also causes damage to roads through increased erosion and rainfall run-off ratios (Piyasiri 2009). The loss of multipurpose reservoir capacity as a result of sedimentation is a major off-site effect. This loss has a major adverse impact on power generation and on the irrigation of agricultural land in the dry zone outside Sri Lanka's Central Highlands.

Land degradation has been recognized as the most serious environmental problem especially in the highlands of the country. A high population density, presently over 300 per km, and a lack of off-farm livelihood opportunities has led to excessive highland exploitation (Ministry of Environment 2010). In the higher elevations in Sri Lanka's Central Highlands, particularly due to the cultivation of tobacco, potato, and vegetable crops and the construction of roads and highways, soil erosion has been taking place to a considerable extent. The degrees of soil erosion extend up to massive land sides. When the same land is planted continuously with the same crop, soil fertility tends to gradually decrease while the application of fertilizes and agro-chemicals further challenge the soil. Due to tillage and pulverizing soil clods, sudden heavy rains, soil erosion, landslides, and the deposition of clay and silt, the reservoirs designed to supply water to hydropower in the region are under great risk. According to the H.R. Wallinford Limited completed hydrographic survey (Hewawasam 2010), sedimentation rates in small reservoirs of the Upper Mahaweli catchment for the period from 1985 to 1993 in the Polgolla barrage has reduced its storage capacity to 56% over a 17 year period. Rantambe reservoir has reduced its storage capacity to 72% during a three year period after impoundment (Table 15.2).

Table 15.2. Original capacity, surveyed capacity and percentage of loss of original capacity due to siltation of hydropower reservoirs impounded in the upper Mahaweli Catchment.

Name of the reservior	Year of the impoundment	Original capacity (Million m³)	Year of the hydrographic survey	Surveyed capacity (Million m³)	The loss of original capacity due to siltation
Kotmale	1985	176.770	1990	184.640	-
Polgolla	1976	5.271	1993	2.794	44%
Victoria	1985	717.530	1993	713.080	01%
Rantembe	1991	10.950	1994	7.900	28%

Source: Wallingford 1995.

Population growth and urbanization

Sri Lanka's Central Highlands mainly cover four administrative districts: Kandy, Nuwara Eliya, Matale, and Badulla. The total population of these four main districts is around 3.3 million and its population density is around

395 per km² (Population Survey 2011). The high population density and sustained efforts to improve living standards have exerted tremendous pressure on the natural environment of the highlands. Unplanned urban population growth has exerted pressure on the Central Highlands and water resources in the cities as well as peripherals with impacts on sewage disposal, waste management and environmentally related health problems. The high proportion of the poor and the growing population, combined with an unequal distribution of benefits from natural resources, make a sustainable development in the Sri Lankan Highlands quite a challenging task. According to the independent evaluation group study on the Mahaweli Development Program, a substantial rudiment of the population, who disliked leaving their original villages, migrated to the higher elevation areas. This worsened the living standard there as they only have access to miniscule land holdings prone to environmental risk.

Multipurpose development program

Due to the geographical configuration with a rain fed central hill zone, the upper catchment of all the major rivers of Sri Lanka are situated and the area enjoys a high hydropower potential. In 1968, to get the best benefits from such a massive water resource, a major, multi-purpose development plan named Mahaweli Development Program (MDP) (Fig. 15.5) was initiated with the help of UNDP and FAO funding and was expected to cover a period of 30 years (Peiris 2006). The Mahaweli River, Sri Lanka's largest river (325 km) with an annual discharge of 7,650 million m³ also has by far the largest catchment area (10,327 km²) covering one sixth of the country (NSF 2000). The Mahaweli River Development Program is the largest integrated rural development multi-purpose program ever undertaken in Sri Lanka and was based on the water resources of Mahaweli and six allied river basins. The main objectives were to increase agricultural production, hydropower generation, employment opportunities, settlement of landless poor and flood control. The program, originally planned for the implementation over a 30-year period, was brought to acceleration in 1979, with incorporation of Mahaweli Authority (The World Bank 2012). At present, the country gains benefit from this project and major hydropower potential will be fully developed in the Upper Mahaweli Catchment (UMC) mainly to generate hydropower, which contributes to about 40–50% of the total hydropower production in the country and also sustains 90% paddy and other crops cultivation in the low land of Sri Lanka (Tolisano 1993).

ACCELERATED MAHAWELI DEVELOPMENT PROJECT AREA

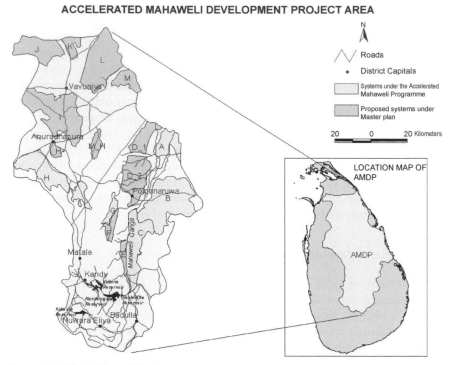

Figure 15.5. The Mahaweli Project.

Impacts of Central Highlands Sri Lanka

Disturbances of the ecologically sensitive areas in Central Highlands in Sri Lanka

Although Highlands of Sri Lanka have an extraordinarily attractive scenery with a very high value of biodiversity, nature conservation receives hardly any attention and is often disappointing (Tolisano et al. 1993, IEG 2012). Due to encroachment on the environmentally sensitive areas, biodiversity is at risk. Cultivation and construction activities along hill slopes are also risky (Gunatillake 2008, Wickramagamage 1990).

Exploiting the environmentally sensitive locations has a history going back about 15,000 year. However, during the early stages, when land was used for food production and residential purposes, these disturbances were relatively low. The upper catchments areas of the major rivers are situated in the Highlands of Sri Lanka were stripped of the natural vegetation to make way for the plantation agriculture during the British colonial period. The plantation area grew from 19 km² to 2,500 km² within less than a century without concerning the natural equilibrium of the watershed (Wickramagamage 1990). The land value of forested land per acre was significantly lower (13–65 pounds sterling) compared to the value of areas cleared for cultivation (100–500 pounds sterling). The consequences were immediately visible in the form of landslides, heavy soil erosion, soil fertility decline and reduction in crop yields siltation of low lying areas, frequent flooding, drying out of streams, etc. During the British colonial period Highlands of Sri Lanka were also used to provide for plantations and roads, railway tracks, holiday homes, etc. There was not enough concern about the ecological role of forests for water balance, soil fertility, erosion prevention, and as habitat. British hunters killed exceedingly large numbers of wild animals, which led to the threat of several species, e.g., the Sri Lankan elephant. A single hunter reported that he alone had killed more than 6,000 wild elephants (Baker 1853). The third stage of invading the environmentally sensitive Central Highlands started after Sri Lanka's independence in 1948. The multipurpose Mahaweli Development program was part of this stage. The attention to ecosystem and biodiversity conservation in the submerged areas and to the impact on dried up areas was inadequate. According to the report An Environmental Evaluation of the Accelerated Mahaweli Development Program, the current data on wildlife population characteristics, migration and dispersal patterns, and habitat requirements are not sufficient to guide management decision making. Habitat conditions in many of the designated protected areas were degraded over time, and restoration or enrichment measures were insufficient to support the existing wildlife populations. In addition to that some environmental and cultural values were destroyed (Tolisano et al. 1993, IEG 2012, Bulankulama 1992). Before the Mahaweli project started there were 257 bird species, 50 mammalian species, and more than 20 reptile species in addition to amphibians and some others. Most of them are extinct now due to the destruction of lower lying areas of the evergreen forest and riparian ecosystems (Tolisano et al. 1993). The inhabitants of the Mahaweli project area were sheltered in newly developed areas, and given a fixed extent of land. During heavy rain periods, they were exposed to diseases spread by mosquitoes. Increased population pressure lead to the exploitation of remaining forest areas for shifting cultivation or for attaining firewood (IEG 2012) and the day to day problems they face are created due to substantial tension and stress (Tolisano et al. 1993, Furset 1994) (Fig. 15.6).

Figure 15.6. Utilization of the sensitive hill slopes (Breuste 2010).

Cultivation and construction activities along hill slopes

The process of soil erosion in Sri Lanka began in the 19th century with the expansion of human settlements and the cultivation of upland rain fed crops. It was aggravated by the changes in land-use patterns during the British administration. The upper catchment areas of major rivers are located in the Central Highlands of Sri Lanka were stripped of natural vegetation to make way for plantation agriculture such as coffee and tea. Land clearing continued even after independence primarily for the establishment of human settlements and for agriculture. During the past five decades, land under human settlements has doubled, while the land brought under crops other than tea, rubber, coconut, and paddy has increased by 250% (Yogaratnam 2010). On the other hand, total land under forests and wildlife and nature reserves has declined by 40%, while land taken up by tea plantations and rubber has fallen by 35 and 25% respectively (Hewawasam 2010). In this situation on-site soil loss rates, particularly in the upper catchment area continue to be greater than soil replacement rates. Other causes are depletion of soil nutrients, damage to physical and chemical properties of the soil, and the reduction in the soils' capacity to retain moisture. Cropping patterns are often inappropriate for soil and microclimatic conditions of the site;

productivity result is reduced and increased reliance on environmentally inappropriate agricultural practices as well. Due to all these causes, massive landslides occur during the rainy seasons. According to the records of the Road Development Authority (Yogaratnam 2010), the total cost of maintaining 1 km of A and B class roads in Nuwara-Eliya District of Central Highlands has increased by almost 350% during the past five years (Yogaratnam 2010). According to studies conducted by Senanayake on factors associated with the occurrence of landslides in Sri Lanka, human activities in the cultivation of tea, rubber, coconut, paddy and vegetables contribute to 35, 20, 10, 13, and 8% respectively (Senanayake 1993, cited by Hewawasam 2010) (Table 15.3).

Table 15.3. Extent of different land use types and estimated soil erosion rates in the UMC (Upper Mahaweli Catchment).

Land use type	Area (km²)	Soil loss (t km⁻² y⁻¹)	Bedrock erosion rate[1] (mm ky⁻¹)
Dense forest	356.6	100	37
Degraded forest and scrubs	435.7	2500	925
Degraded grasslands	141.9	3000	1110
Poorly managed seedling tea	3454.8	5200	1924
Seedling tea with some conservation	252.7	1500	555
Vegetative-propagated tea	114.9	200	74
Paddy	285.7	300	111
Home gardens	537.7	100	37
Shifting cultivation and tobacco	484.6	7000	2590
Market gardens	163.6	2500	925

[1]Converted into corresponding bedrock erosion considering density as 2.7 g per cm³
Source: Hewawasam 2010

Water stream mismanagement

Many rivers and streams flow through catchment areas that are densely populated regions in Central Highlands and are negatively impacted by urbanization. The most consistent and pervasive effects are channeling due to settlements, solid waste mismanagement, stream pollution, and an increase in water discharge. In Akurana at Pinga Oya River (Tributary of Mahaweli), the stream is being channelized nearly 3 km without proper planning and solid waste issues create massive risks (Dissanayake 2002, Dissanayake 2009, Mahees 2009). In Geli Oya at Mahaweli River, similar effects as in Pinga Oya and in Meda Ela have been registered (Dayawansa 2008), where the effects of urbanization described above brought about changes to the stream channel width, the water quality and the stream load. Also, the river banks were smoothened by the construction of concrete walls.

The reduction of the channel width has reduced the channel capacity; the floral coverage and diversity in the stream corridors shows a decreasing pattern towards the town resulting in an increased risk of flooding. Meda Ela and Akurana experienced unprecedented floods in the recent past as a direct result of the stream channel encroachment for the construction of residential and commercial buildings. Recent severe floods caused damage to the properties of Akurana. The river flows into the stagnant water body of the Polgolla barrage resulting in the aggradations of the lower reaches. Increased garbage and environmental pollution are major issues in the urban environments of the Sri Lankan Highlands. Kandy, Matale and Nuwara Eliya are some of the municipalities which are suffering from increased garbage pollution owing to a lack of proper dumping or recycling methods. As of today, infrastructure for garbage collection is lacking in most municipal areas. This has increased uncontrolled scattering and dumping of garbage everywhere in the urban and suburban areas in the highlands, as there is a high potential of water pollution threats due to garbage accumulation. For example, solid wastes getting into water ways at the higher watershed areas lead to serious situations like overflow floods and reduction of water storage capacity of reservoirs associated with hydro power generation. Also, the occurrence of water-borne and water associated diseases is increasing and reports of dengue epidemics in the central region points out the emerging challenges.

The increased garbage quantity also causes slower water flow in many drainage channels in the Sri Lankan Highlands and provides breeding places for disease vectors such as rats and mosquitoes. Pinga Oya, Nanu Oya, Geli Oya, Kandy and especially Meda-Ela are the best examples of polluted streams due to solid waste. Most of the solid waste transfer points are located close to the most sensitive locations such as water ways, road sides, schools, and so on, which poses several health risks (Dissanayake, unpubl. data 2012). Open dumping sites (e.g., Gohagoda in Kandy) cause pollution of ground and surface-water sources (Fig. 15.7). Open burning of waste without any government regulation is widespread in the country and causes bad smell and air pollution in neighborhoods. It contributes to atmospheric pollution and may cause serious health problems. The river water is also contaminated with fertilizer and pesticides, sewerage, and other types of waste from the residential and commercial establishments located on either side of the channel. This causes the spreading of water-borne diseases (Dissanayake 2009a, Dissanayake 2009b, Piyasiri 2009). Although Sri Lanka has adequate rules and regulations with regard to environment conservation the implementation mechanism has weaknesses that prevent a proper implementation.

Figure 15.7. Mismanagement of solid waste tributary in Mahaweli River (photo: Dissanayake 2012).

Adaptations

Environmental conservation

Environmental degradation is the major issue in Central Highlands of Sri Lanka, therefore individuals, organizations, and government programs are trying to introduce several conservation methods such as planting vegetation, contour plowing, maintaining the soil pH, soil organisms, crop rotation, watering the soil, salinity management, terracing, bordering from indigenous crops, no-tilling farming methods and also home gardens. A home garden is a piece of land around the dwelling with clear boundaries and it has a functional relationship with its occupants related to economic, biophysical, and social aspects (Weerakoon 2011). Kandyan home gardens can be identified in Central Highlands Sri Lanka has a valuable, diversified, and sustainable ecosystem; it exhibits the geometric relationships of trees, light attenuation in canopy layers, multiple functions and interactions that occur in limited areas outside natural forest with human co-relations. The canopy stratification minimizes rain drop impact and soil erosion is reported to be negligible. This special forest management practice, primarily inherited as family resource, and its knowledge transfer of ownership from parents to children, means that management has retained continuity (Halladay and

Gilmour 1995). The history of cultivating trees and crops in home gardens, social tree planting, protecting and managing forests, appreciating wildlife, and sustaining the beauty of nature, especially in the Central Highlands, go back more than 25 centuries (Nianthi 2010). Today, the traditional knowledge of agro-forestry is being developed and expanded with the objective of improving living standards, especially among the rural communities not only in the Central Highlands but also all over the country.

National legal enactments and international agreement

We have to acknowledge the proclamation of the Central Highland of Sri Lanka by the UNESCO and Sri Lanka is bound to preserve it. An important cultural place in the highland is the city of Kandy, which was identified as a sacred Buddhist site that has a long history reaching back more than 2,500 years. This allows a global cooperation in conserving this area. Apart from that, a substantial number of legal enactments and international agreements are available as authorities related to the environment that have made efforts to protect the diversity of Sri Lanka's Central Highlands (Table 15.4).

The substantial number of legal enactments to dispel all undesirable aspects due to political interferences, implementation of imposed legal enactments had been delayed or not done at all. If the legal enactments were enforced, the present problems related to environmental, social, cultural, and economic aspects could have been avoided.

Environment conservation organizations

Sri Lanka has an adequate network of government and non-governmental organizations focusing on environment conservation. According to the Sri Lanka Environmental Journalist Forum (SLEJF), there are 7,000 non-governmental organizations with an interest in environmentally friendly activities. Government institutions directly contributing to environmental protection are the Wild Life Environment and Forest Conservation Department, the Central Environment Authority, the Geological Survey and Mines Bureau, the Wildlife Trust of Sri Lanka and the Mahaweli Authority; they protect the highland environmental resources and values (SLEJF 2004).

Other government authorities and 700 other non-governmental organizations additionally provide services related to influencing, encouraging, and promoting sustainable development. They communicate environmental issues and build awareness among the public not only in Sri Lanka's Central Highlands but also in the whole country.

Table 15.4. Central Highland Protection—Related Regulation and Legal Enactments in Sri Lanka.

1. National Environmental Act No. 47 of 1980. (Amendment) No 56 of 1988/No. 53 of 2000.
2. National Environmental (Upper Kotmale Hydro-Power Project- Monitoring) Regulation No. 1 of 2003. Gazette Notification Number 1283/19 dated 10th April 2003.
3. Regulation for Prohibition of use of Equipment for exploration, mining and extraction of Sand & Gem. Gazette Notification Number 1454/4 dated 17th July 2006.
4. Regulation for Prohibition of use of Cultivation of annual crops in high gradient area. Gazette Notification Number 1456/35 dated 4th August 2006.
5. Order under Sections 24(C) and 24(D) to declare Gregory Lake as an Environmental Protection Area. Gazette Notification Number 1487/10 dated 5th March 2007.
6. Order under Section 24(C) and 24(D) to declare Knuckles Environment Protection Area. No. 1507/9 dated 23rd July 2007.
7. Land Ownership Act of 1840 (by British Rulers).
8. Crown Land Encroachment Ordinance of 1840 (by British Rulers).
9. Irrigation Ordinance of 1856 (by British Rulers).
10. Forest Ordinance of 1907 (by British Rulers).
11. Land Development Ordinance of 1935 (by British Rulers).
12. Soil Conservation Acts No. 25 and 29 of 1951 (General Regulations) and 1953 (Special Regulations).
13. Agrarian Services Act of 1959.
14. Water Resources Act of 1964.
15. State Land Ordinance.
16. National Water Supply and Drainage Board Act of 1974.
17. Land Grant Act of 1979.
18. Mahaweli Authority of Sri Lanka Act No. 23 of 1979.
19. National Environment Act of 1980 amended in 1988.
20. Irrigation Ordinance of 1990.
21. Soil Conservation (amended) Act No. 24 of 1996.
22. Scheduled waste (hazardous waste) management regulations 2008—Gazette Notification No. 1534/18 dated 01.02.2008.
23. Regulation for Prohibition of manufacture of polythene or any product of 20 micron or below thickness. Gazette Notification Number 1466/5 dated 10th October 2006.

Copyright: CEA 2012, Jayakody 2012.

Environmental education

In the recent past, there has been an increasing awareness of environmental education from school to university degree level both in Sri Lanka and in the larger regional setting of South Asia. Also governmental and non-governmental organizations are trying to provide informal environmental education, especially for farmers and households. To reach the global environmental targets, Sri Lanka faces big challenges but also huge opportunities. Managing Sri Lanka's Central Highlands ecologically will be an important contribution to both global and national environmental targets.

Conclusion

According to the above mentioned factors and figures it is evident that the highland of Sri Lanka has a unique value of natural beauty and the impacts of its socioeconomic values. There are several failures which have to overcome with the support of international and local community to preserve the heart of the Central Highlands of Sri Lanka.

References

Abeywickrama, W. 2012. Hydropower Generation in Sri Lanka. Daily News online [online]. <http://www.dailynews.lk/2012/10/10/main_News.asp> (accessed: 10/10/2012).

Baker, S.W. 1853. The Rifle and Hound in Ceylon [e-book]. <http://ebookstore.sony.com/ebook/samuel-w-baker/the-rifle-and-hound-in-ceylon/_/R-400000000000000309274> (accessed: 17/10/2012).

Conservation International. 2012. Western Ghats and Sri Lanka—Species. <http://www.conservation.org/where/priority_areas/hotspots/asia-pacific/western-ghats-and-sri-lanka/pages/biodiversity.aspx> (accessed: 18/11/ 2012).

Dayawansa, N.D.K. 2008. Assessment of changing pattern of the river course and its impact on adjacent riparian areas highlighting the importance of multi-temporal remotely sensed data. Conference on Remote Sensing, Colombo, Sri Lanka.

Deraniyagala, S. 1992. The Prehistory of Sri Lanka; an ecological perspective (Revised ed.), Archaeological Survey Department of Sri Lanka, Colombo, Sri Lanka.

Dissanayake, D.M.L. 2002. Settlement growth land use changes and its consequences in Akurana town, Sri Lanka. M. Phil. Thesis, Norwegian University of Science and Technology, Norway.

Dissanayake, D.M.L. 2009a. Disturbances affecting Stream corridors. The proceedings of the 1st National Geographic Conference, University of Peradeniya, Peradeniya, Sri Lanka.

Dissanayake, C.B. 2009b. The multidisciplinary approach to water quality research. Economic Review 35: 16–22.

Furset, R. 1994. Women's Health behavior in Mahaweli System Sri Lanka. M.S. Thesis, Norwegian University of Science and Technology, Trondheim, Norway.

GoldenSriLanka.com. 2011. Hill Country—Nuwara Eliya. <http://goldensrilanka.com/places/hill-country/> accessed: (18/11/2011).

Gunatilleke, N.R. and S. Pethiyagoda. 2008. Biodiversity in Sri Lanka. Journal of the National Science Council of Sri Lanka, Peradeniya, Sri Lanka 36: 25–62.

Halladay, P. and D.A. Gilmour (eds.). 1995. Conserving Biodiversity outside Protected Areas. The Role of Traditional Agro-ecosystems.IUCN-The world conservation Union, Gland, Switzerland. <http://books.google.at/books?hl=en&lr=&id=DNAhLtU3xzYC&oi=fnd&pg=PA11&dq=Conserving+Biodiversity+outside+Protected+Areas.+The+Role+of+Traditional+Agroecosystems&ots=tZXKw0B5yQ&sig=0wM61gHMhuAfhoRkU0gcIUgn8AM> (accessed: 23/01/2013).

Hewawasam, T. 2010. Effect of land use in upper Mahaweli catchment area on erosion landslides and siltation in hydropower reservoirs Sri Lanka. Journal of the National Science Council of Sri Lanka 38(1): 3–14.

Holsinger, M. 2002. History of Ceylon Tea. <www.historyofceylontea.com/articles/thesis.html> (accessed: 17/01/2012).

MaddumaBandara, C.M. 2000. National Resource of Sri Lanka: Water Resources of Sri Lanka. National Science Foundation, Colombo, Sri Lanka.

IEG. 2012. Sri Lanka: Mahaweli Ganga Development. <http://lnweb90.worldbank.org/oed/oeddoclib.nsf/DocUNIDViewForJavaSearch/0D869807701D1EEE852567F5005D8903> (accessed: 31/10/2012).

Mahees, M.T.M. 2009. Political economy of water pollution in Pinga Oya, Mahaweli River. Symposium Proceedings of the Water Professional Day. Faculty of Agriculture, University of Peradeniya, Peradeniya, Sri Lanka.

[MESL] Ministry of Environment Sri Lanka. 2010. Strategies to combat climate change in Sri Lanka. Ministry of Environment, Colombo, Sri Lanka.

Ministry of Environment. 2010. <http://www.environmentmin.gov.lk/about.htm> (accessed: 27/10/2012).

[ME and NR] Ministry of Environment and Natural Resources. 2007. Sri Lanka strategy for sustainable development. Ministry of Environment and Natural Resources, Colombo, Sri Lanka.

[MF and E] Ministry of Forestry and Environment. 1999. Biodiversity Conservation in Sri Lanka: A Framework for Action. Ministry of Forestry and Environment, Colombo, Sri Lanka.

Nianthi, R. 2010. Climate change adaptation and agroforestry in Sri Lanka. *In*: R. Shaw, J.M. Pulhin and J.J. Pereira (eds.). Climate Change Adaptation and Disaster Risk Reduction: An Asian Perspective. Community, Environment and Disaster Risk Management. Emerald Group Publishing Limited, UK 5: 285–305.

[NSF] National Science Foundation. 2000. Natural Resources of Sri Lanka. National Science Foundation, Colombo, Sri Lanka.

Perera, B.M.A.O. and M.C.N. Jayasuriya. 2008. The dairy industry in Sri Lanka: current status and future directions for a greater role in national development. Journal of National Science Foundation Sri Lanka 36: 115–126.

Peiris, G.H. 2006. Sri Lanka: Challenges of the New Millennium. Kandy books, Kandy, Sri Lanka.

Piyasiri, S. 2009. Surface waters, their status and management. Economic Review 35: 16–22.

Ranasinghe, R. and R. Deshyapriya. 2010. Analyzing the significance of tourism on Sri Lankan Economy; An economic analysis. University of Kelaniya, Kelaniya, Sri Lanka.

Ratnasiri, J. and A. Anandacoomaraswamy. 2008. Climate Change and Vulnerability: Vulnerability of Sri Lankan Tea Plantations to Climate Change. Earthscan, London, UK.

SLEJF. 2004. Sri Lanka Directory of Environmental NGOs. <http://www.environmentaljournalists.org/images/Sri_Lanka_Directory_of_Environmental_NGO.pdf> (accessed. 30/12/2011).

Tolisano, J., P. Abeygunewardene, T. Athukaeala, C. Davis, W. Fleming, I.K. Goonesekara, T. Rusinow, H.D.V.S. Vattala and I.K. Weerewardene. 1993. An environmental evaluation of the accelerated Mahaweli Development Program: Lessons learned and donor opportunities for improved assistance Project report.Dai, Bethesda, USA.

UNESCO. 2012. Central Highlands of Sri Lanka. <http://whc.unesco.org/en/list/1203> (accessed: 31/10/2012).

Wallingford, H.R. 1995. Sedimentation Studies in the Upper Mahaweli Catchment, Sri Lanka. HRWallingford Ltd., Oxon, UK.

Weerakoon, L. 2011. Present Situation of Home Gardens in Sri Lanka. Responses to Home Garden Column. <http://www.island.lk/index.php?page_cat=article-details&page=article-details&code_title=22684> (accessed: 29/12/2011).

Wickramagamage, P. 1990. A man's role in the degradation of soil and water resources in Sri Lanka: A historical perspective. Journal of the National Science Council of Sri Lanka 18(1): 1–16.

Yogaratnam, N. 2010. <http://www.dailynews.lk> (accessed: 29/12/2011).

Yogaratnam, N. 2011. Environmental issues and plantation management. Daily News online: <http://www.dailynews.lk> (accessed 29/12/2011).

16

High Tatra—The Challenges of Natural Disaster Recovery and Complex Changes

Maros Finka* and *Tatiana Kluvankova*

INTRODUCTION: SETTING THE CONTEXT

Natural systems of High Tatras

The High Tatras mountains belong to the most valuable ecosystems in Slovakia, a small Central European country. Nearly 60% of its surface is over 300 m, 15% over 800 m and 1% over 1,500 m above the sea level. Around 40% of the Slovak territory is covered by forest ecosystems creating a valuable part of the nature of Slovakia.

The High Tatras represent the first European cross-border national park, now-a-days one of the nine Slovak national parks along with 14 large scale protect areas and 19 special protection areas of Natura 2000 in Slovakia, creating 36% from total the state area. Despite a large proportion of protected areas, the vulnerability of the ecosystems against increasing extreme climate events (drought, floods, wind), global warming, abandonment of traditional management of meadows and pastures, expansion of invasive plants, fragmentation of habitats, etc. is rather high. Practices in the management

SPECTRA+ Centre of Excellene, Slovak University of Technology in Bratislava Vazovova 5, 81243 Bratislava, Slovak Republic.
   Email: tana@cetip.sk
* Corresponding author: maros.finka@stuba.sk

of nature conservation in Slovakia could be characterized by traditional approaches, with static conservation of ecological values. Its implementation has had a rather negative impact on the stability of valuable ecosystems in certain cases, especially in semi natural or cultural ecosystems requiring active management.

Traditional state environmental policy concerned mostly human health and was largely fragmented and administrated by forest and agriculture authorities. Significant improvement in the general public approach to nature and environment protection in Slovakia has been prevalent after the velvet revolution in 1989. The most important driver of this improvement has been establishment of environmental administration in 1995 and EU integration, in particular the implementation of EU Framework directives in nature conservation. Hovewer, not all aspects can be seen in a purely positive way. The biodiversity governance in Slovakia is subordinated to regional administrations and state nature conservancy, which lack adequate coordination of competencies and tasks. Participation of non state actors on planning and decision making is also a challenge as the absence of an accountability mechanism and practice for non-representative participation is prevalent (Kluvánková–Oravská et al. 2009). Involvement of land owners, land-users of protected land on decision making and protection of the ecosystems and natural disasters prevention strategies is missing, despite the increasing pressure of global climate change that would require coordination of competences and implementation of prevention measures.

As reported in The Fourth National Communication of the Slovak Republic on Climate Change (Szemesova et al. 2007), the average annual temperature of the air increased by about 1.1°C in Slovakia in the last century and the precipitation decreased by about 5.6%. As reported in this document there was a rapid decrease of more than 10% of total precipitation in southern Slovakia and on the other hand the increase up to 3%, in the north and northeast of Slovakia were documented. A significant decrease in the relative humidity of the air of up to 5% and a decrease in snow coverage over the whole territory were recorded. Recently a significant increase of local weather extreme events like local heavy rains and thunderstorms with high daily precitation volume and strong winds have been observed, bringing higher risks of local floods, buildings and forest destructions, etc. Local and regional droughts caused by long periods of relatively warm weather have been recorded in last two decades. The spruce ecosystems have been influenced by the bark beetle expansion as well as fires caused by extremely hot and dry periods (Szemesova et al. 2007).

The High Tatras mountains are one of the smallest alpine range in the world, nevertheless, they are the highest and the only mountain range of alpine system in the Slovak Republic, and the highest part of the Carpathians stretching in a large bow-shape from Vienna-Bratislava metropolitan region in the west, along the borders of Slovakia with Czech Republic, and Poland and the Ukraine to Romania in the southeast.

The territory of the High Tatras spans 260 km², although the length of the main mountain ridge from west to east is only 26.5 km. The stretch of mountains from the northest to the southest point is appoximately 17 km long. The state border of Slovakia and Poland runs along the main mountain ridge. There are 25 peaks surpassing 2500 metres above the sea level, among them the Gerlachovsky peak as the highest point of the Carpathians with a height of 2655 m above the sea level. Numerous side mountain ridges spreading out of the main High Tatras ridge are lined by valleys with mountain streams and lakes formed by the past glacier activities. The largest and deepest of them is VelkeHincovopleso (53 m deep, 0,20 km²). The High Tatras are the starting place for waters flowing to both the Baltic and Black Seas.

Mountain ecosystem of the High Tatras is unique and one can still find some of the last remnants of virgin mountain forests in Europe. The mixture of acidic and alkaline rocks, colorful geomorphological formations, lakes, springs and rivers frame the co-existence of broad scale unique mountain species, including some which are endemic to the area. The natural ecosystems in the High Tatras are characterized by the so-called elevational gradation; the species composition changes with increasing elevation and the vegetation can be divided in quite distinct vegetation zones. More than 1,300 recorded plant species makes the High Tatras an important biodiversity centre all over the world. Their ecosystem includes the whole range of large European predators such as bear, fox, lynx, marten wild cat and wolf.

In contrary to the natural structure of forests characterized by a mosaic of wood-vegetation in different stages of a natural development cycle, the dominant part of the forests in High Tatras mountains actually comprises many large tracts of even-aged, spruce-(*Piceaabies*) dominated, dense forest stands with no structural diversity. These large tracts were artificially planted following recurring natural disasters (windbreak and windfall followed by bark beetle damage) in the past in the same way that commercial wood-producing plantations are established (Crofts et al. 2005, Vyskot et al. 2007).

Social systems of High Tatras

These unique ecosystems of the High Tatras were for more than half of century institutionally protected as the first European cross-border national park founded in 1948. The Tatras National Park (Tatransky Narodny Park —TANAP), called High Tatras National Park as well in the Slovak territory was joined with the Tatras National Park (Tatranski Park Narodowy) in

1954 on the Polish side. The area of the National Park on the Slovak side covers 738 sq km (182,360 acres) and, in addition, the buffer zone area is 307.03 sq km (75,867 acres). National Nature Reserves with the highest level of protection cover 51% of the territory. UNESCO has acknowledged the uniqueness of the Tatras and accepted them as a part of international biosphere reserves in 1993. The Tatras National Parks, being a model of international cooperation in the field of environmental protection were identified as the core area of European importance of the Pan-European Ecological Network under the Council of Europe's Pan-European Biological and Landscape Diversity Strategy. The territory of Natura 2000 network covers 86% of the National Park (Crofts et al. 2005).

The territorry of High Tatras has served people by fulfilling a variety of functions, starting with forestry and agriculture, and continuing with silver and gold mining, fishing and hunting. An important role played High Tatra at the start of XX century as an alpine respiratory health resort, later an attractive platform for recreation and sports often expanding into the core zone of natural ecosystems. The years of human interaction with the nature in High Tatras have left visible footprints on original natural ecosystems transforming them to socio-ecosystems with high natural and cultural values on one side, but increasing its vulnerability on the other side. A unique combination of natural values with outstanding cultural heritage in High Tatras is attracting more than three million visitors a year (before destruction in the windstorm in 2004) offering the possibilities for sports, health, education, etc.

On November 19, 2004, the HighTatras Mountains were hit by the storm with the winds reaching a speed of up to 173 km/hour, unprecedented in Central Europe. The storm completely destroyed 13,000 hectares of forest —a third of the total area of the High Tatras National Park. Approximately three million cubic metres of soft wood were damaged (Toma 2009).

The increased number of visitors has put pressure on development and expansion of tourist infrastructure, including open-air sport facilities. Although the main reason for visiting the High Tatras has been, for years, hiking, skiing and other open-air recreational activities, more and more activities such as horseback riding, river rafting, golf and thermal pools have been included. In this context the contstruction of buildings and other infrastructure is considered to be the most significant pressure, and a threat to natural and cultural values of the High Tatras. They have brought many negative impacts on the whole national park environment, although at the start concentrated mostly on the mountain settlements along the main transport axis connecting tourist centres Strbske Pleso and Tatranska Lomnica, including the road and tram line. The high number of visitors and rising spatial demand of their activities causes the radiation of the disturbance to the broader surrounding natural environment, such as erosion, noise and disturbance to wildlife, with a resulting decrease in the numbers of chamois and marmots for example.

To reduce negative impacts the Tatras National Park Administration has implemented a set of different organizational, institutional and conceptual measures. The High Tatras became a natural laboratory for investigating natural phenomena, including the anthropogenic impact on natural components and possibilities for active protective and prospective interventions. The implementation of these measures and activities driven by public administration is contextual and represents only one part of the process in the reaction to the challenges resulting from the global changes demonstrated by the natural disaster in 2004. Another important part of the recovery process is related to the activities of the stakeholders motivated primarily by economic and other interests, rather than purely nature protection. The question is, to which extend are these activites in harmony with the protection of the ecosystem values or even whether there does exist such mechanisms which can be complementary to the public administration's protective interventions.

Natural Disaster as a Driver of Recovery Strategies

Dilemma over the forest management regime

The natural events like storms associated by forest damage are recurrent events in the mountains. The storm on November 19, 2004 affecting around 120 sq km of forest ecosystems at altitudes between 700 m to 1350 m above the sea level. The storm damaged not only very susceptible spruce monocultures, but also damaged to some extent mixed forests, including close-to-nature stands, believed to have higher resistance against wind damage. The spatial range of this storm was influenced by the coincidence of the natural event with the effects of long time human intervations towards changes in the ecosystems. There were primarily damaged forest stands with the spruce monocultures, homogeneous in their age (around 100 years), which had been affected previously by smaller windstorms in the past and which were artificially replanted mainly by spruce. The planting of spruce on unsuitable sites, the utilization of non-autochthonous planting material and too low-intensity thinning led to a very high susceptibility of these stands against wind (Toma 2009). But the storm also revealed many additional hidden problems, mainly in the tourist centres and settlements, such as illegal waste disposals, poorly maintained buildings and infrastructural facilities, illegally rebuilt buildings and devastated green areas.

The disaster in 2004 can be understood as the expectable natural phenomenon catalyzing the transformation of existing very sensitive ecosystems towards more sustainablity and less vulnerablity, and at the same time a challenge for the management of human activities in the High Tatras.

This natural disaster attracted a lot of attention around the whole of Slovakia and Europe. The media, the broad public, politicians, specialists, environmental activists felt the necessity to contribute to the recovery from this hard trial, to the removal of the negative consequences, and to support all the necessary steps which are urgently required for the restoration of the environment in the High Tatras.

In 1995 the responsibilities of the nature conservation in High Tatras National Park shifted to a new actor, the State Nature Conservancy, represented by the National Park Administration. However, it acts only as an advisory body with decision-making powers subordinated to the Ministry of the Environment (Kluvánková–Oravská et al. 2009). The key decision-making body in the area of forest management with direct competencies and legal responsibility to act remains state forest administration represented by State Forest Company. Thus High Tatras NP is under the management and responsibilities of two state agencies.

The position of State Nature Conservancy and State Forest Company varies significantly. The two state actors are guided by different management plans, forest classification coming from a dual regulatory system (forestry regulation vs. nature protection regulation). The discordance between those regulations has been causing difficulties and coordination problems in the management of High Tatras. Most critically is the incompatibility of protection regimes. Forestry management practices, based on production that is in contrast to the ecosystem approach of conservationists allowing natural renewal and leaving dead wood in the ecosystem. Contrasting positions for the type of management and competences in decision making evolved in open conflicts and communication failure.

The windstorm re-opened discussion on the division of competencies and future strategic plans for the High Tatras management. Immediately after the disaster discussion on the future organization and strategic development (including dealing with consequences of the windstorm) started. The community of specialists, representatives of responsible state bodies, municipalities, forest owners and users and other stakeholders in the High Tatras area took part in an international conference held in Zvolen in the Spring 2005 focused on the policy options for storm damage management in protected areas. This conference was organized under the auspices of the Slovak Republic Government, FAO and ECE UN. The key issue became again the question of the type of management, in particular on the type of forest regime, reflected already in the research projects 'Processing the windstorm calamity', 'Forest protection project', 'Forest fire protection project' and 'Revitalizations of forest ecosystems in the High Tatras region' (Jankovič 2007, Toma 2009). The Governmental Committee for the Renewal and Development of the High Tatras was established and took the decision to divide affected forest into three zones: recreational zone, core park zone both to be replanted by native species and two nature reserves and NATURA 2000 sites protected under the EU Habitat Directive (Tichá and Kôprová valleys) both to be left for natural evolution, with no management activities (pick up of deadwood or pesticide treatment of insects).

Management regime of those two reserves (pick up of deadwood and insect treatment), has become a dramatic and problematic issue as of the competing positions of the forestry and nature conservation communities. Forest management, in order to prevent insect outbreak, requires the intervention on deadwood withing a six month of a natural wind storm. Therefore the State forest administration intiated the collection of deadwood in the reserves, arguing that, spruce monocultures particularly weakened after a devastating storm, are an ideal habitat for bark beetles, potentially damaging a neighbouring forest also, which is not under the full protection regime. The risk of bark beetle was considered higher than the potential damage done to the ecosystems by collecting deadwood, as these ecosystems were seriously affected by the storm. In contrast the position of nature conservation was to follow international treaties and EU Habitat Directive, arguing that no interventions should be taken against bark beetles in natural ecosystems as they are already in the process of consolidation by natural succession. Evidence of new forest plants and wild animals was documented (nesting wild birds and young trees) and thus any action would most likely harm the environment. This position was also supported by International Union for Conservation of Nature (IUCN), World Wildlife Fund (WWF) and a later by the Director General of the Environment of the European Commission (DG Environment) indicating that management activities planned for Ticha valley are not in compliance with NATURA 2000 principles and the EU initiated infringement against the Slovak government.

The attempt to harmonize contradicting legal regulation and management practices for forestry and nature conservation in protected forest has not been sucessful. The consequence of this conflict between two parties responsible for the forest management in the Tatras National Park created a barrier for coevolution of two existing regimes and their adaptation to the post-disaster situation.

Intense or sustainable development?

The natural windstorm also created a platform for rapid development of commercial activities in the National Park area. After recouping first shock from the devastation, by the windstorm destroyed landscape, developers intensified demand for the expansion of development projects offering supportive public infrastructure such as transport infrastructure and water supply systems, etc. However these projects are often in contrast with the real market demand and interest of the protection of natural and cultural values of the area. This is a result of being placed in the context where the former socialistic regime eroded principles of environmental protection by ideological legacy and the absence of the market allowed states to be the only regulatory body, often resulting in a de facto open access resource regime (Kluvánková–Oravská et al. 2009). Implementation of the market into the environmental protection in such context creates a number of sectoral coordination problems.

At first weak economic potential at local level has been a challenge for powerfull investors, who are capable of interacting with communal decision making but following their own interests only. A typical example is the expansion of skiing resorts, where the economics of skiing in the High Tatras is rather poor, especially as the variability of the snow fall necessitates the production of artificial snow and existing services are not competitive in the market (Beták et al. 2005). It is assumed that societal costs of such projects would dramatically exceed possible benefits in the public sphere, such as water regime biodiversity and landscape quality or provision of drinking water for the Vysoke Tatry (High Tatras) County.

This situation gave rise to the pressure to develop and to approve new land-use documentation for the High Tatras County. The existence of the regulative document based on the multilateral consensus about the harmonization of different interests in the space is especially important for the areas with outstanding natural and cultural values. The lack of up-dated version of such documents since 1999 was used as the argument in the discussions regarding the frameworks for its approval in 2010 before the existence of new zonation reflecting the outputs from the recovery strategy after the disaster in 2004. The land-use plan defines the proportion of greenery in built-up areas, the density, character and other parameters of the built-up structures, addresses the traffic situation and defines functional use of land. Only on the basis of a land-use plan at the municipal level there is the possibility to extend the built-up area boundaries beyond current limits.

The critics of the civic and environmental activists expressed their apprehensions regarding the not properly limited developments of multi-storey apartment houses, open possibilities for the limit-less decisions of the developers, the change of rural mountain character of the settlements and extensive development of sports infrastructure outside the built-up areas.

The elaborated land-use plan proposal, based on a study from 1999, could mirror only a set of older documents, e.g., the definition of three zones following the designation of the Biosphere Reserve by UNESCO: core, buffer and transition. These are not adequate as the buffer zone throughout is very narrow and there is no buffer or transition zones in the west and north east of the Tatras National Park.

The new proposal of the zonation has not yet been approved, as there still remain disagreements between responsible ministries the Ministry of Environment and the Ministry of Agriculture, between state administration and environmental activists as well as between the National Park Administration and the private land and forest owners on the principles of zoning being imposed on private land. This relates to the requirments of private owners, demanding compensation for the claimed reduction in their owner rights.

This problem was identified by IUCN mission to the Tatras National Park in April 2005 (Crofts et al. 2005) concluding that there is no clear authority and specifically no overall management authority for the Park. There is no comprehensive strategy or management plan, nor a formal provision for the preparation of such documents, the current approved zonation system is inadequate, and the layers of government from national to municipal level results in confusion and inconsistency to all bodies involved. In addition, the IUCN mission identified the highly polarized views about the future management and use of the key parts of the area, including its core, following the windstorm of November 2004. This problem has yet to be solved, although the mission underlined the necessity to cope with the pressures of different interests by supplementary and compatible management arrangements facing the tendencies to enlarge the built-up areas of the settlement units belonging to the town of Vysoke Tatry towards the core zone of the national park and to meet the demands for large recreation areas as part of the town Vysoke Tatry. These demands include new huts, hotels, and transport facilities in the high mountain environment with the highest degree of protection (Crofts et al. 2005).

The necessity to look after new modes of recovery management in the affected areas safeguarding the fulfillment of the nature protection tasks, sustainable ecological, social and economic development in the High Tatras Mountains is determined by the specific land ownership situation in the area. In the High Tatras Nationa Park only 52% of the land is owned by the state, the rest (48%) is privately owned. The proposal for new zoning in the National Park as the base for the implementation of appropriate problem specific management across areas with different ownership mode, prepared by the Ministry of Environment in agreement with the Ministry of Agriculture, remains unapproved even eight years after the disaster, mostly due to the resistance of the land owners.

One of the basic problems is a disagreement between the National Park Administration and the private forest owners on the principle of zoning being imposed on private land as the private owners demand compensation for the claimed reduction in their rights. But the contradiction can be seen between The State Forest Company and the National Park Administration as well, as the state forest administration prefers to retain more land in the B zone allowing an intervention approach in the hope of changing its present unnatural situation to a more natural one. In accordance with the IUCN mission report (Crofts et al. 2005) this debate raises an important issue of principle which divides the two organizations; whether it is reasonable to have an unnatural forest ecosystem in the core area of the park and leave nature to its own devises, or whether to have human intervention to give a greater chance of success of renaturalization.

New Modes of Recovery Strategies

A key critical factor in the case of the High Tatras consolidation is the seen misfit between natural and social systems accelerated by the natural disaster (windstorm) that occurred in 2004. Thus we propose a concept of social and ecological dynamics to highlight human dependence on the capacity of ecosystems to generate essential services, and the vast importance of ecological feedbacks for societal development. The Socio-Ecological System (SES) represents such interconnection. SES include societal (human) and ecological (biophysical) subsystems in mutual interactions (Gallopin 1991). SES concept places humans within nature and focuses on the way in which interconnections between people and their biophysical contexts produce complex adaptive systems.

Both social and ecological systems contain units that interact interdependently and each may contain interactive subsystems as well. A social system includes economy, actors and institutions in mutual interaction (Kluvánková–Oravská 2009). Ecological systems include self-regulating communities of organisms interacting with one another and with their environment (Berkes et al. 2003). Adaptive governance implies establishing compatibility between ecosystems and social systems by creating efficient social norms and rules that are capable to manage systems in an effective and sustainable way. The connectivity pattern within and between social and ecological systems plays an important role in designing effective institutions for sustainable resource use (Gatzweiler and Hagedorn 2002). Resilience, the concept originally used by ecologists in their analysis of population ecology of lands and animals and in the study of managing ecosystems (Folke 2006), is seen as a possible concept to study socio-ecological dynamics. Innovative is the capacity to absorb shocks while maintaining function and provides components for renewal and reorganization following disturbance and sustains capacity for adaptation and learning (Holling 1973, Carpenter et al. 2001, Folke 2006). A resilient ecosystem has the capacity to withstand shocks and surprises and, if damaged, to rebuild itself. In a resilient socio-ecological system, the process of rebuilding after disturbance promotes renewal and innovation. Without resilience, ecosystems become vulnerable to the effects of disturbance that previously could be absorbed.

Resilience in Forest Management

As indicated previously land ownership in the Slovak Republic, is diverse. In forest land 25.5% belongs to the collective private owners—urbars. They constitute a form of self-governed historical land co-ownership regime mainly of forested land and pastures usually within one village. The name originally referred to a register of serfs' properties and their respective duties towards a feudal lord (Štefanovič 1999), created in 18th century for the use of feudalists' pastures and forests for their own purposes. Gradually, serfs were freed from their obligations towards landlords. However, they continued to use pastures and forests and they paid a rent to the landlord in return. After the abolition of serfdom in 1848 those pastures and forests were transferred to them in the form of common property from aristocratic landlords (in 1853) or later (beginning of 20th century) were bought at stock market as a number of aristocracies bankrupted. Property in the urbar is inherited from parents to children. Share in the urbar can be sold only with the approval of the assembly, giving priority to existing members. Two land reforms undertaken in the 20th century significantly affected ownership of urbars. First, undertaken at the establishment of Czechoslovakia (1918) enabled expansion of urbars by transfer of ownership from the aristocracy, however the second disconnected operation of urbars for more than 40 years by nationalization of private land by the communist government (1948). In particular nationalization (1945–1990), when land was in the hands of the state interrupted the inheriting process resulting in significant land share fragmentation and reduced the sizes of individual shares to sometimes less than 1 ha. Urbars were re-established in the process of land re-nationalization initiated in 1993. Urbars are currently regulated by Slovak Law on Land Associations (No. 181/1995).

In the High Tatras urbars have a significant ownership share in particular within the core zone of the national park and natural reserves, including the Ticha valley NATURA 2000 site. However, whoever has the ownership, experience shows that the success of management depends on good consultative arrangements and communications between the managing authority and the other owners, and on satisfactory financial arrangements for the management of the national park.

To undertake managerial responsibilities, community rules for harvesting, replanting and self management were developed over time. Most significantly was the forest degree of Maria Teresa-Austro-Hungarian imperator, issued in 1767 to manage wood as a strategic resource for the mining industry, but also protected forest from overexploitation by overgrazing, illegal timber and inappropriate land-use changes (Nozicka 1956). This document served as management guidelines for forest industry since 1770. The guidelines contained 55 management rules for harvesting and forest revitalization designed to maintain forest quantity and quality for the long term. These include age of the trees permitted for timber, harvesting techniques, harvesting (rotation) and forest revitalization calendars, measures to protect wood from mechanical damage and soil against erosion, duty of registration of type and quantity of timber. The degree also contains regulations for inspections, planting, guidelines for flood protection, regulation of housing and fire protection. Division of the responsibilities and rights were also regulated. Each co-owner of urbar had a duty to participate in the management according to the size of the shares and having the right to collect an annual benefit from the land.

Today urbars operate on ten-year programs controlled by the state forest authority. Timber, replanting and other activities are planned for this period and each subject has a certain flexibility to decide on the strategy for each year. Such a system enables flexibility of decisions to reflect external social and natural shocks, for example timber price decline, wind blow damaging forest and others. Social equity is also used as a reason behind decisionmaking. A number of urbars also use a regular self monitoring mechanism to control harvesting process and an internal sanction system, mainly in the form of gradual exclusion of the rules violators from group benefits. External sanctions are imposed by governmental authorities to regulate forest use.

The main decision-making body is an assembly of owners, which takes place once a year, and adopts an annual harvesting strategy and approved budget. It also delegates all day-to-day decisions to the economic committee, consisting of elected and professional members (Act no. 181/1995 on Land Associations). Members of urbars participate in modifying the operational rules and have the opportunity to contribute to creation of the rules which define their rights and duties. Following Ostrom (1990) they are more likely to create arrangements that are mutually acceptable and adaptable to changes. Urbars can adopt voluntary monitoring of members or other forest users, willing to invest private costs into informal sanctioning which has been found as an effective low-cost control also previously reported by Ostrom et al. (1994). Flexibility, self-governance and local experience helps create conditions for the renewal of long-lasting institutions that have demonstrated their ability of adaptation to external factors.

Given formal and informal arrangements for operation of urbars discussed above, it is possible to declare that urbars are nested within an existing forest organizational structure. However the connection of self-governing urbars within existing forest management and governance structure which is fully hierarchical is problematic. Due to historical management rules in use, self-governance and equal cost benefit sharing urbars can be seen as long lasting and resilient socio ecological systems with high adaptive capacity to external shocks.

Conclusions

The post-disaster development in Tatras National Park is an example of how natural and social systems imposed to global changes are becoming vulnerable to external as well as internal disturbances. Global climate changes may seriously affect the systems capacity to absorb external and internal disturbances and the adaptation capacity of natural and social systems without fundamental changes of their quality. In our study, the windstorm that devastated forest ecosystems of Tatras National Park accelerated existing conflicts and institutional threats in resource management and created several social challenges for forest management and spatial development in the region.

Urbars as specific form of self-governing regime combined with collective ownership was identified as a perspective resource regime to cope with unpredictable disturbances and complexity of global changes. Urbars' can be seen as long surviving institutions for sustainable forest management also under the free market competition and global changes, as they represent rather flexible collaborative management modes combining traditional forms of management following sustainability goals, as well as the economically oriented behaviours of the owners.

In particular it has has been found that operation of urbars determine ecosystem dynamics and sustainable use of forest resources as an attribute of economic profit. In addition institutional structure of urbars increase internal system stability and reduces vulnerability against external shocks. Urbars are thus seen as more resilient than individual private or state property resource regimes.

Furthemore, as the urbars represent historical ties between joint owners —residents creating a dominant part of local communities and forest ecosystems, they seemed to be crucial for continuity and development of local identity, and their participation in management and governance contributes to the stability and sustainability of the landscape. These aspects, as important aspects of landscape values, may be further research challenges.

References

Act no. 181/1995 on Land Associations.

Act no. 326/2005 on Forests.

Berkes, F., J. Colding and C. Folke (eds.). 2003. Navigating Social–Ecological Systems: Building Resilience for Complexity and Change. Cambridge University Press, Cambridge.

Betak, J., L. Bizikova, J. Hanušin, M. Huba, V. Ira, J. Lacika and T. Kluvankova-Oravska. 2005. Smerom k trvalo udržateľnému tatranskému regiónu: (nezávislá štúdia strategického charakteru) — súhrn. Bratislava: REC Slovensko and STUŽ/SR with the support of Fond Tatrypri Nadácii Ekopolis.

Carpenter, S.R. and L.H. Gunderson. 2001. Coping with collapse: ecological and social dynamics in ecosystem management. BioScience 51: 451–457.

Crofts, R., M. Zupancic-Vicar, T. Marghescu and Y. Tederko. 2005. Report from the IUCN Mission to Tatra National Park, Republic of Slovakia, Gland: IUCN.

Folke, C. 2006. Resilience: The emergence of a perspective for social–ecological systems analyses Global Environmental Change 16/2006: 253–267.

Gallopin, G.C. 1991. Human dimensions of global change: linking the global and the local processes. International Social Science Journal 130: 707–718.

Gatzweilder, F. and K. Hagedom. 2002. The evolution of institutions in transition. International Journal of Agricultural Resources, Governance and Ecology 2/2002: 37–58.

Halada, L. and Z. Izakovičová. 2011. Multi-scale interactions between disturbances and ecological and socioeconomical changes — case study High Tatra Mts. (Slovakia), ESI LTER Europe (http://www.lter-europe.net/projects/ecosystem-services-initiative-esi/ESI%20Tatra%20Mts%20Slovakia%20v4.pdf).

Holling, C.S., D.W. Schindler, B.W. Walker and J. Roughgarden. 1995. Biodiversity in the functioning of ecosystems: an ecological synthesis. *In*: C. Perrings, K.-G. Mäler and C.S. Folke. Biodiversity Conservation: Problems and Policies, Kluwer Academic Publisher, Dordrecht, pp. 44–83.

Jankovič, J., S. Celer, V. Caboun, P. Fleischer, K. Gubka, P. Hlavac, I. Chromek, J. Juleny, M. Kamensky, M. Koren, E. Krizová, J. Liska, A. Majlingova, J. Marhefka, R. Rasi, I. Rizman, M. Saniga, M. Schwarz, P. Spitzkopf, M. Suskova, P. Szarka, S. Smelko, L. Smelkova, I. Stefancik, P. Toma, A. Tucekova and J. Vladovic. 2007. Projekt revitalizácie lesných ekosystémov naúzemí Vysokých Tatier postihnutom veternou kalamitou dňa 19. 11. 2004, Zvolen: Národné lesnícke centrum.

Kluvankova-Oravska, T., V. Choborova, I. Banaszak, L. Slavikova and S. Trifunovova. 2009. From government to governance for biodiversity: the perspective of central and eastern European transition countries. *In*: Environmental Policy and Governance. 3: 186–196.

MPRV SR. 2011. Správa o lesnom hospodárstve v Slovenskej republike za rok 2010 – Zelená správa, Bratislava: Miniserstvo pôdohospodárstva a rozvoja vidieka SR.

Nozicka, J. 1956. Přehled vývoje nasich lesu, Statni zemedelske nakladatelstvi, Praha.

Ostrom, E. 1990. Governing the Commons: The Evolution of Institutions for Collective Action. Cambridge University Press, Cambridge.

Szemesova, J., J. Balajka, M. Lapin, J. Mindas, P. Stastny and D. Thalmeinerova. 2005. The Fourth National Communication of the Slovak Republic on Climate Change, Bratislava: Ministry of Environment of the Slovak Republic and Slovak Hydrometeorological Institute.

Štefanovic, M. 1999. Urbárske právo. *In*: Ekonomický poradca podnikateľa 6. Poradca podnikateľa spol. s.r.o.

Šulek, R. 2007. Urbarska sústava na území Slovenska do roku 1918. pp. 163–172. *In*: Acta Facultatis Forestalisno. 2. Lesnícka fakulta Technickej univerzity vo Zvolene.

MZP SR. 2010. Zámer na vyhlásenie zón Tatranského národného parku, Bratislava: Ministerstvo životného prostredia SR.

Toma, P. 2009. Strategic intentions of wind calamity management on protected territories, paper on International workshop, Strbske Pleso May 2009.

Svajda, J. 2006. Rangers in Tatra National Park, 5th International Ranger Federation congress, Stirling.

Vyskot, I., J. Schneider, P. Kupec, J. Fialova, A. Melicharova and D. Smitka. 2007. Wind calamity damages to sanitary-hygienic and social-recreational functions of forest in High Tatras National Park. *In*: J. Rožnovský, T. Litschmann and I. Vyskot (eds.). Klimalesa. Sborní kreferátů z mezinárodní vědecké konference, ČHMÚ, Praha.

17

Local Responses to Global Change: Community Alternatives for 'Good Living' in Latin America

David Barkin

INTRODUCTION

Mountains dominate the landscape in the Americas. In Latin America, the mountain ranges from the Sierras in Mexico to the Andes in the Sur are home to innumerable ethnic groups descendent from the pre-Colombian peoples many of whom were forced to retreat from the lower reaches by colonial expansion or more recent forms of expulsion, including expropriation, theft or chicanery. During the course of the past centuries, most of these diverse communities have learned to co-exist with the extraordinary biodiversity that has survived and been managed with an understanding and a sophistication that we are just beginning to comprehend. The multiple processes of conservation and reproduction has produced the rich quilt of diverse ecosystems that we observe today, the result of a complex intertwining of mechanisms to systematically acquired knowledge of how natural systems function and cultural practices that organized societies to

Xochimilco Campus, Universidad Autónoma Metropolitana in México City, Calzada del Hueso 1100, Villa Quietud, 04960 Coyoacan, DF MEXICO.
Email: barkin@correo.xoc.uam.mx

live harmoniously in them; of course, there were numerous exceptions of communities that were unsuccessful in managing their environments or who succumbed to challenges from other social groups throughout the region known as the Americas, even before the arrival of the European conquerors.

This chapter will examine the ingredients that went into developing constructive strategies to facilitate the survival of the hundreds of ethnic groups that continue to inhabit the highlands of Latin America. Many of these productive and cultural building blocks have endured until today, undergoing a continuing process of adaption, the result of culling of old practices that proved to be obstacles to survival and of integrating new discoveries, accumulating knowledge, and the direct appropriation of contributions from other communities and other social systems, including the very peoples who might be threatening them.[1]

Perhaps the point of departure for this analysis is the changing configuration of the world in which they operate, one that has evolved from membership in regional empires, like the Aztec, Mayan, and Incan confederations, to a complex and variegated process of integration into nation-states and the international economy. Without going into an historical discussion of these developments, this chapter offers a suggestive examination of the concerted efforts by communities throughout the region to maintain their identities, to develop mechanisms to assure increasing degrees of autonomy in the face of intensifying efforts to integrate them into the ranks of the poorest people in national societies and global markets. Far from isolated efforts by individual communities, the current efforts by myriad communities throughout the region reflect explicit attempts to forge alliances and propose collective strategies for strengthening their organizations to better enable them to confront the external forces of homogenization. As such, this proposal for understanding the activities of the mountain communities in Latin America offers a direct counterpoint to much of the literature that examines growth and 'development' in terms of society's progress, or lack of such progress, in advancing along the road of capitalist development and fuller participation in global markets; this literature attaches little importance to the serious collateral damage that causes great harm to the human and natural settings in which it occurs, increasing processes of social exclusion and reducing natural resilience while the State abdicates its historic responsibilities to assure some modicum of balance among distinct groups.[2]

Communities throughout the Latin American highlands are increasingly militant in the face of these problems, defending their right for self-determination, for local empowerment, to thwart the dynamics of globalization. In many places this takes the form of a demand for autonomy, a claim that they receive the political guarantees to govern themselves and, in the process, determine the way in which they organize their economy—the productive system and processes of exchange with others—as well as their political and social structures while devoting the energies and resources necessary to assure the balanced operation of their ecosystems; in current parlance, they are seeking to manage and guide their social metabolism to assure their own well-being and that of their surroundings.[3]

This set of demands and aspirations has not emerged as an abstract socio-political platform, but rather as the logical outcome of the political efforts of communities throughout the region to assert their rights as unique societies, with a cultural heritage and traditions that have been strengthened as a result of the systematic patterns of oppression which they have had to resist. This self-conscious assertion of identities is accompanied by moves to strengthen their collective capacities to govern themselves and their environments; in the process, they are reevaluating the contributions of their traditional activities, their inherited knowledge, and their ecosystems to their well-being while also exploring ways to design and implement new strategies for the future.[4] These new strategies are focused on reversing the historical pattern of social exclusion of these peoples and their regions in the very conception of nation, implementing new productive schemes that promote local material progress and the control of surplus by the communities for channeling into new productive ventures and social projects to improve material well-being and social integration (solidarity) while also assigning resources for programs of environmental conservation and rehabilitation.

In conceptual terms, perhaps the most important contributions of this regional impetus to assert the significance of a renewed participation by the 'first nations' in the political determination of their ways of life in their regions and the discovery of the significance of the communal nature of their societies for internal governance and the (non-pecuniary) revalorization of their ideological/spiritual/cultural heritage for the definition of their decision-making processes and the choices they make. In Meso-America, this combination of factors has been defined as 'communality'—a collective determination of the full panoply of factors that comprise the definition and management of a society, with the significant difference that among these indigenous groups the process evolves explicitly and amid lengthy and profound discussions because of their contentious relationships with the dominant institutions that continuously attempt to demean and even expunge their uniqueness.[5]

Analysis in this chapter, incorporates the proposals of diverse indigenous and peasant groups for their own organization of the rural production process as part of their efforts to ensure the viability of their mountain environments. Their collective commitments to an alternative framework for production and social integration, based on basic principles of social and political organization, offer a realistic but challenging strategy for local progress. These principles, widely agreed upon in broadly based consultations among the communities, are: Autonomy, Solidarity, Self-Sufficiency, Productive Diversification, and Sustainable Management of Regional Resources. Their emphasis on local (regional) economies (rather than larger units) and the use of traditional and agro-ecological approaches in production and the integrated management of ecosystems are the basis for their guarantee of a minimum standard of living for all their members and a corresponding responsibility to participate, thus eliminating the phenomenon of unemployment. An integral part of this approach is the explicit rejection of the notion that people in rural communities conceive of themselves exclusively as farmers, or even as resource managers; rather, in these societies, it is more revealing to understand their decisions as the result of a complex allocation of their time among numerous activities of individual and collective benefit.

Communality

There are a number of fundamental conceptual principles underlying the organization of the societies involved in constructing structure capable of moving towards the 'good life' (buen vivir) discussed in the Latin American literature that are facilitating their efforts to transcend the problems of poverty in their social reality, with a concomitant commitment to productively incorporating all their members into socially useful forms of participation. In the case of Mexico, these principles have been codified by a number of 'organic intellectuals' who have been actively involved in a process of innovation as a part of the process for the consolidation of social capacities in their communities, a self-conscious process of organization contributing to strengthen tradition (Díaz 2007, Martínez Luna 2010). They have coined the category 'communality' to encompass these principles, that include: 1) Direct or participative democracy that defines and ratifies the political, cultural, social, civil, economic and religious aspects of society; 2) The organization of community (collective) work; 3) Community possession and control of land; 4) A common cosmology, including the notion of the Earth as mother (Pachamama) and a respect for community leadership; and 5) A common history and language. This development reflects an epistemological contribution that incorporates the appropriation of nature in a dramatically different way than that conceived by the dominant institutions of the western project of 'civilization' that is embedded in most development programs (Diaz 2007: 38–39).

Communality, in this sense, is not simply the aggregation of individual interests into a collective whole as suggested in the historical notion of 'social contract' (Hobbes, Locke, Kant) that should lead to a 'just society'. The institutions are ineffective because they are backed by:

> An agreement in which each person adheres to the contract to safeguard his own individual interest; if the contract, the political association, does not safeguard them, the individual has the right to break the contract, because (s)he agreed to the arrangement in terms of an egotistical interest, and thus if it does not respond in these terms, the individual may refuse to continue abiding by the contract (Villoro 2003: 48–49).

In contrast, in the context of a peasant association adhering to the principles discussed above, a contract is one that:

> I accept the contract, on the understanding that I am committed to the well-being of the group as a whole, even if it might advance against my own particular interests; therefore, I continue to respect the terms of the contract. Democracy is, in this sense, a political association in which, at the same time, is an ethical agreement, because it is the way in which a public group can guarantee the freedom of everyone in the group, while also remaining a guarantor of autonomy (Villoro 2003: 49).

Communality, then, is a complex composite concept, one that embodies the totality of the collective commitment to individual welfare in the context of an individual commitment to collective well-being. It is an implicit arrangement that goes beyond the limits of material considerations to accept a different responsibility to the community and to its ecosystem, an obligation ground in tradition, in cosmology, to respect the community within its environment.

Although a product of the very specific conditions of the struggles in the highlands of Oaxaca to reclaim their forest resources,[6] the doctrine of communality is increasingly recognized as relevant for understanding the many local struggles for self-governance, for autonomy in the management of social organization and for the right to decide on the best uses for the resources over which the people involved in these struggles. As such, the doctrine is a direct challenge to inherited notions of the sovereignty of the nation-state, of the unquestioned right and ability of national governments to decree the disposition of the nation's resources without reference to the considerations of the local peoples.

Without going into more detail, we wish to simply ensure that there be no illusion about the singularity of the Oaxacan version of this conceptual approach. Similar approaches are evident in the current efforts among the Andean peoples to codify and operationalize the heritage of the Sumak Kawsay (*'buen vivir'* or 'living well'), the explanations of the Zapatistas of their own developments (*mandar obedeciendo* or govern by obeying), the struggles of the Wixarrica (Huichol) against the depredatory expansion of Canadian mining (in their sacred site, Wirikuta), or the myriad other manifestations of peoples throughout the Americas to defend their customs, their territories, their societies, indeed their very existence (e.g., the Abya Yala among the Kuna of Panamá).

In this regard, the introduction of the concept of 'living well' as a guiding principle in the new constitution of Ecuador reflects this search for alternative rationalities to inform national policy. In the mountainous regions of Latin America there is a long history of this realization for the need of a different approach to constructing a path towards well-being: the early contributions (1920s) of the Peruvian José Carlos Mariátegui (1971) were inspired in this search "that does deny the material and intellectual contributions of modernity, but challenges them on an ethical plane." Bringing the discussion up to the present, it is clear that for us today this approach to development recasts the problem, rejecting the need to 'overcome' backwardness and the notions of growth and accumulation. The challenge is to organize a society's human, material and natural resources to assure greater measures of equality and a meaningful process to guarantee the sustainability of its natural patrimony. By abandoning an anthropocentric vision of social processes, the new ethic of development subordinates economic objectives to ecological balance, human dignity and social well-being. The resulting productive systems involve a mixed and solidarity economy in which social and intergenerational justice are assured by diversified social and cultural institutions that attend to the basic needs of food sovereignty and control of natural resources.[7]

Responses to the Development Conundrum

Much Latin American thinking and practice to improve the plight of highland peoples demonstrates that the 'development conundrum' is not one that can be resolved by simply raising savings rates or promoting local productive ventures; increased international investment has proven so devastating that a new phrase sums up the critical theoretical work examining and explaining the process of plundering the environment, exacerbating inequalities and even disenfranchisement: accumulation by dispossession (Harvey 2003). Although many honest practitioners focus on enhancing human capabilities (Sen 2002), and generating new ones to overcome the obstacles to individual achievement, recent experience demonstrates that the monopolistic exercise of power in the political area and the operation of markets cannot be tamed through the well-intentioned creation of institutions, even when they are committed to promoting individual opportunities and enabling the 'free and equal' interplay of social forces (Hill 2007).

As an alternative, and in the face of the intransigence of national economic policy makers, many marginal societies are moving to strengthen their collective capacities to govern themselves, shape their productive systems, and manage their environments; in the process they are reevaluating the contributions of their traditional activities to their well-being while also exploring ways to forge new strategies for their advancement and the protection of their ecosystems. They are focusing on inherited knowledge and their inherited productive systems as sources of wealth and means to consolidate their cohesiveness as societies, while incorporating the latest technological and scientific advances, becoming able innovators and managers and creating new governance capacities consistent with the demands for negotiating with regional, national and international institutions. Their proposals for creating viable strategies require local control of geographic and political space, involving alliances among peoples searching for new responses to the global forces of marginality and exclusion.

A new post-development policy framework is being formulated to help design the appropriate responses for reversing past tendencies of impoverishment and environmental destruction. Traditional organizations and knowledge systems are generating innovative forms of collaboration and production, of political consolidation and social collaboration. The experiences briefly mentioned below offer an attempt to explain why it is necessary to expand beyond the improvement of individual capabilities and the exercise of individual freedoms, if societies are to liberate themselves from the globalized straitjackets imposed by international economic integration with its imperatives of 'free' trade and markets. Although individual improvement and self-betterment continue to be significant, we focus on the primacy of collective determinations of the worth of their activities and the focus on collective entitlements, assuring the viability of community processes for individual participation.

This approach to community welfare requires new forms of collaboration and the forging of alliances that colonial practices and national politics discouraged or even prevented for decades if not centuries. The reasons for this repression are rooted in the autonomy that such practices might have generated for the communities, along with the possibility that the beneficiaries could reduce or even prevent the capture of surplus by outside investors or political groups. The methods used by these communities to produce the goods needed for their survival and reproduction reflect an accumulation of knowledge of the workings of the natural world through the centuries; they developed interesting and innovative solutions to complex problems, appreciated by local communities worldwide and codified into religious and lay traditions that were passed on through the ages in sacred texts, by story tellers, or keepers of 'the word.' In sharp contrast to the practitioners employed by the international development institutions and the academic community that perpetuates the doctrines that guide them, people interested in these communities are learning to appreciate the significance of the conservation and deepening of the traditional knowledge and understanding of their surroundings developed by communities on the basis of other rationalities and cosmologies that are contributing to their well-being and the conservation of their ecosystems; in some cases research teams are now collaborating with these communities to expand their options with innovations based on new research consistent with traditional approaches.[8]

As in many parts of the world, in Mexico there are numerous mountain communities actively involved in efforts to escape from the dynamics of social and economic marginality (Borrini-Feyerbend et al. 2010, Martin et al. 2011, Tauli-Corpuz et al. 2010). National socio-economic policy and programs of national 'integration' systematically impoverished them as part of modernization programs that continue to proclaim international integration and free trade as the most effective path to 'development' (Barkin 2000). In response, many communities are actively involved in resisting programs for local and regional development, constructing alternative projects for local advance. They are reclaiming parts of their history and inviting others to join them in integrating the best of state-of-the-art practice as part of an effort to strengthen their societies, to join them in forging new structures that will promote a meaningful form of sustainability, assuring enduring patterns of equality and an informed process of ecosystem management for rehabilitation and conservation. In what follows we offer several examples in which university based teams were able to interact with these communities to strengthen their collective projects. This experience is based upon the idea that people codify their knowledge systems in such a way as to attempt to manage their environments and produce the goods they need for their own well-being and for improving their conditions.[9]

Innovation to Maintain Tradition

In our collaborations with local communities, we are concerned with identifying projects that contribute to promoting community solidarity and welfare in consonance with environmental equilibria. Our contribution, as outside researchers, is to identify untapped or ill-used resources that can be better mobilized to promote community objectives. In some cases, as will become obvious, we explore the significance of institutions that have atrophied or assert new values that we consider can contribute to community objectives; two such examples are promoting gender equality and the restoration of backyard animal husbandry. By their very nature, these marginalized communities are generally situated in fragile environments, many in mountain settings; the decisions about the directions for social and productive innovation are particularly sensitive to natural conditions, placing a greater responsibility on the outside collaborators in their interactions with the community and their suggestions for innovation. In the examples that follow, the processes that accompanied these interventions, highlighting the elements that contributed to the development of a theoretical approach briefly outlined towards the end are emphasized.

1) For centuries backyard animal husbandry has been a central element for peasant societies around the world. Transnational corporations have systematically undermined this strategy by imposing new technologies that make small-scale family units unviable. Genetic selection produced new breeds of poultry and hogs better suited to factory-like conditions for reproduction and fattening, displacing traditional breeds that are more efficient in processing household and small-farm waste streams, but required more time before they could be marketed. In our search for sustainable regional resource management strategies, we discovered that avocados lowered blood-serum cholesterol levels in hogs (and people); people in some communities were penalized when their hogs grazed in local orchards because the butchers in the region paid less for those animals because they lacked a layer of fat![10] By introducing small modifications in traditional diets, we helped the communities to produce meat lower in fat and encourage backyard hog raising as a complementary and profitable activity to strengthen the regional economy and the role of women as a new social force. Now, some 10 years later, it is clear that the approach has been accepted and the main obstacle to its full implementation will be the need for people from the region to advise and supervise the quality of the diets and the conditions in which the pigs are raised in the backyard stalls.

   In retrospect, the proposed innovation proved relatively easy to implement because the design fits into the existing structure of village life and political organization. Although based on a declining activity (hog raising), the proposed changes were clear to all participants who clearly understood the relationship between diet and animal nutrition; its commercial logic was also compelling, especially within today's precarious rural economy, because of the price premium their low-fat pork commanded. Because of the focus on an activity that women have historically managed and the declining presence of men who must seek work elsewhere, the project struck a particularly responsive chord. Furthermore, with a growing awareness of the need to improve sanitary conditions as a result of improving channels of information and concern about health, the project also stimulated discussion of environmental issues, like water quality and sewage disposal and treatment (Baron and Barkin 2001).

An interesting development emerging from the work on 'lite' pork was the discovery of the nutritional qualities of *verdolaga* (purslane in English), rich in Omega-3, which could be valuable for feeding to hens to produce "enriched" eggs. The plant can be readily incorporated into the diet of laying hens, displacing the fatty omega-6 from the egg yolks, to produce a product that will have less of an impact on the cholesterol of consumers. This program is a logical follow-on to the hog project in the central highlands, harnessing a concern for the integrity of ecosystems in order to introduce a new activity that promises to generate new sources of income for the participants.

These experiences offer a singular window on the development process. Rather than concentrating on individuals and their capacities to participate effectively in regional governance activities, the approach implemented here joins the search for more productive activities with strategies for increased collective capacities to implement programs for the sustainable regional resource management. The communities are becoming active promoters of community programs that increase participation in productive diversification. The significant feature of this process is the relationship between individual initiatives and collective decisionmaking that sanctions and integrates the activities into the collective strategy for regional progress.

2) Indigenous societies, pushed into the mountains by successive waves of productive expansion by conquerors, now find themselves heirs to valuable resources in the headwaters of river basins, resources required for urban-industrial development. Along with problems of global climate change and other ecological phenomena, the lack of water is becoming particularly serious, leading to a desperate search for solutions to mitigate the crisis. Many recent proposals for 'sustainable production', based on individual economic rationality and a liberal development discourse, advance a 'modern' development strategy in which corporations and governments alike do not go beyond a process of 'green washing' corporate activity (Escobar 1995, Leff 1995, Utting 2002). The sustainability discourse frequently camouflages a capitalist rationale and is mixed with a large measure of bio-colonialism, a strategy in which indigenous and peasant communities in regions of mega-diversity do not participate, except as ecological informants and as objects to be rescued.

More recently, however, this strategy also inspires alternative approaches, based on the local appropriation of these concepts by people conscious of the wealth of inherited knowledge that can be used to ameliorate environmental problems. These alternative discourses and perspectives have been used mostly in the southern countries, in part by theorists, intellectuals or practitioners working directly with peoples who express their demands in terms of territorial defense, alternative development, autonomy, sustainability and self-sufficiency (Escobar 1995, Barkin 1998, Toledo 2000). The more successful of these proposals are designed from the local point of view, where the inhabitants become the protagonists of the recovery and preservation of their resources.

These Mexican projects draw on a long history of struggle by different social groups and reflection by southern thinkers who have promoted alternative approaches to sustainability. The basic tenets of this work are: a) the active participation of the local population in the design and implementation of the plans and programs, so that they generate a capacity for self-management and a recuperation of social institutions and cultural identity; and b) the enhancement of the ecological diversity as part of a program that contributes to diversifying the local economic base (Bonfil 1996). Thus, sustainability itself is a complex set of ideas that is understood differently as people assimilate the lessons into their own individual ethos. From the market perspective, the model enables the 'guardians of the forests' to refrain from joining the low-waged labor force[11]—a transitory opportunity, concentrated in the development poles—so that they can become protagonists of their own sustainable regional development.

An example to rebuild a watershed involved an effort to reverse deforestation and compensate for the excessive withdrawals resulting from a mega-tourist development. An environmental rehabilitation program invited the communities to recover their life styles, reinforcing local institutions and diversifying the productive structure, rejecting the standard paternalistic and clienteles' approach; it had three objectives: a) to reconstruct and conserve the region's basins and forests; b) to use the ecosystems in a sustainable manner; and c) to join the inhabitants of the coast of Oaxaca in their efforts to recover their dignity (Barkin and Pailles 2000, 2002). It helped diversify the rural economy by introducing alternative productive systems to raise incomes and strengthen local institutions, blending traditional knowledge systems for conservation with modern production techniques. If the project had not considered the enormous potential of traditional knowledge in ecosystem management, the project would have encountered greater resistance from the local communities, as is common in most projects designed by official development agencies in central offices. This approach has become part of a broadly successful effort throughout Mexico that is now widely recognized as the most wide-spread and consolidated model of community forestry in the world, leading to a diminished dependence on extraction of trees as a large number of different activities in the forest environment assure new sources of income and resources while improving local abilities for self-government and ecosystem management (Merino and Robson 2006).[12]

3) The Mixtecan peoples are an impoverished group living in a desolate region in east-central Mexico, which suffered from centuries of over-exploitation. A group of young people from the National University (UNAM) proposed collaboration to the 100,000 people living in the region to implement an ecosystem rehabilitation program based primarily on water and land management techniques. During the ensuing quarter-century an ambitious series of projects, grouped around an umbrella concept: Water Forever,[13] have improved conditions in the 3.5 million acre region, on the basis of a program firmly anchored in community mobilization and training, based on attacking the problem of water scarcity, caused by three factors: population increase, inadequate natural resources management and unequal access to water, and most especially the over drawing of water supplies by a small number of people legitimated through corrupt power structures.

Deforestation and surface erosion were the main problems to be tackled. A wide variety of land management projects, including dikes, terraces and dams were implemented so that people from all of the communities could participate. The cumulative effect of the hundreds of these small efforts was to substantially increase the arable land under cultivation and to increase the volumes of water available for agricultural production, the animals and the communities. The approach has demonstrated its viability as a mechanism for promoting local capabilities far beyond those of constructing public works and increasing the region's productive potential. The long-term vision and the emphasis on local capacity building for project implementation has transformed the organization into a sort of substitute local government agency, with its own management structure, fleet of vehicles and heavy construction equipment, planning and engineering departments and even a geographic information systems laboratory.

Similar community management projects are springing up throughout Mexico. Community forest management projects now encompass more than one-half of the nation's wood resources, where local groups are developing their own production programs and complementing the protection programs with ecotourism, artisan production, water bottling and the sale of environmental services. Most importantly, these programs are examples of the way in which people are learning to appreciate the value of their inherited cultural traditions and enriching them with techniques and lessons from the current era.

Conclusion

Throughout Latin America mountain people are rediscovering and updating their traditional cosmologies along with their knowledge systems to develop unique proposals for harnessing their material, human and natural resources to improve their quality of life and ensure the protection of their ecosystems. Even in Mexico City, at more than 2,000 meters above sea level, dozens of community groups have organized to take advantage of their resources to strengthen their societies and political structures and raise their living standards. A 2,500 hectare degraded forest has been set aside as an ecotourism site and nature preserve where monthly tens of thousands of visitors are treated to a unique set of hiking and biking trails and nature talks which inform and entertain while employing more than 200 members of the community. A pre-Colombian amphibian, the *Axolotl*,[14] has become a charismatic attraction in the 'floating gardens' of Xochimilco, when one community decided to abandon the crass commercialism in favor of a tour that explains how the complex ecosystem can be managed to provide a variegated cornucopia of fruits, vegetables and small animals that protect the environment and provide for the economic well-being of the people.

The anecdotal and quite selective recounting of a small selection of local development initiatives presented in this chapter cannot do justice to the breadth of activities undertaken by millions of Mexicans and people elsewhere in the world implementing local development strategies on the margin of and in place of unsatisfactory market-based solutions. They are reclaiming cultural mechanisms for organizing productive structures responsive to local needs and strengthening traditional governance organs while creating a new generation of local cadre concerned with their societies' quality of life and the health of their ecosystems; in the process, they are transforming market relations with the outside world, replacing the commercial partners with fair trade institutions and other 'niche' marketers that protect them against unequal exchange.

As mentioned in the introduction to this chapter, we have systematized this set of experiences into a strategy for sustainable regional resource management based on five principles that is being replicated by communities throughout the Third World. Individual communities are moving beyond their local confines to build alliances within and among regions and ecosystems, garnering political and social power to defend themselves; at the same time, as we have pointed out in this chapter, sympathizers and scholars are recognizing the significance of these efforts and their gradual coalescing into movements. Reiterating, the five basic principles of this strategy mentioned above are: autonomy, solidarity, self-sufficiency, productive diversification and sustainable regional resource management.

But it is no longer possible for us or for the communities with which we intend to collaborate to continue assuming that we can design our lives around a consumption scheme and market system like that developed in the western world or continue to depend on a productive apparatus like the one that is decimating our ecosystems and provoking global warming. It is not sufficient for us to learn from the models of 'good behavior' of the communities trying to defend their páramos, their forests, their sacred mountains; we must join them in searching for strategies to build a 'good life', incorporating the principles handed down through the generations among the Andean peoples, and recently coming to the fore in the World Conference of the Peoples on Climate Change and the Rights of Mother Earth, convened in Tiquipaya, Bolivia in March 2010 (Acosta 2010).[15]

If there is one lesson that can be drawn from these experiences, it is for tradition to survive it must become a living process, a resource that is constantly renewed to assure its currency and its value. In Mexico, people that now comprise more than one-quarter of the population, find that indigenous epistemologies are truly a building block for constructing alternatives to globalization: transforming into reality the realization that: Many other worlds are under construction by peoples around the globe right now.

References

Acosta, Alberto. 2010. Only by imagining other worlds, this one will be changed. Thoughts about Good Living. Sustentabilidades, No. 2 (Text in Spanish.).

Barkin, David. 1998. Wealth, Poverty and Sustainable Development. Mexico City: Editorial Jus. (available free from: http://econwpa.wustl.edu/eprints/dev/papers/0506/05060003.pdf).

Barkin, David. 1999. The economic impact of ecotourism: conflicts and solutions in highland Mexico. pp. 157–172. In: Pamela Godde, Michael F. Price and F.M. Zimmerman (eds.). Tourism and Development in Mountain Regions. Cab International, London.

Barkin, David. 2000. Overcoming the Neoliberal Paradigm: sustainable popular development. Journal of Developing Societies XVI (1): 163–180.

Barkin, David (ed.). 2001. Innovaciones Mexicanas en el Manejo del Agua. Mexico City: Universidad Autónoma Metropolitana.

Barkin, David and Carlos Paillés. 2002. NGO-collaboration for ecotourism: a strategy for sustainable regional development in Oaxaca. Current Issues in Tourism 5(3): 245–253. (http://www.planeta.com/planeta/99/0499huatulco.html).

Barkin, David and Carlos Paillés. 2000. Water and forests as instruments for sustainable regional development. International Journal of Water 1(1): 71–79. (http://www.cabi-publishing.org/focus/socio_economics/).

Barón, Lourdes and David Barkin. 2001. Innovations in indigenous production systems to maintain tradition. pp. 211–219. *In*: Cornelia Flora (ed.). Interactions between Agroecosystems and Rural Human Community. CRC Press, Miami, FL.

Berkes, Fikret and Carl Folke. 1998. Linking Social and Ecological Systems: Management Practices and Social Mechanisms for Building Resilience. Cambridge University Press, Cambridge, UK.

Boianovsky, Mauro. 2010. A view from the tropics: Celso Furtado and the theory of economic development in the 1950s. History of Political Economy 42(2): 221–266.

Bonfil Batalla, Guillermo. 1996 [1987]. México Profundo: Reclaiming a Civilization (Translated by Philip A. Dennis). University of Texas Press, Austin.

Borrini-Feyerbend, Grazia, M. Taghi Farvar, Barbara Lassen, Gary Martin, Juan Ernesto F. Ráez-Luna, Carlos Riascos de la Peña and Stan Stevens. 2010. Bio-cultural diversity conserved by indigenous peoples and local communities—examples and analysis. Teheran: ICCA Consortium and Cenesta for GEF SGP, GTZ, IIED and IUCN/CEESP.

Brenner, Ludger and Hubert Job. 2006. Actor-Oriented management of protected areas and ecotourism in Mexico. Journal of Latin American Geography 5(2): 7–27.

Diaz, Floriberto. 2007. Comunidad y comunalidad, En: Sofia Robles H., y Rafael Cardoso J. (comps.), Floriberto Díaz. Comunalidad, energía viva del pensamiento. México: UNAM, pp. 34–50.

Escobar, Arturo. 2010. Una minga para el postdesarrollo: lugar, medio ambiente y movimientos sociales en las transformaciones globales. Lima: Programa Democracia y Transformación Global; Universidad Nacional Mayor de San Marcos. Facultad de Ciencias Sociales.

Escobar, Arturo. 1995. Encountering Development: The Making and Unmaking of the Third World. Princeton University Press, Princeton, NJ.

Esteva, Gustavo. 2001. The meaning and scope of the struggle for autonomy. Latin American Perspectives 28(2):120–148.

Frey, Scott. 2000. Environment and Society Reader. Allyn and Bacon/Longman, Boston.

Fuente, Mario and David Barkin. 2011. Concesiones forestales, exclusión y sustentabilidad. Desacatos (37). 93–110.

Funtowicz, Silvio and Jerry Ravetz. 1993. Science for the post-normal age. Futures 25: 739–755.

Gabriel, Leo. 2007. La tautonomy: autonomies multiculturelles en Amérique Latine et ailleurs. L'Harmattan, Paris.

Harvey, David. 2003. The 'New' Imperialism: Accumulation by dispossession. pp. 63–87. *In*: Leo Pantich and Collin Leys (eds.). Socialist Register 2004: The New Imperial Challenge. Monthly Review, New York.

Hill, Marianne. 2007. Confronting power through policy: on the creation and spread of liberating knowledge. Journal of Human Development 8(2): 259–282.

Illich, Ivan. 1973. Tools for Conviviality. Calder and Boyers, London. Available at: http://clevercycles.com/tools_for_conviviality/.

Klooster, Daniel. 2000. Institutional choice, community, and struggle: a case study of forest co-management in Mexico. World Development 28(1): 1–20.

Leff, Enrique. 2006. Aventuras de la Epistemología Ambiental. Siglo XXI, México.

Leff, Enrique. 1995. Green Production: Toward an Environmental Rationality. Guildford, NY.

Martin, Gary, J. Gary, Claudia Camacho Benavides, Carlos Del Campo García, Salvador Anta Fonseca, Francisco Chapela Mendoza and Marco Antonio González Ortíz. 2011. Indigenous and community conserved areas in Oaxaca, Mexico. Management of Environmental Quality 22(2): 250–266.

Martínez Luna, Jaime. 2003. Comunalidad y Desarrollo. México: Conaculta y Campo, A.C.

Mathews, Andrew Salvador. 2003. Suppressing fire and memory: environmental degradation and political restoration in the Sierra Juárez of Oaxaca, 1887–2001. Environmental History 8(1): 75–108.

Merino Pérez, Leticia and Jim Robson (eds.). 2006. Managing the Commons: Indigenous Rights, Economic Development and Identity. The Ford Foundation, NY.

Pohl Christian, Stephan Rist, Anne Zimmermann, P. Fry, G.S. Gurung, F. Schneider, C.H. Speranza, Boniface Kiteme, S. Boillat, E Serrano, Gertrude Hirsch Hadorn and Urs Wiesmann. 2010. Researchers' roles in knowledge co-production: experience from sustainability research in Kenya, Switzerland, Bolivia and Nepal. Science and Public Policy 37(4): 267–281.

Ravetz, Jerry and Silvio Funtowicz. 1999. Post-normal Science—an insight now maturing. Futures 31: 641–646.

Rist, Gilbert. 2008. The History of Development: From Western Origins to Global Faith, 3 ed. Zed Books, London.

Rist, Stephan, Freddy Delgado Burgoa and Urs Wiesmann. 2003. The role of social learning processes in the emergence and development of Aymara land use systems. Mountain Research and Development 23(3): 263–270.

Sánchez Vázquez, Adolfo. 2003. Filosofía de la Praxis. Siglo XXI, México.

Sen, Amartya, K. 2002. Development and Freedom. Anchor Books, NY.

Tauli-Corpuz, Victoria, Victoria, Leah Enkiwe-Abayao and Raymond de Chavez. 2010. Towards an Alternative Development Paradigm: Indigenous Peoples' Self-Determined Development. Tebtebba Foundation, Baguio City, Philippines.

Toledo, Víctor Manuel and Nicolás Barrera Bassols. 2008. La Memoria Biocultural: La Importancia Ecológica de Las Sabidurías Tradicionales. Icaria, Barcelona.

Toledo, Víctor Manuel. 2000. La Paz en Chiapas: Ecología, luchas indígenas y modernidad alternativa. UNAM y Quinto Sol, Mexico.

Utting, Peter. 2002. The Greening of Business in the South: Rhetoric, Practice and Prospects. Zed Press, London.

Venkateswar, Sita and Emma Hughes. 2011. The Politics of Indigeneity: Dialogues and Reflections of Indigenous Activism. Zed Press, London.

Wolf, Eric. 1982. Europe and the People without History. University of California Press, Berkeley, CA.

Endnotes

1. Eric Wolf (1982) was particularly insightful about this process, observing in one of his seminal books, that if tradition is to survive, along with the people who value it, community leaders must carefully guide their societies through a continuing process of change, guarding the most valuable elements of their cultural heritage while replacing others with changes and acquisitions from the outside.

2. For a detailed discussion of this process in Mexico, see Barkin 1990.

3. There is an explosive discussion of the significance of autonomy in Latin America. For a brief introduction see Gabriel 2007 as well as the website: http://www.latautonomy.org/.

4. This position is intimately related to a profound discussion of a functional in contrast to an ethical dimension of sustainability, particularly in Latin America. Here we refer to the redimensioning of the problem of social and productive exclusion analyzed in the regional literatures as *praxis* (Sánchez Vázquez 2003), "dialogue of knowledge

systems" (Leff 2006), or "traditional knowledge" (Toledo y Bassols 2008).

5. Academic analyses of this problem of the unique character of these peoples and their conflictual relationships with the dominant institutions are assuming an important role in the discussions of their possibilities for developing different paths for achieving material and cultural well-being consistent with environmental balance. An outstanding example of this is Venkateswara and Hughes (2011).

6. For two detailed studies that offer an historical review of these struggles in Mexico, consult Klooster (2000), Mathews (2003), and Fuente and Barkin (2011).

7. There is an abundant literature tracing the significance of this 'heterodox' approach to development and tracing its roots to Indo-American thinking over the past centuries. For a critique of the Western development model in English see Rist (2008). A prominent Latin American development critique, Celso Furtado, enjoys broad respect for his thoughtful contributions (Boianovsky 2010), while other insightful contemporary thinkers, strongly influenced by the work of Ivan Illich (1973), include Arturo Escobar (2010) and Gustavo Esteva (2001).

8. For more information on this work see, for example: Funtowicz and Ravetz (1993); and Ravetz and Funtowicz (1999). Some collections of analytical articles and case studies can be found in: Frey (2000) and Berkes and Folke (1998). For discussions of the way in which research methodologies are being transformed to attend to this new vision of collaboration, see, for example, Pohl et al. 2010, Rist et al. 2003. A different approach is evident in the experiences documented by the Forum of Indigenous and Community Conserved Areas (http://www.iccaforum.org) (Borrini-Feyerbend et al. 2010).

9. This approach is elaborated at greater length in Barkin (1998) as well as in Toledo (2000).

10. A collaborator in this project also used this information to develop a treatment for arteriosclerosis!

11. A note on the role of wage labor in this strategy is in order. Although many members of the communities described in this paper search for work in the capitalist firms and receive wages, including some who migrate abroad in search of higher incomes, the ability of the community to offer them a refuge, a place to return to where their livelihood and that of the family is guaranteed, gives them a measure of freedom that most workers do not have; the proletarian relationship, with its accompanying process of alienation, depends to a large extent on the lack of alternatives for people who must accept the wage-labor accord or face starvation.

12. Unfortunately, well meaning efforts like the current world-wide push to promote REDD (Reduction of Emissions from Degradation and Deforestation) programs through symbolic payments to communities for their conservation efforts are not likely to be more effective than the market oriented programs to offer "payments for environmental services" that removed communities around the world from their native environs in the name of conservation. The tragic history of the Protected Reserve of the Monarch Butterfly in Mexico is a particularly poignant example of good intentions transformed into environmental tragedy (Barkin 1999, Brenner and Job 2006).

13. For a more detailed description of this and other local initiatives for water management in Mexico see Barkin (2001). The group responsible for this program, *Alternativas y Procesos de Participación Social*, is directed by Raúl Hernández Garcíadiego y Gisela Herrerías Guerra. See the website: http://www.alternativas.org.mx.

14. This amphibian has the unique ability to regenerate its limbs, should one or more be lost, and therefore has become a subject for intense research in recent years in centers in the Americas and Europe; an introductory description and analysis can be found in *Wikipedia*.

15. http://www.indigenousclimate.org/index.php?option=com_content&view=article&id=108:declaration-of-the-latin-american-indigenous-forum-on-climate-change&catid=40:ip-declarations&lang=es&Itemid=0. Similar efforts are being proposed in many other places, but should not be confused with those being proffered by organizations trying to propose that this can be achieved under capitalism.

18

Projected Climate Change and Variability and Their Impact on the Canadian Rocky Mountains

Kaz Higuchi[1],* and Fiona Joncas[2]

INTRODUCTION: GLOBAL TRENDS IN CLIMATE—EARTH IS GETTING WARMER

Climate has changed in the historical past, sometimes very dramatically (like the Little Ice Age), and will continue to change in the future. However, since the onset of the Industrial Revolution in the mid-1800s, we have been conducting (unwittingly or not) a global geophysical experiment by emitting into the atmosphere an exponentially increasing amount of greenhouse gases, mainly carbon dioxide (CO_2) originating from fossil fuel combustion (coal, oil, natural gas, etc.), short circuiting the natural global carbon cycle. Another major anthropogenic source of CO_2 is the deforestation activities, mainly in the tropical regions. Based on the basic understanding of the radiative physics, the addition of these greenhouse gases has resulted in an overall increase in the global surface temperature, amplified by an increase in atmospheric water vapour through positive hydrological feedback. "There is unequivocal evidence that Earth's lower atmosphere, ocean, and land surface are warming; sea level is rising; and snow cover,

---

[1] Faculty of Environmental Studies, Graduate Program (Geography Department), York University, Toronto, Ontario, Canada.
[2] Disaster and Emergency Management Program, York University, Toronto, Ontario, Canada.
* Corresponding author

mountain glaciers, and Arctic sea ice are shrinking. The dominant cause of the warming since the 1950s is human activities." (AMS Statement on Climate Change 2012).

Figure 18.1 shows a line graph of the globally-averaged land-ocean temperature anomaly from 1880 to 2010, relative to the WMO (World Meteorological Organization) recommended average temperature climate reference period of 1951 to 1980. Black line connecting solid square symbol shows the annual mean and the solid red line is the corresponding 5-year running mean. The green vertical bars indicate uncertainty estimates.

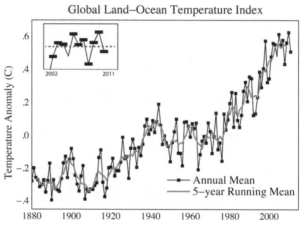

Figure 18.1. Source: http://data.giss.nasa.gov/gistemp/graphs/.

Figure 18.2 shows a globally observed increase in the annually averaged surface temperature during the decade 2000–2009, compared to the WMO recommended reference period of 1951 to 1980. The most significant warming, shown in red, was in the high-latitude Arctic region.

Despite the inter-annual and inter-decadal variability, the observation of the surface temperature of Earth points unequivocally to a warming trend, with about 0.8°C increase from 1901 to 2010, the period which includes a relatively rapid increase of about 0.5°C over the 1979–2010 period. With over 100 years of instrumental temperature record on a global basis, the 10 warmest years have occurred since 1997. Global climate models indicate that these warming trends are projected to continue as CO_2 and other greenhouse gases (such as CH_4 and N_2O) are continually emitted into the atmosphere from human activities (IPCC-AR4 2007). From the meteorological measurements from stations located in the valley floors of the central Canadian Rockies (Banff, Yoho and Jasper National Parks), it has been found that an increase of around 1.4 to 1.5°C in annual temperature occurred from 1888 to 1994 (Luckman 1998, Luckman and Kavanagh 2000).

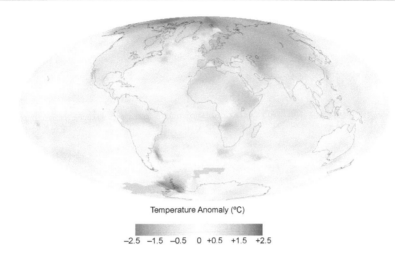

Temperature Anomaly (°C)

-2.5 -1.5 -0.5 0 +0.5 +1.5 +2.5

Figure 18.2. Source: NASA Earth Observatory Image of the Day (http://earthobservatory. nasa.gov/IOTD/view.php?id=42392).

This general increase was primarily influenced by the winter temperature which also contributed significantly to the inter-annual variability of the annual temperature.

Climate change over North America

Figure 18.3 shows projected surface annual mean temperature increases for five different regions in North America (IPCC-AR4 2007) by 21 AOGCMs (Atmosphere-Ocean General Circulation Models) relative to the base period of 1901 to 1950. Black lines indicate averaged observed temperature from 1906 to 2005, while the red envelopes indicate the spread of simulated results by the climate models. Projections from 2001 to 2100 by the models are shown by the orange envelopes for the A1B emission scenario (one of the A1 scenarios describing a very rapid economic growth, with a rapid introduction of new and efficient technologies, and the global population peaking in mid-21st century and declining thereafter). The three vertical bars at the end of the orange envelope denote projected average changes for the period 2091–2100 for different emission scenarios, including the A1B (orange). For the western North American region (WNA), the projected temperature increase ranges from around 2°C to nearly 8°C, depending on the model and the emission scenario. Norgues-Bravo et al. (2007) suggested that a mountainous region will likely experience a projected temperature increase of 2 to 3 times greater than that observed during the 20th century. The projected warming in the mountainous regions could be larger in

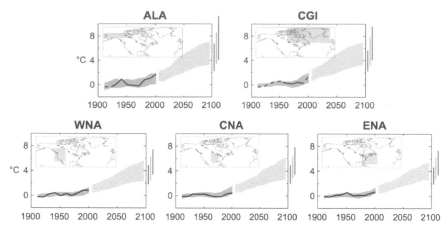

Figure 18.3. Source: Figure 11.11 of IPCC-AR4 (2007), Chapter 11 of Working Group 1.

winter over elevated areas due to snow-albedo feedback. This feedback effect is poorly modeled by AOGCMs due to their insufficient horizontal resolution to resolve the rapidly varying and rising topographic nature of the mountains.

Figure 18.4 shows seasonal variation in the temperature and precipitation changes averaged over 21 climate models driven by the A1B emission scenario. Change is from the periods 1980–1999 to 2080–2099. The top and middle rows show temperature and precipitation changes for annual, winter (DJF—December, January, February) and summer (JJA —June, July, August). The bottom row shows the number of models out of 21 that projected precipitation increases. As indicated in the figure, there is less seasonal variation in the temperature increases along the west coastal regions of North America, due to the influence of the oceans.

Climate scientists usually have a higher confidence in projecting changes in temperature than in other meteorological variables like precipitation. However, in the context of climate change, changes in temperature and precipitation constitute the most important impact on our environment, including socio-economic systems. The second row in Fig. 18.4 shows the seasonal variation in the precipitation change projected by the models.

With the displacement of the westerlies northward, in conjunction with the intensification of the Aleutian Low over the Gulf of Alaska, the climate models project a relatively large percentage increase in the winter precipitation over the northern high-latitude regions, including the Canadian Rockies in Yukon and northern British Columbia (IPCC-AR4, WG-1 2007). This fractional increase is enhanced on the western windward slopes of the Rockies (as an orographic precipitation). As indicated in the figure, the summer season shows a relative decrease in precipitation

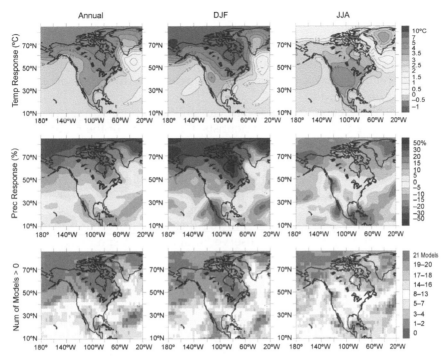

Figure 18.4. Source: Figure 11.12 of IPCC-AR4 (2007), Chapter 11 of Working Group 1.
Color image of this figure appears in the color plate section at the end of the book.

(as much as 15 to 20%) over the southern areas of the Canadian Rocky Mountains. It should be noted that there is less confidence in climate change projections by a global climate model at regional and local scales than, say at an overall continental scale. But it becomes extremely problematic for extreme climatic events at a regional scale.

In the recent past, we have noted that extreme weather events (collectively described as climatic variability) have usually produced disastrous impacts on our socio-economic systems. IPCC-AR4 (2007) has summarized the results of many studies that indicate, under a warming climate, we will see significant changes in temperature and precipitation extremes in the 21st century in many regions of the world. That is, the climate will become more variable. Intensity, Duration and Frequency (IDF) of extreme weather events, such as the heat wave, will increase. Projected amplification of the hydrological cycle by the climate models will lead to intensification of precipitation events (snow and rain) over some regions while other regions will face prolong drought periods.

Although global and regional climate models simulate a general decrease in snow depth in the Rocky Mountains with global warming

(Snyder et al. 2003, Giorgi et al. 2001), a study carried out by Leung et al. (2004) showed an increase in extreme precipitation of up to 10% for the period 2040 to 2060 in winter, particularly in the northern Rockies. With increases in temperature at higher elevations, the probability of rain-on-snow events increases, resulting in an increase in extreme runoff causing flooding downstream.

Simulating climate change and variability in mountainous regions is a challenging task for climate scientists. Complexity of the mountainous topography and surface characteristics can lead to a large spatial and temporal variability in climatic regimes and responses. The density of observations, in time and space, is nowhere near the required level to describe accurately the rapid changes in meteorological variables such as temperature and precipitation in a complex mountainous topography that varies quickly in space. Not only do these changes occur horizontally, but also vertically. It has been observed that climate response is elevation-dependent (Rangwala and Miller 2012).

Although mountains constitute an important component of the climate system, particularly in terms of the regional hydrological cycle, there have only been few model simulations to examine how mountain regions will respond to climate change, due mainly to the fact that the present global and regional climate models do not possess sufficiently adequate spatial resolution to resolve the topographical details of the mountains, as well as the detailed characteristics of the surface, such as various ecosystems (vegetation, snow/ice, etc.), that are needed to properly represent such important feedback mechanisms as snow/ice albedo feedback.

Climate Change and Variability in the Canadian Rocky Mountains

As stated earlier, globally-averaged surface temperature rose nearly 0.5°C from 1979 to 2010. It has been shown that a greater increase in daily minimum temperatures at night than daily maximum temperatures during the day has contributed more to the observed global temperature rise (Karl et al. 1991, Easterling et al. 1997, Zhai and Pan 2003). There have been several studies however, that suggest greater rates of warming have taken place in mountain regions (as compared to lower elevation surfaces), with temperature change as a function of elevation (Diaz and Bradley 1997, Penderson et al. 2010).

Understanding how climate in mountain regions will change with global warming is important because of its significant impact on environment (including humans) not just within the mountains, but also on lowlands downstream where human population and socio-economic activities are concentrated. Over 50% of the rivers in the world originate from glaciers in mountain regions. It has been suggested that 30 to 50% of the existing mountain glacier mass worldwide could disappear by 2100 (Beniston 2003). In the Canadian Rockies, three major rivers (Fraser River, Columbia River and Saskatchewan River) provide freshwater to millions of Canadians for domestic, agricultural, industrial and power generation usage, as well as for tourism. Based on a literature review of studies that looked at climate change in mountains, Rangwala and Miller (2012) lists the following possible impacts due to increasing temperature and precipitation changes:

a) a general decrease in winter and spring snowpack that could lead to changes in the seasonal streamflow pattern that could lead to reduced summer flows,

b) shifts in precipitation [in kind (snow or rain) and amount] could alter the nature of the hydrological cycle that would impact freshwater availability,

c) increased evaporation from small water bodies and would lead to severe summer drying,

d) ecosystem changes that would include intensity and frequency of wildfires, plant mortality due to drought and pest, intrusion of non-native flora and fauna species, shift of the present ecosystems to higher altitudes, and composition of aquatic life.

Increased warming will also result in a reduction of snow cover that could result in early seasonal runoff and increased albedo. Indeed, how snow and ice distribution in time and space will change under global warming will have a significant impact on the mountain hydrological cycle. A modeling study by Leung et al. (2004) indicates that areas experiencing snowfall at the present time will increasingly experience rain. It has been estimated that the snowline on mountain slopes will rise on average about 150 m for every 1°C rise in temperature (IPCC-AR4 2007). Although results vary and are often contradictory in studies examined by Rangwala and Miller (2012), they concluded that a majority of studies they examined "suggest an elevation-dependent climate response in both observations and climate models", with warming rates greater at higher elevations.

Simulating climate change and variability in mountainous regions is a challenging task for climate scientists. The density of observations, in time and space, is nowhere near the required level to describe accurately the rapid changes in meteorological variables such as temperature and precipitation in a complex mountainous topography that varies quickly in space. Not only do these changes occur horizontally, but also vertically. It has been observed that climate response is elevation-dependent (Rangwala and Miller 2012).

There have only been few model simulations to examine how mountain regions will respond to climate change, due mainly to the fact that the present global and regional climate models do not possess sufficiently adequate spatial resolution to resolve the topographical details of the mountains, as well as the detailed characteristics of the surface, such as various ecosystems (vegetation, snow/ice, etc.), that are needed to properly represent such important feedback mechanisms as snow/ice albedo feedback. But also in the case of precipitation, the influence topography has on precipitation process is not adequately represented in models, not only in terms of spatial resolution but also in terms of the basic physics.

In addition to the overall trend in climate change on a multi-decadal scale, the west coast of North America is also influenced by ENSO (El-Niño and Southern Oscillation), one of the prominent low-frequency variability modes of the atmosphere. ENSO is a quasi-periodic oscillation that switches between El Niño (the warm SST anomaly phase of the tropical Pacific Ocean) and La Nina (the cold SST anomaly phase of the tropical Pacific Ocean) with an average period of about 4 years (ranging from 2 to 7 years) (MacMynowski and Tziperman 2008). ENSO is a manifestation of a set of complicated processes of nonlinear air-sea feedback interactions, and it produces inter-annual and decadal variability in temperature and precipitation patterns of the western portion of Canada (Zhao et al. 2012, Bonsal and Shabbar 2011, Shabbar et al. 1997, Shabbar and Khandekar 1996) that has significant impact on the mountain ecosystems and freshwater supply downstream. See Fig. 18.5. Yet, at the present time, "despite considerable progress in our understanding of the impact of climate change on many of the processes that contribute to El Niño variability, it is not yet possible to say whether ENSO activity will be enhanced or damped, or if the frequency of events will change" (Collins et al. 2010).

Simulations of the present climate by the models in the mountain regions are relatively unrealistic and perhaps not so reliable because of inadequate understanding of some of the physical processes that generate the observed climate and the lack of sufficient spatial resolution to represent even those processes we understand well. The models are also deficient in realistically reproducing the characteristics of ENSO which has significant impact on the Canadian Rockies. However, because of the importance of the impact of climate change in the Rockies will have, not only within the mountain regions, but also downstream in the Canadian prairies (affecting the socio-economic and ecosystem functions there), we need to address these impacts and explore possible consequences based on available observational evidence and existing climate model projections that are based on our understanding of physical mechanisms that operate in the mountainous regions (IPCC-AR4 2007).

**TYPICAL JANUARY-MARCH WEATHER ANOMALIES
AND ATMOSPHERIC CIRCULATION
DURING MODERATE TO STRONG
EL NIÑO & LA NIÑA**

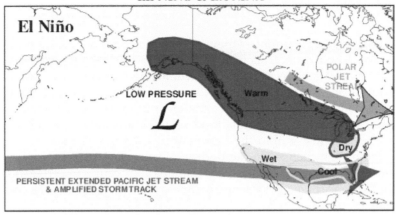

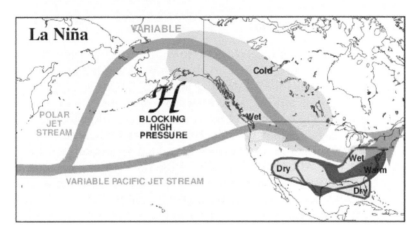

Figure 18.5. During El Nino phase of the ENSO cycle, warmer-than-usual temperature regime dominates the Canadian Rocky Mountain regions. The opposite is true during the La Nina phase, producing above-normal amount of precipitation in the southern regions of the Canadian Rocky Mountains (Source: National Oceanic and Atmospheric Administration (NOAA): http://www.cpc.noaa.gov/products/analysis_monitoring/ensocycle/nawinter. shtml).

Climate Change Impact in the Canadian Rocky Mountains

A summary of the final agreement which came out of the UN Conference on Environment and Development (UNCED) that took place in 1992 states, in part, that:

"Mountains are important sources of water, energy, minerals, forests and agricultural products and areas of recreation. They are storehouses of biological diversity, home to endangered species and an essential part of the global ecosystem."

(UN Earth Summit 1992)

The projected warming of the climate will impact various aspects of the mountain environment, such as on hydrological cycle involving changes in glaciers and snow, ecosystem biodiversity and species distribution, and natural hazards such as landslides. Changes in these natural environmental conditions in response to climate change and variation will then impact on human health, settlements, food security and other socio-economic activities, not just within the mountainous regions but also in regions affected by the changes in the hydrological cycle in the mountains. But where, how and by how much these changes will occur is uncertain because of the existence of many local microclimate regimes caused by the complex mountainous topography.

Glacial retreat and thinning

One of the most visible impacts of climate warming is the gradual decrease in the size of the existing glaciers. Although there are not comprehensive glacier inventory data for the Canadian Rockies, Luckman (1998) suggests that the existing data indicate an average loss of about 25% in area during the 20th century. He notes that the "data make no allowance for glacier thinning and therefore may considerably underestimate volumetric loss" (Luckman 1998). The loss is influenced by changes in both temperatures and precipitation. Although the precipitation data suffer from spatial and temporal, but mostly spatial, limitation in the Rockies, they show decadal variations (influenced by the low-frequency variability modes, such as the Pacific-North America (PNA) Oscillation caused by ENSO), peaking in the mid-20th century (Shabbar et al. 1997). In general terms, Oerlemans and Fortuin (1992) showed that the sensitivity of glacier volume changes to temperature is a function of precipitation. Using seasonal and annual mass balance measurements, Dyurgerov (2003) showed increases in both summer and winter glacier balances are related more to temperature and precipitation, respectively. These increases suggest that the hydrological cycle is being intensified by glaciers.

A relatively more recent and comprehensive analysis and discussion of changing glacier characteristics in western North America under climate change, and the impact on the hydrology, geomorphic hazards and stream water quality due to retreating and melting glaciers was carried out by Moore et al. (2009). The authors have noted that the recent glacial retreat and/or mass loss are accompanied by statistically detectable declines in late-summer glacier-fed streams. For example, glaciers located in the northwest British Columbia and southwest Yukon regions have shown very little terminal retreat but have experienced mass loss due mainly to thinning, causing increases in streamflows fed by the glaciers in these regions.

Melting of glaciers and permafrost, and precipitation more in the form of rain instead of snow, will likely cause various natural hazards such as soil erosion, landslides, debris and mud flows (Evans and Clague 1997). These hazards will pose significant risk to populations living downstream in the lowland regions.

Impact on hydrological cycle and streamflows

Overall, the glaciers in the Canadian Rockies have been retreating consistently over the last 30 years or so. With negative mass balance and volume loss, these glaciers will continue to lose ice and shrink. Some small glaciers have already disappeared, or will soon do so. This situation will likely accelerate in the future under climate warming, producing a pronounced variation in the present seasonal flow patterns of streams fed by the glaciers. Loukas et al. (2002), for example, examined what would happen to the seasonal flow pattern of Illecillewaet River in the Columbia Mountains of British Columbia. By reducing the glacier (water source for the river) by 1/3 and using a hydrological model driven by climate change model data for the period 2008 to 2100, they found, relative to the 1970–1990 level, a slight increase in the streamflow in spring (May and June) and a decrease from July to September.

Kienzle et al. (2012) examined how streamflow in the Cline River watershed would change under the IPCC projected increases in temperatures and precipitation over western Alberta. The Cline River watershed contributes over 40% of the North Saskatchewan River streamflow at Edmonton, Alberta. About 60% of the mean annual Cline River streamflow comes from snowmelt, while about 8% comes from glacier melt. Kiengle et al. (2012) found that increases in mean annual precipitation of about 1.2, 11, and 16.6% would be there in the 2020s, 2050s and 2080s, respectively, leading to "a clear shift in the future hydrological regime, … with significant higher streamflow between October and June, and lower streamflow in July-September." Although nearly 70% of the streamflow in the July-September period can come from glacier melt, it was not included in their streamflow estimates due to "current inadequate simulation of future glacial melt." Thus it is likely that the July-September decreases in the streamflow are over-estimated. Earlier peak streamflow was found to be caused by the earlier melting of the snowpack (due to higher spring temperatures), combined with increased spring rainfall.

Increased streamflow and retreating glaciers (exposing thitherto unexposed ground) can lead to an increase in downstream flooding, sediment transport, increased erosion (e.g., land slides) and change in water quality (Moore et al. 2009). These changes in streamflow regimes have significant socio-economic and ecological impacts. Water resource is essential for such activities as hydroelectric power generation (e.g., Columbia River) and agricultural production.

Impact on ecosystems and socio-economic activities

Significant changes in the existing climatic conditions will alter the complex interaction between the 'natural' environment and human socio-economic activities, such as agriculture, forestry and tourism. Hydrological cycle in terms of water supply and demand will mediate that interaction in many mountain regions.

Sensitivity of potential effects of climate change on water resource for crops in the Okanagan Basin in the interior region of British Columbia was studied by Neilsen et al. (2006). The Okanagan Valley is located in the southern interior of British Columbia, its mild and dry climate influenced by the rain shadow effects from the mountains on the west and a number of lakes (Lake Okanagan being the largest in the valley). The agricultural land usage is characterized by a complex distribution of orchards, vineyards and pasture/forage. Okanagan agriculture is almost entirely dependent on irrigation, drawing water from Okanagan River and lakes, tributary streams and groundwater. Nearly 75% of the necessary water supply comes from tributary streams (Neilsen et al. 2006). It is projected that by 2080, the tributary streams will be reduced by roughly 15 to 30% of the current flow. A combination of decreasing water supply and increasing demand due to increasingly longer growing season (in addition to the demand from increasing population) will increase the occurrences of drought and thus put agriculture production in the Okanagan Valley at extreme risk. Vulnerability of the grape industry in the Okanagan Valley is discussed by Belliveau et al. (2006).

Another major impact of climate change concerns mountain ecosystems and trees. Ecological systems in a mountainous region are sensitive to climate change due to the existence of various species niches adapted to corresponding microclimatic habitats produced by rapid changes in temperature and precipitation regimes, both vertically and horizontally. This produces high biodiversity, "often with sharp transitions (ecotones) in vegetation sequences, and equally rapid changes from vegetation and soil to snow and ice" (Beniston 2003). Luckman (1998) for example noted "small but significant vegetation changes ... are taking place at upper econtonal boundaries (where rapid step-like changes in vegetation types occur) in response" to the overall warming trend in the 20th century. The treeline is perceived as most sensitive to climate change, with the upslope migration being a typical response to warming at higher elevations. However, an investigation at three sites near the Columbia Icefield, Luckman and Kavanagh (1998) showed that the nature of treeline response to climate change depends on local environmental conditions and 'the strategies adopted by individual species', as well as to the genetic diversity of the species in their ability to adapt to rapid environmental change and to be able to achieve wide dispersion of seeds. For example, the treeline of *Abies lasiocarpa* on the north-facing slope has not moved for the last 400 years. On the other hand, the treeline of *Picea engelmannii* on the south-facing slope has migrated upslope by seedling establishment. They have noted that other factors, such as soil and groundwater level, are important in influencing changes in plant species distribution over time. In addition, intensity and extent of forest fires due to frequent drought and high temperatures (as well as due to diebacks caused by insect infestation) can also alter the ecological succession of plant species. Referring to Ebata (2004), Hamann and Wang (2006) mentioned that the pine beetle infestation in British Columbia has become an epidemic, with over 4×10^6 ha destroyed. This is partially caused by increasingly warmer winter minimum temperatures in the Rockies. Ordinarily, colder winter temperatures would kill much of the insect larvae residing inside the tree barks. Harvell et al. (2002) showed that the warming of winter temperatures will cause increased probability in the wintertime survival of various pathogens, as well as causing increased severity in forest disease outbreaks. Henderson et al. (2002) suggested two possible ways in which a forest can change in response to climate change, one by slow and cumulative decline or by catastrophic change, such as a major forest fire. Hamann and Wang (2006) also found that: (1) Tree species at the northern range limits have migrated northward at a rate of 100 km per decade, (2) Hardwood species are generally not affected by climate change, and (3) many of the important conifer species will eventually disappear as their suitable habitat shrinks under climate change. These changes will have a profound impact on the forest industry in British Columbia.

Every year the Rocky Mountains experience a number of forest fires, endangering human lives and socio-economic activities of urban communities (such as Kelowna, British Columbia) located within mountainous regions. As stated earlier in the chapter, IPCC-AR4 (2007) has projected prolonged periods of summer drought with climate warming. It is reasonable to assume therefore, that such a change will cause an enhancement in forest fire frequency and intensity, producing increased risk to major population centres. A combination of summer dryness and insect infestation (such as the pine beetle) would produce increasingly favourable conditions for major forest fires.

Another socio-economic acitivity in the Rocky Mountains which will likely be significantly affected directly and indirectly by climate warming is recreation and tourism. Activities such as skiing, snowmobiling and snowshoeing that depend directly on snow will likely be negatively affected by less snow (Scott and Jones 2005), although if the wintertime temperature remains below zero, snow-making machines can partially offset the impact. Wintertime tourism in areas like Banff and Jasper are significantly at risk. However, such a negative impact might be offset, at least partially, by increased summertime activities (such as mountaineering and hiking) due to longer summer seasons. Recreational activities that are indirectly affected include those that are influenced by landscape changes. This has been referred to as the 'capital of nature' by Krippendorf (1984). Such activities include hunting, fishing and sightseeing.

As noted earlier, changes in the hydrological cycle in the mountains have a significantly important impact on people living downstream lowland areas. In Canada, that basically covers much of the western prairie regions. Much of this impact will be felt through changes in water availability. As stated by Henderson and Sauchyn (2008), "Water impacts our health and well-being, food production, infrastructure, energy production, forestry, recreation, and communities large and small." A significant portion of water (via melting glaciers and snow) from Rockies is channeled to the prairies through Saskatchewan River. Large urban centres like Edmonton depend on the quality and amount of water to sustain human activities. The river also feeds much needed water for agricultural food production. As noted earlier, changing climate in the Rockies will change the water supply and quality streaming out of the mountains. Continued shrinkage of glaciers with consequent decreases in streamflow will compound the increased water shortages already observed during drought years (Akinremi et al. 1999). This situation will become more acute as water demand increases due to rising food production and population. As we have seen with the Cline River watershed study by Kienzle et al. (2012), the streamflow goes through a seasonal variation that can be accentuated by changes in snow and glacier melt. It was noted that increased streamflow and retreating glaciers (exposing thitherto unexposed ground) can lead to an increase in downstream flooding, sediment transport, increased erosion (e.g., land slides) and change in water quality (Moore et al. 2009). It has been reported that a dry condition can lead to increased pathogens and toxic chemicals in water supplies (Charron et al. 2004). Flooding can also lead to an increase in water-borne diseases. A high variation in streamflow can also create small stagnant water bodies such as ponds along a river that can harbour mosquito larvae and increased incidents of West Nile virus outbreaks. Another mosquito-related disease like malaria cannot be reasonably ruled out.

Adaptation to Climate Change in the Rockies

Climate change in the Canadian Rocky Mountains will affect nearly all aspects of various ecosystems and related social and economic activities of humans. The relationship between the 'environment' and 'us' is a function of highly nonlinear processes that change dynamically as we respond in a feedback loop to a continous change (gradually or suddenly) in environmental conditions forced externally by climate warming. Any sustainable and flexible adaption strategy needs to incorporate this dynamic interactive relationship. What makes this process difficult is that future projections of changes in temperature and precipitation, as well as of the dynamics of glacier retreat, are filled with uncertainties, some of which are contributed by not knowing or deciding what our 'optimal' adaptation choices are. This is why adaptive management is a crucial tool in adapting to climate change. The approach is characterized by constant monitoring of (1) various environmental changes in the landscape and (2) the effectiveness of the adaptive strategy to achieve the desired goal. Since adaptation is basically local and regional in scale, any scientifically-based strategy must be tempered with social, economic, cultural and ethical uniqueness of each region affected by climate change. Involvement of stakeholders in any successful adaptive strategy is essential in post-normal science, balancing the needs of socio-economic development and the conservation of the environment.

Hansen et al. (2010) noted in a report on the Yellowstone to Yukon Conservation Initiative that a "responsive climate adaptation strategy will require a fundamental shift in conservation planning" that includes "strong coalitions among divers stakeholders." They have identified the Y2Y (Yellowstone to Yukon) region as the most appropriate spatial scale for providing, among other things, "sufficient space and connectivity for biological and ecological shifts" that could take occur under rapid climate change. The Y2Y region should be able to "conserve ecological diversity" and to "maximize the range of bioclimate variability ..., and to protect natural disturbance regimes such as fire and floods that sustain its ecological communities." The Y2Y organization has also initiated efforts to reduce intrusion of roads, railways and pipelines that cause habitat fragmentation and loss, as well as prevention of the spread of human communities in ecologically sensitive areas. In this process, the Y2Y Conservation Initiative has included various stakeholders composed of "individuals, conservation groups, business, government agencies, Native American Tribes, First Nations, and ecoregional coalitions."

Quality and quantity of water is the most important natural variable we depend on for sustaining our lives. The prairie regions of Canada depend on water from the Rockies for "health and well-being, food production, infrastructure, energy production, forestry, recreation, and communities large and small" (Henderson and Sauchyn 2008). For agriculture, one can reduce water demand by implementing more efficient irrigation methods, as well as irrigating after sunset to reduce evapotranspiration loss. Water recyling for industrial and private consumption is a technology that is already here and can be implemented without delay. Better water management practices include conscious change in personal water usage habits, such as reduction in watering lawns and washing vehicles.

As noted above, any successful adaptation strategy to climate change needs to include discussions with stakeholders right from the initial stages of the strategy development. The importance of this was highlighted by a study conducted by Cohen et al. (2006) in developing water management in the Okanagan Valley region in the interior British Columbia. They carried out four case studies in the Valley, each exploring decision-making processes on different aspects of water management. We already noted earlier, the sensitivity of water resource for agriculture in the Okanagan Valley under climate warming. The Cohen et al. (2006) study explores various general adaptation strategies for water management that involved stakeholders representing fisheries, watershed protection, agriculture, and local water management practices. The study identified five key components constituting a framework for developing an integrated approach to assessing climate change impact on water resource and adaptation strategy development. These are (Cohen et al. 2006):

1) Climate change scenarios: downscaling of global climate change scenarios to the regional level,
2) Hydrological scenarios: using a hydrological model to assess impact of climate change to watershed hydrology,
3) Water supply and demand: developing future demand scenarios for municipalities and irrigated agriculture,
4) Adaptation options: exploring previous management experience and potential future approaches for augmenting water supply and/or reducing water consumption, and
5) Adaptation dialogue with stakeholders: learning about regional perspectives on adapting to climate change.

The dialogue component is important and for it to be conducted successfully, Cohen et al. (2006) identified three essential elements: (1) a clear identification of the dialogue's objectives, (2) an understanding of the climate change and adaptation on a regional scale, and (3) a construction of trust among the participants (researchers and stakeholders) through exchange of information and shared learning. Continuous dialogue among the participants is essential to keep the adaptation strategy flexible and sustainable in the face of complex dynamic nature of climate change and large uncertainties associated with regional climate change projections.

Conclusion

The temporal and spatial density of observations in mountain regions is inadequate to describe accurately the rapid changes in meteorological variables (such as temperature and precipitation) that would occur in a complex topography with steep slopes. Inadequacies of the sparse and scattered observational stations become even more acute when we realize that there is a potential vertical gradient in the nature of the meteorological response to climate change. Many of the existing meteorological stations are few and far in between, and are mostly located in the mountain valleys.

Simulating climate change and variability in mountainous regions is a challenging task for climate scientists. Complexity of the mountainous topography and surface characteristics can lead to a large spatial and temporal variability in climatic regimes and responses. There have only been a few model simulations to examine how mountain regions will respond to climate change, due mainly to the fact that the present global and regional climate models do not possess sufficiently adequate spatial resolution to resolve the topographical details of the mountains, as well as the detailed characteristics of the surface, such as various ecosystems (vegetation, snow/ice, etc.), that are needed to properly represent such important feedback mechanisms as snow/ice albedo feedback. Simulating the details of the existing mountain climate requires a significantly more computation power and resource than is presently available. Also, there are not enough observational data to verify with confidence any model simulation output.

However, understanding how climate in mountain regions will change with global warming is important because of its significant impact on environment (including humans) not just within the mountains, but also on lowlands downstream where human population and socio-economic activities are concentrated. Over 50% of the rivers in the world originate from glaciers in mountain regions. It has been suggested that 30 to 50% of the existing mountain glacier mass worldwide could disappear by 2100. In the Canadian Rockies, three major rivers (Fraser River, Columbia River and Saskatchewan River) provide freshwater to millions of Canadians for domestic, agricultural, industrial and power generation usage, as well as for tourism. We showed results from two case studies, one from the Cline River watershed in western Alberta and another one from the Okanagan River watershed in the interior British Columbia. In both of these cases, impact of climate change on the seasonal variation of streamflow is important in agriculture and power generation, as well as in tourism. The forest industry in British Columbia will be significantly impacted from the ecosystem redistribution of tree species, with many of the important conifer species quickly losing their habitats. A lot more research is required to understand how the interaction between the changes in meteorological variables under climate change and those factors which influence the biodiversity of the environment and the nature of human socio-economic activities.

References

Akinremi, O.O., S.M. McGinn and H.W. Cutforth. 1999. Precipitation trends on the Canadian Prairies. Journal of Climate 12: 2996–3003.

AMS Statement on Climate Change. 2012. An Information Statement of the American Meteorological Society (Adopted by AMS Council 20 August 2012).

Belliveau, S., B. Smit and B. Bradshaw. 2006. Multiple exposures and dynamic vulnerability: Evidence from the grape industry in the Okanagan Valley, Canada. Global Environmental Change 16: 364–378.

Beniston, M. 2003. Climatic change in mountain regions: a review of possible impacts. Climatic Change 59: 5–31.

Charron, D.F., M.K. Thomas, D. Waltner-Toews, J.J. Aramini, T. Edge, R.A. Kent, A.R. Maarouf and J. Wilson. 2004. Vulnerability of waterborne diseases to climate change in Canada. Journal of Toxicology and Environmental Health, Part A 67: 1667–1677.

Cohen, S., D. Neilsen, S. Smith, T. Neale, B. Taylor, M. Barton, W. Merritt, Y. Alila, P. Sheperd, R. McNeill, J. Tansey, J. Carmichael and S. Langsdale. 2006. Learning with local help: expanding the dialogue on climate change and water management in the Okanagan region, British Columbia, Canada. Climatic Change 75: 331–358.

Collins, M., S.-I. An, W. Cai, A. Ganachaud, E. Guilyardi, F.-F. Jin, M. Jochum, M. Lengaigne, S. Power, A. Timmermann, G. Vecchi and A. Wittenberg. 2010. The impact of global warming on the tropical Pacific Ocean and El Nino. Nature-Geoscience 3: 391–397.

Diaz, H.F. and R.S. Bradley. 1997. Temperature variations during the last century at high elevation sites. Climatic Change 36: 253–279.

Dyurgerov, M. 2003. Mountain and subpolar glaciers show an increase in sensitivity to climate warming and intensification of the water cycle. Journal of Hydrology 282: 164–176.

Eastering, D.R., B. Horton, P.D. Jones, T.C. Peterson, T.R. Karl, D.E. Parker, M.J. Salinger, V. Razuvayev, N. Plummer, P. Jamason and C.K. Folland. 1997. Maximum and minimum temperature trends for the globe. Science 277: 364–367.

Ebata, T. 2004. Current status of mountain pine beetle in British Columbia. pp. 52–56. In: P.T.L. Shore, J.E. Brooks and J.E. Stone (eds.). Mountain Pine Beetle Symposium: Challenges and Solutions. Natural Resources Canada, Canadian Forest Service, Pacific Forestry Centre, Information Report BC-X-399, Victoria, British Columbia, Canada.

Evans, S.G. and J.J. Clague. 1997. The impact of climate change on catastrophic geomorphic processes in the mountains of British Columbia, Yukon and Alberta. pp. 1–16. In: E. Taylor and B. Taylor (eds.). The Canada Country Study: Climate Impacts and Adaptation, Volume I: Responding to Global Climate Change in British Columbia and Yukon. British Columbia Ministry of Environment Lands & Parks and Environment Canada, Vancouver, British Columbia, Chap. 7.

Giorgi, F., B. Hewitson, J. Christensen, M. Hulme, H. Von Storch, P. Whetton, R. Jones, L. Mearns and C. Fu. 2001. Regional climate information—evaluation and projections. pp. 583–638. In: J.T. Houghton et al. (eds.). Cambridge Climate Change 2001: The Scientific Basis. Contribution of Working Group I to the Third Assessment Report of the Intergovernmental Panel on Climate Change University Press, Cambridge, United Kingdom and New York, NY, USA.

Hamann, A. and T. Wang. 2006. Potential effects of climate change on ecosystem and tree species distribution in British Columbia. Ecology 87: 2773–2786.

Hansen, L., G. Tabor, C.C. Chester, E. Zavaleta, L. Graumlich, E. Rowland and R. Hebda. 2010. Making adaptation happen: a climate adaptation agenda for the Y2Y region. pp. 53–61. *In*: L. Graumlich and W.L. Francis (eds.). Moving Toward Climate Change Adaptation: The Promise of the Yellowstone to Yukon Conservation Initiative for addressing the Region's Vulnerabilities. Yellowstone to Yukon Conservation Initiative, Canmore, Alberta.

Harvell, C., C. Mitchell, J. Ward, S. Altizer, A. Dobson, R. Ostfeld and M. Samuel. 2002. Climate warming and disease risks for terrestrial and marine biota. Science 296: 2158–2162.

Henderson, N., T. Hogg, E. Barrow and B. Dolter. 2002. Climate Change Impacts on the Island Forests of the Great Plains and the Implications for Nature Conservation Policy. Prairie Adaptation Research Collaborative, Regina, Saskatchewan, pp. 116.

Henderson, N. and D. Sauchyn. 2008. Climate change impacts on Canada's Prairie Provinces: a summary of our state of knowledge. Parks Canada, Report 08-01, pp. 20.

Kienzle, S.W., M.W. Nemeth, J.M. Byrne and R.J. MacDonald. 2012. Simulating the hydrological impacts of climate change in the upper North Saskatchewan River basin, Alberta, Canada. Journal of Hydrology 412-413: 76–89.

Krippendorf, J. 1984. The capital of tourism in danger. Reciprocal effects between landscape and tourism. pp. 427–450. *In*: E.A. Brugger, G. Furrer, B. Messerli and P. Messerli (eds.). The Transformation of Swiss Mountain Regions. Haupt Publishers, Bern.

Leung, L.R., Y. Qian, X. Bian, W.M. Washington, J. Han and J.O. Roads. 2004. Mid-century ensemble regional climate change scenarios for the western United States. Climatic Change 62: 75–113.

Little, J.L., K.A. Saffran and L. Fent. 2003. Land use and water quality relationships in the lower Little Bow River watershed, Alberta, Canada. Water Quality Research Journal of Canada 38: 563–584.

Luckman, B.H. 1998. Landscape and climate change in the central Canadian Rockies during the 20th Century. Canadian Geographer 42: 319–336.

Luckman, B.H. and T. Kavanagh. 2000. Impact of climate fluctuations on mountain environments in the Canadian Rockies. Ambio 29: 371–380.

MacMynowski, D.G. and Eli Tziperman. 2008. Factors affecting ENSO's period. Journal of the Atmospheric Sciences 65: 1570–1586.

Moore, R.D., S.W. Fleming, B. Menounos, R. Wheate, A. Fountain, K. Stahl, K. Holm and M. Jakob. 2009. Glacier change in western North America: influences on hydrology, geomorphic hazards and water quality. Hydrological Processes 23: 42–61.

Neilsen, D., C.A.S. Smith, G. Frank, W. Koch, Y. Alila, W.S. Merritt, W.G. Taylor, M. Barton, J.W. Hall and S.J. Cohen. 2006. Potential impacts of climate change on water availability for crops in the Okanagan Basin, British Columbia. Canadian Journal of Soil Science 86: 921–936.

Oerlemans, J. and J.P.F. Fortuin. 1992. Sensitivity of glaciers and small ice caps to greenhouse warming. Science 258: 115–117.

Pederson, G.T., L.J. Graumlich, D.B. Fagre, T. Kipfer and C.C. Muhlfeld. 2010. A century of climate and ecosystem change in Western Montana: what do temperature trends portend? Climatic Change 98: 133–154.

Rangwala, I. and J.R. Miller. 2012. Climate change in mountains: a review of elevation-dependent warming and its possible causes. Climatic Change 114: 527–547. DOI 10.1007/s10584-012-0419-3.

Scott, D. and B. Jones. 2005. Climate change and Banff National Park: implications for tourism and recreation. Department of Geography, University of Waterloo, Waterloo, Ontario. Report prepared for the town of Banff, Alberta, pp. 25.

Shabbar, A., B. Bonsal and M. Khandekar. 1997. Canadian precipitation patterns associated with the Southern Oscillation. Journal of Climate 10: 3016–3027.

Shabbar, A. and M. Khandekar. 1996. The impact of El Niño-Southern Oscillation on the temperature field over Canada. Atmosphere-Ocean 34: 401–416.

Snyder, M.A., L.C. Sloan, N.S. Diffenbaugh and J.L. Bell. 2003. Future climate change and upwelling in the California Current. Geophysical Research Letters 30: 1823. DOI:10.1029/2003GL017647.

Zhai, P. and X. Pan. 2003. Trends in temperature extremes during 1951–1999 in China. Geophysical Research Letters 30. DOI:10.1029/2003GL018004.

Zhao, H., K. Higuchi, J. Waller, H. Auld and T. Mote. 2012. The impacts of the PNA and NAO on annual maximum snowpack over southern Canada during 1979–2009. International Journal of Climatology. DOI:10.1002/joc.3431.

19

Environmental and Socio-Economic Changes in the Rural Andes: Human Resilience and Adaptation Strategies

Christoph Stadel

INTRODUCTION

Since centuries, rural communities in high mountains had to cope with multiple changes of a harsh environment and fragile livelihoods. In turn, farmers and pastoralists have utilized the full range of natural resources in an optimal and often sustainable way, with the primary objectives of diversifying their economic base and of minimizing the livelihood risks. This has been achieved by a long-standing tradition of experiences and learning processes, by tenacious and inventive methods of resilience, and by skillful adaptation strategies (Gade 1999).

Rural landscapes in the Andes are characterized by an impressive diversity of natural environments and by multiple resource assets. This is particularly the case in the tropical realm where the ecological altitudinal zones of the *tierra caliente*, the *tierra templada*, the *tierra fria* and the *tierra helada*

Department of Geography & Geology, University of Salzburg, Hellbrunner-Str. 34, A 5020 Salzburg, Austria.
Email: christoph.stadel@sbg.ac.at

offer a remarkable range of agricultural potential. This is complemented by a multitude of topographic, hydrographic and climatic niches. The landscape is further overlaid by a cultural mosaic and by the recent impact of political ecology parameters. These agricultural 'archipelagos' and 'overlapping patchworks' (Zimmerer 1999) of human utilization also represent fragile and potentially vulnerable regions exposed to environmental risks and to socio-economic or political threats. The external natural and human stimulators or stressors affecting the environment and people are complemented by new challenges and changes within the rural societies. Thus, farming and human settlement in the Andes have millennia tradition of an 'open' rural system, constantly adapting to an array of changing environmental, political, cultural, social and economic conditions (Stadel 2003b, Stadel 2008).

Environmental Changes

As is the case in many other high mountains, the Andes are susceptible to widespread and frequently occurring natural risks and hazards. This has resulted, in specific geographic and temporal contexts, to conditions of vulnerability and criticality for rural communities, at times even to social and economic disasters. Particularly harmful have been the suddenly occurring earthquakes and volcanic eruptions which often set in motion other geomorphic processes. A notable example of such events in recent times has been the earthquake in the Santa Valley of Peru in 1970 which broke off a large part of the rock and glacier summit face of the Huascarán and buried the settlement of Yungay causing the sudden death of some 20,000 people. Since that time, the old townsite of Yungay was not rebuilt, with new Yungay being founded nearby, but in a presumably safer location. In the Cordillera Blanca, glacier ice, rock and debris material have sometimes also fallen into high mountain lakes, generating huge spillovers and flashfloods further below. Another devastating disaster example was the 1985 eruption of the Nevado del Ruiz in Colombia which triggered a massive *lahar* in the Langunilla Valley at its base and devastated the town of Armero which was never rebuilt but left as a memorial site.

Extraordinary weather and climate events are also deeply affecting the life of Andean people and entailing major changes in Andean environments and societies. Whereas sudden weather events tend to have rather short-term impacts in the form of excessive rains and flooding, droughts, frost, snowfall or hail, periodic weather changes generated by anomalies of the currents of the Pacific Ocean, called *El Niño-* and *La Niña*-events, generally have vaster and longer-lasting repercussions. The abnormally warm *El Niño* generates excessive rains in the Pacific coastal plains of Ecuador and Peru and the adjacent western Cordillera, while the *Altiplano*-region of Peru and Bolivia experience, under these conditions, a deficit of precipitation.

In contrast, when the cold *La Niña*-current reaches the Pacific shore, the reverse situation can be observed in the Andean realm. It is obvious that this pattern leads to temporary changes and adjustments of field cultivation and pastoralism.

The current debate focuses on the impact of 'global warming' of mountain environments. In the tropical Andes, the recent high rates of glacier melting have short-and long-term repercussions on the environment and society. If the smaller snow- and ice-caps of mountains are disappearing, indigenous people are experiencing a spiritual loss. A further grave social and economic consequence is the diminution or disappearance of melt water, most acutely felt during the dry seasons, with shortages becoming increasingly problematic for drinking water supplies, sanitation and irrigation-based agriculture. In the case of larger glaciated regions, this may not yet be a problem; but on a longer term, the situation may also become critical. The threat of a looming water scarcity will put many valley- and lower elevation regions with a perennial cultivation of products for national and global markets at risk. A similar problem may develop in the case of seasonally irrigated *Altiplano* pastures, especially for the wool-based economy of lamas and alpacas. The growing competition for water will likely also exacerbate the socio-economic disparities between 'water-rich' landowners and households on the one hand, and the 'water-poor' small farmers and marginalized households. It will furthermore accentuate the water distribution debate opposing urban and rural water requirements. The water issue will entail various adaptation strategies for Andean agriculture; it will further highlight regional development programs with a focus on a wise and sustainable use of water, and will necessitate a new 'generation' of agrarian reforms with the objective of a more equitable access to and distribution of rural water.

Socio-economic Changes

While until recently global changes in mountain regions have been closely linked to climate changes, it is argued here, that cultural, social, economic, and political changes affecting even formerly remote mountains, may have at least an equally profound and often more immediate impacts. Under the influence of ubiquitous economic and social modernization processes of the rural population, the agrarian structures, for the most part, have undergone profound transformations. Indigenous communities, in particular, are facing the challenges of remaining attached to their traditions, while at the same time being confronted with the opportunities and challenges of innovations and modernization (Stadel 2003). The new imperatives of capital- and market orientation promoted by governments, wealthy landowners and large corporations have changed the production forms, economic strategies and agricultural techniques of many farms. Whereas in the past the major objective of most small farmers has been to minimize environmental and economic risks by traditional strategies of agrarian diversification and a high degree of self-sufficiency, the prevalent new outlook of agriculture is characterized by a greater degree of specialization, higher financial investments, new agricultural technologies and an enhanced profitability. While these objectives are primarily pursued by the larger and productive farms, they also have a 'demonstration effect' for many smaller family farms which in turn may become 'islands of sustainability' (Bebbington 1997). This can be seen, for instance, in the production of milk, butter and cheese, specialized fruit, vegetables, or sheep and alpaca wool for regional, national and international markets; in the proliferation of greenhouses for a cultivation of flowers, and also of numerous poultry stables. In the piedmont regions of the eastern *Cordilleras* in Colombia, Peru and Bolivia, a legal and illegal cultivation of coca represents for many farmers more lucrative economic opportunities than other forms of agriculture. The efforts of the United States and occasional interventions by national governments to promote economic alternatives were largely unsuccessful. Table 19.1 gives a summary of the components of the major changes of agriculture and rural communities.

The Agrarian reforms since the 1960s abolished in most parts the feudal system of the traditional *haciendas* and granted thousands of former landless peasant small parcels of privately owned land. However, the Agrarian reforms in most countries later stalled, or were even annulated, the small farmers often received an inadequate financial and technical support, and also suffered from deficient rural infrastructures and services. In addition, a subsequent subdivision and fragmentation of land holdings aggravated the rural situation. Especially young people had to look for seasonal agricultural employment generally in the lowland plantations, or they sought work in non-agricultural activities, in mines, oilfields, and mainly in cities. Many rural people also emigrated, sent remittances to their remaining family members or made investments in their home villages. Modernization, progress and genuine development still remain largely restricted to local or at best regional niches of affluence and to a small number of actors and 'market players'. For the most part, the prevailing *minifundio* agriculture offers rather limited perspectives and forces the "neoliberal unravelling of the agrarian reform".

Rural Resilience and Adaptation Strategies

The concept of resilience focuses on the resistivity of people-environment–systems. It relates to a flexible adaptation of people to changes of natural

Table 19.1. Changing socio-economic conditions for Andean agriculture and communities.

INCREASED MARKET AND PROFIT ORIENTATION, IMPACT OF GLOBALIZATION

- CONVERSION OF LAND USE FROM SUBSISTENCE ORIENTATION TO MARKET ORIENTATION WITH THE CULTIVATION OF EXPORT-ORIENTED PRODUCTS
- INTENSIFICATION OF AGRICULTURAL LAND USE, OFTEN WITH INCREASED APPLICATION OF IRRIGATION, IMPLEMENTS AND MECHANIZATION
- OFTEN AGRICULTURAL LAND AMALGAMATIONS BY PRIVATE LAND OWNERS, COOPERATIVES, CORPORATIONS AND GOVERNMENTS
- POTENTIALLY DETERIORATING LAND USE AT THE OUTSKIRTS OF THE COMMERCIAL FARMS

STRENGTHENING OF COMMUNITY SUBSISTENCE AGRICULTURE WITH REGIONAL MARKET ORIENTATION

- SUSTAINABLE, TO THE LOCAL CONDITIONS ADAPTED LAND USE
- AGRICULTURAL SYSTEMS AND TECHNIQUES BASED ON PROVEN TRADITIONAL PRACTICES AND ON ECOLOGICALLY AND CULTURALLY ADAPTED METHODS
- AGRICULTURAL LAND USE PROTECTING ENVIRONMENTAL INTEGRITY AND CAUTIOUS RESOURCE USE (LAND, SOIL, WATER, VEGETATION)
- EMPHASIS ON NATIVE CROPS AND CONSERVATION OF BIODIVERSITY
- PRODUCTION AND PROCESSING OF NICHE PRODUCTS FOR UP-SCALE MARKETS

IMPACTS OF MODERNIZATION AND OF NEW TECHNOLOGIES

CHALLENGING NEW INTERFACES BETWEEN CONSERVATION, AGRICULTURE/FORESTRY AND TOURISM

- CONFLICTING OR COMPLEMENTARY INTERESTS AND RESULTING LAND USE
- TRANSITION/BUFFER ZONE BETWEEN AGRICULTURAL LAND USE AND BIOSPHERES, BLENDING AGRICULTURAL PRODUCTION/FOREST USE AND CONSERVATION, OR SHARPLY DELINEATED EXCLUSIVE ECONOMIC AND CONSERVATION ZONES

URBAN GROWTH AND RURAL INVASION

- INVADING URBAN LAND USE WITH TENTACLES OF LAND SPECULATION, INTENSIFICATION OF URBAN MARKET ORIENTED HORTICULTURE, OR 'BLIGHTED' AGRICULTURAL LAND
- CONVERSION OF AGRICULTURAL LAND TO TRANSPORTATION INFRASTRUCTURES (HIGHWAYS; AIRPORTS), SHOPPING CENTERS; INDUSTRIAL PARKS, RECREATION COMPLEXES
- LAND USE CHANGES RESULTING FROM FOREIGN REMITTANCES

IMPACT OF AGRICULTURAL AND RURAL REFORMS

ALTERNATIVE RURAL ECONOMIC ACTIVITIES

- ECOTOURISM WITH A CONSERVATION OF ECOLOGICALLY IMPORTANT OR VULNERABLE AREAS COMBINED WITH ENVIRONMENTALLY COMPATIBLE TOURISTIC INFRASTRUCTURE
- AGRO TOURISM WITH A MAINTENANCE OF AGRICULTURAL LAND USE COMBINED WITH 'SOFT' TOURISTIC INFRASTRUCTURES

Compiled by C. Stadel

and human conditions, in particular to their coping and management of vulnerabilities, risks and crises (Bohle 2008, Janssen and Orstrom 2006).

Over many generations, the rural communities in the Andes have demonstrated remarkable and diverse forms of human resilience and adaptive capacities which have been an integral component of Andean culture (*lo Andino*) (Gade 1999, Stadel 2001). The *campesinos* have always successfully resorted to their traditional wisdom and knowledge in coping with their mountain environment (*saber andino*). Well documented are in particular the concepts and strategies of complementarity (*complementaridad*) and reciprocity (*reciprocidad*), both socially and economically rooted in their community (*comunidad*). Over many generations, Andean people have also developed complex strategies of land and water management as evidenced in particular by the sophisticated irrigation systems and terracing of agricultural fields. Vogl (1990) and others have documented and described the importance of the diverse traditional agricultural techniques from the perspective of an ecological and socio-economic adaptation of rural communities to dynamic natural environments and changing human conditions.

The cultural heritage of the Andean people was always founded by a deep respect of nature and by an attachment to the native soil and local natural resources. In the quest for optimizing the diverse resources of a highly diverse mountain realm and of minimizing the risks of a fragile and vulnerable environment, the complementary strategy has a spatial and multi-sectoral dimension of agricultural production. By a 'vertical control' (Murra 1975) of different ecological zones (*mitimagkuma*), Andean communities are utilizing the potentials and resources of diverse *archipiélagos verticales*. At each altitudinal level and in the various topographic niches, specific crops are cultivated, often in combination with animal husbandry or pastoralism. Furthermore, multiple forms of economic exchanges are carried out between the altitudinal zones, as well as between highlands and lowland areas. In this respect, accessibility and the role of trails, roads and today highways, have always been of crucial importance and have greatly impacted on or even altered the traditional altitudinal land-use system (Allan 1986). In addition, the cultural environment, the age and nature of the settlement process, the availability of irrigation water and water rights, land tenure and inheritance traditions, alternative economic opportunities, the adherence to traditions versus on openness towards modernizations and new technologies and the influence of external actors, are important factors influencing the rural land-use system. The second major 'pillar' of rural resilience and livelihoods is the concept of reciprocity. Upto now, it has been a fundamental socio-economic support system of rural communities and an effective protection against environmental and economic vulnerabilities, and also a communal principle of solidarity and equity (Rist 2000).

The recent environmental, economical, cultural and political changes and impacts have resulted in new challenges for rural families, but they have also generated new 'windows of opportunities'. Both the new potential problems and new challenges require shorter- and long-term adaptation strategies. Both the field cultivation and pastoral economies are generally becoming more intensified and market-oriented. In line with a modified 'staple-economy theory', the destinations of the agricultural products are no longer limited to nearby periodic markets (*ferias*) and regional service centers, but they are increasingly oriented towards more distant large national cities and world markets. These national and global scopes of agriculture though are primarily restricted to the larger, modern and capital-intensive farm operations with a high financial and technological input. Ownership and management of these 'agro-businesses' are often external, but employing a predominant local labor force. These agricultural enterprises are preferably established in environmentally favored sites, and in areas with a good accessibility. The overwhelming majority of farmers, however, remain poor, and marginalized *minifundistas* are eking out a precarious existence 'in the shadow of globalization'. Thus the former *hacienda–minifundio* dichotomy has become substituted by an equally sharp and unjust contrast between a new and dynamic 'agro-capitalism' and a largely stagnant and self-sufficient small-scale agriculture. The latter is in many cases affected by an out-migration, generally of younger and more dynamic people, 'eroding' the communities of their precious human resources. A temporary migration to seek employment on large farms, in mines, oilfields or in urban jobs may provide the families with much needed additional income, but it could also lead to a social alienation of the migrants from the families they left behind. A permanent emigration, in turn, has generated cash flows of remittances (*remesas*) to families and relatives and to some investments in the home villages. Some of these investments have been valuable 'seed funds' for local and regional development projects; some others though have merely generated 'show pieces' without a sustainable development spin-off. Table 19.2 summarizes major forms and strategies of the adaptation of agriculture and rural communities to the changing environmental and human conditions in the Andean realm.

Today, national governments, NGOs and large international conservation agencies are attempting to protect the Andean ecosystems and their land-, forest- and water resources, and by promoting different types of parks and protected areas. As laudable as these efforts may be, they can also be problematic for the local rural communities. In most cases, these environments have been an integral part of the living- and economic spaces of Andean people. An exclusionary natural ecology perspective, ignoring the cultural traditions and needs of the local population, cannot justify externally imposed natural conservation strategies. Successful, well

Table 19.2. Adaptation forms and strategies of Andean agriculture and rural Communities.

ADAPTATION FORMS AND STRATEGIES OF AGRICULTURE	ADAPTATION TO:	ADAPTATION FORMS AND STRATEGIES OF RURAL COMMUNITIES
• Seasonal and spatial shifts of agricultural calendar • Altitudinal complementarity of agricultural systems ("Vertical Control") • Niche- and mosaic agriculture • Terracing, contour ploughing • Traditional and modern irrigation techniques • Application of fertilizers & pesticides • Field and crop rotation • Polyculture and intercropping • Frost alleviation and protection techniques, freeze-/drying techniques (e.g., "*chuño*" preparation) • Agropastoral systems, animal husbandry • Agroforestry systems • Land degradation protective measures • Adjustments in the type and number of herding animals • Revival of traditional crops, introduction of new cultigens and products • Production and marketing cooperatives • Revival of "*saber andino*"	• Altitude • Relief • Water supply • Soil conditions • Climate and climate change • Distance & accessibility • Cultural traditions • Modernization, new technologies • Variable demographic conditions and migration patterns • Land tenure • Subsistence versus market requirements • Rural and urban labour market • Government and NGO-Programmes • International/global actors and institutions	• Adapted location for communities • Adaptation of communities to environmental condition and changes • Protection of communities from environmental risks and hazards • Revitalization of traditional knowledge • Sustainable use of local natural resources • Principle of reciprocity • Mobilization of local human resources and potentials (e.g., tourism, industry) • Exploration of new economic opportunities • Alternative new—agricultural employment • Commuting to urban centers, seasonal migration, permanent outmigration • Enhanced supply and quality of drinking and irrigation waters • Enhancement of health and sanitation • Rural empowerment, social mobilization, *autogestión*

Compiled by Stadel 2011

integrated 'people and biosphere' concepts and strategies, for instance in the Sajama National Park in Bolivia, demonstrate that one might blend, in a climate of genuine partnership, respect and dialogue, the objectives of both protecting nature and securing sustainable livelihoods.

With its attractive scenery and environmental and cultural qualities, the Andes have become a major recreational space for urban populations, and a destination for national and international tourism. As is the case in other mountain regions of the world, this can be beneficial for local populations and a stimulus for economic development. The massive influx of recreationists and tourists and the ensuing new infrastructures may, however, also be a disturbing factor for the Andean environments and societies. Examples for both types of impacts can be observed in the vicinity of the metropolises of Caracas, Bogotá, Lima or Santiago de Chile; or in the case of major international tourism flows along the *Avenida de los Volcanes* in Ecuador, the *Cordillera Blanca* or the *Valle Sagrado* in Peru, or on the shores of Lake Titicaca in Bolivia and Peru. On the other hand, 'gentle' and sustainable forms of tourism and tourism management, involving and benefitting local populations, can help rural people without unduly disturbing the environment and cultural traditions. This potential exists particularly in various forms of an eco- or agro-tourism, as the example of the Huaraz region in the *Cordillera Blanca* demonstrates.

Apart from the economic dimension of resilience and adaptation measures, but also linked to these, are the social and political components of coping strategies. A central pillar has to be the active involvement and participation of all segments of the local civil society, and a mobilization of the knowledge, experiences and energies of the rural human resources. This 'Local Development and Democratic Participation Model' (Rhoades 2006) is based on a preparedness to 'listen' to the voices of all local stakeholders and on a sharing and mutual learning process in coping with the challenges of a changing Andean environment and society. He argues earlier (Rhoades 2000) that 'local voices and visions' be effectively integrated into the 'Global Mountain Agenda'. An effective participation of local families and associations requires a genuine 'enablement' and 'empowerment' of all stakeholders, including the politically underprivileged groups of women, minorities and isolated and poor people. Participatory forms of planning and development encompass a wide range of objectives, ranging from biodiversity conservation (Llambie et al. 2005), to a strengthening of small-scale agriculture, to education and health improvements, and a fostering of alternative forms of employment and income generation. Underlying this approach is the concept of environmental and social justice and the strife to implement the development postulate of '*El Buen Vivir*', even for the geographically marginal areas and underprivileged people. The author's experience shows that this 'listening' and consensus-building approach may be difficult and time consuming and will require a skillful and patient mediation approach.

The local political participation concept rejects an uncritical application of external intervention strategies and supports a response to the challenges and development options 'with identity' (Rhoades 2006). In particular, the value and wisdom of the ancestral indigenous knowledge has to play a pivotal role in assessing appropriate resilience and adaptation measures. Laurie et al. (2005) plead for a 'professionalization of indigenous knowledge' in a quest for a new type of 'ethno-development'. In an ethno-ecological research project in southern Ecuador, Pohle (2008) has underlined the importance of indigenous concepts of biodiversity management as an essential contribution to sustainable land development. She concludes that "the protection of biodiversity is intimately bound up with the protection and promotion of cultural diversity, particularly when the 'hot spots' of biodiversity coincide with those of cultural diversity" (Pohle 2008: 103). In a similar vein, Rist (2007) emphasizes the importance of bio-cultural diversity for a sustainable endogenous development. Under the right circumstances, this does not ignore the importance and potential usefulness of external expertise, advice and support. Andean people, throughout their long history, have always been exposed to external influences. In some instances, they have attempted to resist these influences from the outside and continued to cling to their traditional practices and livelihoods. In other cases, they have adopted innovations, new techniques and alternative economic options to adapt to the challenges and chances of a changing environment. It is generally recognized that rural communities can greatly benefit from the active support of 'Western' scientists and technicians, from governments, NGOs and large institutions in their local resilience efforts, adaptation strategies and sustainable development and management programs (Apffel-Marglin 1998, Neubert and Macamo 2002).

Conclusion

Andean societies have always coped with environmental, economic, cultural and political changes. Often these exogenous forces have disturbed traditional livelihoods, required new adaptations, and also changed Andean landscapes and societies; according to Bebbington (2000), the Andes are characterized by frequent 'livelihood transitions and place transformations'. At times, they also resulted in new economic outlooks, acculturations, migration patterns, and different spatial and societal disparities. While the various agrarian reforms during the last half century attempted to achieve a more equitable land distribution and new opportunities for *Sierra* farmers in new colonization areas, the current impact of 'agro-capitalism' and market-orientation has created a situation, in which wealthy and powerful private and corporate stakeholders are the winners, with a majority of rural people remaining at the periphery of this form of development.

In this context, the resilience and adaptive capacities of rural communities are precarious, and their outcome is difficult to assess. It appears though, that a sustainable securing of the livelihoods depends both on successful endogenous factors and favorable exogenous influences and interventions. Within the rural societies an intensified education and training (*capacitación*) and a mobilization and empowerment of people to remain 'in control' of their environment, and to find their own forms of resilience, adaptive strategies and development alternatives, will be imperative. However, these efforts have to be 'accompanied' and supported by an external financial, infrastructural, technical and political assistance. The ultimate objective of this partnership support has to be a sustainable and equitable improvement of the livelihoods for the majority of *campesinos*, rather than the often prevailing aim of maximizing outputs and profits for a small minority of stakeholders, without much concern for a responsible environmental stewardship. Stadel (2006) has pleaded for a *campesino-oriented development* and has suggested a conceptual model of 'Sustainable *Campesino* Communities' on the basis of favorable intrinsic and extrinsic influences (Fig. 19.1).

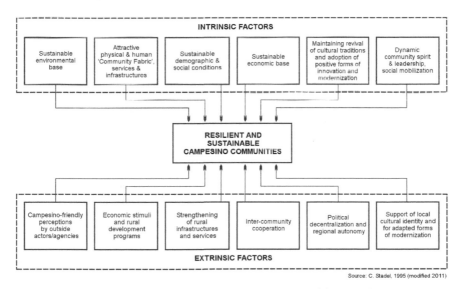

Source: C. Stadel, 1995 (modified 2011)

Figure 19.1 Intrinsic and extrinsic factors of resilient and sustainable campesino communities
Source: C. Stadel 1995 (modified 2011).

References

Allan, N.G.R. 1986. Accessibility and altitudinal zonation models of mountains. Mountain Research and Development 6(3): 185–194.

Apffel-Marglin, F. (ed.). 1998. The Spirit of Regeneration: Andean Culture Confronting Western Notions of Development. Zed Books, London.

Bebbington, A. 1997. Social capital and rural intensification: local organizations and islands of sustainability in the rural Andes. The Geographical Journal 163(2): 189–197.

Bebbington, A. 2000. Reencountering development: livelihood transitions and place transformations in the Andes. Annals of the Association of American Geographers 90: 495–520.

Bebbington, A. 2001. Globalized Andes? Livelihoods, landscapes and development. Ecumene 8(4): 414–436.

Bohle, H.-G. 2008. Leben mit Risiko—Resilience als ein neues Paradigma für die Risikowelten von morgen. pp. 435–441. In: C. Felgentreff and T. Glade (eds.). Naturrisiken und Sozialkatastrophen. Berlin and Heidelberg.

Gade, D. 1999. Nature and Culture in the Andes. University of Wisconsin Press, Madison.

Janssen, M.A. and E. Orstrom. 2006. Resilience, vulnerability and adaptation. IHDP Newsletter 1: 10–11.

Kay, C. 1999. Rural development: from Agrarian reform to neoliberalism and beyond. pp. 272–304. In: R.N. Gwynne and C. Kay (eds.). Latin America transformed. Globalization and Modernity. Arnold, London and New York.

Laurie, N.D., R. Andolina and S.A. Radcliffe. 2005. Ethnodevelopment: social movements, creating experts and professionalizing indigenous knowledge in Ecuador. Antipode 37(3): 470–496.

Llambie, L.D., J.K. Smith, N. Pereira, A.C. Pereira, F. Valero, M. Monasterio and M.V. Dávila. 2005. Participatory Planning for Biodiversity Conservation in the High Tropical Andes: Are Farmers Interested? Mountain Research and Development 25(3): 200–205.

Murra, J.V. 1975. El control vertical de un máximo de pisos ecológicos en la economía de las sociedades andinasIn. Instituto de Estudios Peruanos (ed.). Formaciones económicas y políticas del mundo andino. Lima 59–115.

Neubert, D. and E. Macamo. 2002. Entwicklungstrategien zwischen lokalem Wissen und globaler Wissenschaft. Geographische Rundschau 54(10): 12–17.

Pohle, P. 2008. Indigenous and local concepts of land use and biodiversity management in the Andes of Southern Ecuador. pp. 89–105. In: J. Löffler and J. Stadelbauer (eds.). Diversity in Mountains Systems (=Colloquium Geographicum 31). Asgard-Verlag Sankt Augustin. Bonn.

Rhoades, R. 2000. Integrating Local Voices and Visions into the Global Mountain Agenda. Mountain Research and Development 20(1): 4–9.

Rhoades, R. (ed.). 2006. Development with Identity. Community, Culture and Sustainability in the Andes. CABI Publishing, Wallingford, OX and Cambridge, MA.

Rist, S. 2000. Linking ethics, and the market: campesino economic strategies in the Bolivian Andes. Mountain Research and Development 20(4): 310–315.

Rist, S. 2007. The importance of bio-cultural diversity for endogeneous development. pp. 76–81. In: B. Harverkorst and S. Rist (eds.). Endogeneous Development and Bio-Cultural Diversity. The interplay of worldview, globalization and locality. COMPAS series on Worldviews and Sciences 6. Berne-Leusden.

Stadel, C. 2001. "Lo Andino": andine Umwelt, Philosophie und Weisheit. Innsbrucker Geographische Studien 32: 143–154.

Stadel, C. 2003a. Indigene Gemeinschaften im Andenraum—Tradition und Neu-orientierung. pp. 75–88. In: B. Eitel et al. (eds.). Naturrisiken und Naturkatastrophen—Lateinamerika (=HGG-Journal 18). Heidelberg.

Stadel, C. 2003b. L'agriculture andine: traditions et mutations. pp. 193–207. *In*: CERAMAC (ed.). Crises et mutations des agricultures de montagne. CERAMAC, Clermont-Ferrand.

Stadel, C. 2006. Entwicklungsperspektiven im ländlichen Andenraum. Geographische Rundschau 58(10): 64–72.

Stadel, C. 2008. Agrarian diversity, resilience and adaptation of Andean agriculture and rural communities. pp. 73–88. *In*: J. Löffler and J. Stadelbauer (eds.). Diversity in Mountain Systems (=Colloquium Geographicum 31). Asgard-Verlag Sankt Augustin. Bonn.

Vogl, C.R. 1990. Traditionelle andine Agrartechnologie. Beschreibung und Bedeutung traditioneller landwirtschaftlicher Techniken des peruanischen Andenraums aus der Sicht ökologischer und sozio-ökonomischer Anpassung. PhD thesis. Vienna.

Zimmerer, K.S. 1999. Overlapping patchworks of mountain agriculture in Peru and Bolivia: toward a regional-global landscape model. Human Ecology 27(1): 135–165.

20

Climate Change and the Rocky Mountains[#]

James M. Byrne,[1,*] *Daniel Fagre,*[2] *Ryan MacDonald*[3]
and *Clint C. Muhlfeld*[4]

[1] Professor, Ph.D., University of Lethbridge, Alberta, Canada.
[2] Research Ecologist, Ph.D., USGS Northern Rocky Mountain Science Center, Glacier National Park, University of Montana, USA.
[3] Research Associate, Ph.D., University of Lethbridge, Alberta, Canada.
[4] Research Ecologist & Research Assistant Professor, Ph.D., USGS Northern Rocky Mountain Science Center, Glacier National Park, The University of Montana, Flathead Lake Biological Station, USA.
[#] Rocky Mountains image courtesy Garth Lenz.
[*] Corresponding author

INTRODUCTION

For at least half of the year, the whole of the Rocky Mountains are shrouded in snow that feeds a multitude of glaciers. Snow and ice melts into rivers that have eroded deep valleys with rich aquatic and terrestrial ecosystems.

The Rocky Mountains are the major divide on the continent. Melt water from a multitude of glaciers, snowfields, and rainfall feeds major river systems that run to the Pacific, Atlantic and Arctic oceans. The Rockies truly are the water tower for much of North America, and part of the Alpine backbone of North and South America. But for purposes of this chapter, we consider the Rocky Mountains of the Canadian provinces of Alberta and British Columbia, and the states of Montana, Idaho, Wyoming and Colorado.

The altitude of the Rocky Mountains condenses the weather, climate and ecosystems of thousands of kilometers of latitude into very short vertical distances. On a good day, a strong hiker can journey by foot from the mid-latitude climates of the great plains of North America to an arctic climate near the top of Rocky Mountain peaks—ecologists refer to the bio-climate zone above tree line as alpine tundra since it bears such similarities to the tundra ecosystem of the far north. Rapid changes in elevation typical of mountain terrain create some of the most diverse ecosystems in the world, but it is those rapid changes in microclimate and ecology that make mountains sensitive to climate change. The energy budget in mountains varies dramatically not only with elevation but with slope and aspect. A modest change in the slope of the terrain over short distances may radically change the solar radiation available in that location. Shaded or north facing slopes have very different microclimates than the same elevations in a sunlit location, or for a hill slope facing south. The complexities associated with the diverse mountain terrain of the Rockies compound complexities of weather and climate and create diverse, amazing ecosystems, which are home to a diversity of plant and animal species that have adapted to persist in these changing environments for millennia.

The discussion in this chapter addresses the impacts of climate change in the Rocky Mountains. Climate change is real and ever present, and the role of each of us in changing the climate is also real and present. The Rocky Mountains are a vast and complex region that is valuable both for resources and ecosystems. The Rockies cannot provide the valuable resources we need and treasure, particularly clean plentiful water, unless we protect and conserve mountain ecosystems. Hopefully the discussion in this chapter of the major changes ongoing in the Rocky Mountains due to climate change will help understand the impacts of climate change and add to the collective will in society to minimize this change in future.

Hydrological Response to Climate Change

The Rocky Mountains play a critical role in governing the hydrological regime in large regions of Canada and the United States, largely due to seasonal snowpacks (Schindler and Donahue 2006, Stewart 2009). Snow processes are governed by topography and vegetation as these factors determine the interaction between the land surface and the atmosphere. Orographic effects on precipitation result in greater total accumulation of snowpack in high elevation mountain regions relative to low-lying areas (Dixon et al. 2013). Vegetation plays a key role in partitioning precipitation into transpiration, evaporation, sublimation, and soil moisture storage, controlled primarily by the interception of precipitation and water uptake through root systems (Pomeroy and Schmidt 1993, Storck et al. 2002, Villegas et al. 2010). Changes in vegetation as a function of natural (e.g., pests, wildfire) and human (e.g., timber harvest) disturbance; therefore, have important implications for the hydrologic cycle of mountain environments (Winkler and Boon 2010, Burles and Boon 2011, Pomeroy et al. 2012).

The response of mountain hydrology to climate is highly season-specific, driven by snowmelt providing recharge for surface and groundwater supplies which is released at specific periods during the year (Stewart et al. 2004). The spring freshet period provides the majority of the annual streamflow for many regions across western North America (Barnett et al. 2008). Storing this spring freshet is, therefore, relied heavily upon to sustain dense human populations in low-lying areas. Agriculture, municipal, and industrial development all rely on capture and storage of the historically-reliable spring freshet. As changes in the timing and magnitude of spring freshet occur, the management of these systems will present a challenge for decades to come.

A number of studies have already documented hydrological changes in snow dominated regions, including earlier snowmelt onset (Cayan et al. 2001, Mote et al. 2005, Stewart et al. 2004, Stewart 2009, Clow 2010) as a result of warmer winter and spring air temperatures. Streamflow in rivers fed by the eastern slopes of the Rocky Mountains has been significantly declining over the last century (Rood et al. 2005, St. Jacques et al. 2010). Streamflow regimes in systems fed by the western slopes of the Rocky Mountains are experiencing increased proportions of winter rain (Knowles et al. 2006), resulting in higher winter flow conditions. The reduction of late season streamflow as a result of decreased glacial runoff has also been observed (Marshall et al. 2011). Recent work has shown that up to 60% of climate-related hydrological trends over the Western United States from 1950 to 1999 are in fact human-induced (Barnett et al. 2008).

Snow accumulation in the Rocky Mountains is expected to decline with continued atmospheric warming (Beniston et al. 2003, Hamlet and

Lettenmaier 1999, Lapp et al. 2005, Larson et al. 2011, MacDonald et al. 2011) as the proportion of rain to snow increases, altering the timing and magnitude of snowmelt contributions to streamflow in mountainous regions (Barnett et al. 2005). Shifts in the timing of snowmelt runoff are likely to result in lower late-season streamflow conditions in mountain regions (Huntington et al. 2012). Winter flooding is likely to increase due to a higher proportion of winter rainfall (Goode et al. 2013) and mountain glaciers will continue to be depleted in a warmer climate (Marshall et al. 2011). The effect of climate change is likely to be compounded by enhanced landscape disturbance due to wildfire and pest outbreaks (Flannigan et al. 2005, Hicke and Jenkins 2008, Littell et al. 2009), significantly affecting water supply from snow-dominated regions. The anticipated reduction in available water for both human and ecosystem needs poses a substantial challenge for future water resource management in Rocky Mountain ecosystems (Harma et al. 2011).

A particularly important feature of Rocky Mountain rivers is that they have large contributions of groundwater to streamflow (Ward 1994). During the summer groundwater-dominated mountain streams are very important for the survival of aquatic organisms. This is largely due to the fact that at high temperatures fish may experience reduced growth associated with depleted oxygen, appetite, and enzyme efficiency. In extreme cases, fish could even die when thermal tolerances are exceeded. Therefore, the temperature of the water, particularly cool water associated with groundwater upwelling during summer controls habitat selection of many fish (Power et al. 1999). Cunjak (1996) suggests the protection against extreme abiotic conditions such as ice, reduced oxygen and streamflow fluctuations during winter periods are also critical to fish survival. Groundwater upwelling enables fish to overwinter in streams which would not otherwise contain suitable habitat in the absence of the relatively warm, oxygenated water provided by groundwater (Brown et al. 2011). Changes in groundwater regimes of the Rocky Mountains as a function of climate change have important consequences for ecosystems and suggest this sensitive water-resource must be accounted for when adaptively managing these systems.

Snow and the Terrestrial Ecosystem

Although glacier recession and glacial meltwater directly impact hydrology and aquatic ecological interactions, it is the seasonal snowpack that exerts the predominant influence on alpine terrestrial ecosystems. Thus, the 50-year downward trend in snow water equivalent of mountain snowpacks, the earlier spring melt (up to four weeks), and the shift toward more rain, even during winter, all have consequences for terrestrial vegetation in the Rocky Mountains (Pederson et al. 2011). At the highest elevations, species like the pygmy poppy (*Papaver pygmaeum*), a threatened endemic in the Rocky Mountains of northwestern Montana and southwestern Alberta, are distributed near and below snowfields and are dependent on snowmelt in late summer for adequate moisture. Because the soil water holding capacity is low at these elevations, the continued loss of snow and ice is expected to severely reduce the abundance of the pygmy poppy. Lesica (2012) reported loss of snow deposits, drying of tundra and local declines of half the alpine species he monitored. Similarly, studies of snowbed vegetation (i.e., vegetation that grows later in the season as pockets of late-lying snowmelt) in Colorado and elsewhere suggest comparable dependencies of alpine flora on persistent snow both for moisture and nutrients. Logan Pass, in Montana, has a suspended wetland in the alpine zone that is fed by meltwater from surrounding ice and snowfields and contains a rich flora (Lesica 2002) that includes species on the edge of their ranges. This unique biodiversity hotspot has been monitored for climate change impacts since 1988. Lescia and McCune (2004) found losses and reductions in abundance consistent with a warming trend and earlier snowmelt. To date, there has been no monitoring or estimates of change in the water balance for Logan Pass but diminished snowpack is likely to affect the numerous springs that allow this concentration of plants to persist.

Snowpacks accumulate atmospherically deposited nitrogen during the cold season for as much as nine months and then slowly release nitrogen and other nutrients with snowmelt during the growing season (Bowman and Seistadt 2001). These nutrient inputs have large impacts on plant growth and dynamics in the mostly inorganic environment in which alpine flora exist. This fertilization effect may diminish in the future with the reduced presence of snow in early summer when plant uptake peaks. However, Baron et al. (2009) have documented additional nitrogen release from retreating glaciers and rock glaciers as sediments are exposed during warmer and drier summers. In some mountain areas atmospheric nitrogen deposition has increased enough to become a problem for aquatic systems, so there may be a compensatory effect if summer rainfall can deliver sporadically what used to be provided continuously by melting snowfields. These interacting effects of nitrogen deposition and changing climate are spatially and temporally variable but are important to document for projecting how alpine tundra vegetation will fare in the future.

Alpine vegetation on mountain summits is also likely to be susceptible to warming air temperatures and diminished snow. Diaz and Eischeid (2007) showed that the climate envelope that is unique to alpine tundra has decreased in area by 73% in two decades and is likely to disappear in the western U.S. in the near future. Plants in alpine tundra may find refugia by dispersing to other mountain aspects or thermally sheltered microtopographic features (Scherrer and Korner 2011). But summit species, identified as distinct communities from alpine tundra plants (Malanson and Fagre 2013), will have few, if any, options since they already occupy the highest elevations. In Europe, where there are summit vegetation records going back a century, some plant species have disappeared from some mountain ranges and others are now found much higher in elevation than before (Parolo and Rossi 2008, Pauli et al. 2007). It is reasonable to expect the same trend in the Rocky Mountains of North America.

The interface between alpine tundra and high-elevation forests, the Alpine Treeline Ecotone (ATE), is another area where reduced snow persistence and rising temperatures have started to profoundly change the dynamics of mountain ecosystems. Although the position and form of the ATE is thought to be climatically controlled, and therefore changes in it are an indication of climate warming, as much as half of the treelines are affected by geomorphology (e.g., debris flows, rockfall, landslides), snow avalanches, and occasional fires in the northern Rocky Mountains (Butler et al. 2007). Nonetheless, ATEs have increased in biomass, canopies have become denser, and the mean position of trees have moved upward in North America and the Rocky Mountains and this has been attributed to climate change (Malanson et al. 2007, Holtmeier and Broll 2007, Klasner and Fagre 2002). One consequence is that more snow is now retained at the ATE with taller and more complex canopy architecture, potentially affecting alpine hydrology and seedling establishment. Another, however, is that there is more fuel continuity and fires from downslope can now be carried into the alpine when dry conditions exist. Westerling et al. (2006) indicated that the largest increases in fire frequency for the past six decades have been in mid-elevation forests above 2,100 m. As trees become established further upslope there is a corresponding reduction in the area of alpine tundra that is also host to many of the Rocky Mountains' charismatic wildlife species, such as mountain goats (*Oreamnos americanus*) and bighorn sheep (*Ovis canadensis*). In addition to loss of habitat area, in the future when trees may cover passes, there could be reduced migration between sub-ranges due to predator avoidance, and consequent genetic isolation.

Another consequence of enhanced montane tree growth related to changes in snow is the in-filling of subalpine meadows. Deep snow that accumulates in meadows and persists to mid-summer or later can suppress tree seedling establishment but has less effect on establishment and vitality of meadow forbs and grasses. These forbs and grasses provide valuable food resources and high quality habitat for insects, birds, many burrowing mammals, ungulates and bears. Some of the forest/meadow patterns in upper elevation basins are alternating ribbons of trees and open meadow that are largely controlled by underlying bedrock topography with snow accumulating in the depressions between ridges (Butler et al. 2003). This pattern provides a large edge effect, which maximizes the value to wildlife.

Since the mid-20th century, our examination of historic photographs and repeats photography has shown that many meadows have become smaller as newly established young trees flourish and that this meadow reduction is coincident with the multi-decadal reduction in snowpack. As tree seedlings invade the meadows and canopies close with maturing trees, the loss in functional wildlife habitat is much greater than the absolute areal loss of meadows would indicate.

Climate change and other stressors

The fate of whitebark pine (*Pinus albicaulis*) illustrates the role of multiple stressors, or stress complexes, on the future persistence of keystone species in a warming world. These alpine trees colonize high ridges and other areas with formidable climates and, because of their snow-retaining broad crowns, modify conditions enough to affect alpine hydrology and allow other tree species to establish. Whitebark pine seeds also provide a vital high calorie food source for grizzly bears (*Ursus arctos*) and other animal species. Whitebark pine seed dispersal is largely dependent on Clark's Nutcracker (*Nucifraga columbiana*), a medium-sized bird related to jays, that caches seeds across broad reaches of the mountains (McKinney et al. 2009). Because not all the seed caches are consumed, these birds uniquely re-establish whitebark pine populations after disturbance, such as wildfire, has eliminated the pines from part of their range. However, beginning last century, an introduced pathogen, white pine blister rust (Cronartium ribicola), has decimated whitebark pine in large parts of its range and prompted attempts to find rust resistant whitebark pine as part of a restoration effort. Although the significant damage to whitebark pine populations by the blister rust is not directly climate related, increased threats from widespread bark beetle outbreaks (Logan et al. 2010) and high-elevation fires that have increased with warming temperatures has diminished the prospects of whitebark pine recovering robustly. In parts of the Greater Yellowstone Ecosystem, up to 95% of cone-bearing whitebark pine has died from beetle outbreaks and climate change is likely to have exacerbated such outbreaks (Logan et al. 2010). Another potential stressor on whitebark pine is that Clark's Nutcracker populations have seemingly dropped in areas where more than 90% of whitebark pine has disappeared, such as the Crown of the Continent Ecosystem in northern Montana and southern Alberta and British Columbia, because the birds are dependent on the whitebark pine seed crop. Because Nutcracker's seed dispersal is essential to long-term maintenance of whitebark pine distribution, their population reduction or local extirpation would be yet another obstacle to whitebark pine recovery (McKinney et al. 2009). Thus, whitebark pine, already negatively affected by one stressor, the blister rust, now must contend with several other stressors that directly or indirectly result from climate change.

Snow-dependent wildlife vulnerabilities

Over millennia many wildlife species have adapted to snow regimes in the Rocky Mountains to improve their survival, reproduction and persistence. As snowpacks diminish in depth and extent, melt earlier in the spring, and change thermal characteristics as they become warmer, these fine-tuned characteristics may lose their advantages or even become maladaptive. The snowshoe hare (Lepus americanus), for instance, rapidly molts to change the color of its fur to white as winter approaches. While effective against a snow background, it serves to attract predators against a snowless forest background, indicating that timing this pelage change to patterns of snowfall is critical. Pederson et al. (2010) documented that snowmelt has shifted two weeks earlier in western Montana. Mills et al. (2013) recorded numerous instances of snowshoe hares remaining partially white in early spring well after snow had melted. If the pelage change is triggered by day-length changes, as many biological phenomena are, then climate change-driven speeding of spring snowmelt will result in a long-term seasonal mismatch for this organism and mismatched hares die at higher rates (Mills et al. 2013). Projected reductions in average duration of snowpack suggest that there will be an additional 40–69 snowless days by the end of the century and hares will be mismatched four to eight times more often.

A major predator of the snowshoe hare is the lynx (Lynx canadensis), so hare population declines are expected to adversely affect lynx. Lynx also share the adaptation to snow of having extra-large feet to provide support when moving across snow, which provides a competitive edge for the lynx over other medium-sized carnivores such as coyotes (Canis latrans). With a less snow-dominated environment becoming a new reality in much of the Rocky Mountains, wildlife managers have to consider whether the loss of the lynx's competitive advantage will result in their population decline. Significant resources have gone into establishing lynx in many areas of its current core habitat but the degree to which they can adapt to climate change is speculative.

Landscape Scale Disturbance

Many climate change effects are direct, such as the facilitation of seedling establishment due to reduced snowpack persistence, whereas other effects are indirect. Perhaps the most transformative indirect effect is climate change's role in altering disturbance processes such as wildland fire and massive insect outbreaks. In the past several decades, wildland fire in the Rocky Mountains has increased in frequency and scale, particularly in Colorado, Wyoming and Montana where significantly fuel loads have had longer warm seasons to dry out (Westerling et al. 2006). Although part of the more active fire seasons and potential for future increases in fires are due to past forest management and fire suppression policies, longer growing seasons have contributed tree growth and additional fuel as well. Fires are now four times more frequent and burn more than six times the annual average area of past years (Westerling et al. 2006). The average fire season now lasts 78 days longer than the long-term mean. By 2050 the area burned by wildland fires is expected to double in the northern U.S. Rocky Mountains, especially in August, due to higher temperatures (Yue et al. 2013). Increased fire frequency also means that a larger percentage of the mountain landscape has had recent fires and this affects numerous ecosystem dynamics such as soil instability, debris flows, canopy interception of snow, changed albedo and stream turbidity. With hotter and longer summers projected in the next decades, Rocky Mountain ecosystems are likely to become more fire dominated and cause significant social disruption and financial costs. But larger and more frequent fires will also transform many mountain areas because the current dominant vegetation will not necessarily re-establish after fires, even decades later. Many forests can persist for centuries after the climatic conditions under which they were established have changed. Climate warming that has already occurred in the Rocky Mountains means that once a catastrophic wildland fire has burned an area sufficiently to remove most of the vegetation, the formerly dominant tree species may not be able to re-establish and species adapted to warmer conditions and more frequent fire will be the beneficiaries. These landscape scale transformations will, in many cases, constitute thresholds that have been crossed and the new vegetation communities and dynamics will be permanent from a human perspective.

Although episodic forest insect outbreaks are a common feature of mountain ecosystems in the Rocky Mountains, the significantly larger scale and length of recent outbreaks has exceeded the norm. Logan et al. (2010) and others have tied warmer winters to larger overwintering beetle populations, faster lifecycles for some species, and more prevalent drought-stressed trees less able to defend against beetle infestation. The mountain pine beetle killed between 2–3.3 million hectares of mostly lodgepole pine forests in the western USA in a 13-year outbreak. In the Rocky Mountains it appears that beetle species have benefitted the most from fewer extreme cold events during winter but other forest insects such as spruce budworm have also increased in prevalence.

Aquatic Biota in the Rocky Mountains

Climate change is rapidly altering the structure and function of aquatic ecosystems across the Rocky Mountains of North America. Although the effects of climate change vary across a variety of spatial and temporal scales, warming trends and climate model simulations indicate that stream habitats will become warmer, have more variable thermal and hydrologic regimes, and will be more susceptible to stochastic disturbances such as flooding, wildfire and drought (Jentsch et al. 2007). Combined with existing stressors of habitat loss and invasive species, climate-induced changes in precipitation and air temperature are predicted to have significant effects on the distribution, abundance and phenology of many aquatic species (Walther et al. 2002, Parmesan and Yohe 2003). This is particularly true for cold-water, stenothermic species (e.g., fish and invertebrates) inhabiting mountainous streams. These species show strong temperature and flow-related range contractions and thus are potentially the most threatened groups of species due to impending climate change (Milner et al. 2001, Parmesan 2006). As climate warming rapidly progresses throughout the Rockies, aquatic biota must adapt in place, shift to track suitable habitats (i.e., climatic niches) or be extirpated.

Salmonids

Native salmonids (e.g., trout, char and salmon) are a group of fishes with considerable ecological and socioeconomic significance in the Rocky Mountains. Salmonid fishes are especially vulnerable to the effects of environmental change in freshwater systems because they require cold, connected and high-quality habitats, which are easily fragmented by changes in thermal and hydrologic regimes (Williams et al. 2009, Haak et al. 2010). Moreover, many populations have experienced large reductions in historically occupied habitats over the past century, owing to anthropogenic activities such as habitat degradation, fragmentation and invasions of non-native fishes (Paul and Post 2001). Consequently, many remaining populations are small in numbers and isolated in headwater streams, where they are at risk of extirpation due to environmental (e.g., flooding and wildfire) and demographic stochasticity and loss of genetic variability (Peterson et al. 2008). Rapid water temperature increases throughout the Rocky Mountains combined with these existing stressors have already made it possible to see how a changing climate has been affecting streams habitats with potential consequences for fish populations (Isaak et al. 2012). Climate changes, therefore, could directly or indirectly lead to increasing habitat and population fragmentation and accelerated decline of extant populations inhabiting the Rocky Mountain region. These environmental and biological changes, however, will not be uniform across aquatic ecosystems in the Rockies.

A number of bioclimatic models have been developed for native salmonids in the Rocky Mountain region—all of which forecast substantial reductions in thermally-suitable habitats during the 21st century (Keleher and Rahel 1996, Williams et al. 2009, Haak et al. 2010). These models congruently show that as water temperatures continue to warm and exceed species' physiological thresholds, populations will become increasingly fragmented and will retreat into colder headwater habitats. For example, using an upper temperature threshold of 22°C as a constraint for cold water trout and char (Keleher and Rahel 1996), models predict that an increase of 5°C in mean air temperature would reduce the amount of thermally-suitable salmonids habitat by 70% across the Rocky Mountain region. Wenger et al. (2011b) forecasted the climate warming effects of increased temperatures and altered flows on four interacting species of trout across the interior western United States and predicted a 47% decline in total suitable trout habitat by 2080. In the interior Columbia River basin, models predict that bull trout (*Salvelinus confluentus*) may lose 18–92% of thermally suitable natal habitat due to climate warming over the next 50 or more years (Rieman et al. 2007). Jones et al. (2014) used a spatial hierarchical modeling framework to predict stream temperatures and estimated that if average air temperatures were to rise 3°C by 2050, bull trout would potentially lose 58% of foraging, migrating and overwintering areas and 36% of spawning and rearing habitat in the Flathead River system, Canada and USA.

Although increased stream temperatures will be one of the primary consequences of a rapidly changing climate, increased disturbances and altered flow regimes will also have a significant impact on the persistence of native salmonids and aquatic biota. Reduced snowpack, earlier and more rapid spring runoff, increased floods, and drought all pose additional stressors to native trout populations and aquatic organisms in western North America (Poff 2002, Williams et al. 2009, Haak et al. 2010). In the western United States, decreased snowpack and earlier spring runoff have been linked to increases in the frequency and severity wildfire (Westerling et al. 2006), which are causing extensive habitat changes and, in some cases, direct mortality of trout and other native fishes. Interactions of stochastic disturbances with fragmentation appear to be the greatest threats to the persistence of small, isolated Colorado River cutthroat trout (*O. c. pleuriticus*) populations (Roberts et al. 2013). Similarly, decreasing summer habitat volumes and increasing drought and wildfire disturbances are perhaps the greatest extirpation risks for remaining Rio Grande cutthroat trout (*O. c. virginalis*) populations, which are already highly fragmented and confined to small (<10 km) headwater streams above natural and anthropogenic barriers (Zeigler et al. 2012). Reduced snowpack and earlier spring runoff decreases summer base flows and available habitat, which will further accelerate increasing water temperatures during a thermally stressful time for many aquatic organisms.

Climate-induced changes in flow and temperature regimes may also have direct effects on salmonid reproduction and recruitment across Rocky Mountain streams. Native salmonids are either spring spawners, adapted to deposit eggs in the streambed on the descending limb of the snowmelt hydrograph (e.g., cutthroat trout) or fall spawners, which deposit their eggs during the fall and emerge the following spring (e.g., bull trout) (Northcote 1997). In either case, salmonids may be especially sensitive to increased flood events because high flows can scour incubating eggs from gravel substrates or wash away newly emerged fry. In some regions, winter floods are projected to continue increasing due to warming temperatures causing precipitation to shift from snow to a snow-rain mix. Likewise, more rapid spring snowmelt runoff may reduce recruitment of recently emerged juveniles, creating an additional environmental stressor that these fish must respond to.

Native fishes have persisted for millennia in these cold mountainous landscapes that have undergone tremendous geologic and hydrologic change shaped by continental glaciation, flood, wildfire, and periods of extreme temperature warming. These differences in climatic regimes and environmental conditions have allowed salmonids to adapt to a diversity of aquatic habitats (e.g., streams, rivers, lakes, ponds) through highly plastic life histories, which are primarily linked to temperature and flow. Shifts in the timing and magnitude of ecologically important flows and temperatures are clearly evident across the Rocky Mountain region. For example, migratory cutthroat trout and bull trout spawn and rear in natal headwater streams, but move downstream to grow and mature in rivers (i.e., fluvial) or lakes (i.e., adfluvial) (Rieman and McIntyre 1995, Muhlfeld and Marotz 2005, Muhlfeld et al. 2009). These unique evolutionary behaviors sustain the genetic diversity of populations and are critical for the persistence of salmonid metapopulations in complex riverscapes. If climate warming trends continue, however, headwater streams used for spawning and early rearing will be at greater risk of extirpation by stochastic perturbations, whereas downstream habitats used for foraging, migrating and overwintering may become thermally and hydrologically unsuitable, resulting in the loss of migratory life history forms, genetic diversity and headwater source populations. Thus, the migratory nature of salmonid fishes may make them particularly susceptible to extirpation or decline in a rapidly warming world. However, the diversity of life history strategies found in these fish may enable some populations to adapt through phenotypic or genetic means or move to track suitable habitats to complete their life cycle. Generally, the adaptive capacity of aquatic organisms is still relatively unknown, but recent empirical evidence suggests that salmonid fishes can rapidly adapt via natural selection to climate warming (Crozier et al. 2011, Kovach et al. 2012).

Interspecific differences in life histories and thermal niches suggest that the effects of changing climate patterns are unlikely to be consistent across species, populations and ecosystems. For example, bull trout may be more sensitive than cutthroat trout to the direct effects of climate warming (Wenger et al. 2011a). Bull trout have optimal temperatures that are substantially lower than those of other salmonids, so spawning and rearing habitats are constrained to a patchwork of cold headwater habitats across river networks (Selong et al. 2001). Additionally, bull trout are fall spawners, which are particularly susceptible to frequent high winter flows (Wenger et al. 2011a). Increasing temperatures and flood risks pose serious threats to this declining species. In some cases, however, future stream temperature warming may benefit trout living in the coldest environments of the Rockies. Al-Chokhachy et al. (2013) linked future stream temperatures with fish growth models to investigate how changing thermal regimes could influence the future distribution and persistence of Yellowstone cutthroat trout in the Greater Yellowstone Ecosystem. For high-elevation populations, there were significant increases in fish growth attributable both to warming of cold water temperatures and to extended growing seasons. However, these benefits could be offset by the interspecific effects of corresponding growth of sympatric, non-native fishes currently occupying lower elevation streams. Finally (Crozier and Zabel 2006), demonstrated that Chinook salmon (*Oncorhynchus tshawytsha*) have widely variable responses to temperature throughout the Salmon River basin in Idaho, resulting in very different projections of population size and probability of persistence across the basin.

In addition to the direct effects on habitat quality, climate warming will indirectly affect native salmonids and aquatic biota by exacerbating interactions with non-native, invasive species. Climate changes may facilitate the expansion of invasive species or modify interactions among native species through competition, predation, disease and introgressive hybridization (Rahel et al. 2008). In many cases, stream temperature increases are exacerbating biotic interactions by facilitating expansion of non-native species into native fish habitats (Wenger et al. 2011a). For example, many populations of native cutthroat trout have been displaced from natal habitats through competitive interactions with introduced brook trout, brown trout (*S. trutta*), and rainbow trout. Laboratory experiments and natural studies show that non-native salmonids may have a competitive growth advantage at warmer temperatures. In the Northern Rockies, recent summer temperature increases and wildfire disturbances appear to be allowing introduced rainbow trout (*Oncorhynchus mykiss*) distributions to expand upstream, thereby enhancing the spread of hybridization with native westslope cutthroat trout (*O. clarkii lewisi*) populations (Muhlfeld et al. 2009, Isaak et al. 2012).

Aquatic invertebrates

Aquatic invertebrates are also important to regional biodiversity in Rocky Mountain aquatic ecosystems. They are useful biological indicators of climate-induced changes in mountain ecosystems because they are integral components of aquatic food webs and their distributions and abundances are strongly linked to temperature and stream flow gradients (Milner et al. 2001). Species-specific distributions are likely to be profoundly affected as temperatures continue to rise beyond species' thermal limits and precipitation patterns change potentially causing some streams to become intermittent. Elevational changes in the timberline associated with climate change will also affect species' distributions and the structure and function of stream ecosystems.

Cold-water stenothermic invertebrate species inhabiting alpine stream environments in the Rocky Mountains are especially vulnerable to climate-induced warming and snow loss. Recent data show increased magnitude and rate of warming with extensive loss of glaciers and snowpack throughout the Rocky Mountains. For example, the loss of glaciers in Wateron-Glacier International Peace Park is iconic of the combined impacts of global warming and reduced snowpack; 125 of the estimated 150 glaciers existing in 1850 have disappeared, and the remaining 25 are predicted to be gone by 2030 (Hall and Fagre 2003). Changes in the hydrological cycle associated with snow and ice loss are likely to warm perennial streams and some may transition to intermittent flows. These changes threaten the stability of sensitive ecosystems that provide critical habitat for alpine-restricted stream invertebrates facing increased risks of extinction, such as the rare caddisfly *Allomyia bifosa* (Hauer et al. 2007) and the endemic meltwater stonefly *Lednia tumana* (Muhlfeld et al. 2011), *Lednia tumana*, discussed further under case studies, is the first invertebrate species that warrants the protections of the U.S. Endangered Species Act (ESA) due to climate-change-induced glacier loss. Genetic diversity in alpine invertebrates may be particularly sensitive to temperature warming because many species are undergoing rapid range shifts in these environments (Jordan et al. In-review).

Case Studies of Response to Climate Change in the Rocky Mountains

Glaciers and climate change

Mountains are the water towers of western North America, supplying 85% of the freshwater people use and providing sustained release of water through the summer that is essential for ecosystem processes (USGCRP 2000). Glaciers contribute significantly to these sustained releases. Arguably one of the most visible impacts of a changing climate in the Rocky Mountains is the shrinkage of alpine glaciers and perennial snowfields. Unlike seasonal snowpacks that can vary substantially from year to year, glaciers track changes over decades and are a better indicator of the direction and pace of climate change. These glaciers provide direct indications of the pace of changes to broader mountain environments because they integrate temperature, precipitation and other climate variables. They are relatively easy to map and inventory compared to other ecosystem components (e.g., soil carbon), are well distributed in the Rocky Mountains, and are understood by managers and the public alike as proxy signals of changing water balance in the mountains. Their shrinkage has direct consequences for total water available to humans and ecosystems but it is their regulatory effect on timing of water release that is a critical ecological service. Glacier meltwater provides a significant component of late summer baseflow in mountain streams, keeping streams viable, but also keeps many mountain streams cold in late summer when seasonal snowpacks are long gone. This is essential for aquatic species of special concern, such as bull trout (*Salvelinus confluentus*), that require cold water. Retreating glaciers alter stream chemistry by exposing bedrock surfaces that release various forms of nitrogen that in turn act to fertilize oligotrophic alpine lakes and streams (Fountain et al. 2012). Shrinking glaciers can also create a variety of hazards to downstream human communities and are a source of significant ecological disturbance (Moore et al. 2009).

The Rocky Mountains have hundreds of glaciers, largely clustered near the continental divide and often small by global standards. In the U.S. the two largest concentrations are in the Wind River Range of Wyoming, where Gannett glacier is the single largest glacier south of Canada, and the Lewis and Livingstone ranges of Glacier National Park, Montana. In Canada, alpine glaciers are larger and valley glaciers flow from icefields such as the Columbia Ice Field north of Banff National Park and the Wapta Ice Field in Banff National Park (Ommanney 2002).

During the past century, virtually all the glaciers in the Rocky Mountains have become smaller (Figs. 20.1 and 20.2). In Glacier National Park, an estimated 100 km^2 of ice cover that existed in the late 19th century (Key et al. 2002) has been reduced to just 16 km^2 by 2005 (Fagre and McKeon 2010).

Jackson Glacier_1911_Elrod_UofM archives

Jackson Glacier_2009_Fagre USGS

Figure 20.1. Repeat photographs of Jackson Glacier showing the dramatic ice reduction since 1911.

At least 150 glaciers existed at the end of the Little Ice Age (Carrera 1989) but only 25 glaciers larger than 0.1 km² exist now (Fagre and McKeon 2010). These are likely to be gone in decades with estimates from models ranging from 2030 (Hall and Fagre 2003) to 2080 (Brown et al. 2010) as the end dates depend on rates of future climate change. What is clear is that this area will be devoid of glaciers without a reversal in the current upward trend in temperatures. Fountain (2007) has indicated that the glaciers of Glacier National Park have shrunk more quickly than other areas of the western U.S. (66% vs. 25–40%) but a recent paper (Fountain et al. 2012) shows that the higher elevation glaciers in Wyoming are now shrinking at comparable rates. Pochop et al. (1990) had earlier reported that glaciers in the Wind

Sperry_c1920_Elrod_UM

Sperry_mid_2008_McKeon_USGS

Figure 20.2. Photos of Sperry Glacier, Montana in 1920 and 2008 demonstrate the extensive ice loss typical of glaciers in the Rocky Mountains in the last 100 years.

River Range had shrunk by more than 30% since 1950 and that they were half the size than when first photographed in the 1890s. Gannett glacier alone lost 50% of its volume since 1920 and is expected to disappear by 2050 at current rates of retreat. Further south in Colorado, several very small glaciers continue to exist because snow avalanches and wind-deposited snow contribute enough mass to compensate for shrinking due to warming temperatures (Hoffman et al. 2007).

In Canada, the same trends are evident. The Peyto Glacier in Banff National Park, Alberta, has been monitored since 1896. Peyto has lost 70% of its volume in that time, and had only five positive mass balance years between 1970–2003 (Comeau et al. 2009). The Athabasca glacier, the largest in the Rocky Mountains, has receded more than 1.5 km in the last 125 years and lost more than 50% of its volume as it continues to drain the Columbia Ice Field. Because it is close to the Ice Fields Parkway, a major tourist destination, it is likely the most visited glacier in North America and has accelerated in its retreat since 1980. The melting glaciers of the Rocky Mountains clearly signal changes at the topographic apex of mountain ecosystems that cascade downward and throughout other system components.

Booth (2011) modeled glacier ice mass balance for all glaciers in the headwaters of the North Saskatchewan watershed along the Continental divide in Jasper National Park, Alberta. That study used a series of climate scenarios representing the likely range of future climate for that region as forecast by a number of global circulation models. Their results presented in Fig. 20.3, show that under all future scenarios glacier ice mass in the northern Rockies will likely continue to shrink to very modest volumes isolated in high mountain locations.

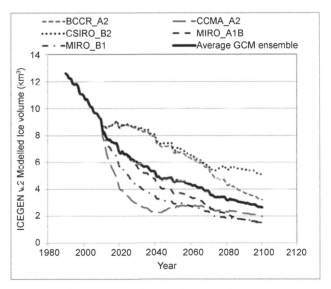

Figure 20.3. Expected decline in ice mass through 2100 in the North Saskatchewan headwaters in Jasper National Park, Alberta for a range of possible future climate scenarios (Booth 2011). The lesser rates of decline will occur if and only if humanity is successful in substantial decreases in greenhouse gas emissions. The worst climate scenarios shows most of the alpine glaciers ice mass will be gone by 2040.

Glaciers and rare alpine invertebrate—Lednia tumana

Alpine aquatic species are important to regional biodiversity in the Rocky Mountains. One species that may be particularly vulnerable to climate change is the meltwater stonefly *Lednia tumana*—a macro invertebrate

Figure 20.4. Lednia tumana—a macro invertebrate species endemic to the Waterton-Glacier International Peace Park (WGIPP) area (Canada and USA) and a recent candidate for listing under the U.S. Endangered Species Act due to climate-change-induced glacier loss (Credit: Joe Giersch, USGS).

species endemic to the Waterton-Glacier International Peace Park (WGIPP) area (Canada and USA) and a recent candidate for listing under the U.S. Endangered Species Act due to climate-change-induced glacier loss. During 14 years of research, *L. tumana* was found to inhabit a narrow distribution, restricted to short sections (~500 m) of cold, alpine streams directly below glaciers, permanent snowfields and springs. Bioclimatic models suggest that climate change threatens the potential future distribution of these sensitive habitats and the persistence of *L. tumana*, with the species predicted to lose over 80% (current ~23 km², future ~4.5 km²) of its potential current range under future warming induced glacier and perennial snowfield loss. These projected reductions in suitable habitat suggest that *L. tumana* may be one of the first known macro invertebrates to become extinct due to recent climate change. Mountain top aquatic invertebrates that exhibit severe climate-related range-restrictions are ideal early-warning indicators of thermal and hydrological modification that may be associated with climate warming in mountain ecosystems.

Bull trout

The bull trout (*Salvelinus confluentus*—Fig. 20.6) is a native char to northwestern North America, including portions of the Rocky Mountains in Canada and the United States, which requires large, ecologically diverse and connected coldwater river and lake habitats for persistence (Fig. 20.5). Bull trout populations have declined throughout much of their native range, and

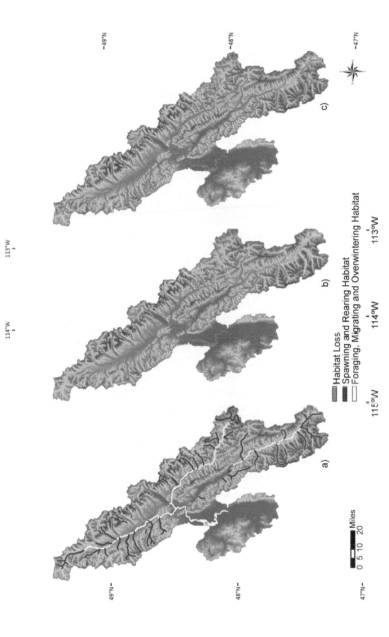

Figure 20.5. Current bull trout habitat, and habitat loss associated with 2059 and 2099 climate simulations in the upper Flathead River system, United States and Canada (from Jones et al. 2013).

Color image of this figure appears in the color plate section at the end of the book.

Figure 20.6. Spawning bull trout in the Wigwam River, Canada (Top credit: Joel Sartore). Bottom: juvenile bull trout and westslope cutthroat trout inhabit a pool in Ole Creek, Glacier National Park (Credit: Jonny Armstrong).

Color image of this figure appears in the color plate section at the end of the book.

the species is listed as threatened species under the US Endangered Species Act and a blue-listed species in Canada. Declines are primarily the result of habitat degradation, fragmentation, overharvest, non-native invasive species, and climate change. Having one of the lowest upper thermal limits and growth optima of all salmonids in North America, the bull trout is an excellent indicator of warming temperatures and modified hydrologic regimes in stream networks. Natal spawning and rearing habitats are often fragmented and constrained to the coldest headwater streams (<13°C; Jones et al. 2013). As air and water temperatures continue to increase, the lower portions of stream networks, which are used by bull trout for foraging, migrating and overwintering habitat, may become thermally unsuitable and headwater spawning and rearing streams may become isolated because of increasing thermal fragmentation during summer (Isaak et al. 2010, Jones et al. 2013; Fig. 20.5). Bull trout are also fall spawners, which may explain why populations typically fare poorly in streams with frequent high winter flows and suggests that the recent increases in winter flood risks across portions of the Rocky Mountains are a cause for concern. Currently, conservation efforts are focusing on maintaining natural connections and a diversity of coldwater habitats over a large spatial scale to conserve the full expression of life history traits and processes influencing the natural dispersal among populations.

Freshwater algae invades with warming waters

Climate warming may also promote expansion of invasive plants in aquatic ecosystems. *Didymosphenia geminata* (*didymo*) is a freshwater alga (Fig. 20.7) native to North America that has recently spread to lower latitudes and warmer waters, and increasingly forms large blooms that cover streambeds

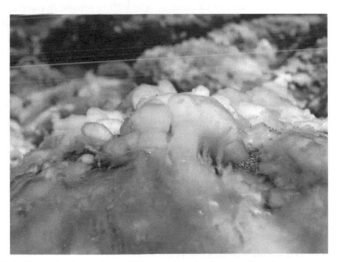

Figure 20.7. *Didymosphenia geminata* (didymo) is a freshwater alga that may cause dramatic changes in the diversity and abundance of aquatic insect species when algae blooms occur in warming Rocky Mountain streams and rivers. Such losses would impact the entire food web (Credit: Joe Giersch, USGS).

Color image of this figure appears in the color plate section at the end of the book.

and modify the composition of macro invertebrate communities. *Didymo* is currenlty present in an estimated 64% of streams in Glacier National Park (GNP). *Didymo's* abundance is positively related to summer stream temperatures in GNP, which are likely to continue increasing over the coming decades, thereby increasing the extent and severity of didymo blooms (Schweiger et al. 2011). Didymo may be a useful indicator of thermal and hydrological modification with climate warming across the Rocky Mountain region.

North Saskatchewan River

Snowpack from watersheds on the eastern slopes of the Canadian Rocky Mountains is an important source of water for the western prairie provinces of Alberta, Saskatchewan and Manitoba (Schindler and Donahue 2006). The North Saskatchewan basin is important because it is subject to the 1969 Master Agreement on Apportionment, which dictates that Alberta must allow 50% of the annual natural flow to enter the province of Saskatchewan, while maintaining minimum in-stream flow requirements (North Saskatchewan Watershed Alliance 2005). This presents a challenge for water managers because reservoir operations on the upper North Saskatchewan are at risk as a result of decreased reservoir reliability under future climate change (Minville et al. 2009, Minville et al. 2010).

MacDonald et al. (2012) demonstrated that snow accumulation and ablation in the North Saskatchewan basin are sensitive to climate change. Climate warming is likely to result in an upward shift in elevation of the zero degree isotherm, with a transition to more precipitation occurring as rain than snow annually. The average maximum SWE is not likely to change substantially under future conditions because winters will remain cold. However, the timing of spring snowmelt onset is likely to change under a range of future climate scenarios (Fig. 20.8). An earlier snowmelt onset over large portions of the North Saskatchewan basin will likely reduce late season water availability for humans and ecosystems. Reductions in available water will force adaptive management strategies to be implemented in order to avoid negative consequences for water users.

Adaptive Management in the Rocky Mountains

Comprehensive assessments of regional and local climate trends and trajectories will be integral for assessing potential impacts of climate warming in Rocky Mountain ecosystems (Pederson et al. 2010). The single largest source of uncertainty is simply how much and how fast the Earth's climate will warm. Additional inconclusiveness exists about how large-scale changes in the atmosphere will be realized at regional and local scales.

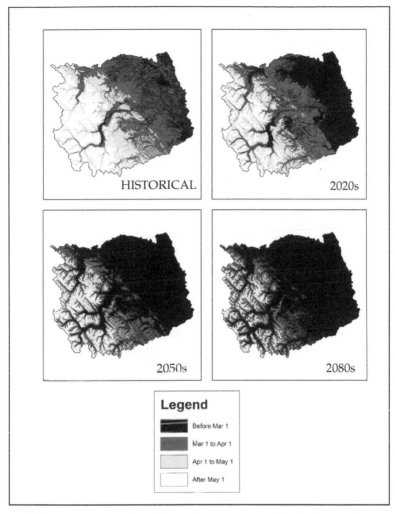

Figure 20.8. The 30-year mean date of complete snow removal for the historical, 2020, 2050, and 2080 periods in the North Saskatchewan watershed Canada. Where complete snow removal does not occur (i.e., higher elevations), the date of minimum SWE is used (from MacDonald et al. 2012).

Understanding the interactions between climate shifts and existing stressors is important to identifying which species and ecosystems are likely to be affected by projected changes and why they are likely to be vulnerable. Vulnerability assessments can be used to identify populations and habitats at risk for conservation management. Although broad-scale assessments are valuable for predicting species' vulnerabilities to climate warming and for raising awareness within the scientific community, they generally lack the spatial resolution (e.g., population-level) and predictive accuracy that managers need to make informed decisions on where and when to implement management actions (e.g., stream restoration and habitat protection). More recently, advances in geostatistical modeling of stream systems have greatly improved temperature predictability by using spatial data to explain variation across heterogeneous river networks in the Rockies (Isaak et al. 2010, Jones et al. 2013). These types of fish population and habitat models are currently being used to help direct conservation strategies at local (i.e., stream reach) scales aimed to improve population resilience and resistance against future changes through adaptive management.

Adaptation planning has emerged as a powerful management tool to help people and natural systems prepare for and cope with the current and projected impacts of climate change and other important cumulative stressors, such as habitat loss and invasive species. Climate change requires altering traditional approaches to conservation and natural resource management, which are focused on short term response variables, to conservation and restoration goals focused toward longer time periods (e.g., several decades) and larger scales (e.g., landscapes and biogeographical areas). This includes adaptation analyses that account for an increasingly unknown future. Adaption planning may include conservation measures to reduce deleterious effects or to take preventative measures to slow the impact and rate of climate change. As such, climate change adaptation planning is rapidly becoming the primary lens for conservation and natural resource planning and management to develop approaches that minimize risk for increasingly different and uncertain future changes.

Climate adaptation planning requires assessing the vulnerability of aquatic species, habitats and ecosystems to future climate change scenarios. Accordingly, management options are identified and implemented to reduce sensitivity and exposure to existing and future stressors, increasing resiliency and adaptive capacity across large spatial scales. In some cases, climate change may result in the expansion of suitable habitats, but for many coldwater dependent species, these changes are likely to restrict and further reduce suitable habitats to headwaters streams, resulting in highly fragmented habitat networks. For migratory salmonids, conserving the connectivity, size and extent of existing high quality habitats will be an important conservation strategy, as well as helping to identify where to undertake habitat restoration within a river network and which methods would best improve population resilience against future changes.

Conclusion

Global climate change is one of the greatest problems facing humanity now and that problem will continue to confront us throughout this century.

This chapter has provided specific discussion on how climate change will affect the Rocky Mountains of western North America. The complex terrestrial and aquatic ecosystems in the Rocky Mountains have evolved within the climate since the last great glaciation ended 10,000 years ago. Human-induced climate change will bring weather extremes on both the short and long-term to the Rocky Mountains. On average, those of us who depend upon the Rocky Mountains for resources will have to adjust to shorter winters and longer summers that will lead to a myriad of changes. Climate scientists are quite certain that greater variability will bring greater challenges for ecosystems and for society.

This chapter dwelt primarily with changes to the ecosystems of the Rocky Mountains. Snow and ice are critical components of hydrology and ecology, and snow and ice will be radically changed as the climate warms. Expect much greater variability in soil moisture conditions in forests and valleys that will lead to longer and more intense droughts than we have experienced in the past. Forest fires will be much more common, leading to whole scale land-cover changes in many areas. These changes will have dramatic impacts on terrestrial and aquatic ecosystems. In wet years we will experience extreme floods, like the province of Alberta suffered in summer 2013, with damages now estimated at US$5–$6 billion. The damages require years for recovery, and in some places extensive, costly flood protection will be needed. The same is true of Colorado Rocky Mountain regions that suffered through the worst fire year ever in the spring of 2013, and just a few months later the northern Rockies of Colorado experienced unprecedented, terrible flooding.

Humanity will adjust to climate change as best we can. The same is true of aquatic and terrestrial ecosystems in the Rocky Mountains. Where possible, plant and animal species will migrate in search of suitable habitat. Those fragile species already inhabiting areas near the tops of mountains may have nowhere to go. Aquatic species may be challenged by mega-floods, droughts where low flows makes survival difficult, or even impossible; and changing water temperatures will make current habitats less functional for indigenous species and allow new species to invade.

Adaptive management may offer some solutions but often the adaptation needs of humanity and society are in conflict with those of ecosystems. We will have to make very careful decisions, recognizing that the value of the resources we gain from the Rocky Mountains is based substantially on the value of ecosystems that keep the mountains a viable, healthy environment. The challenges of climate change in the Rocky Mountains are already quite overwhelming in many areas and will only continue to build. Humanity has to take immediate steps to both mitigate the impacts of climate change and develop adaptation techniques, policies and plans that will protect the interest of ecosystems and people. Failure to protect ecosystems adequately will lead to serious decline in the quality of life for all of us.

References

Barnett, T.P., J.C. Adam and D.P. Lettenmaier. 2005. Potential impacts of a warming climate on water availability in snow-dominated regions. Nature 438: 1–7.

Barnett, T.P., D.W. Pierce, H.G. Hidalgo, C. Bonfils, B.D. Santer, T. Das, G. Bala, A.W. Wood, T. Nozawa, A.A. Mirin, D.R. Cayan and M.D. Dettinger. 2008. Human-induced changes in the hydrology of the western United States. Science 319: 1080. DOI: 10.1126/science.1152538.

Baron, J.S., T.M. Schmidt and M.D. Hartman. 2009. Climate-induced changes in high elevation stream nitrate dynamics. Global Change Biology 15: 1777–1789.

Beniston, M. 2003. Climatic change in mountain regions: a review of possible impacts. Climatic Change 59: 5–31.

Brown, J., J. Harper and N. Humphrey. 2010. Cirque glacier sensitivity to 21st century warming: Sperry Glacier, Rocky Mountains, USA, Global and Planetary Change 74(2): 91–98.

Brown, R.S. and W.A. Hubert. 2011. A primer on winter, ice, and fish: what fisheries biologists should know about winter ice processes and stream-dwelling fish. Fisheries 36: 8–26.

Booth, Evan. 2011. Modelling the effects of climate change on glaciers in the Upper North Saskatchewan River Basin. Master of Science thesis, Department of Geography, University of Lethbridge.

Burles, K. and S. Boon. 2011. Snowmelt energy balance in a burned forest plot, Crowsnest Pass, Alberta, Canada. Hydrological Processes 25(19): 3012–3029.

Butler, D.R., G.P. Malanson and D.B. Fagre. 2007. Influences of geomorphology and geology on alpine treeline in the American West-more important than climatic influences? Physical Geography 28: 434–450.

Butler, D.R., G.P. Malanson, M.F. Bekker and L.M. Resler. 2003. Lithologic, structural, and geomorphic controls on ribbon forest patterns in a glaciated mountain environment. Geomorphology 55: 203–217.

Carrara, P.E. 1989. Late Quaternary Glacial and Vegetative History of the Glacier National Park Region, Montana. U.S. Geological Survey (USGS) Bulletin 1902. Denver, CO., pp. 64.

Cayan, D.R., S.A. Kammerdiener, M.D. Dettinger, J.M. Caprio and D.H. Peterson. 2001. Changes in the onset of spring in the western United States. Bulletin of the American Meteorological Society 82: 399–415.

Clow, D.W. 2010. Changes in the timing of snowmelt and streamflow in Colorado: a response to recent warming. J. Climate 23: 2293–2306.

Comeau, L.E.L., A. Pietroniro and M.N. Demuth. 2009. Glacier contribution to the north and south Saskatchewan rivers. Hydrological Processes 23(18): 2640–2653.

Crozier, L.G., M.D. Scheuerell and Richard W. Zabel. 2011. Using time series analysis to characterize evolutionary and plastic responses to environmental change: a case study of a shift toward earlier migration date in sockeye salmon. American Naturalist 178(6): 755–773.

Crozier, L.G. and R.W. Zabel. 2006. Climate impacts at multiple scales: evidence for differential population responses in juvenile Chinook salmon. Journal of Animal Ecology 75(5): 1100–1109.

Cunjak, R.A. 1998. Physiological consequences of overwintering in streams: the cost of acclimitization? Canadian Journal of Fisheries and Aquatic Sciences 45: 443–452.

Diaz, H. and J.K. Eischeid. 2007. Disappearing Òalpine tundraÓ Koppen climatic type in the western United States. Geophysical Research Letters 34: L18707, 1–4.

Dixon, D., S. Boon and U. Silins. 2013. Watershed-scale controls on snow accumulation in a small montane watershed, southwestern Alberta, Canada. Hydrological Processes. DOI: 10.1002/hyp.9667.

Fagre, D.B. and L.A. McKeon. 2010. Documenting disappearing glaciers: repeat photography at Glacier National Park, Montana, USA. pp. 77–88. *In*: R.H. Webb, D.E. Boyer and R.M. Turner (eds.). Repeat Photography: Methods and Applications in the Natural Sciences. Island Press, Covelo, CA.

Flannigan, M.D., B.D. Amiro, K.A. Logan, B.J. Stocks and B.M. Wotton. 2005. Forest fires and climate change in the 21st century. Climatic Change 11: 847–859.

Fountain, A.G. 2007. A century of glacier change in the American West. EOS Trans. of the American Geophysical Union 88(52).

Fountain, A.G., J.L. Campbell, E.A. Schuur, S.E. Stammerjohn, M.W. Williams and H.W. Ducklow. 2012. The disappearing cryosphere: Impacts and ecosystem responses to rapid cryosphere loss. BioScience 62: 405–415.

Fountain, A.G., H.J. Basagic IV, C. Cannon, M. Devisser, M.J. Hoffman, J.S. Kargel, G.J. Leonard and K. Thorneykroft. 2012. Glaciers and perennial snowfields of the American Cordillera, USA. *In*: J.S. Kargel, G.J. Leonard, M.P. Bishop, P. Blondel, C. Horwood, A. Kaab and B. Raup (eds.). Global Land Ice Measurements from Space. Praxis-Springer (Publ.), London (in press).

Goode, J.R., J.M. Buffington, D. Tonina, D.J. Isaak, R.F. Thurow, S. Wenger, D. Nagel, C. Luce, D. Tetzlaff and C. Soulsby. 2013. Potential effects of climate change on streambed scour and risks to salmonid survival in snow-dominated mountain basins. Hydrological Processes 27(5): 750–765.

Haak, A.L., J.E. Williams, D. Isaak, A. Todd, C.C. Muhlfeld, J.L Kershner, R.E. Gresswell, S.W. Hostetler and H.M. Neville. 2010. The Potential Influence of Changing Climate on the Persistence of Salmonids of the Inland West. U.S. Department of the Interior, U.S. Geological Survey, Open-File Report 2010–1236.

Hall, M.H.P. and D.B. Fagre. 2003. Modeled climate-induced glacier change in Glacier National Park, 1850–2100. BioScience 53(2): 131–140.

Hamlet, A.F. and D.P. Lettenmaier. 1999. Effects of climate change on hydrology and water resources in the Columbia basin. Journal of the American Water Resources Association 35: 1597–1623.

Harma, K.J., M.S. Johnson and S.J. Cohen. 2011. Future water supply and demand in the Okanagan basin, British Columbia: a scenario-based analysis of multiple, interacting stressors. Water. Resour. Manage. DOI 10.1007/s11269-011-9938-3.

Hauer, F.R., J.A. Stanford and Mark S. Lorang. 2007. Pattern and process in northern Rocky Mountain headwaters: ecological linkages in the headwaters of the Crown of the Continent. Journal of the American Water Resources Association 43(1): 104–117.

Hicke, J.A. and J.C. Jenkins. 2008. Mapping lodgepole pine stand structure susceptibility to mountain pine beetle attack across the western United States, Forest Ecology and Management 255: 1536–1547.

Hoffmann, M.J., A.G. Fountain and J.M. Achuff. 2007. 20th-century variations in area of cirque glaciers and glacierets, Rocky Mountain National Park, Rocky Mountains, Colorado, USA. Annals of Glaciology 46: 349–354.

Holtmeier, F.-K. and G. Broll. 2007. Treeline advance—driving processes and adverse factors. Landscape Online 1: 1–33.

Huntington, J.L. and R.G. Niswonger. 2012. Role of surface-water and groundwater interactions on projected summertime streamflow in snow dominated regions: an integrated modeling approach. Water Resources Research 48: W11524. DOI: 10.1029/2012WR012319.2012.

Isaak, D., C. Muhlfeld, A. Todd, R. Al-Chakhady, J. Roberts, J. Kershner, K. Fausch, S. Hostetler. 2012. The Past as a Prelude to the Future for Understanding 21st Century Climate Effects on Rocky Mountain Trout. Fisheries 37: 542–556. HYPERLINK "http://www.fs.fed.us/rm/boise/AWAE/projects/stream_temp/downloads/2012_PastAsPreludeForClimateEffectsOnRockyMountainTrout.pdf".

Isaak, D.J., C.H. Luce, Bruce E. Rieman, David E. Nagel, Erin E. Peterson, Dona L. Horan, Sharon Parkes and Gwynne L. Chandler. 2010. Effects of climate change and wildfire on stream temperatures and salmonid thermal habitat in a mountain river network. Ecological Applications 20(5): 1350–1371.

Jentsch, A., J. Kreyling and Carl Beierkuhnlein. 2007. A new generation of climate-change experiments: events, not trends. Frontiers in Ecology and the Environment 5(7): 365–374.

Jones, L.A., C.C. Muhlfeld, L.A. Marshall, B.L. McGlynn and J.L. Kershner. 2014. Estimating Thermal Regimes of Bull Trout and Assessing the Potential Effects of Climate Warming on Critical Habitats. River Res. Applic. 30: 204–216. doi: 10.1002/rra.2638.

Keleher, C.J. and F.J. Rahel. 1996. Thermal limits to salmonid distributions in the rocky mountain region and potential habitat loss due to global warming: a geographic information system (GIS) approach. Transactions of the American Fisheries Society 125(1): 1–13.

Key, C.H., D.B. Fagre and R.K. Menicke. 2002. Glacier retreat in Glacier National Park, Montana. pp. J365–J381. *In*: R.S. Jr. Williams and J.G. Ferrigno (eds.). Satellite Image Atlas of Glaciers of the World, Glaciers of North America - Glaciers of the Western United States. U.S. Geological Survey Professional Paper 1386-J. United States Government Printing Office, Washington D.C., USA.

Klasner, F.L. and D.B. Fagre. 2002. A half century of change in alpine treeline patterns at Glacier National Park, Montana, USA. Arctic, Antarctic and Alpine Research 34: 49–56.

Knowles, N., M.D. Dettinger and D.R. Cayan. 2006. Trends in snowfall versus rainfall in the western United States. Journal of Climate 19: 4545–4559.

Kovach, R.P., A.J. Gharrett et al. 2012. Genetic change for earlier migration timing in a pink salmon population. Proceedings of the Royal Society B-Biological Sciences 279(1743): 3870–3878.

Lapp, S., J.M. Byrne, S.W. Kienzle and I. Townshend. 2005. Climate warming impacts on snowpack accumulation in an alpine watershed: a GIS based modeling approach. International Journal of Climatology 25: 521–526.

Larson, R.P., J.M. Byrne., D.L. Johnson, M.G. Letts and S.W. Kienzle. 2011. Modelling Climate Change Impacts on Spring Runoff for the Rocky Mountains of Montana and Alberta I: Model Development, Calibration and Historical Analysis. Canadian Water Resources Journal 36:17Đ34.

Lesica, P. 2002. Flora of Glacier National Park. Oregon State University Press, Corvallis, Oregon, USA.

Lesica, P. 2012. Monitoring Rare Arctic-Alpine Plants at the Periphery of their Range in Glacier National Park, Montana. Final Report to Glacier National Park, pp. 14.

Lesica, P. and B. McCune. 2004. Decline of arctic-alpine plants at the southern margin of their range following a decade of climatic warming. Journal of Vegetation Science 15: 679–690.Ê

Littell, J.S., Donald McKenzie, David L. Peterson and Anthony L. Westerling. 2009. Climate and wildfire area burned in western U.S. ecoprovinces, 1916–2003. Ecological Applications 19: 1003–1021.

Logan, J.A., W.W. MacFarlane and L. Willcox. 2010. Whitebark pine vulnerability to climate-driven mountain pine beetle disturbance in the Greater Yellowstone Ecosystem. Ecological Applications 20: 895–902.

MacDonald, R.J., J. Byrne, S. Boon and S.W. Kienzle. 2012 Modelling the potential impacts of climate change on snowpack in the North Saskatchewan River watershed, Alberta. Water Resources Management 26(11): 3053–3076.

MacDonald, R.J., J.M. Byrne, S.W. Kienzle and R.P. Larson. 2011. Assessing the potential impacts of climate change on mountain snowpack in the St. Mary River catchment, Montana. Journal of Hydrometeorology 12: 262–273.

Malanson, G.P. and D.B. Fagre. 2013. Spatial contexts for temporal variability in alpine vegetation under ongoing climate change. Plant Ecology 214: 1309–1319.

Malanson, G.P., D.R. Butler, D.B. Fagre, S.J. Walsh, D.F. Tomback, L.D. Daniels, L.M. Resler, W.K. Smith, D.J. Weiss, D.L. Peterson, A.G. Bunn, C.A. Hiemstra, D. Liptzin, P.S. Bourgeron, Z.

Shen and C. Millar. 2007. Alpine treeline of western North America: Linking organism-to-landscape dynamics. Physical Geography 28: 378–396.

Marshall, S., E.C. White, M.N. Demuth, T. Bolch, R. Wheate, B. Menonuos, M.J. Beedle and J. Shea. 2011. Glacier water resources on the eastern slopes of the Canadian Rocky Mountains. Canadian Water Resources Journal 36: 109–134.

McKinney, S.T., C.E. Fiedler and D.F. Tomback. 2009. Invasive pathogen threatens bird-pine mutualism: implications for sustaining a high-elevation ecosystem. Ecological Applications 19: 597–607.

Meddens, A., J. Hicke and C. Ferguson. 2012. Spatiotemporal patterns of observed bark beetle-caused tree mortality in British Columbia and the western United States. Ecological Applications 22: 1876–1891.

Merritt, R.W., K.W. Cummins et al. 1984. The Ecology of Aquatic Insects. Praeger Publishers, New York.

Mills, L.S., M. Zimova, J. Oyler, S. Running, J. Abatzoglou and P. Lukacs. 2013. Camouflage mismatch in seasonal coat color due to decreased snow duration. Proc. National Academy of Sciences 110: 7360–7365.

Milner, A.M., J.E. Brittain, Emmanuel Castella and Geoffrey E. Petts. 2001. Trends of macroinvertebrate community structure in glacier-fed rivers in relation to environmental conditions: a synthesis. Freshwater Biology 46: 1833–1847.

Minville, M., F. Brissette, S. Krau and R. Leconte. 2009. Adaptation to climate change in the management of a Canadian water-resources system exploited for hydropower. Water Resources Management 23: 2965–2986.

Minville, M., F. Brissette, S. Krau and R. Leconte. 2010. Behaviour and performance of a water resource system in Quebec (Canada) under adapted operating policies in a climate change context. Water Resources Management 24: 1333–1352.

Moore, R.D., S.W. Fleming, B. Menounos, R. Wheate, A. Fountain, K. Stahl, K. Holm and M. Jakob. 2009. Glacier change in western North America: influences on hydrology, geomorphic hazards, and water quality. Hydrological Processes 23: 42–61.

Mote, P.W., A.F. Hamlet, M.P. Clark and D.P. Lettenmaier. 2005. Declining mountain snowpack in western North America. Bulletin of the American Meteorological Society 86: 39–49.

Muhlfeld, C.C. and B. Marotz. 2005. Seasonal movement and habitat use by subadult bull trout in the upper Flathead river system, Montana. North American Journal of Fisheries Management 25: 797–810.

Muhlfeld, C.C., J.J. Giersch, F. Richard Hauer, Gregory T. Pederson, Gordon Luikart, Douglas P. Peterson, Christopher C. Downs and Daniel B. Fagre. 2011. Climate change links fate of glaciers and an endemic alpine invertebrate. Climatic Change 106(2): 337–345.

Muhlfeld, C.C., T.E. McMahon, Matthew C. Boyer and Robert E. Gresswell. 2009. Local habitat, watershed and biotic factors in the spread of hybridization between native westslope cutthroat trout and introduced rainbow trout. Transactions of the American Fisheries Society 138(5): 1036–1051.

Muhlfeld, C.C., T.E. McMahon and Durae Belcer. 2009. Spatial and temporal spawning dynamics of native westslope cutthroat trout, Oncorhynchus clarkii lewisi, introduced rainbow trout, Oncorhynchus mykiss, and their hybrids. Canadian Journal of Fisheries and Aquatic Sciences 66(7): 1153–1168.

North Saskatchewan Watershed Alliance. 2005. State of the North Saskatchewan Watershed Report 2005: A foundation for collaborative watershed management. North Saskatchewan Watershed Alliance, Edmonton, Alberta. 202 pages.

Northcote, T.G. 1997. Potamodromy in salmonidae-living an dmoving in the fast lane. North American Journal of Fisheries Management 17(4): 1029–1045.

Ommanney, C.S. 2002. Satellite image atlas of North America, Glaciers of Canada: History of glacier investigations in Canada, edited by D.o.t.I.G. Survey, U.S. Government Printing Office, Washington D.C., pp. J27–82.

Parmesan, C. 2006. Ecological and evolutionary responses to recent climate change. Annual Review of Ecological Systems 37: 637–669.

Parmesan, C. and G. Yohe. 2003. A globally coherent fingerprint of climate change impacts across natural systems. Nature 421(6918): 37–42.

Parolo, G. and G. Rossi. 2008. Upward migration of vascular plants following a climate warming trend in the Alps. Basic and Applied Ecology 9: 100–107.

Paul, A.J. and J.R. Post. 2001. Spatial distributions of native and nonnative salmonids in the streams of the eastern slopes of the Canadian rocky mountains. Transactions of the American Fisheries Society 130: 417–430.

Pauli, H., M. Gottfried, K. Reiter, C. Klettner and G. Grabherr. 2007. Signals of range expansions and contractions of vascular plants in the high Alps: observations (1994–2004) at the GLORIA master site Schrankogel, Tyrol, Austria. Global Change Biology 13: 147–156.

Pederson, G.T., L.J. Graumlich, D.B. Fagre, T. Kipfer and C.C. Muhlfeld. 2010. A century of climate and ecosystem change in Western Montana: what do temperature trends portend? Climatic Change 98(1-2): 133–154.

Pederson, G.T., S.T. Gray, T. Ault, W. Marsh, D.B. Fagre, A.G. Bunn, C.A. Woodhouse and L.J. Graumlich. 2011. Climatic controls on the snowmelt hydrology of the northern Rocky Mountains. J. Climate 24: 1666–1687.

Peterson, D.P., B.E. Rieman, Jason B. Dunham, Kurt D. Fausch and Michael K. Young. 2008. Analysis of trade-offs between threats of invasion by nonnative brook trout (Salvelinus fontinalis) and intentional isolation for native westslope cutthroat trout (Oncorhynchus clarkii lewisi). Canadian Journal of Fisheries and Aquatic Sciences 65: 557–573.

Pochop, L., R. Marston, G. Kerr, D. Veryzer, M. Varuska and R. Jacobel. 1990. Glacial icemelt in the Wind River Range, Wyoming. pp. 118–124. In: Watershed Planning and Analysis in Action, Symposium Proceedings of IR Conference, Durango, CO. July 9–11, 1990.

Poff, N.L. 2002. Ecological response to and management of increased flooding caused by climate change. Philosophical Transactions of the Royal Society of London Series a-Mathematical Physical and Engineering Sciences 360(1796): 1497–1510.

Pomeroy, J.W. and R.A. Schmidt. 1993. The use of fractal geometry in modelling intercepted snow accumulation and sublimation. Proceedings of the Eastern Snow Conference 50: 1–10.

Pomeroy, J., X. Fang and C. Ellis. 2012. Sensitivity of snowmelt hydrology in Marmot Creek, Alberta, to forest cover disturbance. Hydrological Processes 26: 1891–1904.

Power, G., R.S. Brown and J.G. Imhof. 1999. Goundwater and fish-insights from northern North America. Hydrological Processes 13Ê: 401–422.

Rahel, F.J., B. Bierwagen and Y. Taniguchi. 2008. Managing aquatic species of conservation concern in the face of climate change and invasive species. Conservation Biology 22(3): 551–561.

Rieman, B.E. and J.D. McIntyre. 1995. Occurrence of bull trout in naturally fragmented habitat patches of varied size. Transactions of the American Fisheries Society 124(3): 285–296.

Rieman, B.E., D. Isaak, Susan Adams, Dona Horan, David Nagel, Charles Luce and Deborah Myers. 2007. Anticipated climate warming effects on bull trout habitats and populations across the interior Columbia River basin. Transactions of the American Fisheries Society 136(6): 1552–1565.

Roberts, J.J., K.D. Fausch, Douglas P. Peterson and Mevin B. Hooten. 2013. Fragmentation and thermal risks from climate change interact to affect persistence of native trout in the Colorado River basin. Global Change Biology 19(5): 1383–1398.

Rood, S.B., G.M. Samuelson, J.K. Weber and K.A. Wywrot. 2005. Twentieth-century decline in streamflows from the hydrographic apex of North America. Journal of Hydrology 81: 1281–1299.

Scherrer, D. and C. Korner. 2011. Topographically controlled thermal-habitat differentiation buffers alpine plant diversity against climate warming. J. Biogeography 38: 406–416.

Schindler, D.W. and W.F. Donahue. 2006. An impending water crisis in Canada's western Prairie Provinces. Proceedings of the National Academy of Sciences, USA 103: 7210–7216.

Schweiger, W., I.W. Ashton, Clint C. Muhlfeld, Leslie A. Jones and Loren L. Bahls. 2011. The Distribution and abundance of a nuisance native alga, Didymosphenia geminata, in streams of Glacier National Park: Climate drivers and management implications. Park Science 28(2): 78–81.

Selong, J.H., T.E. McMahon, Alexander V. Zale and Frederic T.B. Arrows. 2001. Effect of temperature on growth and survival of bull trout, with application of an improved method for determining thermal tolerance in fishes. Transactions of the American Fisheries Society 130(6): 1026–1037.

St. Jaques, M.J., D.J. Sauchyn and Y. Zhao. 2010. Northern Rocky Mountain streamflow. Geophysical Research Letters 37(6).

Stewart, I.T. 2009. Changes in snowpack and snowmelt runoff for key mountain regions. Hydrological Process 23: 78–94.

Stewart, I.T., D.R. Cayan and M.D. Dettinger. 2004. Changes in snowmelt runoff timing in western North America under a business as usual climate change scenario. Climatic Change 62: 217–232.

Storck, P., D. Lettenmaier and S.M. Bolton. 2002. Measurement of snow interception and canopy effects on snow accumulation in a mountainous maritime climate, Oregon, United States. Water Resources Research 38. doi:10.1029/2002WR001281.

USGCRP (U.S. Global Change Research Program). 2000. Climate change and America: Overview Document. A Report of the National Assessment Synthesis Team. Washington, D.C.

Villegas, J.C., D.D. Breshears, C.B. Zou and D.J. Law. 2010. Ecohydrological controls of soil evaporation in deciduous drylands: how the hierarchical effects of litter, patch and vegetation mosaic cover interact with phenology and season. Journal of Arid Environments 74: 595–602.

Walther, G.R., E. Post, Peter Convey, Annette Menzel, Camille Parmesan, Trevor J.C. Beebee, Jean-Marc Fromentin, Ove Hoegh-Guldberg and Franz Bairlein. 2002. Ecological responses to recent climate change. Nature 416(6879): 389–395.

Ward, J.V. 1994. Ecology of alpine streams. Freshwater Biology 32: 277–294.Ê

Wenger, S.J., D.J. Isaak, Jason B. Dunham, Kurt D. Fausch, Charles H. Luce, Helen M. Neville, Bruce E. Rieman, Michael K. Young, David E. Nagel, Dona L. Horan, Gwynne L. Chandler. 2011a. Role of climate and invasive species in structuring trout distributions in the interior Columbia River Basin, USA. Canadian Journal of Fisheries and Aquatic Sciences 68: 988–1008.

Wenger, S.J., D.J. Isaak, Seth J. Wenger, Daniel J. Isaak, Charles H. Luce, Helen M. Neville, Kurt D. Fausch, Jason B. Dunham, Daniel C. Dauwalter, Michael K. Young, Marketa M. Elsner, Bruce E. Rieman, Alan F. Hamlet and Jack E. Williams. 2011b. Flow regime, temperature, and biotic interactions drive differential declines of trout species under climate change. Proceedings of the National Academy of Sciences 108: 34, 14175–14180.

Westerling, A.L., H.G. Hidalgo, D.R. Cayan and T.W. Swetnam. 2006. Warming and earlier spring increase western US forest wildfire activity. Science 313(5789): 940–943.

Westerling, A., H.G. Hidalgo, D.R. Cayan and T.W. Swetnam. 2006. Warming and earlier spring increase western U.S. forest wildfire activity. Science 313: 940–943.

Williams, J.E., A.L. Haak, Helen Neville and Warren Colyer. 2009. Potential consequences of climate change to persistence of cutthroat trout populations. North American Journal of Fisheries Management 29: 533–548.

Winkler, R. and S. Boon. 2010. The effects of mountain pine beetle attack on snow accumulation and ablation: a synthesis of ongoing research in British Columbia. Streamline Watershed Management Bulletin 13(2): 25–31.

Yue, X., L. Mickley, J. Logan and J.O. Kaplan. 2013. Ensemble projections of wildfire activity and carbonaceous aerosol concentrations over the western United States in the mid-21st century. Atmospheric Environment 77: 767–780.

Zeigler, M.P., A.S. Todd and Colleen A. Caldwell. 2012. Evidence of recent climate change within the historic range of Rio Grande cutthroat trout: implications for management and future persistence. Transactions of the American Fisheries Society 141(4): 1045–1059.

21

Climate Change Impacts On The Northwestern Andean (NOA) and Western (Cuyo) Andean Regions of Argentina

H.D. Ginzo[1,]* and *A. Faggi*[2,3]

INTRODUCTION

In the last decades Argentine has actively participated in the elaboration and implementation of international policies leading towards the care of the global environment. The country has been enthusiastically involved in the international process that crystallized in the elaboration of the UNFCCC[1]; signed the Convention in 1992, ratified and enforced it in 1994. In doing so Argentina committed herself to fulfill the ultimate objective of the Convention, which is the stabilization of greenhouse gas concentrations "… at a level that would prevent dangerous anthropogenic (human induced) interference with the climate system." Argentina accepted this commitment taking into account its "… common but differentiated responsibilities, and their specific national and regional development priorities, objectives,

---
[1] Instituto del Clima, Academia Argentina de Ciencias del Ambiente, Av. Santa Fe 1145, C1045ABF C.A. Buenos Aires, Argentina.
   Email: hdginzo@arnet.com.ar
[2] CONICET-MACN, A. Gallardo 470, C1405DJR C.A. Buenos Aires, Argentina.
   Email: afaggi@macn.gov.ar
[3] UFLO, Nazca274, C1406AJO C.A. Buenos Aires, Argentina.
* Corresponding author

and circumstances." Argentina chaired the international negotiations that materialized into the Kyoto Protocol to the Convention in 1995.[2] The country is an observer in the World Bank's Forest Carbon Partnership Facility, a program aimed at financing the implementation of REDD+ projects in developing countries. Because of its commitment to the principles of the Convention, Argentina has submitted two national communications to the UNFCCC together with corresponding inventories of emissions and removals of Greenhouse Gases (GHG), and one review of the first national communication. In both communications Argentina presented diverse studies which could be the foundation for devising national policies aimed to confronting the likely impacts of climate change. Two of those are particularly relevant to the present chapter. The first national communication (RA 1997) contains a study on the vulnerability of oases located in both the NOA and Cuyo regions; the second communication (RA 2007) contains climate scenarios for the whole country.

In this chapter we assess the relevant information on regional vulnerabilities, mitigation and adaptation actions to plausible impacts of Climate Change on two mountain regions in Argentina: one in the Northwest (NOA[3] region), and the other in the West (Cuyo[4] region) of the country (Fig. 21.1). The NOA and Cuyo regions are closely related to the septentrional and the mid sections of the Argentinean Andean Ridge. Roveta (2008) has aptly dealt with climate change impacts in the meridional section of the Argentinian Andes.

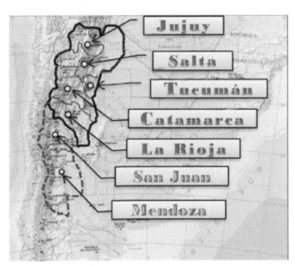

Figure 21.1. Partial physical map of Argentina showing the boundaries of the NOA region (continuous thick black contour line) and the Cuyo region (segmented thick gray contour line). Labels show the names of the provinces in each region (Map obtained from IGN (2011).

Brief Description of the NOA and Cuyo Regions

The contiguous area of Argentina is 2,791,810 km² (IGN 2011); after Brazil, it is the largest country in South America. Its wedge-like shape stretches along about 3,900 km in the N-S direction; climate zones range from subtropical in the North to tundra in the South. As a consequence of this varied climate the country shows large diversity in biological and landscape elements. Argentina has been divided geographically in five regions, three of which— Patagonia,[5] Cuyo and NOA—are closely related to the Andean Ridge.

The NOA and Cuyo regions share the High Andes, Puna, Hill and Valley Scrubland ecoregions. The NOA region in addition contains the *Yungas*, which is absent from the Cuyo region. The latter contains the Flatland and Plateau Scrubland ecoregion, which is not in the NOA region (Burkart et al. 1999).

Water (Rivers)

Both NOA and Cuyo have a lot of rivers varying in size from brooks to rivers. Some of these rivers directly reach the Atlantic Ocean, like the Colorado River at the end of a 1,000-km journey. Other rivers such as the *Juramento-Salado* River (which has its source in the Andes and discharges into the *Paraná* River) and the *Desaguadero* River (with springs also in the Andes, and flows into the Colorado River) indirectly flow into the Atlantic Ocean.

Some other rivers make endorheic basins and discharge in either lakes or salt lakes (as in *Puna*) or diffuse underground through permeable rock, as many rivers in the Cuyo region do. These are mostly fed with melt water from glaciers. In the NOA region, however, rivers are mostly fed by rain- and snowfall. All these rivers, no matter their sources, are markedly seasonal: flow rates peak in the summer months, in coincidence with annual higher temperatures and snowmelt, and are minimal or null in the winter.

Rivers are essential to technologically advanced agriculture and the industrial processing of grapes, olives, stone and pip fruits, forage crops, vegetables, etc., since these crops rely on irrigation with either surface or groundwater. In the NOA region there is enough rainfall for dryland farming in areas to the east of the *Yungas* where the mean annual precipitation is over 500 mm.yr^{-1}.

Drivers of change: major human activities in the region

Agriculture, tourism and mining are relevant sources of income to the NOA and Cuyo regions, all these activities are strongly dependent on water availability. In this chapter agriculture is taken in its broadest sense; i.e., it includes farming, silviculture, aquaculture and any other production system based on the raise and care of a living organism (FAO 2009).

Agriculture

This part only refers to industrial agriculture, which is characterized by being highly mechanized, intensively fertilized, weeded, and pest- and disease-ridden with chemical products, mostly harmful to the environment and wildlife. The other agriculture, more akin to the principles of agroecology (Altieri 1996), is here considered as an ecosystem service and it is denominated 'food and fiber'. Industrial agriculture thrives in the valleys belonging to the Hill and Valley Scrubland, and in those in between *Puna* and the *Yungas*. In the drier (western) areas in NOA agriculture is economically feasible exclusively under irrigation. The region shared 28% of the national irrigated area in 2002 (INDEC 2011).

a) NOA

In this region the provinces with the largest proportion of farmland area under irrigation in 2002 were Catamarca and La Rioja. The system of gravitational (or flood) irrigation is the most common in the region; it was used in 68% of the irrigated area compared to either sprinkler or localized irrigation, which were each used in 16% of the irrigated areas. Gravitation irrigation was mostly used in Salta and Jujuy; sprinkler irrigation was the main system in Tucumán and Catamarca, and localized irrigation was the preferred system in La Rioja.

Each province had a preferred combination of cropping systems under irrigation (Table 21.1).

Table 21.1. Cropping systems (grayed cells) which took up with at least 75% of the area under irrigation in the NOA region in 2002 (data from INDEC 2011).

Cropping system	Province				
	Jujuy	Salta	Tucumán	Catamarca	La Rioja
sugarcane	▒	▒			
tobacco	▒	▒			
vegetables			▒		
orchards			▒	▒	▒
cereals		▒	▒		▒
oilseeds		▒		▒	
forages		▒		▒	

b) Cuyo

The most important agricultural production system is winemaking. Until the 1970s this industry was almost exclusively oriented to satisfy the domestic market. From that time onwards some multinational enterprises completely updated technical, institutional and commercial procedures and processes to put the industry on a firm footing to play in the international arena of high quality wines. No less important was the parallel development of ancillary commercial concerns like winery- and gastronomy-oriented tourism (Richard-Jorba 2008).

Almost 100% of the agricultural area in each of the provinces of San Juan and Mendoza was under irrigation in 2002 (INDEC 2011). About 87% of that area was irrigated gravitationally, and 12.3% was under localized irrigation; sprinkler irrigation was not used in the Cuyo region.

Gravitational irrigation predominated over localized irrigation in Mendoza and the converse is true in San Juan.

Tourism

National and international tourism is a well-developed industry in both regions; adventure tourism and sport fishing are some of the most important amenities there (Mintur 2011).

In the provinces of Jujuy and Salta (NOA region) the ruins of pre-Columbian cultures, and fortifications and buildings dating back to the time of the Spanish conquest and colonization are major tourist attractions (Bergesio and Montial 2010). The beauty of the landscape together with a dry and sunny weather from March to November attracts tourists interested in trekking, sightseeing, and leisure to the whole NOA region. In the province of La Rioja the Talampaya National Park offers hills with multicolored walls and interesting indigenous carvings (Mintur 2011). In the province of Jujuy the Quebrada de Humahuaca (Humahuaca gorge) is the most important tourism destination; it is also an UNESCO World Heritage Site.

In Cuyo there are two main attractions for both domestic and foreign tourists: geology and wineries; some of the latter provide lodging and high-quality gastronomy. Vineyards and wineries of international renown have led to the development of rural, adventure, cultural, agricultural and ecological tourism (Pastor 2010). In the province of San Juan the Valley of the Moon (Ischigualasto National Park), an eerie landscape of wind-carved hills that also is a repository of plant and animal fossils.

We take the beauty of the regional landscapes as the ecosystem service *par excellence*, because it is nature in its full manifestation which attracts people to gaze at it. Both the NOA and Cuyo regions have particularly distinct landscapes; the former region, however, has a larger diversity of climes and biomes than the former. On the NOA and Cuyo regions there are no data adequately disaggregated (or at least data that can be used as a basis) to assessing the impacts of touristic activities on both the local and the global biophysical and biological environments.

Mining

In the NOA and Cuyo regions mining is a widespread extractive activity, particularly in the High Andes and Puna ecoregions. Several minerals are extracted from the ranges in the High Andes: gold, silver and copper (NOA and Cuyo), lead and zinc (NOA). Sodium, lithium, borate and potassium salt deposits are also common in the High Andes (Minería 2009).

Open-pit surface mining is the most common extractive procedure used in Argentina (Comelli et al. 2010). It is considered harmful to the environment, particularly in the case of metal ore mining where large amounts of removed rock are washed with corrosive and/or toxic chemicals for extracting the mineral of interest (Brenning and Azócar 2010). This technology uses large amounts of water, which on the one hand poses an environmental risk and, on the other hand, competes with its use in agriculture and human consumption. Mining not only contaminates water because of acid rock-drainage; mining in the vicinity of glaciers degrades the permafrost and creates instability in the glacier, eventually leading to its complete loss (Brenning and Azócar 2010). Other impacts of open-mining are related to aesthetics, noise, air quality (dust and pollutants), vibration, water discharge and runoff, subsidence and process-wastes.

Oil and petrochemical industries

Oil wells in production are in the provinces of Jujuy and Salta (NOA) and in the province of Mendoza (Cuyo). The average (2004–2008) reserve of those wells was just about 13% of the national average for that period. Petrochemical industries are present in Cuyo but not in NOA.

Typical water use in oil refineries ranges from 246 to 341 liters per barrel of oil (Sandy 2005). In refineries water is basically used for steam production, cooling service and removal of water-soluble inorganic compounds (utility water). About 48% of the total use of water is for cooling; this is consumptive use of water and it returns to the atmosphere by evaporation. In the petrochemical industry most of water is also used for cooling. Wastewater is generally reused in steam systems, but after some purification. Water use in oil exploration is negligible except in the case of enhanced oil recovery.

The use of water in the oil and petrochemical industry might compete with other uses under a scenario of global warming-induced water stress if the recent decadal (1999–2010) decline (IAPG[6]) in oil production from NOA (47,600 $m^3.yr^{-1}$) and Cuyo (80,500 $m^3.yr^{-1}$) were definitely reversed.

Food and fiber

This part deals with the traditional agriculture, i.e., cropping and livestock rising by local people for their own consumption and for local markets.

The High Andes and Puna are among the poorest regions in South America (Lichtenstein and Vilá 2003). Diverse aboriginal populations practice subsistence farming and livestock on public lands. Local populations tend herds of both wild—vicuna and guanaco—and domesticated—alpaca and lama—American camelids that live on hilly country in both the NOA and Cuyo regions. All of these camelids are sources of meat and hair. Lamas are mostly raised for their meat. Its hair has some demand but it is not as valuable as vicuna's. The best quality fiber was US$ 1.93 per kilogram paid to the herder in 2007. The lama population was about 161,400 heads in 2002 (Minagri 2011). About two-thirds of it was in the province of Jujuy, in northwestern Argentina, mostly in the Puna ecoregion. The most valuable of these is vicuna, also prized for its sacral symbolism as a property of Mother Earth (Pachamama) (Lichtenstein and Vilá 2003). Goats are also raised for their milk and meat on natural pastures covering hill slopes in both the NOA and Cuyo regions.

Farming depends mainly on rainfall since the aboriginal people constitute a landless peasantry group and, therefore, they are practically excluded from participating in local irrigation schemes. Altitude, harsh weather, destructive frosts, scarcity of water, droughts, soil erosion and overgrazing constrain agriculture and economic growth. These restrictions are enhanced by the absence of production systems adapted to a xeric environment, long distances, difficulty to reach markets, and low demand for regional products. Lack of economic opportunities encourages the emigration of locals to rural and urban conditions at lower altitudes in quest of better living conditions (Lichtenstein and Vilá 2003).

Terraces and peatlands are substrata used by indigenous peoples for raising crops in the High Andes ecoregion. Aymara and Atamaqueño people drain peatlands and strip away their surface layers in order to use the underlying organic mineral soil for farming (Squeo et al. 2006). Terracing is an example of an indigenous adaptive technology developed over millennia for growing native species in a particularly harsh arid environment of the High Andes (FAO 2009). Aboriginal people grow many native food plant species on these terraces. Terraces were made not only for facilitating cultivation work, but to collect water from sporadic rains and snowmelt running off mountain slopes (FAO 2009).

It is plausible that the ecosystems in the ecoregions High Andes, Puna and both Scrublands, which provide basically food and fiber for local indigenous populations, can be severely disrupted by unexpected biophysical perturbations and human interventions on them. For example, local herders to raise livestock on natural grasslands in hills lying about the Calchaquí valleys extend along the provinces of Jujuy, Salta, Tucumán and Catamarca. Itinerant herders move about their livestock looking for better forage; frequently they increase animal load beyond the carrying capacity of a grassland, triggering an overgrazing episode that promptly leads to the inception of erosive processes in communities prone to them because they grow on structurally fragile soils (Molinillo 1993).

Vulnerabilities to Climate Change

The climatic trends detected over most of the country during the latest four decades are likely consequences of global climate change (RA 2007). Those trends have been more evident in some regions than in others, presumably because high-quality climatic data was unevenly distributed over the country's area.

Seven major climate vulnerabilities to plausible changes in temperature and precipitation were identified for the country, some of which are relevant to the NOA and Cuyo regions. Each of these regions, however, shows particular vulnerabilities to those climate variables (Table 21.2).

As to the future, changes in the global climate could have the following impacts on NOA and Cuyo in the interval 2020–2040 (Table 21.3).

Plausible Biological, Physical, Economic and Social Impacts

Water is the regional resource to be mostly impacted by global warming. National scenarios (RA 2007) show a clear connection between the availability of water in the NOA and Cuyo regions and the increase in temperature. As mountain rivers supply water for agriculture, energy generation, and settlement needs, climate change is very likely to have large economic and social impacts. The efficacy of regional mitigation measures to avoid water stresses is difficult to assess in view of the complexity of the climate system and the dearth of dedicated studies.

Increased warming is likely to enhance the aridity of the region by increasing evaporation, which in turn would make soils dryer and reduce the levels of groundwater reserves. A relatively large part of the NOA region High Andes and Puna and all of Cuyo are under high to very high water stress as measured by the (water) withdrawal to availability ratio,

Table 21.2. Replication of major countrywide vulnerabilities to climate change in mountain-related ecoregions in NOA and Cuyo regions.

Climate effect	Vulnerability	
	NOA	Cuyo
Increase in mean yearly rainfall during the 20th century	Yes—*Puna, Yungas* and the Hill and Valley Scrubland	Yes—both Scrublands
Increase in the frequency of extreme rainfall events	No	No
Increase in air temperature	No	Yes—High Andes (glacier retraction)
Increase in the frequency of heat waves	Yes—heavily-built cities	Yes—heavily-built cities
Decrease in river flow regime	No	Yes—High Andes (glacier retraction)
Increase in river flow variability and in risk of frequent drought spells [1]	Yes	No
On the degree of social vulnerability [2]	Low to high	Low to medium

Table elaborated on the basis of information contained in RA (2007). [1] (Barros and Kullock 2006). [2] Social vulnerability as measured with an index built with variables reflecting demographic, economic and welfare conditions sensitive to either withstand or avoid damages under disaster events (Natenzon et al. 2006).

Table 21.3. Subjective assessment of the probability that national impacts of climate change reflected on the NOA and Cuyo regions in the interval 2020–2040.

Impact	NOA	Cuyo	Drivers
Increase in plant water stress, and its concomitant impact on biodiversity, ecosystem productivity and agricultural production.	likely	unlikely	Increased evapotranspiration driven by increasing air temperatures.
Decrease in the flows of major regional rivers, which translates into decreases in agricultural production, electricity hydrogeneration, and urban quality of life.	unlikely	likely	Decrease in precipitation (rain and snow) over the Andean Ridge between parallels 30° S and 36° S. Persistent retreat of the Andean glaciers, particularly to the south of parallel 30° S. Increase in the altitude of the 0°C-isotherm south to parallel 30° S.

Table elaborated on the basis of information contained in RA (2007). In one of the probability scales used by the IPCC in its publications, a likely event is one with >66% chance of occurring, while an unlikely event is one with a <33% chance (Bates et al. 2008).

which is larger than 0.4 (Bates et al. 2008). In a scenario of future scarcity of water in both regions, reasonable mitigation measures should preferably be independent of the use of water or alternatively be water-thrifty at least.

The observed decrease in precipitations (snowfall and rainfall) over the Cuyo region together with the likely 1°C warming and its consequent increase in evapotranspiration will certainly strain the supply of water to the irrigated croplands spread along the foothills of the Andes. The absolute amount of water available for them is expected to decrease, which would increase the risk of reducing crop yields (RA 2007).

The use of groundwater instead of surface water (from rivers) is not a solution to water scarcity because groundwater has the same sources as surface water and it is, therefore, vulnerable to the decrease of precipitation in the Andean Ridge (Canziani et al. 1995). A modeling exercise of the northern watershed in the Province of Mendoza showed that a reduction of 10% in the supply of surface water under a scenario of climate change with a 40-year timeline would lead to the economic unfeasibility of irrigation because of the increased use of groundwater (Barros and Kullock 2006).

Climate Impacts, Mitigation and Adaptation Issues

In the NOA and Cuyo regions there is a generalized awareness of the regional consequences of climate change. Presently there are no concrete programs or policies to cope with plausible impacts of it, but there are some non-institutionalized actions addressing reasonable mitigation and adaptation avenues to decrease the vulnerability of those regions. A conceptual framework for a national adaptation program has been elaborated using the Adaptation Policy Framework as an imprint (Girardin and Kozulij 2006). There are studies on vulnerability to climate change that could be used to frame regional adaptation policies. One of these was made on glacier dynamics in the Cuyo region (Canziani et al. 1995), where a decrease in river flow is expected to occur in a global warming scenario. The economic implications of this impact are presumed to be huge in view of the national and international relevance of the regional wine industry, which depends on irrigation.

Many of the current adaptation activities pertain to autonomous adaptation, i.e., the adaptation to the variability of climate that takes place in the absence of policies and measures from specific institutions.

1. **Heat-waves in large cities.** The impact of these phenomena can be abated through actions on the 'urban heat-island'. These actions are increasing albedo on urban and built-up surfaces, increasing vegetation cover, decreasing impervious surface area, decreasing anthropogenic heating, increasing structural and natural shading, and green roofs to

Table 21.4. Summary table for climate impacts and mitigation actions.

Climate impact	Mitigation action
Heat waves in large cities	Reduction of the urban-heat island phenomenon (1)
Rising temperature	Prohibition of deforestation and forest degradation (2.a)
	Afforestation of water catchments (2.b)
	Enhancement of carbon stocks and moisture conservation in agricultural soils (2.c)
	Use of firewood instead of fossil fuels (2.d)
	Growing *jatropha*, *canola*, and *sugarcane* for producing biofuels (2.e)

name the most relevant actions (Taha 2011). In view of a plausible future with less water vegetation cover could be expanded with tree species adapted to xeric conditions, like *Prosopis* spp. in the NOA region.

2. **Rising temperature.** Temperature rise can be abated by either restraining the emission of GHG and/or removing them (mostly CO_2) from the atmosphere. Physically[7] the increase in 1°C represents a 7% change in specific humidity. Large-scale warming associated with GHG emissions is a driver of hydrological change. The result is a modification of the frequency, amount, and intensity of precipitation with downstream effects including an increased incidence of extreme dryness, broad-scale increases in soil moisture, a regime shift in global evapotranspiration, and relatively wet areas becoming wetter and drier areas becoming drier (Schwalm et al. 2011). On a regional scale it is reasonable to assume the same broad manifestation of a coupling between the carbon and the water cycles. The following are some actions intended to decrease the impacts of global warming.

   a. Forbidding practices leading to deforestation of native forests and/or their degradation will keep those forests as active CO_2 sinks, i.e., as absorbers of atmospheric CO_2, and also avoid CO_2 emissions from the soil, the burning of harvest waste and frequently the cultivation of the deforested land. Strict enforcement of the native forests law and Argentina's adherence to the UN REDD+ program would allow the provinces of the NOA and Cuyo regions impacting positively on their respective regional hydrologic cycles.

   b. The afforestation of water catchments (mostly in the *Yungas*) would not only increase the quality of water streams but would supply the service of long-term carbon sequestration. This characteristic would make these dedicated plantations sources of emission reduction certificates in the current voluntary carbon market, and in an institutional one created under the umbrella of the UNFCCC (2.a).

c. The dissemination of non-tillage cropping—very common in the country's major cereal growing area—together with the adoption of cover-crops or mulching would keep carbon in the soil and reduce water loss by evaporation. In this way regional warming would be constrained and increasingly scarce water would be better used. In this respect there is relevant research by Uliarte et al. (2006, 2009a,b,c) and Catania et al. (2010) which focused on the effects of species and management to the vegetation between the rows in vineyards can have on the potential of CO_2 fixation and water consumption.

d. Firewood is commonly burnt for domestic and industrial heating in the NOA and Cuyo regions, e.g., drying tobacco leaves in the provinces of Salta and Jujuy. The use of this renewable fuel should be promoted as a substitute of fossil fuels in so far firewood were sustainably obtained, i.e., by avoiding the deforestation and degradation of native forests.

e. Agriculturally marginal lands in arid areas of the provinces of Catamarca and La Rioja have been promoted for planting jatropha (*Jatropha curcas* L.) for producing biodiesel from its oil-rich seeds (Panadero Pastrana 2009). In Mendoza the National Institute for Agricultural Research is trialling the canola crop for producing biodiesel (Silva Colomer 2010); the same is being done in the province of Tucumán in its experiment station.[8] Bioethanol[9] from sugarcane has been produced in the NOA region since 40 years ago, but its blend with petrol is not mandatory yet.

Table 21.5. Current and future adaptation activities/policies.

Adaptation issues	Time frame	Adaptation actions
International trade	Current	Raising awareness (1)
General	Current	Studies on the valuation the ecosystem services, building climate-related databases, etc. (2)
Alternative energy	Current	Solar and biomass (3)
Water use	Future	Sustainable management and governance (4)
Institutional strenghtening	Future	Promotion of research, development and diffusion (5) and Regional system for compiling greenhouse gas inventory (6)
Extreme climatic events	Future	Early-warning systems (7)

1. **Raising awareness (current action).** Raising and promoting awareness of the risks of climate change among the civil society and policymakers is a well-tested approach to adaptation. In this direction the federal agency overseeing the country's wine production and trade has been promoting meetings among stakeholders to inform and discuss the future implications of climate change on the quality of grapes and wines produced in the country, and the present and future relevance of the wine industry carbon footprint for international trade (PwC 2009). In the near future countries could implement diverse trade restrictions and/or duties on high-carbon goods, i.e., goods that even if they had a low carbon signature in the production chains, they could have a large carbon signature due to transport. As Argentina is far away from many wine markets, it should be ready to reduce the carbon footprint of its wines.

2. **Knowledge building (current action).** In the NOA region the Regional Climate Change Network is a binational project between the provincial governments and the universities of Salta and Jujuy (Argentina) and Tarija (P. S. of Bolivia). This network promotes studies on the principles underlying the valuation of environmental services, devise policies of regional adaptation and mitigation to climate change, and building databases for monitoring climate change.[10]

3. **Alternative energies (current action).** There are initiatives in the Puna for heating and cooking with solar energy, and implementation of alternative energies and production of ethanol in agroindustries in Salta and Jujuy (http://www.asades.org.ar/). In the province of San Juan a wind-farm has recently been opened.

4. **Water management and governance (future action).** A fair water pricing system should be established in order to induce farmers to take water as a production cost that might produce negative returns if water were wastefully used. The fair pricing of water should be concomitant with a complete and deep overhaul of both the infrastructure, e.g., run-down and obsolescence of the control, distribution, and drainage infrastructure—and the governance, e.g., lack of coordination among provincial and federal agencies—of the public irrigation systems.

In line with the reasons mentioned above, the future expansion of the irrigated area should be based on the adoption of localized irrigation instead of surface irrigation, because the former is more efficient than the latter (Howell 2003).

The need for a program for capacity building aimed at extension service personnel. These personnel should closely collaborate with farmers to raise their awareness on the consequences of climate change to their production systems, and advise them on technical, managerial and commercial aspects of irrigated agriculture under restricted water availability and increasing air temperature.

The need of a regional agency for planning reasonable policies addressing water conservation, water recycling, and water-use optimization in both productive and consumptive human activities; and implementing them in harmony with federal agencies responsible for the management of national water resources to enhance the efficacy and the efficiency of those policies (Girardin 2006).

5. **Institutional strengthening (future action).** The need of a regional agency to fund interdisciplinary mitigation and adaptation research programs with universities and other public and private research institutions, and foster the participation of stakeholders in various stages of the planning and realization of those programs.

6. **Institutional strengthening (future action).** The need of a regional administrative system for planning, coordinating and compiling greenhouse gas inventories with a view of using them to base mitigation and adaptation regional policies.

7. **Extreme climatic events (future action).** The need of an early warning system for reducing the risk of damages to people and property from extreme climatic events.

Conclusions

The hydrologic regime in the areas of the Cuyo and NOA regions under strong influence of the Andean system of mountains is the most exposed to the regional peculiarities of precipitation and temperature evoked by global warming. Water stored in glaciers and in forested watersheds is expected to decrease in the future thereby putting under stress the wellbeing of the regional population and ecosystems and, in the case of the major exorheic regional rivers, that of life downstream. The population and the economy in the NOA and Cuyo regions are most likely to increase in the future—the demand for water is likely to increase in a plausible scenario of water availability already strained by a regional warming. In that same scenario extreme weather events are likely to occur, adding a further strain to the environment, life and property. To cope with these threats the regions of NOA and Cuyo should take decisive and timely steps to increase their adaptability and resilience to a changing climate. In view of the foregoing, each of NOA and Cuyo regions will need to implement context-specific policies of adaptation and mitigation to the extent of the expected climate change impacts. Most of those would be in line with recent conclusions from a relevant meeting of Latin American countries.[11]

Life and property are under threat in the NOA and Cuyo regions. The regions should not wait for the outcome of present international negotiations under the aegis of the UNFCCC; they should prod the federal government to move forward as well, because a changing climate does not respect provincial political boundaries.

Acknowledgements

The authors are very grateful to Nora Madanes (University of Buenos Aires) and Leónidas O. Girardin (Fundación Bariloche) for their critical revision of the manuscript. We are also indebted to Patricia Perelman (Figs. 5, 7, 10, 14), Viviana Gomez (Fig. 8) and Marcelo Canevari (Fig. 6).

References

Altieri, M. 1996. Indigenous knowledge re-valued in Andean agriculture. ILEIA Newsletter. ILEIA Centre for learning on sustainable agriculture. 12: 1–7.
Barros, V. and D. Kullock. 2006. Resumen Ejecutivo. Informe Final. pp. 1–5. B8 Programa Nacional de Adaptación y Planes Regionales de Adaptación. Informe Nacional de Cambio Climático. Fundación Instituto Torcuato Di Tella. 295 pp.
Bates, B.C., Z.W. Kundzewicz, S. Wu and J.P. Palutikof. 2008. Climate Change and Water. Technical Paper of the Intergovernmental Panel on Climate Change, IPCC Secretariat, Geneva, 210 pp.
Bergesio, L. and J. Montial. 2010. Declaraciones patrimoniales, turismo y conocimientos locales. Posibilidades de los estudios del folklore para el caso de las ferias en la quebrada de Humahuaca (Jujuy-Argentina). Trabajo y Sociedad 14: 19–35.
Brenning, A. and G.F. Azócar. 2010. Minería y glaciares rocosos: Impactos ambientales, antecedentes políticos y legales, y perspectivas futuras. Rev. Geogr. Norte Gd. 47: 143–158.
Brown, A.D., S. Pacheco, T. Lomáscolo and L. Malizia. 2005. Situación ambiental en los bosques andinos yungueños. pp. 53–56 . *In*: A. Brown, U. Martinez Ortiz, M. Acerbi y J. Corcuera (eds.). La Situación Ambiental Argentina 2005. Fundación Vida Silvestre. Buenos Aires. Argentina.
Burkart, R., N.O. Bárbaro, R.O. Sánchez and D.A. Gómez. 1999. Eco-regiones de la Argentina. Administración de Parques Nacionales. 21 pp.
Cabrera, A.L. 1976. Regiones fitogeográficas argentinas. Enciclopedia Argentina de Agricultura y Ganadería. Tomo II, Editorial ACME Buenos Aires, Argentina, 85 pp.
Canziani, O.F., M. del R. Prieto, R.M. Quintela and D. Huggenberger. 1995. Vulnerabilidad de los oasis comprendidos entre 29 °S y 36 °S ante condiciones más secas en los Andes altos. Proyecto ARG/95/G/31 - PNUD – SECYT. 124 pp.
Catania, C.D., S. Avagnina de del Monte, E.M. Uliarte, R.F. Del Monte and J. Tonietto. 2010. El clima vitícola de las regiones productoras de uvas para vinos de Argentina. Revista 13°, 2(17): 20–47.
Comelli, M., M.G. Hadad and M.I. Petz. 2010. Hacia un desarrollo (in)sostenible en América Latina. El caso de la minería a cielo abierto en la Argentina. Argumentos. Revista de crítica social 12: 132–157.
FAO. 2009. comiadventuras 2. Serie Ciencia, Salud y Ciudadanía. Proyecto de Alfabetización Científica. Educación Alimentaria y Nutricional. Revista para el alumno. 2 nivel (3° y 4° grados/años de Educación General Básica/Primaria). Proyecto CP/ARG/3101 (T) "Educación Alimentaria y Nutricional en las escuelas de Educación General Básica/ Primaria". Food and Agriculture Organization of the United Nations. Rome, 34 pp.

Gasparri, N.I., H.R. Grau and E. Manghi. 2008. Carbon Pools and Emissions from Deforestation in Extra-Tropical Forests of Northern Argentina between 1900 and 2005. Ecosystems 11: 1247–1261.

Girardin, L.O. 2006. Jurisdicción para el abordaje de las políticas y medidas. Informe Final. pp. 197–198. *In*: B8 Programa Nacional de Adaptación y Planes Regionales de Adaptación. Informe Nacional de Cambio Climático. Fundación Instituto Torcuato Di Tella.

Girardin, L.O. and R. Kozulj. 2006. Lineamientos de las políticas de adaptación. Informe Final. Section 4.2. pp. 199–209. *In*: B8 Programa Nacional de Adaptación y Planes Regionales de Adaptación. Informe Nacional de Cambio Climático. Fundación Instituto Torcuato Di Tella.

Grau, H.R., M.E. Hernández, J. Gutierrez, N.I. Gasparri, M.C. Casavecchia, E.E. Flores and L. Paolini. 2008. A peri-urban neotropical forest transition and its consequences for environmental services. Ecology and Society 13(1): 35.

Hector, A. and R. Bagchi. 2007. Biodiversity and ecosystem multifunctionality. Nature 448: 188–190.

Howell, T.A. 2003. Irrigation efficiency. pp. 467–472. *In*: Encyclopedia of Water Science. Manuel Dekker, Inc., New York, US.

IGN. 2011. División Política, Superficie y Población de la República Argentina. Instituto Geográfico Nacional de la República Argentina.

INDEC. 2011. Instituto Nacional de Estadísticas y Censos. http:\\www.indec.gov.ar.

IPCC. 2006. Forest Land. 83 pp. *In*: H.S. Eggleston, L. Buendia, K. Miwa, T. Ngara and K. Tanabe (eds.). Volume 4: Agriculture, Forestry and Other Land Use. 2006 IPCC Guidelines for National Greenhouse Gas Inventories, National Greenhouse Gas Inventories Programme, IGES, Japan.

IUCN. 2011. IUCN Red List. International Union for Conservation of Nature. http://www.iucn.org/about/work/programmes/species/red_list/.

Lara, A., C. Little, R. Urrutia, J. McPhee, C. Álvarez-Garretón, C. Oyarzún, D. Soto, P. Donoso, L. Nahuelhual, M. Pino and I. Arismendi. 2009. Assessment of ecosystem services as an opportunity for the conservation and management of native forests in Chile. Forest Ecology and Management 258(4): 415–424.

Lichtenstein, G. and B. Vilá. 2003. Vicuna use by Andean communities: an overview. Mountain Research and Development 23: 198–201.

Manrique, S., J. Franco, V. Núñez and L. Seghezzo. 2011. Potential of native forests for the mitigation of greenhouse gases in Salta, Argentina. Biomass and Bioenergy 35(5): 2184–2193.

Minagri. 2011. Camélidos silvestres. Ministerio de Agricultura, Ganadería y Pesca. República Argentina. http://www.minagri.gob.ar/SAGPyA/ganaderia/camelidos/01_Informaci%C3%B3n/02_Silvestres/index.php[AF6].

Minería. 2009. Minería en números.2009. Secretaría de Minería, Ministerio de Planificación Federal, Inversión Pública y Servicios. República Argentina. http://www.mineria.gov.ar/pdf/mineriaennumeros.pdf.

Mintur. 2011. Ministerio de Turismo de la Nación. http://www.turismo.gov.ar.

Molinillo, M.F. 1993. Is traditional pastoralism the cause of erosive processes in mountain environments? The case of the cumbrescalchaquíes in Argentina. Mountain Research and Development 13(2): 189–202.

Natenzon, C., A. Murgida and M. Ruiz. 2006. Vulnerabilidad Social al Probable CC. pp. 32–73. *In*: Informe Final. Impacto Socioeconómico del Cambio Climático en la República Argentina. Fundación Bariloche.

Panadero Pastrana, C. 2009. Recursos vegetales de desarrollo estratégico con finalidad energética. Proyecto PNEG1412. C. INTA. 26 pp.

Pastor, G.C. and L.M. Torres. 2010.¿Turismo en territorios periféricos? Algunas reflexiones a propósito de un estudio de caso en el "Desierto de Lanalle", Argentina. Estudios y Perspectivas en Turismo 19: 163–181.

PROSAP. II. El PROSAP y el riego. Una herramienta para el desarrollo de las economías regionales y la generación de empleo rural . Servir al Agro. pp. 15–93. Programa de Servicios Agrícolas Provinciales. Ministerio de Agricultura, Ganadería y Pesca de la Nación.

PwC. 2009. Efectos del cambio climático sobre la industria vitivinícola de Argentina y Chile. Estudio sobre los impactos y las medidas de adaptación en un escenario de calentamiento global hacia el año 2050. Price Waterhouse & Co. Buenos Aires and Santiago. 84 pp.

Sandy, T. 2005. Water Reduction and Reuse in the Petroleum Industry. Petroleum Environmental Research Forum. CHM2Hill. http://www.perf.org/pdf/sandy.pdf.

Schwalm, C.R., C.A. Williams and K. Schaefer. 2011. Carbon consequences of global hydrologic change, 1948–2009. J. Geo. Res. 116: G03042.

Seghezzo, Lucas, José N. Volante, José M. Paruelo, Daniel J. Somma, E. Catalina Buliubasich, Héctor E. Rodríguez, Sandra Gagnon and Marc Hufty. 2011. Native Forests and Agriculture in Salta (Argentina). The Journal of Environment and Development 20: 251–277.

Silva Colomer, J. 2010. Colza bajo riego en Mendoza. Doc IIR-BC-INF-06-10. INTA. 2 pp.

Squeo, F.A., B.G. Warner, R. Aravena and D. Espinoza. 2006. Bofedales: high altitude peatlands of the central Andes. Revista Chilena de Historia Natural 79: 245–255.

SRH. 2011. Sistema Nacional de Información Hídrica. Base de Datos Hidrológica Integrada. Subsecretaría de Recursos Hídricos de la Nación. Gobierno de la República Argentina.

RA. 1997. Primera Comunicación Nacional del Gobierno de la República Argentina a la Convención Marco de las Naciones Unidas sobre Cambio Climático. 109 pp.

RA. 2007. 2a. Comunicación Nacional de la República Argentina a la Convención Marco de las Naciones Unidas sobre Cambio Climático. [2nd National Communication of the Argentine Republic to the United Nations Framework Convention on Climate Change]. pp. 200.

RA. 2011. Series Estadísticas Forestales. 2003-2009. Secretaría Secretaría de Ambiente y Desarrollo Sustentable. Jefatura de Gabinete de Ministros. Presidencia de la Nación. 63 pp.

Richard–Jorba, R.A. 2008. Crisis y transformacionesrecientes en la regiónvitivinícolaargentina: Mendoza y San Juan, 1970–2005. Estud. Soc 16: 82–123.

Roa-García, M.C., S. Brown, H. Schreier and L.M. Lavkulich. 2011. The role of land use and soils in regulating water flow in small headwater catchments of the Andes. Water Resources Research 47: 1–12.

Roveta, J.R. 2008. Resilience to Climate Change in Patagonia, Argentina. The Gatekeeper 140. 24 pp.

Taha, H. 2011. Meso-urban modeling in support of heat-island mitigation. Urban Climate News (IAUC) 39: 5–12.

Trumper, K., M. Bertzky, B. Dickson, G. van del Heijden, G. Jenkins, M. Manning and P. June. 2009. The Natural Fix? The role of ecosystems in climatic mitigation. A UNEP rapid response assessment. United Nations Environment Program. UNEP-WCMC. Cambridge, UK, 68 pp.

Uliarte, E.M., R.F. del Monte and C.A. Parera. 2006. Influencia del manejo de suelo mediante coberturas vegetales en el microclima de viñedos bajo riego (cv. Malbec). Le Bulletin de l'OIV 79: 5–22.

Uliarte, E.M., R.F. Del Monte, A. Ambrogetti and M. Montoya. 2009a. Evaluación y elección de diferentes especies de coberturas vegetales en viñedos bajo riego de Mendoza. En: Revista 13°, 2(15): 18–34.

Uliarte, E.M., H.R. Schultz, C.A. Parera, R.F. Del Monte, E.E. Alessandria and M. Bonada. 2009b. Manejo edáfico del viñedo frente al cambio climático: Balance de CO_2 y demanda hídrica en especies cultivadas como cobertura vegetal y su relación con alternativas de labranza. Es. En: XII Congreso Latinoamericano de Viticultura y Enología. Montevideo, Uruguay, 11 al 13 de noviembre de 2009.

Uliarte, E.M., R.F. del Monte, C.A. Parera and S.M. Avagnina de del Monte. 2009c. Influencia del manejo de suelo mediante coberturas vegetales establecidas en el desarrollo vegetativo, producción y características de vinos en viñedos bajo riego superficial (cv. Malbec). Le Bulletin de l'OIV 82: 205–227.

Villagra, P.E., C. Giordano, J.A. Alvarez, J.B. Cavagnaro, A. Guevara, C. Sartor, C.B. Passera and S. Greco. 2011. Ser planta en el desierto: estrategias de uso de agua y resistencia al estrés hídrico en el Monte Central de Argentina. Ecología Austral 21: 29–42.

Endnotes

1. United Nations Framework Convention on Climate Change. It entered into force on 21 March 1994. Referred to as 'the Convention' in what follows.
2. Argentina signed the Protocol in 1998, ratified it in 2001, and enforced it in 2005.
3. Provinces of Jujuy, Salta, Tucumán, Catamarca and La Rioja.
4. Provinces of San Juan and Mendoza.
5. Provinces of Neuquén, Rio Negro, Chubut, Santa Cruz, and Tierra del Fuego.
6. www.iapg.org.ar Instituto Argentino del Petróleo y del Gas.
7. Clausius-Clapeyron equation.
8. www.eeaoc.org.ar.
9. www.tabacal.com.ar.
10. http://cclimaticosalta.blogspot.com/.
11. Conclusions of a regional Technical Meeting on Climate Change impacts, adaptation and development in Mountain Regions. Organised by the Ministerio de Relaciones Exteriores de Chile, the Mountain Partnership, FAO, and the World Bank. 26–28 October 2011. Santiago de Chile.

22

Impact of Global Changes on Mountains: Case Study of Brazil

Sérgio Murari Ludemann

INTRODUCTION: LANDSLIDES AND SLOPE INSTABILITY MECHANISMS

The stability of a slope is related to the shear strength of the soil where a fine balance needs to be maintained between the forces perpendicular to the slope and tangential to the slope. However, at times natural or anthropogenic processes, may interfere with this balance which could induce landslides. This is called the safety factor, which is a relationship between resistant forces and the acting forces.

In simple terms, we can say that the friction and cohesion of the materials are the main strength parameters to be considered. The friction depends on weight and cohesion depends on the conditions of soil moisture. The determination of strength parameters is based on surveys for the identification of different soils that can make a slope and testings, both '*in situ*' and in the laboratory in each of the soils identified in the profile. The strength parameters can also be obtained based on correlations of the soils indicated in surveys and visual inspections with test data available in the literature.

When extreme events occur, such as heavy and prolonged rains, there may be a saturation of the deeper layers of the slope, which can cause a decrease in the cohesion of the local soils and a reduction of the effective weight of the slope because the weight will be supported by water (as a

Ludemann Engenheiros Associados S/S Ltda.
Email: sergio@ludemann.com.br

tub), but the water has no shear strength, reducing significantly the forces of friction and cohesion that resist slipping. On the other hand, the movement of water in the soil causes percolation, which changes the balance of the slope and leads to landslides.

If the slope has vegetation, these landslides are less but if the slope has been cleared for settlements even less intense rainfall events cause the imbalance triggering landslides. Based on this, the inhabitation of slopes should be avoided and already inhabited slopes should be vacated, restoring the original topography and natural vegetation coverage. In many situations, however, it is not possible to evacuate the slopes, it is necessary to urbanize and consolidate existing occupations.

This chapter deals with a criterion for classification and ranking of risks of landslides slopes, part of the actions necessary to urbanization and consolidation of the occupied slopes.

Global Change, Extreme Weather and Prevention

Recent research by the Brazilian Agricultural Research Corporation (EMBRAPA) in conjunction with the University of Campinas—SP (UNICAMP), where they studied 78 meteorological stations all over Brazil, that showed an increase in average annual temperature of 1.5°C since 1968 to date. As a result of this increase in temperature, there was a significant increase in the intensity of rainfall and there has been an increase in the amount of annual precipitation. There has been a poor temporal and spatial distribution for rainfall events with prolonged drought seasons and intense and prolonged rainfall events during rainy season. This leads to increase in the occurence of landslides in the urban areas.

In 1991 UNDRO— United Nations Disaster Relief Organization proposed a model to counter natural disasters based on Prevention and Preparation.

Prevention activities are related to technical and scientific studies that can assess the likelihood of occurrence and magnitude of a natural disaster, establishing mechanisms to prevent loss of not only lives, but also to minimize property losses. The preparation actions are linked to the logistical nature of the confrontation of emergency situations and care of affected populations, and is usually assignments of civil defense agents (Alheiros, M.M. in UFPE 2008). These aspects are very important, but not discussed in this chapter.

Occupation of the Hills and Mapping Geotechnical Risks

In Brazil, as well as in other countries urbanization has been totally unplanned and has usually marginalized poor people. It has also led to a concentration of people in urban cities and delcining rural population —people in search of better jobs (to make more money) and small farmers (either they have lost their lands to big farms or cannot compete with mechanization of large farms) are moving to cities. However, because of this mass migration, cities are becoming populated and there is a lack space for further development. This is giving rise to either illegal settlements or occupation of fragile land (such as hill sides) by low-income strata group in a disorderly urban concentration areas known as slums.

Within the context of social actions needed in areas of low income include: projects involving slum upgrading projects that involve integration of social services for the weaker socio-economic communities; survey of the lack of urban infrastructure (sanitation, transportation, education, etc.), their needs for cultural and social activities and eliminating risks (such as geological or geotechnical in character and include the houses with low structural stability).

The mapping, characterization and prioritization of geological and geotechnical risks in slums and informal settlements including risk of low stability houses in the slums are discussed next.

Currently, slums are the possible alternative housing for a significant proportion of the population of large cities. Since immigrants settle in the empty urban areas left behind in the process of growth of the city, the conditions in slums are unfavorable and are on fragile land which is vulnerable due to erosion and have higher risk of mud/land slides and flooding (especially along the rivers). Most of the time, disjointed specific programs are impelemted after an accident (e.g., landslide) but this is not enough to prevent further disasters. It is necessary to act in a planned way, using preventive interventions with multiple, integrated and prioritized through a rational system of risk control (Carvalho 1994).

The geotechnical risk situations are often related to human intervention, but may be aggravated or even due to the unfavorable geological characteristics for human occupation in a particular location. In this sense, the mapping of geotechnical risk situations is the starting point for an adequate planning of interventions required. The goal of this activity is the development of a comprehensive diagnosis of geotechnical risk situations. Such risk situations may be associated with areas of slopes (including massive landslides on and/or cut slopes in natural soil, landslides in landfills, the outbreak of erosion, falling blocks, etc.) or the low areas (including undermining the foundations of houses and/or restraints built along by the stream, rupture of embankments on soft soils, erosion of embankments, etc.). It is proposed, according to the recommendations made by UNDRO 1991, that this diagnosis is carried out on two scales:

- Scale Zoning applied to all of the existing slums in the city, including the delimitation of risk areas, the identification of active destructive processes and evaluation of the susceptibility (or probability) of accidents. The zoning should provide general information such as the amount of housing subject to different levels of sensitivity and spatial distribution, allowing characterization of the priority sectors for intervention. It is a subsidy for the planning of action lines of the municipal administration, involving the program of information and a social mobilization program to build shelters, scaling human and material resources to be allocated by the government, definition of demand for slum upgrading programs, implementation of works of stabilization, etc.;
- Scale Registration, applied to areas of high susceptibility to landslides, involving the identification and ranking of homes at risk and social characteristics of families. The risk register should provide specific information about the risk level of housing implemented in priority areas of risk, the required conditions, socio-economic characteristics of households and environmental aspects, providing a subsidy for monitoring actions of civil defense, temporary removal of residents during heavy rains, implementation of emergency stabilization works, etc.

Zoning should be performed on the set of slums with a history of accidents associated with geotechnical risks described above for areas of slopes and/or downloaded (wetlands) and aims to characterize, map and prioritize areas of risk. It includes the following activities:

- Characterization of risk sectors (physical delimitation and identification of active destructive processes) involving the following activities:
  - ◦ Identification, from the available data (geological, geotechnical letters, aerial photographs, aerophotogrametric refunds and iso-inclination maps), sectors of the study area which, in its physical configuration, present situations of potential risk;
  - ◦ Research the geological and geotechnical surface (field), to identify constraints and observation of possible indicators of instability associated with possible destructive processes;
  - ◦ Identification of the destructive processes operating in each sector of potential risk, and
  - ◦ Definition of the site risk over a base map, including the entire area under the influence of the destructive processes identified.
- Assessment of the likelihood of destructive events and setting the level of risk (susceptibility) sector. In actual practice this activity in Brazil is used as the method for qualitative risk analysis, in which the experience

of the technicians in charge of the analysis is used to estimate the probability of destructive events and their potential consequences.

- Evaluation of the potential consequences of each sector (number of houses liable to destruction). The consequences are evaluated in terms of number of houses that can be destroyed in case of occurrence of instability processes. In the case of a risk analysis of a qualitative nature, the estimate of the consequences of destructive events is performed through the trial of the charge of technical analysis. The analysis procedure involves:

  ° Evaluation of the possible ways of developing the active destructive process (volumes mobilized, trajectories of the debris, areas of outreach, etc.); and
  ° Estimates of the number of houses in the area with a risk of being destroyed in case of occurrence of various forms of development of the destabilization process.

- Presentation of the results obtained. The results of the risk zoning, which must be synthesized into maps, aerial photographs and tables, include:

  ° The delimitation of areas of risk and its cartographic representation;
  ° The description of the process of destabilization;
  ° The level of susceptibility of the sector and;
  ° The number of houses liable to destruction.

In the qualitative risk analysis, the probability of destructive events (disasters) is assessed subjectively and expressed in literal terms (very high, high, medium or low). Since the probability of occurrence of the destructive process is a function of the time period considered, it is recommended to consider in the risk analysis a period of not less than one year, comprising at least one rainy season. Thus, when estimating the destructive events probability, the technicians in charge of the analysis assess the probability of a destructive event possible during an episode of intense and prolonged rainfall. Other periods of validity of the mapping may be adopted, adjusting for risk assessments of such a period, which should be explicit in the technical report. It is recommended that the study should not include more than a 10 years time period.

The proposed procedure comprises:

- The qualitative evaluation of the probability of occurrence of the destructive process in the course of an episode of intense and prolonged rainfall, held from the topographic features of human interventions, geological and geotechnical characteristics of the underground site, the indicators of instability, of evidence of earlier (past) destructive occurrences of events and interviews with residents and;

- The definition of the level of susceptibility of the sector, qualitative expression of destructive process occurrence probability.

The criteria currently used in the preparation of the diagnostic report for identification and classification of risk situations is based on work by Cerri & Carvalho 'Hierarchy of Risk Situations in slums in São Paulo' (Carvalho 1990), updated based on the criteria most recently practiced and is described below:

- No risk: Stable sectors, the geotechnical point of view, providing adequate security against the cases examined, whose topography has a slope less than 15% whatever the conditions of occupation or even 30% since the houses have adequate containment of the slope;
- Risk 1-Low: Situations which have not been identified and where instability processes maintaining the status quo, do not expect the occurrence of destructive events, but where the steepness of the slopes is between 15 to 30% and the houses are poor and/or do not promote proper containment of the slope, or where the steepness of the slope exceeds 30%, whatever the type of occupation;
- Risk 2-Medium: Situations where the processes of destabilization meet with development potential, but there is no evidence of instability. Low probability of occurrence of destructive events during episodes of intense and prolonged rainfall, implying the possibility of a smaller number of fatal victims, in case of instability, or where the steepness of the slopes is between 30 and 45% and existing buildings are poor houses and/or do not promote proper containment of the slope, or slopes where the inclination exceeds 45%, whatever the type of occupation,
- Risk 3-High: Situations where the processes of instability are early stage or with high development potential and the evolution of processes is likely within the established period of observation (one year). The probability of the occurrence of destructive events during episodes of intense and prolonged rainfall, implys the possibility of a large number of casualties, in case of instability. Places where the steepness of the slopes is between 45 and 60% and existing buildings are poor houses and/or do not promote proper containment of the slope, or slope where the inclination exceeds 60%, whatever the type of occupation;
- Risk 4-Very High: Situations where the destructive processes are in an advanced stage of development. The indicators of instability (cracks in the soil, abatement steps on slopes, cracks in houses or retaining walls, tilted poles or trees, slip scars, etc.) are present, revealing a high probability of destructive events during episodes of heavy and prolonged rains, requiring urgent intervention to prevent loss of life.

Places where the steepness of slopes greater than 60% and existing buildings are poor houses and/or do not promote proper retaining of the slope.

The above criteria prioritizes risk situations involving loss of life compared to those that would tend to produce only material losses.

The delimitation of areas with different risk situations depends on the correct characterization of the processes of instability, also enabling the demarcation of the area potentially subject to its effects. Thus, depending on the mechanism of instability and position of the indicators of instability, it identifies the boundaries of the sector risk, which should include the area directly involved in the initiation process, the passage area of the mass and area of unstabilized deposition of material resulting from the destabilization process (Carvalho 2000).

After completion of the zoning phase, the process of registration of risk situations begins. As stated earlier, the risk register should provide specific information about the risk level of houses in priority areas of risk, defining the necessary interventions, economic and social characteristics of families, environmental issues, providing a subsidy for monitoring actions of civil defense, temporary removal of residents during heavy rains, implementation of emergency stabilization works, etc.

It is interesting to note that in the risk analysis in slums, not only housing is regarded as an element of risk, but also the urban infrastructure, industries, traffic routes for vehicles and pedestrians, and other elements of infrastructure.

At this stage it is necessary to supplement the information gathered during the risk zoning, through field inspections and bibliographic data (maps, charts, etc.), with a campaign-specific geotechnical investigation to measure the soil strength parameters involved in processes of destabilization, confirming or correcting the degree of risk as defined above and providing subsidies for the proposed interventions, either through monitoring or works of geotechnical stabilization and consolidation. With regard to the geotechnical studies at this stage of registration, must be alert to the presence of preparatory and starting agents of risk situations, clearly defining the mechanisms of instability. The preparatory agents may be geomorphologic character (natural slopes, drainage lines, weakness planes, schistosity of rock, etc.), geotechnical character (thickness of the layers and shear strength of the same, the existence of steep cuts, landfills thick launch of trash and debris on the slopes, etc.) or hydraulic character (concentration of runoff, erosion, saturation of the subsoil, infiltration trenches or wells, sewage spills on the slopes, etc.). The starting agents can be heavy rains, cuts in unstable material, embankments on soft soils, concentrated overloads (houses with direct foundation, accumulation of various materials, earthquakes, etc.).

Therefore, a destabilization process depends on the existence of preparatory and starting agents. Such processes can be shallow translational landslides (thicker colluvial soils cover on residual soil), deep landslides (thick landfill in residual soils with low shear strength, etc.), blocks down (towards the courts foliation or weakness planes in rock alteration soil and/ or altered rocks, exhibition plans of fractures in rock, etc.), slow movements colluvial masses, debris flow, etc. Some of these processes can occur abruptly, but in most cases signs of destabilization process before it occurs can be noticed. It is very difficult, however, to preview the time between the identification of instability signs and the destructive process failure.

The identification of preliminary agents, the evaluation of the possibility of starting agents and the anticipated mechanisms of instability should be the main objectives of the risk register.

Areas with very high levels of risk should be vacated and an emergency declared until it can be left to the social care teams to quantify the number of families to be removed in each defined area and position of public agencies setting locations for temporary shelter or permanent families to be removed.

Areas with high level of risk should be examined, each case should be defined by technical analysis that is in charge of monitoring and are subjected to intervention until a final, or whether these areas should also be vacated.

Areas with medium and low levels of risk should be monitored until a plan of intervention and to eliminate the risk factors.

Warning Systems and Monitoring of Slopes

Public agencies spend extensive funds to solve the problems of slopes usually at a stage when it is too late or the slope is at a stage of imminent collapse (Rocha 1986). The solutions to the risk areas should be planned in an integrated manner, seeking to meet whenever possible, not only the elimination of risk but also the feasibility of implementing urban infrastructure (access roads, sanitation, lighting, etc.) thus optimizing the use of public resources.

With this in mind, the greatest challenge facing the issue of geotechnical interventions is when solutions are based only on existing problems and focus on one issue rather than looking at the holistic picture. This can be achieved with proper planning of public expenditures, the development of public policies on land use and soil, the adequacy of relevant legislation and effective enforcement of these policies and laws.

Another major challenge is to change the current philosophy that a geotechnical intervention should be made mandatory by a work of containment and/or stabilization. The city of São Paulo currently has approximately 3,000 local geotechnical risks, involving tens of thousands of residents. It is unthinkable from the standpoint of economic and social development to solve a problem of this magnitude with mass evictions or the construction of massive consolidation works.

Whereas, as previously mentioned, there are several situations where a destabilization process does not occur without presenting early indications, and that the geotechnical accidents occur more frequently in the rainy season (main starting agent), it is possible, therefore, to establish indicators of actual occurrence of destructive events that can be monitored. Local residents can be also be educated to identify these indicators and take appropriate action to safeguard their lives, their families and neighbors.

The implementation of procedures for monitoring places of risk is fundamental to the strategy to reduce damage from geotechnical accidents. It consists in acting on all areas, from past occurrences or specialized technical studies, which have a higher probability of being able to develop some process of instability of the ground.

This type of intervention through monitoring, however implies that Government is ready to take the risk inherent in the situation and this step is taken as an immediate intervention due to lack of funds or till an integrated plan is developed.

To guide the work of monitoring, it is proposed that minimum specifications and criteria be followed in this work. These criteria should be viewed as a suggestion and may be adjusted and supplemented by the contribution of the technical expertise present at the site, here involving not only the geotechnical professionals but also professionals in social work, architecture and urban planning, infrastructure and other specialties involved in the process of slum upgrading and low-income housing developments.

Criteria for monitoring

- All residents located in risk areas will be notified, according to the classification of the risk map prepared in the phases of zoning and risk register, explaining in every case the degree of risk which that resident is subjected to (very high, high, medium or low);
- Will only be capable of monitoring:
  - The locations where the processes do not trigger instability;
  - The places where the processes of instability may manifest itself over time, providing evidence of their possible outbreak;
  - The locations where the destructive power of instability will occur abruptly, is not enough to cause loss of human life by enabling people to leave the scene of the accident unscathed.
- Every resident in the risk area should be given these specifications, and any subsequent adjustments and additions;

- Signs that may indicate an outbreak of the instability phenomenon or the beginning of a destructive process are:
  - ○ For houses located upstream of containment and/or existing slope, plumb and/or bulging of the containment, the appearance of water at the foot of slopes and/or retaining walls, peeling of the surface slope, the appearance of cracks and/or rebates in the embankment amount, the appearance of cracks and/or cracks in walls and/or floors of houses;
  - ○ For houses located downstream of restraints and/or existing slope, plumb and/or bulging of the containment, the appearance of water at the foot of slopes and/or retaining walls, peeling of the surface slope, the appearance of cracks and/or cracks in walls of houses if they are holding their own or are connected to it, rebates in the amount of embankment retaining walls;
  - ○ For houses located on slopes: Poles and/or leaning trees, the appearance of cracks and/or rebates on the surface of the ground along the slope;
  - ○ For houses located on the floodplain: Cracks in masonry and rebates on the floors of buildings, cracks and/or rebates on embankments and retaining walls along the houses, taking off foundations by undermining and erosion of embankments.
- A qualified technical team will train a group of people (community leaders, volunteers, etc.) so that they can coordinate and constantly check the monitoring work done by residents;
- Each resident must constantly observe the emergence of signs that may indicate the process of an outbreak of instability;
- When any of these signs are noticed, the resident and his family must vacate the property immediately and notify the local technical team responsible for technical monitoring;
- Residents should be instructed in the notification, to seek technical assistance from municipal agencies, the name and contact by telephone of technicians to look for in an emergency should be explicitly provided;
- Residents should only re-occupy their homes after the risk is eliminated;
- There should be an undertaking from the Government stating that the protocol resident received these instructions and attempts to fulfill them.

Conclusion

This proposed monitoring will only help in avoiding the possible loss of life and will depend heavily on understanding and cooperation of residents. To minimize the likelihood of a possible failure of this procedure, it is recommended that these monitoring sites are inspected regularly by qualified technicians. The frequency of inspections can vary from one visit every 45 days during the dry season of the year, and a visit every 15 days or less during the rainy season. It is essential at this point of intervention, when only monitoring is done that there is a larger vision or a plan for intervention that includes social, infrastructure and geotechnical issues in addition to urban planning.

Studies could be developed to correlate rainfall and geotechnical accidents, seeking to establish alert levels and rainfall that could serve as a warning for evacuation of critical areas.

References

Carvalho, Celso Santos. 1994. Controle de Riscos Geotécnicos em Encostas Urbanas—X Congresso Brasileiro de Mecânica dos Solos e Engenharia de Fundações 3: 825–831.

Carvalho, Celso Santos. 1997. Processo de Instabilização de Taludes em Maciços Artificiais – 2° Pan-American Symposium on Landslides 2: 901–908.

Carvalho, Celso Santos. 2000. Risco Geotécnico em Favelas—Publicação do IPT—Instituto de Pesquisas Tecnológicas do Estado de São Paulo—Brasil.

Cruz, Paulo. 1981. AMBS—Encontro Técnico sobre Estabilização de Taludes, Editado pela ABMS—Associação Brasileira de Mecanica dos Solos, São Paulo, Brasil.

Da Cunha, Eugênio P.V. and Celso S. Carvalho. 1998. Controle de Riscos Geotécnicos no Projeto de Urbanização de Favelas—XI Congresso Brasileiro de Mecânica dos Solos e Engenharia de Fundações I: 473–479.

Figueiredo, R.B. 1997. Engenharia Social: Soluções para Áreas de Risco na Cidade de São Paulo —2° Pan-American Symposium on Landslides 2: 909–912.

Gonçalves, Renata Ribeiro Do Valle e Assad, Eduardo Delgado. Análise De Tendências De Temperatura Mínima Do Brasil - XVI Congresso Brasileiro de Agrometeorologia 2009 – Belo Horizonte – MG.

Hachich, Waldemar. 1988. Análise Probabilística de Estabilidade—Encontro Técnico sobre Estabilidade de Encostas. Único 49–73.

Hoek, Evert and John Bray. 1981. Rock Slope Engineering, 3rd Edition. E & FN Spon, London, UK.

Lambe, Willian T. and Robert V. Whitman. 1979. Soil Mechanics. John Wiley & Sons Inc., New York, USA.

Rocha, Euler Magalhães da. 1986. O problema das Encostas em Áreas Urbanas—VIII Congresso Brasileiro de Mecânica dos Solos e Engenharia de Fundações V: 125–134.

UFPE. 2008. Universidade Federal de Pernambuco & Ministérios das Cidades: Gestão e Mapeamento de Riscos Socioambientais, Curso de Capacitação, Recife, Brasil.

Index

Color Plate Section

Chapter 1

Photo 2. Nepal, Khumbu. Imja Lake from the air in 2007. The newly regarded major threat to outbreak of the lake is the course of its drainage through the end moraine via the series of small lakes (Photograph kindly supplied by Sharad P. Joshi, ICIMOD, Kathmandu).

Chapter 2

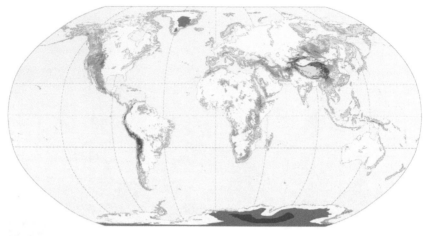

Mountain Zones

	>= 4500 meters
	3500 –4500 m
	2500 –3500 m
	1500 –2500 m and slope > 2 degree
	1000 –1500 m and slope > 5 degree or local elevation range > 300 m
	300 –1000 m and local elevation range > 300 m
	Outline of territory with some mountain areas (below map resolution)

Other

	Non-mountain terrain

Box 1. Some definitions of Mountains and High Mountains.

Source: UNEP 2002.

Chapter 3

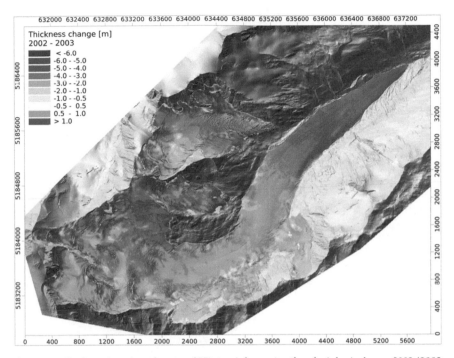

Figure 3.2. Surface elevation change of Hintereisferner for the glaciological year 2002/2003, derived from multi-temporal airborne laser scanner data. Blue indicates positive elevation changes (>0.3 m), white a 'stable' surface and yellow to red colours indicate negative elevation changes. For the respective period, the overall elevation change of Hintereisferner was –3.7 m, elaboration by the authors.

Figure 3.3. Surface elevation change of Reichenkar rock glacier (Austrian Alps), derived from multi-temporal airborne laser scanner data. Blue indicates positive elevation changes, white a 'stable' surface and yellow to red colours indicate negative elevation changes. Note: The area with blue colours are indicating horizontal motion, elaboration by the authors.

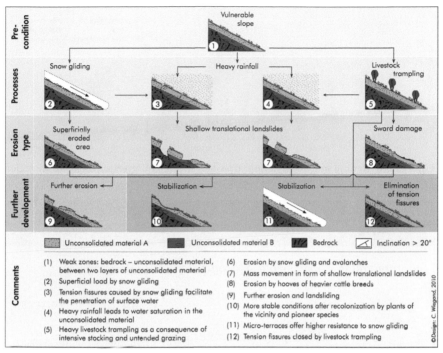

Figure 3.5. Simplified overview of process, erosion types and possible subsequent developments of shallow erosion in high mountains (Source: Wiegand and Geitner 2010: 81).

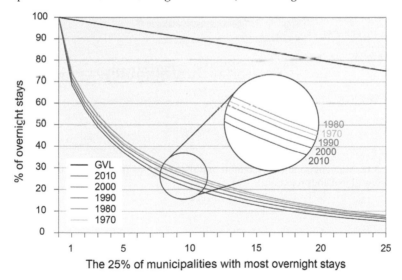

The 25% of municipalities with most overnight stays

Figure 3.10. Increasing concentration of tourism in Austria (Lorenz Curve): GVL stands for the hypothesis that pernoctations are equally distributed to the municipalities. The distance to GVL demonstrates the concentration of tourism on some municipalities. © O: Bender 2011.

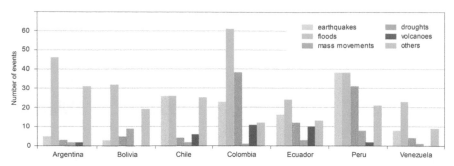

Figure 3.11. Natural and socio-natural disasters in the countries of the Andean region, 1950–2009 (Source: The International Disaster Database 2010).

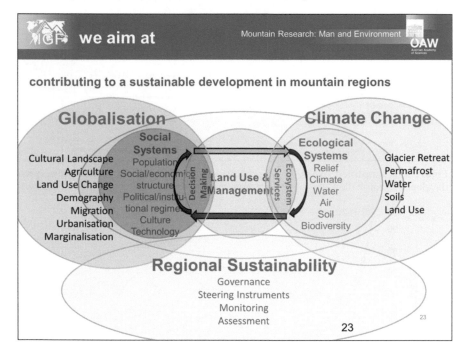

Figure 3.12. Man-environment system in mountain regions (modified by the authors from: GLP, www.globallandproject.org).

Chapter 4

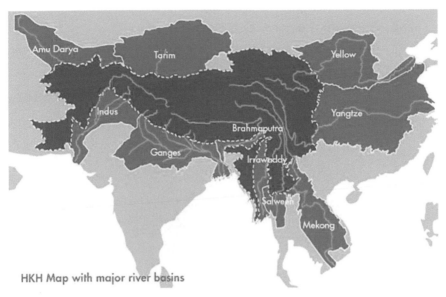

HKH Map with major river basins

Figure 4.3. Hindu Kush Himalaya and Tibetan Plateau with Major River Basins (Source: ICIMOD).

Chapter 6

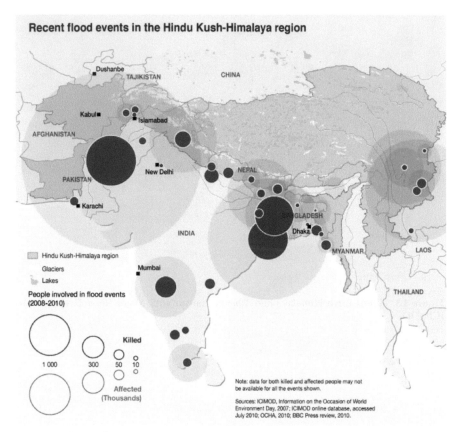

Figure 6.2. Recent flood events in Hindu Kush-Himalaya (Source: Kaltenborn et al. 2010).

Chapter 8

Plate 8.1 Mt. Kenya showing remaining Glaciers.

Chapter 9

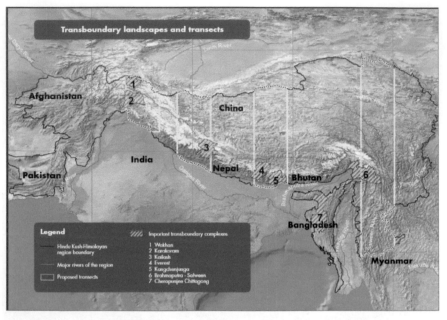

Figure 9.1. Map showing transbounday landscapes and transects in the HKH.

Chapter 10

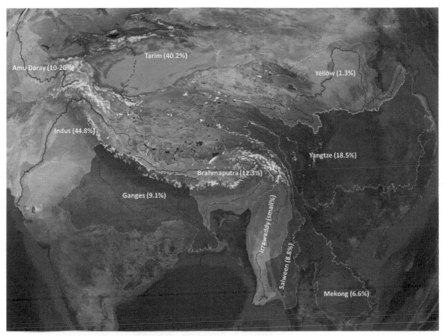

Figure 10.6. Predicted percentage of glacial melts contributing to basin flows in the Himalayan basins. (Data from Xu et al. 2008. Shape files superimposed on background image from ESRI ArcGlobe 10.0 from UNEP 2012a).

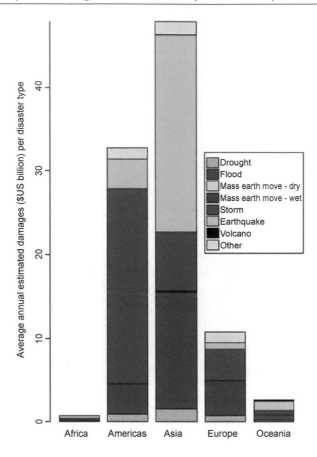

Figure 10.13. Average annual damages (US$ Billion) caused by reported natural disaster 1990–2011. EM–DAT: The OFDA/CRED International Disaster Database—www.emdat. be—Université Catholique de Louvain, Brussels—Belgium (http://www.emdat.be/about).

Chapter 11

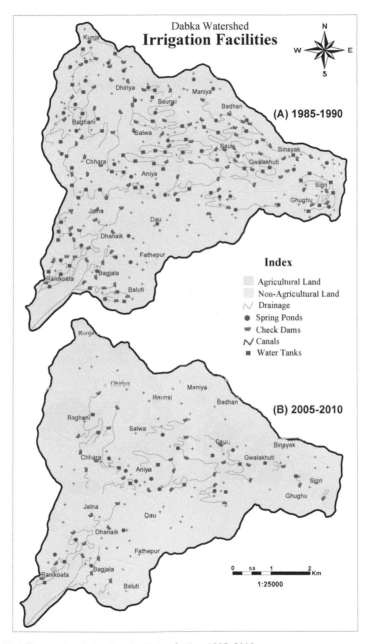

Figure 11.6. Decreasing irrigation facilities during 1985–2010.

Chapter 12

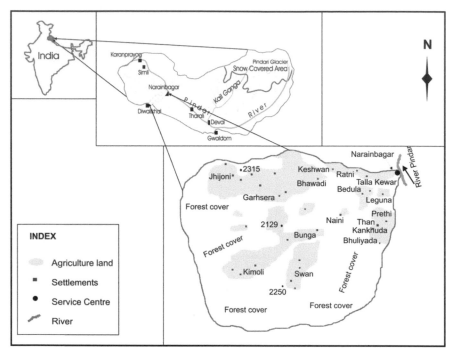

Figure 12.1. Location map of Kewer Gadhera sub-watershed showing recent land-use pattern.

Chapter 13

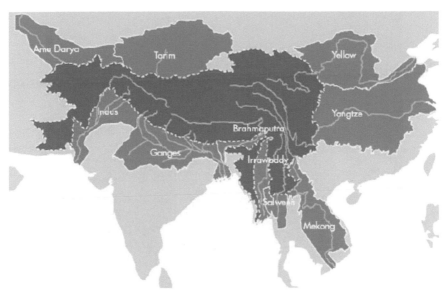

Figure 13.1 Hindu Kush Himalaya region with major river basins.

Chapter 14

Figure 14.2. Recent photo of Dig Thso which drained catastrophically in 1985. The breached dam is still well visible (Photo courtesy of K. Fujita).

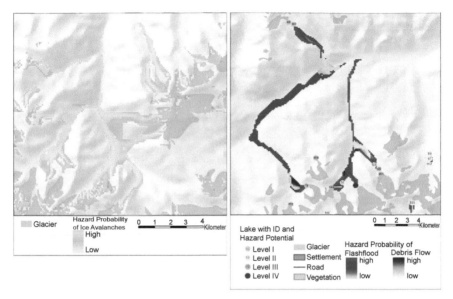

Figure 14.5. Examples from the flow modelling. Left: probability of an area affected by ice avalanches, right: probability of an area affected by flash floods and mudflows (Bolch et al. 2011a, Peters 2009).

Chapter 18

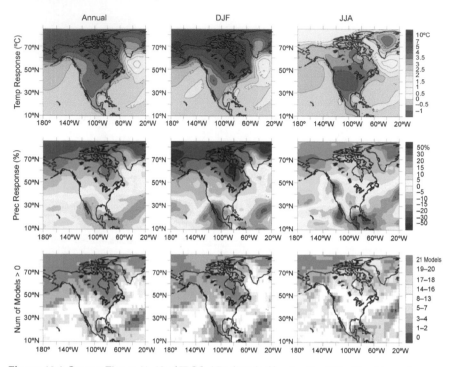

Figure 18.4. Source: Figure 11–12 of IPCC-AR4 (2007), Chapter 11 of Working Group 1.

Chapter 20

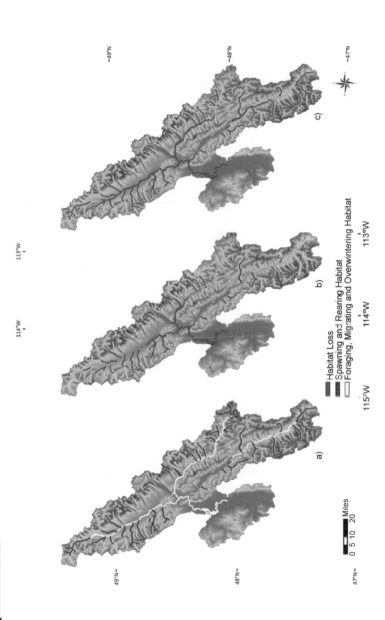

Figure 20.5. Current Bull Trout habitat, and habitat loss associated with 2039 and 2099 climate simulations in the upper Flathead River system, United States and Canada (from Jones et al. 2013).

Figure 20.6. Spawning bull trout in the Wigwam River, Canada (Top credit: Joel Sartore). Bottom: juvenile bull trout and westslope cutthroat trout inhabit a pool in Ole Creek, Glacier National Park (Credit: Jonny Armstrong).

Figure 20.7. Didymosphenia geminata (didymo) is a freshwater alga that may cause dramatic changes in the diversity and abundance of aquatic insect species when algae blooms occur in warming Rocky Mountain streams and rivers. Such losses would impact the entire food web (Credit: Joe Giersch, USGS).

T - #0317 - 071024 - C528 - 234/156/23 - PB - 9780367377908 - Gloss Lamination